PROGRESS IN BRAIN RESEARCH

VOLUME 101

BIOLOGICAL FUNCTION OF GANGLIOSIDES

Other volumes in PROGRESS IN BRAIN RESEARCH

PROGRESS IN BRAIN RESEARCH

VOLUME 101

BIOLOGICAL FUNCTION OF GANGLIOSIDES

Proceedings of Nobel Symposium 83

EDITED BY

LARS SVENNERHOLM *(Göteborg, Sweden)*

ARTHUR K. ASBURY *(Philadelphia, PA, USA)*

RALPH A. REISFELD *(La Jolla, CA, USA)*

KONRAD SANDHOFF *(Bonn, Germany)*

KUNIHIKO SUZUKI *(Chapel Hill, NC, USA)*

GUIDO TETTAMANTI *(Milan, Italy)*

GINO TOFFANO *(Milan, Italy)*

ELSEVIER

AMSTERDAM — LONDON — NEW YORK — TOKYO

1994

ISBN 0-444-81658-5 (volume)
ISBN 0-444-80104-9 (series)

Published by:
Elsevier Science BV
P.O. Box 211
1000 AE Amsterdam
The Netherlands

Library of Congress Cataloging-in-Publication Data

Nobel Symposium (1993 : IBM Nordic Education Centre)
    Biological function of gangliosides : proceedings of Nobel
Symposium 93 / edited by Lars Svennerholm ... [et al.].
        p.   cm. -- (Progress in brain research ; v. 101)
    Includes bibliographical references and index.
    ISBN 0-444-81658-5 (acid-free paper)
    1. Gangliosides--Physiological effect--Congresses.
I. Svennerholm, Lars. II. Title. III. Series.
QP376.P7 vol. 101
[QP752.G3]
612.8'2 s--dc20
[612.8'14]   93-30538
                                                       CIP

Transferred to digital printing 2006
Printed and bound by CPI Antony Rowe, Eastbourne

List of Contributors

Armstrong, D.M., Fidia-Georgetown Institute for the Neurosciences, Georgetown University Medical School, 3900 Reservoir Rd., N.W., Washington, D.C. 20007, USA

Asbury, Arthur K., Dept. of Neurology, School of Medicine, Hospital of University of Pennsylvania, Philadelphia, PA 19104-6055, USA

Beitinger, Heinz, Zoological Institute, University of Stuttgart-Hohenheim, D-7000 Stuttgart 70 (Hohenheim), Germany

Bigner, Darell, Preuss Laboratory for Brain Tumor Research, Duke University Medical Center, Box 3156, Durham, NC 27710, USA

Burger, D., Dept. of Neurology and Laboratory of Neurobiology, Centre Hospitalier Universitaire Vaudois, CH-1011 Lausanne, Switzerland

Casamenti, Fiorella, Dept. of Preclinical and Clinical Pharmacology, University of Florence, Viale Morgagni 65, 50134 Florence, Italy

Costa, Erminio, Fidia-Georgetown Institute for the Neurosciences, Georgetown University Medical School, 3900 Reservoir Rd., NW, Washington, D.C. 20007, USA

Costello, Catherine E., Dept. of Chemistry, Room 56-029, Massachusetts Institute of Technology, 77 Massachusetts Ave., Cambridge, MA 02139, USA

Cuello, Claudio, McGill University, Dept. of Pharmacology and Therapeutics, McIntyre Medical Building, 3655 Drummond Street, Suite 1325, Montreal, Quebec, Canada

Doherty, Patrick, Dept. of Experimental Pathology, UMDS, Guy's Hospital, London Bridge, London SE1 9RT, England

Edelman, Gerald, The Scripps Research Institute, 10666 North Torrey Pines Rd., La Jolla, CA 92037, USA

Fredman, Pam, Department of Clinical Neuroscience, Section of Psychiatry and Neurochemistry, University of Göteborg, Göteborg, Sweden

Garofalo, Lorella, McGill University, Dept. of Pharmacology and Therapeutics, McIntyre Medical Building, 3655 Drummond Street, Suite 1325, Montreal, Quebec, Canada

Gillies, Stephen D., Fuji Immunopharmaceuticals Inc., Lexington, MA, USA

Griffin, John W., Johns Hopkins University, School of Medicine, Baltimore, MD, USA

Guidotti, A., Fidia-Georgetown Institute for the Neurosciences, Georgetown University Medical School, 3900 Reservoir Rd., N.W., Washington, D.C. 20007, USA

Hakomori, Sen-itiroh, The Biomembrane Institute, 201 Elliott Ave W, Seattle, WA 98119, and Dept. of Pathobiology, University of Washington, Seattle, WA 98195, USA

Handgreitinger, Rupert, Universitäts Kinderklinik, University of Tübingen, Tübingen, Germany

Holmgren, Jan, Dept. of Medical Microbiology and Immunology, University of Göteborg, S-413 46 Göteborg, Sweden

Irie, Reiko F., John Wayne Cancer Institute, 2200 Santa Monica Blvd., Santa Monica, CA 90404, USA

Juhasz, Peter, Mass Spectrometry Resource, Dept. of Chemistry, Mass. Inst. of Technology, Cambridge, MA 02139, USA

Kharlamov, A., Fidia-Georgetown Institute for the Neurosciences, Georgetown University Medical School, 3900 Reservoir Rd., N.W., Washington, D.C. 20007, USA

Kiedrowski, L., Fidia-Georgetown Institute for the Neurosciences, Georgetown University Medical School, 3900 Reservoir Rd., N.W., Washington, D.C. 20007, USA

Kuntzer, T., Dept. of Neurology and Laboratory of Neurobiology, Centre Hospitalier Universitaire Vaudois, CH-1011 Lausanne, Switzerland

Körtje, Karl-Heinz, Zoological Institute, University of Stuttgart-Hohenheim, D-7000 Stuttgart 70 (Hohenheim), Germany

Latov, Norman, Dept. of Neurology, College of Physicians and Surgeons, Colombia University, 630 West 168th Street, New York, NY 10031, USA

Ledeen, Robert, Depts. of Neurosciences and Physiology, New Jersey Medical School - UMDNJ, 185 South Orange Ave., Newark, New Jersey 07103-2757, USA

Liberini, Paulo, McGill University, Department of Pharmacology and Therapeutics, McIntyre Building, 3655 Drummond Street, Suite 1325, Montreal, Quebec, Canada

Mahoney, James A., Depts. of Pharmacology and Neuroscience, The Johns Hopkins University School of Medicine, Baltimore, MD 21205, USA

Manev, H, Fidia S.p.A., Via Ponte della Fabbrica, 3/a, 35031 Abano Terme, Italy

Marcus, Donald, Depts. of Medicine, and Microbiology and Immunology, Baylor College of Medicine, Houston, TX 77030, USA

Maysinger, Dusica, McGill University, Dept. of Pharmacology and Therapeutics, McIntyre Medical Building, 3655 Drummond Street, Suite 1325, Montreal, Quebec, Canada

McKhann, Guy, Dept. of Neurology, The Johns Hopkins University, Baltimore, MD 21218, USA

Morton, Donald, John Wayne Cancer Institute, 2200 Santa Monica Blvd., Santa Monica, CA 90404, USA

Mueller, Barbara M., The Scripps Research Institute, Dept. of Immunology, La Jolla, CA, USA

Nagai, Yoshitaka, Tokyo Metropolitan Institute of Medical Science, Bunkyo-ku, Tokyo 113, Japan

Nardelli, E., Dept. of Neurology, University of Verona, Italy

Needham, Leila K., Depts. of Pharmacology and Neuroscience, The Johns Hopkins University School of Medicine, Baltimore, MD 21205, USA

Oderfeld-Nowak, Barbara, Nencki Institute of Experimental Biology, Polish Academy of Sciences, 3 Pasteura, Warszawa, Poland

Paulson, James C., Cytel Corp. and Depts. of Chemistry and Molecular Biology, Scripps Research Institute, 3525 John Hopkins Court, San Diego 92121, USA

Pepeu, Giancarlo, Dept. of Preclinical and Clinical Pharmacology, University of Florence, Viale Morgagni 65, 50134 Florence, Italy

Perreault, Hélène, Mass Spectrometry Resource, Dept. of Chemistry, Mass. Inst. of Technology, Cambridge, MA 02139, USA

Picasso, S., Dept. of Neurology and Laboratory of Neurobiology, Centre Hospitalier Universitaire Vaudois, CH-1011 Lausanne, Switzerland

Polo, A, Fidia-Georgetown Institute for the Neurosciences, Georgetown University Medical School, 3900 Reservoir Rd., N.W., Washington, D.C. 20007, USA

Rahmann, Hinrich, Zoological Institute, University of Stuttgart-Hohenheim, D-7000 Stuttgart 70, Germany

Ravindranath, Mepur H., John Wayne Cancer Institute, 2200 Santa Monica Blvd., Santa Monica, CA 90404, USA

Reisfeld, Ralph A., The Scripps Research Institute, Dept. of Immunology, La Jolla, CA, USA

Riboni, Laura, Dept. of Medical Chemistry and Biochemistry, The Medical School, University of Milan, Via Saldini 50, 20133 Milan, Italy

Rösner, Harald, Zoological Institute, University of Stuttgart-Hohenheim, D-7000 Stuttgart 70 (Hohenheim), Germany

Sandhoff, Konrad, Institut für Organische Chemie und Biochemie der Universität Bonn, Gerhard-Domagk-Str. 1, D-5300 Bonn 1, Germany

Schluep, M., Dept. of Neurology and Laboratory of Neurobiology, Centre Hospitalier Universitaire Vaudois, CH-1011 Lausanne, Switzerland

Schnaar, Ronald L., Depts. of Pharmacology and Neuroscience, The Johns Hopkins University School of Medicine, Baltimore, MD 21205, USA

Seybold, V., Zoological Institute, University of Stuttgart-Hohenheim, D-7000 Stuttgart 70 (Hohenheim), Germany

Siegel, Donald A., The Howard Hughes Medical Institute, The Rockefeller University, Box 279, 1230 York Avenue, New York, NY 10021, USA

Skaper, Stephen D., Fidia Research Laboratories, Abano Terme, 35031, Italy

Steck, Andreas, Dept. of Neurology and Laboratory of Neurobiology, Centre Hospitalier Universitaire Vaudois, CH-1011 Lausanne, Switzerland

Suzuki, Kunihiko, The Brain and Development Research Center, Depts. of Neurology and Psychiatry, University of North Carolina School of Medicine, Chapel Hill, NC 27599, USA

Svennerholm, Lars, Department of Clinical Neuroscience, Section of Psychiatry and Neurochemistry, University of Göteborg, Göteborg, Sweden

Swank-Hill, Patti, Depts. of Pharmacology and Neuroscience, The Johns Hopkins University School of Medicine, Baltimore, MD 21205, USA

Terry, Robert, Dept. of Neurosciences, University of California, San Diego, 9500 Gilman Drive, La Jolla, CA 92093-0624, USA

Tettamanti, Guido, Dept. of Medical Chemistry and Biochemistry, The Medical School, University of Milan, Via Saldini 50, Milan 20133, Italy

Tiemeyer, Michael, Depts. of Pharmacology and Neuroscience, The Johns Hopkins University School of Medicine, Baltimore, MD 21205, USA

Tsuji, Shuichi, Dept. of Glyco Molecular Biology, Glycobiology Research Group of the Frontier Research Program, The Institute of Physical and Chemical Research (REKEN), Wako City, Saitama, 351-01, Japan

Van Echten, Gerhild, Institut für Organische Chemie und Biochemie der Universität Bonn, Gerhard-Domagk-Str.1, D-5300 Bonn 1, Germany

Walsh, Frank, Dept. of Experimental Pathology, UMDS, Guy's Hospital, London Bridge, London SE1 9RT, England

Weng, Nanping, Dept. of Medicine, Baylor College of Medicine, Houston, Texas 77030, USA

Wiegandt, Herbert, Physiologisch-Chemisches Institut, Philipps-Universität, Marburg, Germany

Wikstrand, C.J., Department of Pathology, Duke University Medical Center, Box 3156, Durham, NC 27710, USA

Wroblewski, J.T., Fidia-Georgetown Institute for the Neurosciences, Georgetown University Medical School, 3900 Reservoir Rd., N.W., Washington, D.C. 20007, USA

Wu, Gusheng, Depts. of Neurosciences and Physiology, New Jersey Medical School — UMDNJ, 185 South Orange Avenue, Newark, NJ 07103–2757, USA

Yu, Alice L., University of California, San Diego, CA, USA

Yu, Robert, Dept. of Biochemistry and Molecular Biophysics, Medical College of Virginia, Virginia, Commonwealth University, Richmond, VA, USA

Preface

When given the assignment of organizing a Nobel Symposium on Gangliosides, l recalled the first meeting devoted to gangliosides only, arranged by the late Professor Paul Mandel, of Strasbourg. The meeting was held in a monastery in Mont-Sainte-Odile, Alsace, France and on all later occasions when ganglioside researchers have met afterwards they have remembered the many honest, stimulating lectures and informal discussions held there. A contributing factor to the great success of the meeting and the warm friendship between the participants was an outstanding chef, who introduced us to Alsacian cuisine and wine. More than six years passed before the next ganglioside meeting was held in France. That one was dedicated to Paul Mandel, and I was given the privilege of arranging it. There have been more frequent opportunities for ganglioside researchers to meet more in the last decade, thanks to the generosity of Fidia Research Laboratories. The possibility of meeting and discussing the newest findings has led to exciting progress in this rapidly evolving field. The history of gangliosides has always been linked to the medical field. Ernst Klenk isolated the first gangliosides from cases with Niemann-Pick's and Tay-Sachs' disease. The metabolism of gangliosides was elucidated from studies of the ganglioside lipidoses. An excellent model for the study of the receptor function of gangliosides was the interaction between cholera toxin and GM1-ganglioside. A large number of new gangliosides have been isolated and characterized from carcinomas, and attempts to treat cancer patients have led to intense research into the production of anticarbohydrate antibodies.

When planning the program for the Nobel Symposium, we decided to focus on the role of gangliosides in three areas of medicine in which rapid progress has been achieved during the last decade: cancer, peripheral neuropathies and Alzheimer's disease. Novel ganglioside antigens have been characterized in malignant gliomas, which might be equally important target antigens for immunotherapy as gangliosides GD2 and GD3 have been for melanomas. Rapid progress has been made in understanding the pathogenesis of peripheral neuropathies, and the controversial role of gangliosides is being thoroughly discussed. Recent studies have conclusively shown that the degree of dementia in Alzheimer disease is closely related to the loss of axodendritic arborisation. The ability of gangliosides to protect the neuronal processes and to prevent degeneration is the experimental basis for therapeutic ganglioside administration. The sections on basic ganglioside research were therefore focused on the role of gangliosides in neuronal differentiation and development and their receptor functions and cell surface activities.

When organizing this program, l have had invaluable help and support from my co-editors, who have suggested speakers, chaired the sessions and finally reviewed the manuscripts from their sessions. l wish to express my most sincere thanks to them: Professor Arthur K. Ashbury, Ralph A. Reisfeld, Konrad Sandhoff, Kunihiko Suzuki, Guido Tettamanti and Gino Toffano. In an early phase of the organization we felt that the remarkable progress in cell adhesion had to be communicated to the participants and we therefore asked the inventor and master of the field Nobel Laureate Professor Gerald

Edelman to give a plenary lecture on Molecular Basis of Cell Adhesion, which he generously agreed to do.

The symposium was sponsored by the Nobel Foundation through its Nobel Symposium Fund. IBM graciously defrayed all the local costs by placing its conference center, IBM Nordic Education Centre with all its facilities, at our disposal.

LARS SVENNERHOLM
Göteborg
March 1993

Designation and Schematic Structure of Gangliosides and Allied Glycosphingolipids

Lars Svennerholm

Department of Clinical Neuroscience, Section of Psychiatry and Neurochemistry, University of Göteborg, Göteborg, Sweden

With gangliosides and other higher glycosphingolipids naming problems arise from the complexity of the carbohydrate moiety of these compounds. The systematic names of the oligosaccharides linked to the lipophilic portion, i.e. ceramide, are so cumbersome that they cannot be used for oral communication. This difficulty has been overcome by creating suitable trivial names for the parent oligosaccharides, with the prefixes ganglio-, globo- and lacto- to which the number of monosaccharides in the oligosaccharide is indicated by the suffixes -triaose, -tetraose, etc. (IUPAC-IUB, 1977) (Table I). Klenk (1942) defined ganglioside as a complex glycosphingolipid containing sialic acid. When it was discovered that brain tissue contained several gangliosides (Svennerholm, 1955) with varying number of sialic acids we suggested the generic terms mono-, di-, and trisialogangliosides (Svennerholm and Raal, 1962) which were easily accepted. It was, however, much more difficult to agree about a common system of trivial names or symbols for the various gangliosides and most investigators preferred to introduce their own designation. For more than 10 years most ganglioside conferences devoted time for intense debates about the most appropriate designations of the various gangliosides. Today, the code system suggested by Svennerholm in 1963 and continuously extended (Svennerholm, 1980, 1988) is generally accepted for gangliosides of the ganglio series, while there is still no general agreement about the designation of gangliosides of the lactoseries. The designations suggested in 1963 were based on the findings that brain tissue contained four major gangliosides of the ganglio series with a tetraose chain of neutral sugars. They were designated to belong to the G1 series where G stands for ganglio. The four gangliosides differed with regard to the number of sialic acids where M, D, and T stood for mono-, di-, and tri-sialyl groups. There were two disialogangliosides, GD1a and GD1b, and that one which contained a disialylgroup at the internal galactose was designated b (Table II). When gangliosides of the ganglio series with three sialic acids linked to the internal galactose were detected, they were designated to belong to the c series. When the biosynthesis of the gangliosides of the ganglio series has been finally settled it is evident that the designation of an a, b and c series predicted the crucial role of the sialyltransferases in the ganglioside metabolism (Van Echten and Sandhoff, 1993).

Gangliosides lacking the terminal galactose were given number 2 and when lacking the disaccharide galactosyl-*N*-acetylgalactosamine number 3. We have suggested the symbol cisGM1 for the monosialogangliotetraosyl-ceramide in which the sialyl group is attached to the terminal galactose. Some researchers have unfortunately already used the designation of GM1b for this ganglioside, but b is already used to designate gangliosides with a disialyl group attached to the internal galactose. The name GM1b ought to be abandoned.

Differences in linkage (e.g. $1 \rightarrow 4$ versus $1 \rightarrow 3$) in otherwise identical sequences are in general indicated by 'iso' used as a prefix. We have used it in the nomenclature for the lactoseries gangliosides (Table III). The designation of gangliosides of the lactoseries follows the principal for gangliosides of the ganglio series. We have used the term LM1 for the sialyllactotetraosyl ceramide in which the terminal linkage is $\beta 1 \rightarrow 4$, because it occurs more abundantly in the nervous system and other organs under physiological conditions. Iso LM1 is used for the ganglioside with a terminal $\beta 1 \rightarrow 3$ linkage. The designation paragloboside is still used for the lacto-tetraosylceramide. This term was given to the glycolipids when its chemical structure was unknown, but para means chemically related to or derived from (New Webster's Dictionary, 1992) and it will be equally incorrect to continue to use the name paragloboside for lactotetraosylceramide as to use strandin (Folch-Pi, 1951) for ganglioside. Further, I have also suggested the short symbols LK1 and HexLK1 for the two glucuronic acid-containing glycosphingolipids of the lactotetraose series (L) since they occur in killer cells (K) and will react with the HNK-1 antibody (Table IV).

References

Folch, J., Arsove, S. and Meath, J.A. (1951) Isolation of brain strandin, a new type of large molecule tissue component. *J. Biol. Chem.*, 191: 819–830.

IUPAC-IUB Commission on Biochemical Nomenclature (CBN) (1977) The Nomenclature of Lipids. *Eur. J. Biochem.*, 79: 11–21.

Klenk, E. (1942) Über die Ganglioside, eine neue Gruppe von zucherhaltigen Gehirnlipoiden. *Hoppe-Seyler's Z. Physiol. Chem.*, 273: 76–86.

New Webster's Dictionary and the Thesaurus of the English Language (1992) Lexicon Publications Inc., Daubury, CT.

Svennerholm, L. (1956) Composition of gangliosides from human brain. *Nature.* 177: 524–525.

Svennerholm, L. (1963) Chromatographic separation of human brain gangliosides. *J. Neurochem.*, 10: 613–623.

Svennerholm, L. (1980) Ganglioside designation. In: L. Svennerholm, P. Mandel, H. Dreyfus and P.F. Urban (Eds.) *Structure and Function of Gangliosides. Adv. Exp. Med. Biol., Vol 125.* Plenum Press, New York, pp. 11.

Svennerholm, L. (1988) Immunological and tumoral aspects of gangliosides. In: R.W., Ledeen, E.L., Hogan, G., Tettamanti, A.J., Yates and R.K., Yu. (Eds.) *New Trends in Ganglioside Research,* Fidia Research Series, Vol 14. Liviana Press, Padova, pp. 135–150.

Svennerholm, L. and Raal, A. (1961) Composition of brain gangliosides. *Biochem. Biophys. Acta,* 53: 422–424.

Van Echten, G. and Sandhoff, K. (1993) Ganglioside metabolism: enzymology, topology and regulation, *J. Biol. Chem.*, 268: 5341–5344.

TABLE I

Names and abbreviations of neutral glycosphingolipids

Structure	Trivial name of oligosaccharide	Designation of glycosphingolipid
GalNAcβ1 → 4Galβ1 → 4GlcCer	Gangliotriaose	GgOse₃Cer
Galβ1 → 3GalNAcβ1 → 4Galβ1 → 4GlcCer	Gangliotetraose	GgOse₄Cer
Galα1 → 4Galβ1 → 4GlcCer	Globotriaose	GbOse₃Cer
GalNAcβ1 → 3Galα1 → 4Galβ1 → 4GlcCer	Globotetraose	GbOse₄Cer
GlcNAcβ1 → 3Galβ1 → 4GlcCer	Lactotriaose	LcOse₃Cer
Galβ1 → 4GlcNAcβ1 → 3Galβ1 → 4GlcCer	Lactotetraose	LcOse₄Cer
Galβ1 → 3GlcNAcβ1 → 3Galβ1 → 4GlcCer	isoLactotetraose	isoLcOse₄Cer

TABLE II

Designation and schematic structure of ganglioseries gangliosides of biological interest.

Designation		Schematic structure
Svennerholm (1963, 1980)	IUPAC-IUB (1977)	
GM4	I^3NeuAc-GalCer	NeuAcα2 → 3Galβ1 → 1'Cer
GM3	II^3NeuAc-LacCer	NeuAcα2 → 3Galβ1 → 4Glcβ1 → 1'Cer
GM2	II^3NeuAc-GgOse$_3$Cer	GalNAcβ1 → 4(NeuAcα2 → 3)"Galβ1 → 4Glcβ1 → 1'Cer
GM1	II^3NeuAc-GgOse$_4$Cer	Gal,β1 → 3GalNAcβ1 → 4(NeuAcα2 → 3)Galβ1 → 4Glcβ1 → 1'Cer
cis GM1 (*)	IV^3NeuAc-GgOse$_4$Ger	NeuAcα2 → 3Galβ1 → 3GalNAcβ1 → 4Galβ1 → 4Glcβ1 → 1'Cer
Fuc-GM1	IV^2Fuc,II^3NeuAc-GgOse$_4$Cer	Fucα1 → 2Galβ1 → 3GalNAcβ1 → 4(NeuAcα2 → 3)Galβ1 → 4Glcβ1 → 1'Cer
GD3	II^3(NeuAc)$_2$-LacCer	NeuAcα2 → 8NeuAcα2 → 3Galβ1 → 4Glcβ1 → 1'Cer
GD2	II^3(NeuAc)$_2$-GgOse$_3$Cer	GalNAcβ1 → 4(NeuAcα2 → 8NeuAcα2 → 3)Galβ1 → 4Glcβ1 → 1'Cer
GD1a	IV^3NeuAc,II^3NeuAc-GgOse$_4$Cer	NeuAcα2 → 3Galβ1 → 3GalNAcβ1 → 4(NeuAcα2 → 3)Galβ1 → 4Glcβ1 → 1'Cer
GD1b	II^3(NeuAc)$_2$-GgOse$_4$Cer	Galβ1 → 3GalNAcβ1 → 4(NeuAcα2 → 8NeuAcα2 → 3)Galβ1 → 4Glcβ1 → 1'Cer
Fuc-GD1b	IV^2Fuc,II^3(NeuAc)$_2$-GgOse$_4$Cer	Fucα1 → 2Galβ1 → 3GalNAcβ1 → 4(NeuAcα2 → 8NeuAcα2 → 3)Galβ1 → 4Glcβ1 → 1'Cer
GT3	II^3(NeuAc)$_3$-LacCer	NeuAcα2 → 8NeuAcα2 → 8NeuAcα2 → 3Galβ1 → 4Glcβ1 → 1'Cer
GT1a	IV^3(NeuAc)$_2$,II^3NeuAc-GgOse$_4$Cer	NeuAcα2 → 8NeuAcα2 → 3Galβ1 → 3GalNAcβ1 → 4(NeuAcα2 → 3)Galβ1 → 4Glcβ1 → 1'Cer
GT1b	IV^3NeuAc,II^3(NeuAc)$_2$-GgOse$_4$Cer	NeuAcα2 → 3Galβ1 → 3GalNAcβ1 → 4(NeuAcα2 → 8NeuAcα2 → 3)Gal,β1 → 4Glcβ1 → 1'Cer
GT1c	II^3(NeuAc)$_3$-GgOse$_4$Cer	Galβ1 → 3GalNAcβ1 → 4(NeuAcα2 → 8NeuAcα2 → 8NeuAcα2 → 3)Galβ1 → 4Glcβ1 → 1'Cer
GQ1b	IV^3(NeuAc)$_2$,II^3(NeuAc)$_2$-GgOse$_4$Cer	NeuAcα2 → 8NeuAcα2 → 3Galβ1 → 3GalNAcβ1 → 4(NeuAcα2 → 8NeuAcα2 → 3)Galβ1 → 4Glcβ1 → 1'Cer
GQ1c	IV^3NeuAc,II^3(NeuAc)$_3$-GgOse$_4$Cer	NeuAcα2 → 3Galβ1 → 3GalNAcβ1 → 4(NeuAcα2 → 8NeuAcα2 → 8NeuAcα2 → 3)Galβ1 → 4Glcβ1 → 1'Cer
GP1b	IV^3(NeuAc)$_3$,II^3(NeuAc)$_2$-GgOse$_4$Cer	NeuAcα2 → 8NeuAcα2 → 8NeuAcα2 → 3Galβ1 → 3GalNAc,β1 → 4(NeuAcα2 → 8 NeuAcα2 → 3)Galβ1 → 4Glcβ1 → 1'Cer
GP1c	IV^3(NeuAc)$_2$,II^3(NeuAc)$_3$-GgOse$_4$Cer	NeuAcα2 → 8NeuAcα2 → 3Galβ1 → 3GalNAcβ1 → 4(NeuAcα2 → 8NeuAcα2 → 8 NeuAcα2 → 3)Galβ1 → 4Glcβ1 → 1'Cer

IUPAC-IUB (1977) Commission on Biochemical Nomenclature. The Nomenclature of Lipids. *Lipids*, 12: 455-468.

Svennerholm, L. (1963) Chromatographic separation of human brain gangliosides. *J. Neurochem.*, 10: 613-623.

Svennerholm, L. (1980) Ganglioside designaltion. *Adv. Exp. Med. Biol.*, 125: 11.

(*) This ganglioside has often been designated GM1b. However, b is used in the Nomenclature to design a ganglioside of the ganglioseries with two sialic acids linked to the internal galactose (Svennerholm, 1963).

(**) Terms within brackets ( ) represent branching point in the molecule.

Abbreviations: Gal = galactose, Glc = glucose, GalNAc = N-acetylgalactosamine, GlcNAc = N-acetylglucosamine, Fuc = Fucose, NeuAc = N-acetylneuraminic acid, Cer = ceramide.

TABLE III

Designation and schematic structure of Lactoseries gangliosides of biological interest

Designation		Schematic structure
Svennerholm (1980, 1988)	IUPAC-IUB (1977)	
3'-LM1	IV^{3}NeuAc-nLcOse$_4$Cer	NeuAcα2 → 3Galβ1 → 4GlcNAcβ1 → 3Galβ1 → 4Glcβ1 → 1'Cer
3'-isoLM1	IV^{3}NeuAc-LcOse$_4$Cer	NeuAcα2 → 3Galβ1 → 3GlcNAcβ1 → 3Galβ1 → 4Glcβ1 → 1'Cer
6'-LM1	IV^{6}NeuAc-nLcOse$_4$Cer	NeuAcα2 → 6Galβ1 → 4GlcNAcβ1 → 3Galβ1 → 4Glcβ1 → 1'Cer
Fuc-3'-isoLM1	IV^{3}NeuAc,111^{4}Fuc-LcOse$_4$Cer	NeuAcα2 → 3Galβ1 → 3(Fucα1 → 4)GlcNAcβ1 → 3Galβ1 → 4Glcβ1 → 1'Cer
3',8'-LD1	IV^3(NeuAc)$_2$-nLcOse$_4$Cer	NeuAcα2 → 8NeuAcα2 → 3Galβ1 → 4GlcNAcβ1 → 3Galβ1 → 4Glcβ1 → 1'Cer
3',6'-isoLD1	IV^{3}NeuAc,III^{6}NeuAc-LcOse$_4$Cer	NeuAcα2 → 3Galβ1 → 3(NeuAcα2 → 6)GlcNAcβ1 → 3Galβ1 → 4Glcβ1 → 1'Cer

Svennerholm (1988) Immunological and tumoral aspects of gangliosides. In R.W. Ledeen, E.L. Hogan, G. Tettamanti, A.J. Yates and R.K. Yu (Eds), New Trends in Ganglioside Research, Fidia Research Series 14, Liviana Press, Padova, pp. 135–150. Abbreviations as in Table II.

TABLE IV

Designation and schematic structure of glycosphingolipids related to gangliosides

Designation		Schematic structure
Trivial	IUPAC-IUB (1977)	
Glucosylceramide	GlcCer	Glcβ1 → 1'Cer
Galactosylceramide	GalCer	Galβ1 → 1'Cer
Lactosylceramide (CDH)	LacCer	Galβ1 → 4Glcβ1 → 1'Cer
Trihexosylceramide (CTH)	GbOse$_3$Cer	Galα1 → 4Galβ1 → 4Glc,β1 → 1'Cer
Globoside	GbOse$_4$Cer	GalNAc,β1 → 3Galα1 → 4Galβ1 → 4Glcβ1 → 1'Cer
LA1 lactotetraosylceramide (Paragloboside)	nLcOse$_4$Cer	Galβ1 → 4GlcNAcβ1 → 3Galβ1 → 4Glcβ1 → 1'Cer
isoLA1 isolactotetraosyl-ceramide	isoLcOse$_4$Cer	Galβ1 → 3GlcNAcβ1 → 3Galβ1 → 4Glcβ1 → 1'Cer
LK1 Sulfoglucuronylneolacto-tetraosylceramide (SGPG)	IV^{3}GlcUA(3-sulfate)-nLcOse$_4$Cer	3-sulfateGlcUAβ1 → 3Galβ1 → 4GlcNAcβ1 → 3 Galβ1 → 4Glcβ1 → 1'Cer
HexLK1 Sulfoglucuronyllacto-hexaosylceramide (SGLPG)	VI^{3}GlcUA(3-sulfate)-nLcOse$_6$Cer	3-sulfateGlcUAβ1 → 3Gal,β1 → 4GlcNAcβ1 → 3 Galβ1 → 4GlcNAcβ1 → 3Galβ1 → 4Glcβ1 → 1'Cer

Neolactotetraosylceramide was originally termed paragloboside when its structure was unknown. Para is derived from Greek and means beside, beyond wrong, irregular, and has been used in chemistry to denote a compound of unknown structure. When its structure has been elucidated I suggest that the name should be abandoned. I have suggested the name LK1 and HexLK1 for the two glucuronic acids containing acidic glycosphingolipids of the lactotetraose series since they occur in killer cells and have a lactotetraose core structure.

Contents

Section III — Receptor and Cell Surface Activities of Gangliosides

Section IV — Gangliosides and Cancer

L. Svennerholm, A.K. Asbury, R.A. Reisfeld, K. Sandhoff, K. Suzuki, G. Tettamanti and G. Toffano (Eds.)
Progress in Brain Research, Vol. 101
© 1994 Elsevier Science BV. All rights reserved.

CHAPTER 1

Adhesion and counteradhesion: morphogenetic functions of the cell surface

Gerald M. Edelman

The Scripps Research Institute, 10666 North Torrey Pines Road, La Jolla, CA 92037, U.S.A.

Introduction

It is a commonplace that the plasma membrane and cell surface form a significant boundary for transport and signaling. Before the metazoan transition, these processes took place through substances released or diffusing in the environment, or through brief and specialized contacts with other cells. But evolution of multicellular aggregates leading to the development of multicellular organisms required a new set of important functions. These concern the regulation of morphogenesis by coupling the mechanochemical events that affect cell shape, motion, and adhesion to the control of division and differentiation.

Adhesion of cells to one another or of cells to substrates is central to establishing the sequence of such processes and to achieving the appropriate morphology during development. It is for this reason that a search for molecules mediating or modulating adhesion is particularly significant. In the last 15 years, considerable progress has been made in identifying and characterizing molecules that carry out such functions (Edelman and Thiery, 1985; Edelman et al., 1990; Edelman and Crossin, 1991). These morphoregulatory molecules fall into three families: cell adhesion molecules (CAMs), substrate adhesion molecules (SAMs) and cell junctional molecules (CJMs).

Two main themes have emerged from a study of these molecular families: (1) the interactions of these molecules at the cell surface couple transmembrane control via the cytoskeleton (Edelman, 1976) to spe-

cific external binding events; and (2) distinct place-dependent patterns of expression of these molecules emerge during morphogenesis (Crossin et al., 1985; Crossin et al., 1986; Edelman and Crossin, 1991). To comprehend how these themes can be related to morphogenesis, it is important to understand the mechanism of various binding events affecting cell adhesion, shape, and motion. It is equally important to understand the means by which the genes for morphoregulatory molecules can give rise to place-dependent expression at the cell surface during development.

The cell surface is obviously central to these morphogenetic events. In 1976, I proposed a series of hypotheses on the possible connection between transmembrane control of cell surface receptors and the primary processes of development (Edelman, 1976). The main suggestion was that interactions between the cytoskeleton and the cytoplasmic portions of particular receptors could act to transmit internal cellular states to the cell surface and that, reciprocally, the interactions of such receptors with various ligands could alter the state of the cytoskeleton and affect intracellular signals. The net result of such cell surface modulation, as I called it, was to regulate cell–cell interaction, cell motion, and cell shape, as well as contribute to the control of cell division. Considerable evidence that such modulation actually takes place has now accumulated (Schwartz et al., 1989; Crossin, 1991).

In this brief account, I want to provide a few

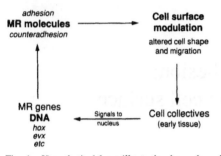

Fig. 1. Hypothetical loop illustrating how place-dependent gene expression of homeobox-containing genes may lead to morphoregulatory (MR) molecule expression. This in turn leads to modulation of the cell surface and alteration of primary processes. The formation of cell collectives (by CAM and SAM interactions) changes nuclear signals for DNA expression leading to expression of new MR molecules and new patterns. (Obviously historegulatory genes are also involved: see (Edelman, 1992) for a description of the complete morphoregulator hypothesis)

examples derived from studies in our laboratory, emphasizing properties of CAMs and their place-dependent expression. Then I will turn to an example of a molecule in the SAM family, cytotactin, that has interesting mixed functions. Interactions with cytotactin can alter cell shape and motion and also inhibit adhesive events. Finally, I want to touch on a central question of morphogenesis—how genes control the place-dependent expression of CAMs and SAMs in the embryo. Knowledge about this important process is quite scanty but recent observations on the promoter regions of CAMs and SAMs and their control by homeobox-containing genes (Jones et al., 1990; Jones et al., 1992a,b; Edelman and Jones, 1992) promise to help us obtain a satisfactory answer to this question.

As a result of studies of this kind, a large shift has occurred in thinking about histogenesis and animal morphogenesis. Instead of the notion of specific place markers coding the interactions of each cell with its neighbors, the notion has emerged of a dynamic multilevel control process coupling gene expression to the mechanochemistry of cell shape and motion (Fig. 1). To understand this morphoregulator hypothesis (Edelman, 1984; Edelman, 1992) more fully, one must know more about the detailed properties of morphoregulatory molecules.

CAMs and Cell Adhesion

Since the early descriptions of N-CAM and L-CAM (Brackenbury et al., 1977; Thiery et al., 1977; Gallin et al., 1983; Yoshida-Noro et al., 1984) it has become evident that there are two main families of CAMs (Table I, see Edelman and Crossin, 1991). The N-CAM superfamily consists of various proteins having a series of extracellular domains homologous to immunoglobulins and to fibronectin type III repeats. The cytoplasmic domains of these molecules vary in length and structure as a result of alternative RNA splicing events. In some cases, these events lead to molecular forms that are bound to the cell surface by phosphatidyl inositol linkages. The binding of members of this family to yield cell–cell adhesion is divalent cation-independent. The second main family of CAMs consists of so-called cadherins, the binding of which is divalent cation-dependent. These molecules share a common cytoplasmic domain but have extracellular domains of different binding specificities. No structural homologies are evident between the N-CAM and cadherin families. During development, however, different cells express members of both families at their surfaces in particular combinations.

In most cases, CAM binding is homophilic, i.e. a given CAM binds to the identical CAM on an apposing cell (Hoffman and Edelman, 1983). In general, therefore, the specificity of cell–cell interaction depends on the ability of each CAM to bind exclusively to itself. This is not universal, however: certain CAMs bind to each other and to cell surface molecules such as integrins in a heterophilic fashion (Marlin and Springer, 1987).

The binding of CAMs depends on direct or indirect interactions of their cytoplasmic domains with the cytoskeleton. The redistribution of CAMs at the cell surface, their ability to bind, and the coupling of that binding to various mechanisms governing cell shape, motion, and sorting all depend on such cytoskeletal interactions. The dependence of cytoskeletal linkage on the cytoplasmic domain of CAMs and the varied consequences of such a linkage have been shown by constructions of deleted or chimeric molecules (Nagafuchi and Takeichi, 1988; Jaffe et al., 1990;

TABLE I

Cell Adhesion Molecules

This table is a composite of the work from numerous laboratories. Individual references may be found in Edelman and Crossin (1991) and Takeichi (1990).

N-CAM Superfamily (Calcium Independent)

Neural	Immune	Miscellaneous	Invertebrates
N-CAM	I-CAM-1	CEA	fasciclin II
Ng-CAM	I-CAM-2	NCA	amalgam
Nr-CAM	CD4	BGP-1	neuroglian
L1/NILE	LAR	pollovirus receptor	DLAR
Contactin/F11		PECAM-1/endo CAM	DPTP
F3		V-CAM	ApCAM
Tag-1/Axonin-1		OB-CAM	
MAG, SMP, DM-GRASP, SC1		PDGF R	
Po		MUC 18	
HT7		DCC	

Cadherin Superfamily (Calcium-dependent)

L-CAM		
uvomorulin/E-cadherin	T-cadherin	EP-cadherin (X.)
N-cadherin/A-CAM	cadherins 4-11	XB-cadherin (X.)
P-cadherin	desmoglein/dgl	U-cadherin (X.)
B-cadherin	desmocollins I&II (dg2/3)	fat gene product (D.)
M-cadherin	PVA (Pemphigus vulgaris antigen)	
R-cadherin		

Ozawa et al., 1990) and their use in transfection experiments (Fig. 2).

Evidence is extensive for the expression of specific combinations of CAMs at particular times and locations during development and tissue regeneration in a given species (Crossin et al., 1985; Edelman and Crossin, 1991). This place-dependent expression begins as early as the two-cell stage in vertebrate development. The tissue-specific combination of particular CAMs and the order of their expression in development appears to be critical in border formation, epithelial-mesenchymal transformation, and cell migration. Moreover, it has been shown that perturbation of CAM expression and binding leads to changes in morphology. Blockade by antibody fragments of the homophilic binding of N-CAM or Ng-CAM results, for example, in disruption of retinal structure and mapping (Buskirk et al., 1980), in alteration of patterns of neurite fasciculation (Hoffman et al., 1986), or in altered patterns of cell motion during cerebellar layering (Chuong et al., 1987). Similarly, super-expression or ectopic expression of cadherins after injection of their RNAs in embryos can lead to distorted morphologies (Kintner, 1988).

As the above brief survey indicates, a dynamic pattern of expression of CAMs occurs at the cell surface. This pattern affects the formation of epithelia and mesenchyme in a definite order that is characteristic of a given tissue during development. Moreover, CAM binding at the cell surface links transmembrane interaction with the cytoskeleton to specific patterns of extracellular binding events. It is not difficult to see how such a process might alter tissue patterns by affecting cell shape and motion. Nonetheless, in most cases, the detailed mechanisms of these transmembrane signalling events remain to be worked out. A particularly promising example is provided by studies of the linkage of cadherins to cytoskeletal ele-

4

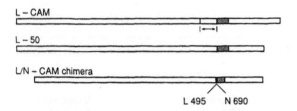

Fig. 2. Chimeric and truncated molecules based on L-CAM and N-CAM structure. Schematic diagrams of L-CAM, L-50 and L/N-CAM are presented. All three molecules have identical extracellular domains with the exception of a 45 amino acid segment, denoted by an arrow in the L-CAM construct, which is missing in the chimeric molecule, L/N CAM. Hatched areas represent transmembrane domains. The junction between L-CAM and N-CAM amino acid sequences is indicated in the Fig., with amino acids numbered as described (Cunningham et al., 1987; Gallin et al., 1987) Cells transfected with the L-50 CDNA (a form with the C-terminal 50 amino acids missing) expressed the molecule at the cell surface but did not bind to each other in short-term assays. Cells transfected with L-CAM CDNA bound in short-term assays and, in long-term sorting assays (Friedlander et al., 1989) they sorted from untransfected L cells and from cells transfected with the chimeric L/N-CAM CDNA. In contrast, cells transfected with the L/N-CAM chimeric CDNA did not sort from untransfected cells although the cells that expressed L-CAM or L/N chimeric CAM bound equally well to each other and to themselves in short-term binding assays. From (Jaffe et al., 1990).

ments by intermediary cytoplasmic proteins, the catenins (Ozawa et al., 1989; Nagafuchi and Takeichi, 1989; Ozawa et al., 1990).

SAMs and Counteradhesion

Substrate adhesion molecules (or SAMs) can also mediate adhesion of cells to the extracellular matrix or to other cells by their binding to cell surface molecules called integrins(Hynes, 1990; Ruoslahti, 1991). A striking characteristic of SAMs is the large number of combinatorial modes of interaction made possible by various molecular sites for binding with each other or with cells (Fig. 3). Besides encouraging adhesion, certain SAMs modulate and even inhibit it.

While it is obvious that cell adhesion can affect cell patterning, it is perhaps less obvious that such patterning may also depend on molecules that are counteradhesive, i.e. molecules that inhibit adhesion. A striking example is the extracellular matrix molecule, cytotactin. Cytotactin (or tenascin) was independently discovered in a number of laboratories (see Erickson and Bourdon, 1989) and goes by a variety of different names. Its structure (Jones et al., 1989) is remarkable in several respects. The molecule (Fig. 4) consists of six polypeptides linked at their amino-termini by disulfide bonds to form a so-called hexabrachion. Each chain consists of a series of EGF repeats, followed by several fibronectin type III repeats (variable in number, depending on alternative RNA splicing), and finally by domains homologous to portions of fibrinogen.

Unlike fibronectin and laminin, cytotactin shows a pattern of expression that is sharply restricted in time and place during development. In this respect, its place-dependent pattern of expression resembles those observed for CAMs; indeed, cytotactin can be seen in patterns that alternate or correlate with those of certain CAMs at various sites such as somites, the developing brain, and developing bones and joints. It is found in the extracellular matrix in the presence of other more generally widespread molecules such as collagen and fibronectin.

One of the striking properties of cytotactin is its effect on cell shape, process extension, and cell motion. Cells plated on substrates of cytotactin generally become rounded even when they adhere (Friedlander et al., 1988; Prieto et al., 1992). An analysis of the detailed binding properties of the various cytotactin domains has been recently carried out in my laboratory (Fig. 4B, Prieto et al., 1992). This work reveals a distinct pattern of cell attachment, flattening, or rounding for each domain that is characteristic of that type of domain. Thus, while EGF-like domains and certain fibronectin type III domains discourage flattening and cell extension, other fibronectin type III and fibrinogen-like domains support these properties.

A good example of the counteradhesive properties of cytotactin is seen in tissue culture experiments on neurite extension and contact. When cytotactin is mixed with fibronectin or laminin and coated in a region of a tissue culture dish adjacent to a region

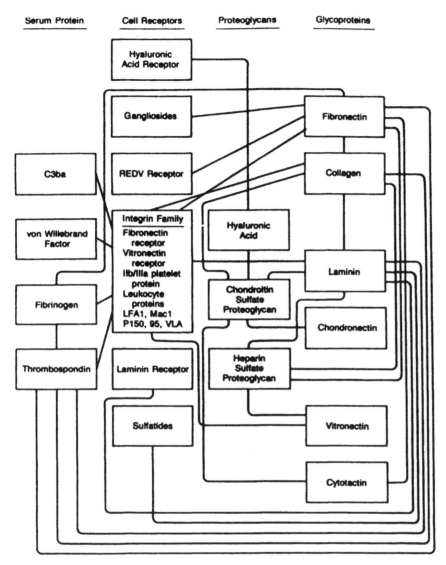

Serum Protein **Cell Receptors** **Proteoglycans** **Glycoproteins**

Fig. 3. Network of ECM proteins, SAMs, and serum proteins capable of interacting with one another and with cell surface receptors. Lines in the diagram indicate possible interactions based on binding sites; actual interactions form subsets that differ in protein type and amount at different embryonic sites. One would expect the modulatory effects of such interactions to reflect the chemistry and structure of the part of the network that is involved at each morphogenetic site. The interactions form the basis for the idea of SAM modulatory networks which can, by altered combinations of interactions, lead to different forms of global and local cell surface modulation. From (Edelman, 1988).

containing fibronectin or laminin alone, and dorsal root ganglia are cultured on that mixed substrate, neurite extension is inhibited in the region containing cytotactin to varying degrees depending on the ratio of cytotactin to the permissive substrates (Fig. 5).

These findings raise the possibility of a mosaic function mediated by the cytotactin molecule that is comprised of adhesive and counteradhesive activities. Both activities may actually operate in vivo. While, in the absence of other interactions, the whole molecule is essentially counteradhesive in tissue culture experiments (Friedlander et al., 1988; Spring et al., 1989; Prieto et al., 1992), evidence of a central role in guiding cell migration has been found in cerebellar

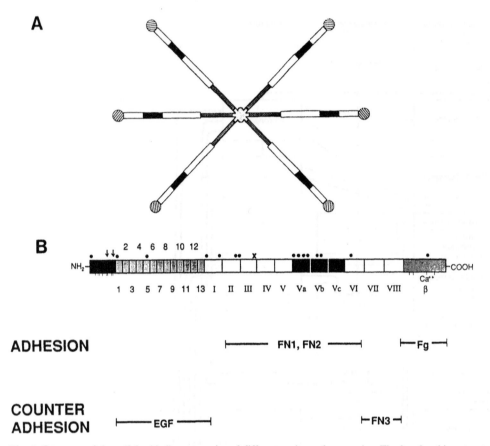

A

B

NH₂— ... —COOH

2 4 6 8 10 12

1 3 5 7 9 11 13   I   II   III   IV   V   Va Vb Vc VI VII VIII   β

Ca⁺⁺

ADHESION

|———— FN1, FN2 ————| |—Fg—|

**COUNTER
ADHESION**

|——— EGF ———| |—FN3—|

Fig. 4. Summary of the cellular binding properties of different regions of cytotactin. The hexabrachion structure of cytotactin is shown schematically in (A). Based on the linear sequence (B) of chicken cytotactin (Jones et al., 1989) shown at the top, regions corresponding to the domains of the protein delineated below were prepared in bacterial expression vectors (Prieto et al., 1992). The isolated domains were tested for their ability to support cell attachment (adhesive) or to prevent cell attachment and flattening on fibronectin (counteradhesive).

tissue slices (Chuong et al., 1987). Blockade of cytotactin binding by specific antibody fragments results in failure of migration of granule cells from the molecular layer into the internal granular layer. The interactions of cytotactin with other matrix proteins such as fibronectin and proteoglycan (Hoffman and Edelman, 1987) may modulate which of the cytotactin domains dominate the functional effects of the binding of the molecule to the cell surface.

A striking correlation between cell rounding and inhibition of cell division has been found when fibroblasts are bound to cytotactin in culture (Crossin, 1991). The effects on cell division appear to be related to the inability of cytotactin treated cells to raise their pH. Such effects have previously been

shown to be due to the action of the Na⁺/H⁺ antiporter reviewed in (Moolenaar, 1986; Grinstein et al., 1989). Despite their coincidence, however, the cell rounding and anti-mitogenic effects of cytotactin binding appear to be independent and parallel events (Crossin, 1991).

Clearly, cytotactin, which shows place-dependent expression like CAMs, is capable of modulating the cell surface. This alters primary cellular processes such as shape change, migration, and division, with resulting changes in cellular patterns. The early observations (Yahara and Edelman, 1973; Edelman, 1976) on the plant lectin concanavalin A as a global modulator of the cell surface affecting cell shape, cytoskeletal events, and cell division thus appear to

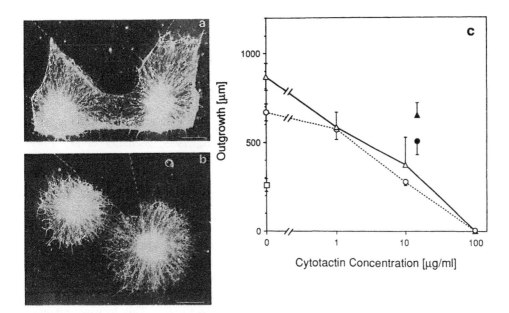

Fig. 5. Effect of purified cytotactin on neurite outgrowth from dorsal root ganglia. *a,b*. Neurite growth at a cytotactin boundary. Micrographs of live explants. Dorsal root ganglia were explanted on a rectangular area incubated with 100 μg/ml laminin (*a*) or fibronectin (*b*) near a vertex of a triangular region on which 100 μg/ml cytotactin was incubated on top of laminin or fibronectin, before blocking with BSA. Dashed lines (*a,b*) indicate the position of two sides of the triangles. The bottom and sides of the outgrowth in (*a*) indicate the limits of the area coated with laminin. Note that neurites avoid the region coated with both cytotactin and laminin (*a*) or cytotactin and fibronectin (*b*). Bar = 1 mm. *c*. Quantitation of neurite outgrowth on mixtures of cytotactin and either laminin or fibronectin. Dorsal root ganglia were explanted into wells and incubated with mixtures of the indicated concentrations of cytotactin and 10 μg/ml laminin (△) or fibronectin (○). □ indicates the outgrowth on BSA. Neurite outgrowth in the presence of soluble cytotactin (10 μg/ml) was determined for ganglia grown in wells coated with 10 μg/ml laminin (▲) or fibronectin (●). From (Crossin et al., 1990).

have a natural counterpart in cytotactin (Crossin, 1991). It is, of course, not at all certain that these different molecules act in the same fashion, and a detailed biochemical comparison of their effects would be useful and revealing.

Potential roles for Gangliosides and Sulfatides in Morphogenesis

The main focus of this conference is on gangliosides and it may seem strange that I have not so far considered their role in morphogenesis. Although very few studies have appeared showing a direct role for these molecules in morphogenesis, evidence is beginning to mount for their ability to modulate CAM and SAM actions. For example, definite evidence exists for the binding of gangliosides and sulfatides to fibronectin and of sulfatides to thrombospondin. Recently (Crossin and Edelman, 1992), we have found that

sulfatides but not gangliosides, can bind to cytotactin (Fig. 6). While no unique carbohydrate specificity was found, the binding showed some specificity and was strongly dependent on divalent cations and was susceptible to increases in ionic strength. The binding is non-linear, with the highest affinity represented by $K_a = 10^8$ M. The evidence suggests that there are multiple binding sites on the cytotactin molecule (Fig. 6). Experiments with antibodies to integrins raise the possibility that, like fibronectin, cytotactin has an integrin receptor. So far, a definitive cell surface receptor for cytotactin has not been identified. It is conceivable that sulfatides may constitute one site for cell binding of cytotactin, particularly in the brain.

Additional observations on the effects of gangliosides in neurite extension mediated by N-CAM (Doherty et al., 1992) suggest that these molecules may modulate the *cis* interactions of CAMs or possi-

8

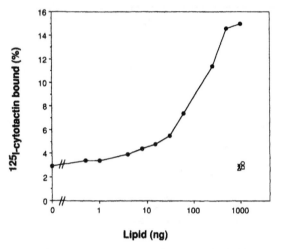

Lipid (ng)

Fig. 6. Solid phase assay of cytotactin-glycolipid binding. Two-fold serial dilutions of glycolipids were dried onto microtiter wells and assayed for ^{125}I-cytotactin binding. The % of input counts of labelled cytotactin is plotted versus the amount of lipid in the coating solution. · – ·, sulfatides; (△)–(△), mixed brain gangliosides; ×–×, lactosylceramide; (□)—(□), galactosylceramide; (○)—(○), glucosylceramide.

bly alter signals transferred from CAM binding events to the cytoskeleton or cell nucleus. It is an intriguing challenge to attempt to define more specific roles for gangliosides in altering CAM and SAM interactions, particularly in connection with modulation of signaling. As indicated in Fig. 1, identification of signaling mechanisms acting in the morphoregulatory loop is particularly fundamental. Recent experiments suggest ways in which that challenge might be met.

Signaling and the gene control of morphoregulatory molecule expression

Signaling events between different tissues can strongly regulate the expression of morphoregulatory molecules, as shown, for example, in the nervous system. Cutting a nerve to a muscle with subsequent Wallerian degeneration leads to a striking re-expression of N-CAM in the affected muscle. At the same time, N-CAM and Ng-CAM levels are reduced in the ventral horn on the side of the affected spinal segment and they are increased in the adjoining dorsal

root ganglia. The patterns revert to normal only 200–300 days later, after Wallerian regeneration is complete (Daniloff et al., 1986).

Similar activity-dependent signalling is seen for cytotactin as it appears in the whisker barrels of the somatosensory cortex in rodents. Cytotactin is normally expressed in the boundary of each whisker barrel (Crossin et al., 1989; Steindler et al., 1989). Ablating a row of whiskers (with destruction of their follicles) leads to a rearrangement of the cortical distribution of cytotactin to follow the fusion of the adjoining barrels that results after removal of input stimuli (Crossin et al., 1989). Thus, the expression of cytotactin in barrel borders depends somehow on the activity of those axons mapping into the barrel.

In order to govern the place-dependent expression of CAMs and of cytotactin, specific signals must affect their local synthesis. This focusses attention on the nature of the genes that regulate the expression of morphoregulatory molecules. The discovery of an effective set of control genes affecting the transcription of such molecules would help explain how particular cellular patterns are actually constrained during morphogenesis. A particularly promising set of candidate genes are the homeobox-containing genes, for these genes and the proteins they specify have been shown to determine regional tissue patterns in a number of species (Kessel et al., 1990). The initial observations on these genes were carried out in *Drosophila* (Lewis, 1978). Subsequently, it has been found that, in vertebrate species, a number of Hox genes that are arranged in four genomic complexes have a similar developmental regulatory role (McGinnis and Krumlauf, 1992). One of the striking features of such genes is that their linear arrangement in the genome reflects their effects along a particular developmental axis: genes located 3′ affect more anterior structures whereas those located in a more 5′ position affect more posterior structures, often in a segmental pattern.

In our studies of cytotactin in the embryo, we were impressed with the sharp anteroposterior sequence of its expression (Crossin et al., 1986; Tan et al., 1987; Prieto et al., 1990). A good example is seen in developing somites (Fig. 7). The molecule and its mRNA

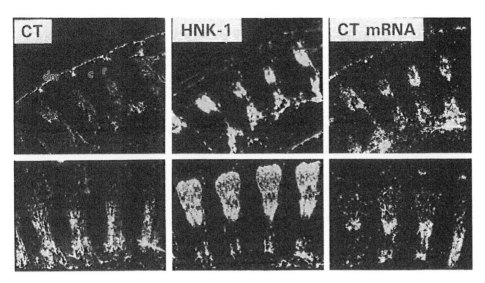

Fig. 7. Immunofluorescence and in situ hybridization of chick embryos of stages 18 and 20. Parasagittal sections of stages 18 (top) and 20 (bottom) were double-labeled with HNK-1 monoclonal antibodies (fluorescein) or cytotactin (CT) polyclonal antibodies (rhodamine). In situ hybridization for cytotactin is shown in parallel sections (CTmRNA). Rostral is to the right in all panels. r, Rostral; c, caudal; dm, dermamyotome. Bar = 240 µm.

is expressed first in the posterior portion of each somite beginning cephalad and proceeding caudad. Later, it is seen only in the anterior sclerotomal portions of the somites at a location in which neural crest cells differentiate into dorsal root ganglia.

This analogy with invertebrates suggested the possibility that homeobox-containing genes may be involved in this process. Analyses of the promoter regions of the genes encoding N-CAM (Hirsch et al., 1990; Chen et al., 1990) and cytotactin (Jones et al., 1990) have in fact uncovered putative sequences for the binding of proteins encoded by homeobox-containing genes. A search was therefore carried out for evidence that the expression of such genes can regulate the transcription of reporter gene constructs driven by the promoters of N-CAM and cytotactin. As shown in Fig. 8, it was found that two *Xenopus laevis* genes, Hox 2.5 and Hox 2.4, that are adjacent to each other in the genomic complex, can differentially modulate mouse N-CAM promoter activity after co-transfection of appropriate constructs into NIH/3T3 cells. A sequence necessary for expression was found in a 47 base pair segment contained in the N-

CAM promoter. When linked to a minimal promoter and a chloramphenicol acetyltransferase (CAT) reporter gene sequence, this segment enhanced basal transcription in 3T3 cells. This activity was greatly increased by co-transfection with Hox 2.5 and it was blocked by co-transfection by Hox 2.4. These results indicate that two neighboring Hox genes can differentially modulate transcription from the promoter of a CAM.

Similar co-transfection experiments using CAT reporter constructs and plasmids driving the expression of the mouse homeobox-containing gene, Evx-1, showed enhanced activity of the cytotactin promoter (Fig. 9). Deletion analysis localized the sequences in the promoter that contributed to the activation to an 89 base pair segment containing a phorbol 12-*O*-tetradecanoate 13-acetate response element (TRE/AP-1 element). Mutation of the TRE/AP-1 sequence abolished the ability of Evx-1 to activate the promoter. The TRE/AP-1 element has been shown to be a target for transcription factors encoded by the *fos* and *jun* gene families. This raises the possibility that cytotactin promoter modulation by Evx-1

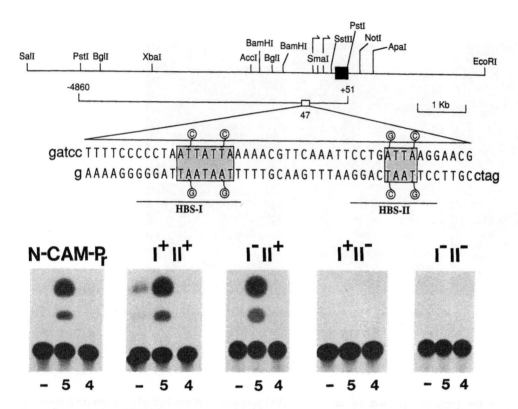

Fig. 8. Co-transfection of NIH 3T3 cells with plasmids driving the expression of Xenopus Hox 2.5 and Hox 2.4 genes modulates the activity of a chloramphenicol acetyltransferase (CAT) reporter gene driven by the N-CAM promoter. A 4.9 kb *Pst* I fragment including the first exon (black box) and the 4860 base pairs of 5~ flanking sequence from the mouse N-CAM gene was cloned upstream of a promoterless CAT gene. This plasmid (N-CAM-Pr) gave no detectable activity when transfected without the Hox gene expression plasmid (N-CAM-Pr, lane -). However, transfection with Hox 2.5 stimulated CAT activity driven by the N-CAM promoter (N-CAM-Pr, lane 5) while transfection with Hox 2.4 had little or no effect (N-CAM-Pr, lane 4). A 47 base pair region (between 512 and 559 base pairs upstream of the ATG codon in the first exon) containing two potential homeodomain binding sites (designated HBS-I and HBS-II) was cloned upstream of a minimal SV40 promoter and a downstream CAT gene. In addition, multiple base pair substitutions were made in the TAAT sequence cores of HBS-I and HBS-II. These plasmids were tested in co-transfection experiments to determine whether the HBSs participate in the modulation of N-CAM promoter activity by Hox gene products and, if so, to determine whether specific HBSs mediated the response. The wild type HBS sequence (HBSI$^+$II$^+$) had significant basal activity in NIH 3T3 cells. Co-transfection with Hox 2.5 elevated expression from basal levels significantly, but co-transfection with Hox 2.4 repressed the basal activity. Mutation of HBSI (HBSI$^-$II$^+$) had no effect on Hox 2.5 activation. However, a mutation either of HBSII (HBSI$^+$II$^-$) or both HBSs (HBSI$^-$II$^-$) completely abolished activation by Hox 2.5. These data suggest that Hox 2.5 gene product activation of the N-CAM promoter is mediated by HBS-II, and that Hox 2.4 is a potent inhibitor of Hox 2.5 activation. From (Edelman, 1993).

may involve a growth-factor signal transduction pathway. Co-transfection experiments (Gruss, P. and Chalepakis, G. et al., unpublished) with paired-box containing genes (Pax genes) have also been shown to modulate the activity of both the N-CAM and cytotactin promoters, extending the observations to yet another family of homeobox-containing genes.

These initial findings with homeobox-containing genes, whose perturbation leads to regional anomalies (Kessel et al., 1990; LeMouellic et al., 1992), provide a clue to understanding the mechanism of place-dependent expression of morphoregulatory molecules. The results (Jones et al., 1992a;b) represent the first findings of targets for these genes aside from other homeobox-containing genes. Before sound conclusions can be drawn, it will be necessary

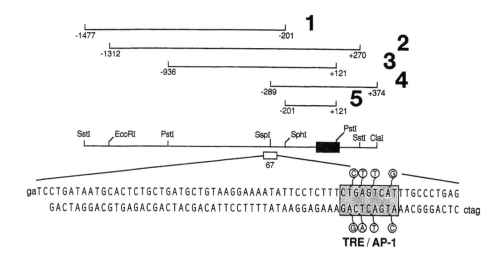

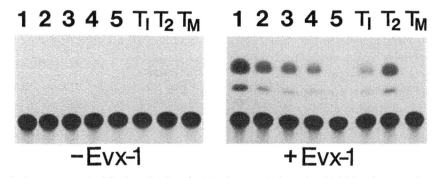

Fig. 9. Sequences required for the activation of cytotactin promoter by a plasmid driving the expression of the homeobox-containing gene, Evx-1. Eight plasmids driving the expression of a chloramphenicol acetyltransferase (CAT) reporter gene were constructed and tested for their expression after co-transfection of NIH 3T3 cells with a plasmid driving the expression Evx-1. A restriction map of part of the first intron, the first exon, and 1477 base pairs of 5~ flanking sequence of the chicken cytotactin gene is shown. Plasmids 1–5 (drawn above) containing the different segments of the promoter region shown were cloned upstream of a promoterless CAT gene and tested for their activity in the absence (−Evx-1) or presence (+Evx-1) of a construct driving the expression of Evx-1. None of these reported constructs yielded detectable CAT expression without the added Evx-1 plasmid (Lanes 1–5, −Evx-1 Panel). However, constructs 1–4 gave high levels of CAT activity, while construct 5 gave no activity, (Lanes 1-5, +Evx-1 Panel). This data indicated that the region between −289 and −201 was required for Evx-1 activation of the cytotactin promoter. This small segment of 89 base pairs was found to contain a TRE/AP-1 sequence (indicated by gray shaded box). Either one copy (T1) or two copies (T2) of a 67 base pair segment from the cytotactin gene 5~ flanking sequence that included the TRE/AP-1 site was inserted upstream of an SV40 minimal promoter CAT gene cassette and tested for activation by Evx-1. Both constructs yielded significant CAT activity in the presence of co-transfected Evx-1 expression plasmid. Mutation of the TRE/AP-1 element with multiple base pair substitutions (indicated by circles) resulted in the loss of Evx-1 activation. These data suggest that the TRE/AP-1 site is involved in the activation of the cytotactin promoter by Evx-1. From (Edelman, 1993).

to confirm these observations in vivo. Nevertheless, if we assume that homeobox containing genes, which are known to be expressed in local patterns, can affect the place-dependent expression of morphoregulatory molecules, we can begin to discern how the mechanochemical effects of these molecules might give rise to specific tissue patterns. One part of the loop proposed in the morphoregulator hypothesis (that shown on the left side of Fig. 1) seems likely to be confirmed. It will still be necessary to show how

12

the linkage of cells by particular combinations of CAMs in the presence of SAMs can lead to new signals to the nucleus (the bottom part of Fig. 1) and result in expression of different combinations of morphoregulatory molecules to yield new histological patterns. The exact nature of these signals remains to be discovered. Presumably, the specific alteration of such signals depends on CAM and SAM mediated cell–cell interactions within a given particular cell collective. Evidence for a role of CAMs and SAMs in such signaling is sparse but is beginning to accumulate (Doherty et al., 1991).

In this brief discussion of the morphogenetic functions of the cell surface, my objective was to point to a multi-level control system linking gene expression to cell surface modulation and vice versa. Adhesion and counteradhesion play pivotal roles in this morphogenetic system. The number of molecular mechanisms involved at each level and their coordination in a place-dependent manner to yield animal form are certainly among the most impressive achievements of natural selection. The subtlety and beauty of these mechanisms are all the more impressive when one considers that they arose as a consequence of selection acting upon the morphology and function of the entire phenotype (Edelman, 1988). Our further understanding of these relationships will not only provide a clearer picture of the connection between evolution and development, but will provide a sound basis for connecting molecular histology to normal and pathological processes.

Acknowledgements

Work from the author's laboratory was supported by USPHS grants HD-09635, DK-04256, NS-22789, and AG-09326.

References

Brackenbury, R., Thiery, J.-P., Rutishauser, U. and Edelman, G.M. (1977) Adhesion among neural cells of the chick embryo. I. An immunological assay for molecules involved in cell-cell binding. *J. Biol. Chem.*, 252: 6835–6840.

Buskirk, D.R., Thiery, J.-P., Rutishauser, U. and Edelman, G.M. (1980) Antibodies to a neural cell adhesion molecule disrupt histogenesis in cultured chick retinae. *Nature (Lond.)*, 285: 488–489.

Chen, A Reyes, A. and Akeson, R. (1990) Transcription initiation sites and structural organization of the extreme 5' region of the rat neural cell adhesion molecule gene. *Mol. Cell. Biol.*, 10: 3314–3324.

Chuong, C.-M., Crossin, K.L. and Edelman, G.M. (1987) Sequential expression and differential function of multiple adhesion molecules during the formation of cerebellar cortical layers. *J. Cell Biol.*, 104: 331–342.

Crossin, K.L. (1991) Cytotactin binding: Inhibition of stimulated proliferation and intracellular alkalinization in fibroblasts. *Proc. Natl. Acad. Sci. USA*, 88: 11403–11407.

Crossin, K.L., Chuong, C.-M. and Edelman, G.M. (1985) Expression sequences of cell adhesion molecules. *Proc. Natl. Acad. Sci. USA*, 82: 6942–6946.

Crossin, K.L., Hoffman, S., Grumet, M., Thiery, J.- P. and Edelman, G.M. (1986) Site-restricted expression of cytotactin during development of the chicken embryo. *J. Cell Biol.*, 102: 1917–1930.

Crossin, K.L., Hoffman, S., Tan, S.-S. and Edelman, G.M. (1989) Cytotactin and its proteoglycan ligand mark structural and functional boundaries in somatosensory cortex of the early postnatal mouse. *Dev. Biol.*, 136: 381–392.

Crossin, K.L., Prieto, A.L., Hoffman, S., Jones, F.S. and Friedlander, D.R. (1990) Expression of adhesion molecules and the establishment of boundaries during embryonic and neural development. *Exp. Neurol.*, 109: 6–18.

Crossin, K.L. and Edelman, G.M. (1992) Specific binding of cytolactin to sulfated glycolipids. J. Neurosci. Res., 33: 631–638.

Cunningham, B.A., Hemperly, J.J., Murray, B.A., Prediger, E.A., Brackenbury, R. and Edelman, G.M. (1987) Neural cell adhesion molecule: Structure, immunoglobulin-like domains, cell surface modulation and alternative RNA splicing. *Science*, 236: 799–806.

Daniloff, J.K., Levi, G., Grumet, M., Rieger, F. and Edelman, G.M. (1986) Altered expression of neuronal cell adhesion molecules induced by nerve injury and repair. *J. Cell Biol.*, 103: 929–945.

Doherty, P., Ashton, S.V., Moore, S.E. and Walsh, F.S. (1991) Morphoregulatory activities of NCAM and N-Cadherin can be accounted for by G protein-dependent activation of L- and N-Type neuronal Ca^{2+} channels. *Cell*, 67: 21–33.

Doherty, P., Ashton, S.V., Skaper, S.D., Leon, A. and Walsh, F.S. (1992) Ganglioside modulation of neural cell adhesion molecule and N–cadherin- dependent neurite outgrowth. *J. Cell Biol.*, 117: 1093–1099.

Edelman, G.M. (1976) Surface modulation in cell recognition and cell growth. *Science*, 192: 218-226.

Edelman, G.M. (1984) Cell adhesion and morphogenesis: The regulator hypothesis. *Proc. Natl. Acad. Sci. USA*, 81: 1460–1464.

Edelman, G.M. (1988) *Topobiology: An Introduction to Molecular*

Embryology. Basic Books, New York.

Edelman, G.M. (1992) Morphoregulation. *Dev. Dynamics*, 193: 2-10.

Edelman, G.M. (1992) Mediation and Inhibition of Cell Adhesion by Morphoregulatory Molecules. In: *Cold Spring Harbor Symp. on Quant. Biol.*, 57: 317–325.

Edelman, G.M. and Crossin, K.L. (1991) Cell adhesion molecules: Implications for a molecular histology. *Annu. Rev. Biochem.*, 60: 155–190.

Edelman, G.M. and Jones, F.S. (1992) Cytotactin: a morphoregulatory molecule and a target for regulation by homeobox gene products. *Trends Biochem. Sci.*, 17: 228–232.

Edelman, G.M. and Thiery, J.-P., (Eds.) (1985) The Cell in Contact —Adhesions and Junctions as Morphogenetic Agents, Wiley, New York.

Edelman, G.M., Cunningham, B.A. and Thiery, J.-P. (Eds.) (1990) Morphoregulatory Molecules, John Wiley & Sons, New York.

Erickson, H.P. and Bourdon, M.A. (1989) Tenascin: an extracellular matrix protein prominent in specialized embryonic tissues and tumors. *Annu. Rev. Cell Biol.*, 5: 71–92.

Friedlander, D.R., Hoffman, S. and Edelman, G.M. (1988) Functional mapping of cytotactin: proteolytic fragments active in cell-substrate adhesion. *J. Cell Biol.*, 107: 2329–2340.

Friedlander, D.R., Mege, R.-M., Cunningham, B.A. and Edelman, G.M. (1989) Cell sorting-out is modulated by both the specificity and amount of different cell adhesion molecules (CAMs) expressed on cell surfaces. *Proc. Natl. Acad. Sci. USA*, 86: 7043–7047.

Gallin, W.J., Edelman, G.M. and Cunningham, B.A. (1983) Characterization of L-CAM, a major cell adhesion molecule from embryonic liver cells. *Proc. Natl. Acad. Sci. USA*, 80: 1038–1042.

Gallin, W.J., Sorkin, B.C., Edelman, G.M. and Cunningham, B.A. (1987) Sequence analysis of a CDNA clone encoding the liver cell adhesion molecule, L-CAM. *Proc. Natl. Acad. Sci. USA*, 84: 2808–2812.

Grinstein, S., Rotin, D. and Mason, M.J. (1989) Na+/H+ exchange and growth factor-induced cytosolic Ph changes. Role in cellular proliferation. *Biochim. Biophys. Acta*, 988: 73–97.

Hirsch, M.-R., Gaugler, L., Deagostini-Bazin, H., Bally-Cuif, L. and Goridis, C. (1990) Identification of positive and negative regulatory elements governing cell-type-specific expression of the neural cell adhesion molecule gene. *Mol. Cell. Biol.*, 10: 1959–1968.

Hoffman, S. and Edelman, G.M. (1983) Kinetics of homophilic binding by E and A forms of the neural cell adhesion molecule. *Proc. Natl. Acad. Sci. USA*, 80: 5762–5766.

Hoffman, S. and Edelman, G.M. (1987) A proteoglycan with HNK-1 antigenic determinants is a neuron-associated ligand for cytotactin. *Proc. Natl. Acad. Sci. USA*, 84: 2523–2527.

Hoffman, S., Friedlander, D.R., Chuong, C.-M., Grumet, M. and Edelman, G.M. (1986) Differential contributions of Ng-CAM and N-CAM to cell adhesion in different neural regions. *J. Cell Biol.*, 103: 145–158.

Hynes, R.O. (1990) *Fibronectins*, Springer-Verlag, New York.

Jaffe, S.H., Friedlander, D.R., Matsuzaki, F., Crossin, K.L., Cunningham, B.A. and Edelman, G.M. (1990) Differential effects of the cytoplasmic domains of cell adhesion molecules on cell aggregation and sorting-out. *Proc. Natl. Acad. Sci. USA*, 87: 3589–3593.

Jones, F.S., Chalepakis, G., Gruss, P. and Edelman, G.M. (1992b) Activation of the cytotactin promoter by the homeobox-containing gene, Evx-1. *Proc. Natl. Acad. Sci. USA*, 89: 2091–2095.

Jones, F.S., Crossin, K.L., Cunningham, B.A. and Edelman, G.M. (1990) Identification and characterization of the promoter for the cytotactin gene. *Proc. Natl. Acad. Sci. USA*, 87: 6497–6501.

Jones, F.S., Hoffman, S., Cunningham, B.A. and Edelman, G.M. (1989) A detailed structural model of cytotactin: Protein homologies, alternative RNA splicing and binding regions. *Proc. Natl. Acad. Sci. USA*, 86: 1905–1909.

Jones, F.S., Prediger, E.A., Bittner, D.A., DeRobertis, E.M. and Edelman, G.M. (1992a) Cell adhesion molecules as targets for hox genes: N-CAM promoter activity is modulated by co-transfection with Hox 2.5 and 2.4. *Proc. Natl. Acad. Sci. USA*, 89: 2086–2090.

Kessel, M., Balling, R. and Gruss, P. (1990) Variation of Cervical Vertebrae after Expression of a Hox-1.1 Transgene in Mice. *Cell*, 61: 301–308.

Kintner, C. (1988) Effects of altered expression of the neural cell adhesion molecule, N-CAM, on early development in *Xenopus* embryo. *Neuron*, 1: 545–555.

LeMouellic, H., Lallemand, Y. and Brulet, P. (1992) Homeosis in the mouse induced by a null mutation in the hox-3.1 gene. *Cell*, 69: 251.

Lewis, E.B. (1978) A gene complex controlling segmentation in *Drosophila. Nature (Lond)*, 276: 565–576.

Marlin, S.D. and Springer, T.A. (1987) Purified intercellular adhesion molecule-1 (ICAM-1) is a ligand for lymphocyte function-associated antigen 1 (LFA-1). *Cell*, 51: 813–819.

McGinnis, W. and Krumlauf, R. (1992) Homeobox Genes and Axial Patterning. *Cell*, 68: 283–302.

Moolenaar, W.H. (1986) Effects of growth factors on intracellular Ph regulation. *Annu. Rev. Physiol.*, 48: 363–376.

Nagafuchi, A. and Takeichi, M. (1988) Cell binding function of E-cadherin is regulated by the cytoplasmic domain. *EMBO J.*, 7: 3679–3694.

Nagafuchi, A. and Takeichi, M. (1989) Transmembrane control of cadherin-mediated cell adhesion: a 94 kDa protein functionally associated with a specific region of the cytoplasmic domain of E-cadherin. *Cell Regul.*, 1: 37–44.

Ozawa, M., Baribault, H. and Kemler, R. (1989) The cytoplasmic domain of the cell adhesion molecule uvomorulin associates with three independent proteins structurally related in different species. *EMBO J.*, 8: 1711–1717.

Ozawa, M., Ringwald, M. and Kemler, R. (1990) Uvomorulin-catenin complex formation is regulated by a specific domain in the cytoplasmic region of the cell adhesion molecule. *Proc. Natl. Acad. Sci. USA*, 87: 4246–4250.

14

Prieto, A.L., Jones, F.S., Cunningham, B.A., Crossin, K.L. and Edelman, G.M. (1990) Localization during development of alternatively-spliced forms of cytotactin by *in situ* hybridization. *J. Cell Biol.*, 111: 685–698.

Prieto, A.L., Andersson-Fisone, C. and Crossin, K.L. (1992) Characterization of multiple adhesive and counteradhesive domains in the extracellular matrix protein cytotactin. *J. Cell. Biol.*, in press:.

Ruoslahti, E. (1991). Integrins as receptors for extracellular matrix. In: Cell Biology of Extracellular Matrix. Plenum Press, New York.

Schwartz, M.A., Both, G. and Lechene, C. (1989) Effect of cell spreading on cytoplasmic Ph in normal and transformed fibroblasts. *Proc. Natl. Acad. Sci. USA*, 86: 4525–4529.

Spring, J., Beck, K. and Chiquet-Ehrismann, R. (1989) Two contrary functions of tenascin: dissection of the active sites by recombinant tenascin fragments. *Cell*, 59: 325–334.

Steindler, D.A., Cooper, N.G.F., Faissner, A. and Schachner, M. (1989) Boundaries defined by adhesion molecules during development of the cerebral cortex: The J1/tenascin glycoprotein in the mouse somatosensory cortical barrel field. *Dev. Biol.*, 131: 243–260.

Takeichi, M. (1990) Cadherins: A molecular family important in selective cell–cell adhesion. *Annu. Rev. Biochem.*, 59: 237–252.

Tan, S.-S., Crossin, K.L., Hoffman, S. and Edelman, G.M. (1987) Asymmetric expression in somites of cytotactin and its proteoglycan ligand is correlated with neural crest cell distribution. *Proc. Natl. Acad. Sci. USA*, 84: 7977–7981.

Thiery, J.-P., Brackenbury, R., Rutishauser, U. and Edelman, G.M. (1977) Adhesion among neural cells of the chick embryo. II. Purification and characterization of a cell adhesion molecule from neural retina. *J. Biol. Chem.*, 252: 6841–6845.

Yahara, I. and Edelman, G.M. (1973) The effects of concanavalin A on the mobility of lymphocyte surface receptors. *Exp. Cell Res.*, 81: 143–155.

Yoshida-Noro, C., Suzuki, M. and Takeichi, M. (1984) Molecular nature of the calcium-dependent cell-cell adhesion system in mouse teratocarcinoma and embryonic cells studied with a monoclonal antibody. *Dev. Biol.*, 101: 19–27.

SECTION I

Structure and Metabolism of Gangliosides

L. Svennerholm, A.K. Asbury, R.A. Reisfeld, K. Sandhoff, K. Suzuki, G. Tettamanti and G. Toffano (Eds.)
Progress in Brain Research, Vol. 101
© 1994 Elsevier Science B.V. All rights reserved.

CHAPTER 2

Ganglioside metabolism: enzymology, topology and regulation

Konrad Sandhoff and Gerhild van Echten

Institut für Organische Chemie und Biochemie der Universität Bonn, Gerhard-Domagk-Str. 1, D-53121 Bonn 1, Germany

Introduction

Glycosphingolipids (GSL) are characteristic components of the outer leaflet of plasma membranes in cell- and species-specific patterns in all eukaryotic cells (Ledeen and Yu, 1982; Svennerholm, 1984; van Echten and Sandhoff, 1989). Gangliosides are a sialic acid-containing group of the GSL which are found concentrated on the surfaces of neurons.

The exact functional role of the GSL and their influence on membrane dynamics are yet to be defined more precisely. However, studies in various laboratories indicate important possibilities. For example, GSL are thought to play a role in cell differentiation and morphogenesis (Hakomori, 1984a). They have been identified as binding sites on the cell surfaces for viruses, bacteria and toxins (Karlsson, 1989; Svennerholm, 1984). Recent evidence demonstrates their involvement in cell type specific adhesion processes (Phillips et al. 1990; Walz et al. 1990). While the GSL are localized mainly on cell surfaces, their biosynthesis and degradation take place in different intracellular organelles. Therefore, a stringent control of GSL metabolism and intracellular transport is required for the maintenance of a balanced GSL pattern in individual cells.

Our current knowledge of ganglioside biosynthesis and their intracellular traffic has been derived almost exclusively from metabolic studies. Whereas ganglioside biosynthesis starts on the membranes of the endoplasmic reticulum (ER) and continues on the Golgi membranes, catabolism occurs after endocytosis in the lysosomal compartment.

Like formation of glycoproteins, biosynthesis of GSL is presumably accompanied by an intracellular vesicle-bound membrane flow of the growing molecules from the ER through the Golgi cisternae to the plasma membrane (for review see Kobayashi et al., 1992; Schwarzmann and Sandhoff, 1990). The involvement of glycolipid binding and/or transfer proteins in the transport of GSL can, however, not be excluded at the present (Watanabe et al., 1980; Tiemeyer et al., 1989).

Biosynthesis of gangliosides in ER-Golgi complex

Glycosphingolipid biosynthesis starts with the condensation of serine and palmitoyl-CoA (Fig. 1), which is catalyzed by the pyridoxal phosphate-dependent serine palmitoyltransferase (SPT), yielding 3-ketosphinganine (Braun and Snell, 1968; Stoffel et al., 1968; Mandon et al., 1991). This ketone is afterwards rapidly reduced to sphinganine by a NADPH-dependent reductase (Stoffel et al., 1968). Although not yet proved there is strong evidence that the introduction of the 4-*trans* double bond occurs only after addition of an amide-linked fatty acid (Ong and Brady, 1973; Merrill and Wang, 1986; Rother et al., 1992). This finding amply indicates that sphingosine is probably not an intermediate of the de novo biosynthetic pathway of the GSL.

All above enzymatic steps involved in the biosyn-

18

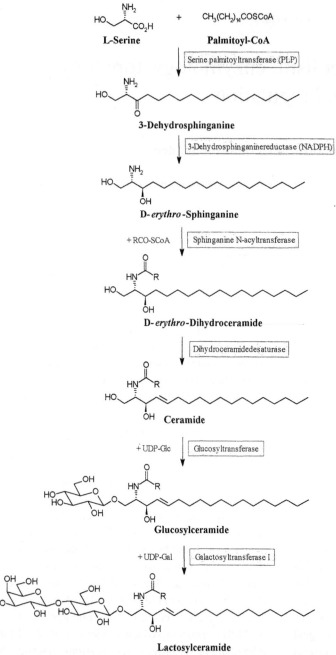

L-Serine   +   CH$_3$(CH$_2$)$_{14}$COSCoA   **Palmitoyl-CoA**

Serine palmitoyltransferase (PLP)

3-Dehydrosphinganine

3-Dehydrosphinganinereductase (NADPH)

D-*erythro*-Sphinganine

+ RCO-SCoA   Sphinganine N-acyltransferase

D-*erythro*-Dihydroceramide

Dihydroceramidedesaturase

Ceramide

+ UDP-Glc   Glucosyltransferase

Glucosylceramide

+ UDP-Gal   Galactosyltransferase I

Lactosylceramide

Fig. 1 Scheme for sphingolipid biosynthesis from serine to lactosylceramide. All enzymatic steps (except desaturation, which is not yet clear) take place on the cytosolic leaflet of ER or Golgi membranes (R = alkylchain).

thesis of ceramide appear to be located on the cytoso-lic face of the endoplasmic reticulum (ER) (Mandon et al., 1992). The following steps of GSL biosynthe-sis, however, are found localized on Golgi mem-branes. Thus the question of ceramide transport from the ER to the Golgi arises. One possibility is ceramide transport via vesicular membrane flow (van Meer, 1989). There is, however, evidence from studies using NBD-ceramide which suggests that non-vesicular transport may be involved (Lipsky and Pagano, 1985a,b; Pagano and Sleight, 1985). Studies on cell free trafficking and sorting of membrane lipids between the ER and the Golgi apparatus also provide similar indications (Moreau et al., 1991). Although the findings concerning the precise localization of ceramide glucosyltransferase within the Golgi and/or a pre-Golgi compartment are diverging, the accessibility of the enzyme from the cytoplasmic side of Golgi vesicles was demonstrated (Coste et al., 1985, 1986; Trinchera et al., 1991; Futerman and Pagano, 1991; Jeckel et al., 1992).

Also, the galactosyltransferase I which catalyzes the formation of LacCer, the common precursor of most GSL families is accessible from the cytosolic side of the Golgi and therefore has the same topology as ceramide glucosyltransferase (Trinchera et al., 1991).

The sequential addition of further monosaccharide or sialic acid residues to the growing oligosaccharide chain, yielding GM3 and more complex gangliosides, is catalyzed by membrane bound glycosyltransferases, which have been shown to be restricted to the luminal surface of the Golgi apparatus (Carey and Hirschberg, 1981; Creek and Morré, 1981; Fleischer, 1981; Yusuf et al., 1983a,b; Trinchera et al., 1991). Therefore, a transfer of the starting material, LacCer, from the cytosolic to the luminal side of the Golgi membranes is required but how this occurs has not been experimentally demonstrated so far.

LacCer, gangliosides GM3, GD3, and GT3 are formed in an early Golgi compartment by different glycosyltransferases and serve as the precursors of more complex O, a, b and c series gangliosides, respectively. As demonstrated for rat liver Golgi, sequential glycosylation of these precursors is catalyzed by a set of rather unspecific glycosyltransferases (Fig. 2). They transfer the same sugar residue to analogous glycolipid acceptors, differing only in the number of neuraminic acid residues bound to the inner galactose of the oligosaccharide chain (Schwarzmann and Sandhoff, 1990).

Thus only one GalNAc-transferase catalyzes the reaction from LacCer, GM3, GD3 and GT3 to GA2, GM2, GD2 and GT2, respectively (Pohlentz et al., 1988; Iber et al., 1992b) and only one single galactosyltransferase is responsible for the formation of GA1, GM1a and GD1b from GA2, GM2 and GD2, respectively (Iber et al., 1989).

Likewise, one single sialyltransferase IV converts GA1, GM1a and GD1b to GM1b, GD1a and GT1b, respectively (Pohlentz et al., 1988) and again one and the same sialyltransferase V is responsible for the reaction GM1b, GD1a, GT1b and GQ1c to GD1c, GT1a, GQ1b and GP1c, respectively (Iber and Sandhoff, 1989; Iber et al., 1992b).

Moreover kinetic studies showed that the glycosyltransferases, which catalyze the first steps of the ganglioside biosynthesis are more specific for their GSL substrates than enzymes forming the more complex gangliosides. Sialyltransferase I and II are quite specific for their respective substrates whereas sialyltransferase IV is able to convert GM1a to GD1a as well as LacCer to GM3 and possibly also GalCer to GM4 (Iber et al., 1991; Iber, 1991), and sialyltransferase V is able to convert GD1a to GT1a as well as GD3 to GT3 and possibly GT3 to GQ3 (Iber, 1991; Iber et al., 1992b).

The observation that sialyltransferase V is able to convert GD3 to GT3 in vitro, could enable mammalian cells to synthesize c-series gangliosides (see Fig. 2) using their normal enzyme equipment. Quite recently ganglioside GP1c was indeed detected in low quantities, additionally to the main a- and b-series gangliosides, in human brain tissue (Miller-Prodraza et al., 1991). It is also possible that it may be formed by sialylation of GQ1b.

It is, however, not yet clear where exactly in the Golgi stack the individual glycosylation reactions take place. If the respective steps of GSL biosynthesis are localized in different Golgi compartments and the biosynthetic process is coupled to a vesicle bound membrane flow through different Golgi compartments of the growing molecules, inhibitors of exocytotic membrane flow should attenuate formation of complex gangliosides. In such a system with radioactive precursor-fed cells, the biosynthetic labeling of

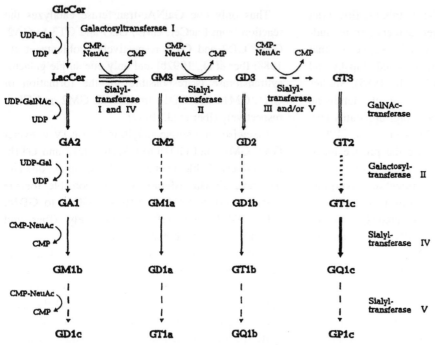

Fig. 2 General scheme for ganglioside biosynthesis (Iber et al., 1992b). All the steps are catalyzed by glycosyltransferases of Golgi membranes. The terminology used for gangliosides is that recommended by Svennerholm (1963).

intermediates before the respective block should be increased and then labeling of complex GSL beyond the drug-induced transport block should be decreased.

One of the first drugs used for this purpose was the cationic ionophore monensin, which was tested on neurotumor cells (Miller-Prodraza and Fishman, 1984), fibroblasts (Saito et al., 1984) and primary cultured neurons (Hogan et al., 1988; van Echten and Sandhoff, 1989). Monensin primarily impedes vesicular membrane flow between the proximal and the distal Golgi cisternae (for review see Tartakoff, 1983). In the presence of this drug incorporation of [^{14}C]galactose into GlcCer, LacCer, GM3, GD3 and GM2 of these cells was significantly increased while labeling of more complex gangliosides, like GM1a, GD1a, GD1b, GT1b and GQ1b decreased remarkably (van Echten and Sandhoff, 1989).

An impressive example of uncoupling ganglioside- and sphingomyelin biosynthesis was observed in the presence of the antibiotic brefeldin A (BFA) (Fig. 3), which causes vesiculation of the Golgi and allows the fusion of cis-, medial- and to some extent also trans elements of Golgi with endoplasmic reticulum (ER) (Fujiwara et al., 1988; Doms et al., 1989; Lippincott-Schwartz et al., 1989, 1990). Under the influence of BFA the biosynthetic labeling of primary cultured neurons with different precursors of GSLs like [^{14}C]galactose, [^{14}C]serine, [^{3}H]sphingosine and [^{3}H]palmitic acid resulted in a drastic reduction of label in gangliosides GM1a, GD1a, GD1b, GT1b, GQ1b and to a smaller extent of sphingomyelin (SM), whereas labeling of GlcCer, LacCer, GM3 and GD3 increased dramatically (van Echten et al., 1990a). Similar results were obtained in CHO cells (Young et al., 1990).

Opposite results, namely an increase of SM biosynthesis after BFA treatment of CHO cells, was, however, very recently reported (Brüning et al., 1992). To clarify this contradiction we pretreated in parallel experiments primary cultured cerebellar neurons and fibroblasts with BFA (1 µg/ml) and then fed

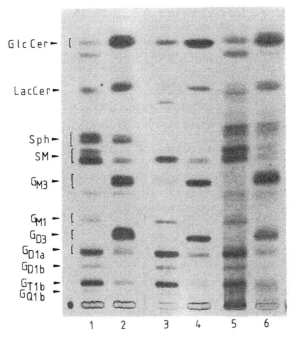

GlcCer —
LacCer —
Sph —
SM —
G_{M3} —
G_{M1} —
G_{D3} —
G_{D1a} —
G_{D1b} —
G_{T1b} —
G_{Q1b} —

1   2   3   4   5   6

Fig. 3 The effect of brefeldin A on the incorporation of radioactive precursors into the sphingolipids of cerebellar cells (van Echten et al., 1990a). Cells were incubated for 3 h in medium containing no addition (lanes 1, 3, 5), or 1 µg/ml BFA (lanes 2, 4, 6). The radioactive precursors were then added to the culture medium: [³H]palmitic acid (lanes 1,2); [³H]sphingosine (lanes 3,4); [¹⁴C]serine (lanes 5,6); after 21 h cells were harvested. Sphingolipids were extracted, purified, separated by TLC and detected by fluorography. The mobilities of standard sphingolipids applied to the chromatogram are shown. (SM-sphingomyelin, Sph–sphingoid base)

the cells with NBD-C6-Cer. As shown in Table I for cultured neurons we reproduced exactly our previously published findings (van Echten et al., 1990a) while for fibroblasts we obtained an almost 2-fold increase of NBD-C6-SM synthesis after BFA treatment (Stotz, 1990). Interestingly, this increase of SM biosynthesis in fibroblasts in the presence of BFA was not observed (on the contrary there was a 40 % reduction of SM formation) when tritiated sphingosine instead of NBD-C6-Cer was administered as a precursor to the cells (Manheller, 1990). Thus caution is indicated in the interpretation of data obtained in different cell systems and especially when different, and most important when unnatural sphingolipid precursors are used.

Taken together our studies with BFA in primary cultured neurons suggest that GM3 and GD3 are synthesized in the early Golgi compartment, whereas complex gangliosides such as GM1a, GD1a, GD1b, GT1b and GQ1b are formed in a late Golgi compartment beyond the BFA induced block. Recent attempts to subfractionate Golgi of rat liver (Trinchera and Ghidoni, 1989) and of primary cultured neurons (Iber et al., 1992a) support this suggestion. However, the experimental data (Iber et al., 1992a) indicate that none of the glycosyltransferases is exclusively localized either in the cis or in the trans Golgi/TGN compartment; rather, it seems more likely that the activity of the 'early' transferases (e.g. sialyltransferase I) continuously decreases, whereas that of the 'later' transferases (e.g. sialyltransferase IV and V) continuously increases from the early to the late Golgi compartments.

It is noteworthy that Matyas and Morré (1987) reported the presence of ganglioside biosynthetic enzymes, though at low levels, also in highly purified fractions of the endoplasmic reticulum of rat liver.

TABLE I

The effect of BFA on sphingolipid biosynthesis in cultured cells.

	Fibroblasts	Cerebellar neurons
	% of controls	
GlcCer	330	350
NBD–GlcCer	140	210
SM	70	47
NBD–SM	175	55
GM3	185	1380
NBD–GM3	300	2000
GM1	50	17
NBD–GM1	n.d.	100
GD3	—	800
NBD–GD3	—	1000

Cells were treated in the absence (controls) or presence of 1 µg/ml BFA. After 3 h 3–[³H]sphingosine (2 µCi/1.2 nmol/ml) or alternatively NBD-C₆-Cer (1µM) was added to the culture medium and cell culture continued for 21 h. Cells were harvested and lipids extracted as described in van Echten et al. (1990a).

Moreover, they propose that gangliosides synthesized in the Golgi cisternae are not only transported to the plasma membrane, but also back to the ER and to other internal endomembranes.

While the topology of the enzymatic steps involved in GSL biosynthesis is still shrouded with mystery, the regulation of their biosynthetic steps is even more poorly understood. A good correlation has been observed, however, between the expression of one certain ganglioside during ontogenesis and the relative activity of the respective glycosyltransferases catalyzing its biosynthesis (Daniotti et al., 1991). This study indicates that the formation of cell- and species- specific GSL patterns on cell surfaces is most likely under transcriptional control of the respective glycosyltransferases (Hashimoto et al., 1983; Nakakuma et al., 1984; Nagai et al., 1986).

Terminal glycosylation of glycoproteins and glycosphingolipids change with development, differentiation and oncogenic transformation (Hakomori, 1984b; Feizi, 1985; Rademacher et al., 1988). In transformed cells it has been demonstrated that alterations in the glycosylation of different glycoconjugates correspond to the respective changes in the expression of the relevant glycosyltransferases (Coleman et al., 1975; Nakaishi et al., 1988a,b; Matsuura et al., 1989;).

Recent studies reported about a successful purification of some glycolipid glycosyltransferases (Honke et al., 1988; Basu et al., 1990; Gu et al., 1990; Yu et al., 1990; Melkerson-Watson and Sweeley, 1991) but further studies are necessary to obtain structural data of these glycosyltransferases and their respective genes for better understanding the regulation of ganglioside biosynthesis at the transcriptional level.

In addition to the genetic level some evidence for an epigenetic regulation of GSL biosynthesis is now available. Very recent studies provide strong evidence that sphingosine, the long chain base backbone of all GSL seems to regulate the first step of sphingolipid biosynthesis in neurons, that is the formation of 3-ketosphinganine catalyzed by serine palmitoyltransferase (SPT) (van Echten et al., 1990b; Mandon et al., 1991). The decrease of SPT activity observed in cultured neurons after feeding sphingosine or azidosphingosine is not caused by a direct feedback inhibition; rather it is a down regulation caused by an unkown pathway possibly involving a sphingosine-binding protein, either triggering a modification of SPT or repressing the transcription of its mRNA or inhibiting its translation. The key regulatory role of SPT was also suggested in studies on the regulation of epidermal sphingolipid synthesis by permeability barrier (Holleran et al., 1991). The authors demonstrated that 5–7 h after epidermal barrier disruption with acetone, incorporation of $(^3H)H_2O$ into sphingolipids as well as SPT activity increased specifically.

A feedback control of GSL biosynthesis has also been suggested, but so far only in in vitro studies. Thus, GM2-synthase of rat liver Golgi was strongly inhibited by GD1a while GD3 synthase was inhibited most effectively by GQ1b, indicating that the regulating steps in the synthesis of a- and b-series gangliosides are preferentially inhibited by their respective products (Nores and Caputto, 1984; Yusuf et al., 1987). Similar effects were recently reported by Shukla et al. (1991) who observed a feedback control of GlcCer-synthase by complex gangliosides in brain microsomes. Another enzyme in the GSL series, GM3 synthase of rat liver Golgi, was also found to be inhibited by gangliosides though at very high concentrations (Richardson et al., 1977). An autoregulation of GD3-synthase by GD3 was very recently reported (Iber et al., 1992b). Feedback control of GSL biosynthesis, however, has not yet been observed in vivo or in cell culture.

We think that the sequence LacCer → GM3 → GD3 → GT3 (Fig. 2) is quite essential for the regulation of ganglioside biosynthesis. According to this model, regulation of the formation of the main ganglioside series in mammals, 'a' and 'b', is linked with the control of GalNAc-transferase and sialyltransferase II, converting GM3 into GM2 ('a'-series) or into GD3 ('b'-series), respectively.

Recently it has been shown that a reversible shift of ganglioside biosynthesis from a- to b-series was achieved by lowering the pH from 7.4 to 6.2 in the culture medium of murine cerebellar cells (Iber et al.,

1990). This effect can be explained by the different pH profiles of the two key regulatory glycosyl-transferases involved. At pH 6.2 sialyltransferase II, a key enzyme in the biosynthesis of b-series, is more active than GalNAc-transferase, the first enzyme in the synthesis of a-series gangliosides, whereas at pH 7.4 the reverse is true.

Suggestions have also been made in favour of possible involvement of extracellular proteins in regulation of GSL biosynthesis. Quiroga and Caputto (1988) purified and characterized an inhibitor of GalNAc-transferase from chicken blood serum. This inhibitor seems to participate in the regulation of ganglioside synthesis directing it either to the formation of GD3 (b-series gangliosides) or GD1a (a-series gangliosides). When added to the culture medium of dispersed chicken embryonic retina cells it affected the ratio GD3 to GD1a and inhibited cell differentiation (Quiroga et al., 1991).

It is interesting that the activity of GD3-synthase (sialyltransferase II) is maintained by different hormones and epidermal growth factor or activated by

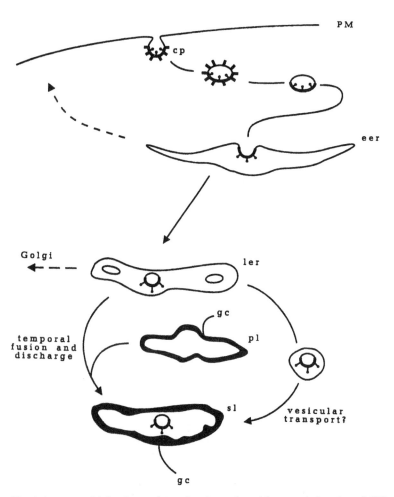

Fig. 4 A new model for the topology of endocytosis and lysosomal digestion of GSL derived from the plasma membrane (Fürst and Sandhoff, 1992). During endocytosis glycolipids of the pm are supposed to end up in intraendosomal vesicles (multivesicular bodies) from where they are discharged into the lysosomal compartment. pm, plasma membrane; cp, coated pit; eer, early endosomal reticulum; gc, glyccalix; ler, late endosomal reticulum; pl, primary lysosome; sl, secondary lysosome; ↑, glycolipid; → proposed pathway of endocytosis of GSL derived from the pm into the lysosomal compartment; and − − → other intracellular routes for GSL derived from the pm.

steroid sex hormones in isolated or cultured rat hepatocytes, respectively, whereas that of GM3-synthase (sialyltransferase I) is not (Mesaric and Decker, 1990 a,b).

Degradation of gangliosides in lysosomes

Components and fragments of the plasma membrane (PM) reach the lysosomal compartment mainly by an endocytotic membrane flow through the early and late endocytotic reticulum (Griffiths et al., 1988). During this vesicular membrane flow molecules are subjected to a sorting process which directs some of the molecules to the lysosomal compartment and others to the Golgi or back to the PM (Wessling-Resnick and Braell, 1990; Kok et al., 1991). It remains, however, an open question whether components of the PM will be included as components into the lysosomal membrane after successive steps of vesicle budding and fusion along the endocytotic pathway. It is thus quite unlikely that components of the lysosomal membrane originating from the PM should be more or less selectively degraded by the lysosomal enzymes.

Alternatively, the observation of multivesicular bodies at the level of the early and late endosomal reticulum (McKanna et al., 1979; Hopkins et al., 1990; Kok et al., 1991) suggests that parts of the endosomal membranes — possibly those enriched in components derived from the PM — budd off into the endosomal lumen and thus form intraendosomal vesicles (Fig. 4). These vesicles enriched in PM components could be delivered by successive processes of membrane fission and fusion along the endocytotic pathways directly into the lysosol for final degradation of their components. Thus glycoconjugates, originating from the outer leaflet of the PM would enter the lysosol as components of the outer leaflet of endocytic vesicles, facing the digestive juice of the lysosomes. This hypothesis is supported by the accumulation of multivesicular storage bodies in Kupffer cells and fibroblasts of patients with a complete deficiency of the *sap* (sphingolipid activator protein) precursor protein (Harzer et al., 1989; Schnabel et al., 1992; Schmid et al., 1992) and by the observation

that the epidermal growth factor receptor derived from the plasma membrane is not integrated into the lysosomal membrane after internalization into lysosomes of hepatocytes (Renfrew and Hubbard, 1991).

Degradation of GSL occurs in lysosomes of mammalian cells by stepwise action of specific acid hydrolases. Several of these enzymes are found to need the assistance of small glycoprotein cofactors, the so called 'sphingolipid activator proteins' (SAPs) (Fürst and Sandhoff, 1992) to attack their lipid substrates with short hydrophilic headgroups.

Since the discovery of the sulfatide activator pro-

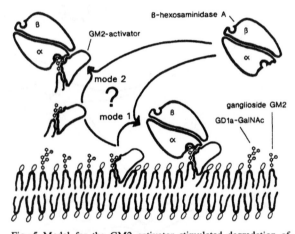

Fig. 5 Model for the GM2–activator stimulated degradation of ganglioside GM2 by human hexosaminidase A (Meier et al., 1991). Water–soluble hexosaminidase A does not degrade membrane–bound ganglioside GM2 in the absence of GM2–activator or appropriate detergents. But it degrades analogues of ganglioside GM2 which contain a short acyl residue or no acyl residue (lyso-ganglioside GM2). They are less firmly bound to the lipid bilayer and more water–soluble than GM2. Ganglioside GM2 bound to a lipid–bilayer, e.g. of an intralysosomal vesicle (see Fig. 4), is hydrolyzed in the presence of the GM2–activator. The GM2–activator binds one ganglioside GM2 molecule and lifts it a few Å out of the membrane. This activator/lipid complex can be reached and recognized by water–soluble hexosaminidase A which cleaves the substrate. However, it is also possible, that the activator–lipid complex leaves the membrane and the enzymatic reaction takes place in free solution. The terminal GalNAc–residue of membrane–incorporated ganglioside GD1a–GalNAc protrudes from the membrane far enough to be accessible to hexosaminidase A without an activator.

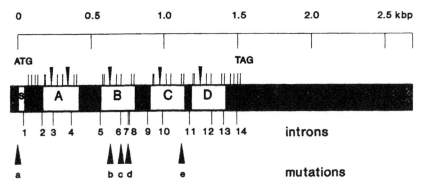

Fig. 6 Structure of the *sap*–precursor cDNA (Holtschmidt et al., 1991). The cDNA of *sap*–precursor codes for a sequence of 524 amino acids (or of 527 amino acids, see Holtschmidt et al., 1991) including a signal peptide of 16 amino acids (termed s, for the entry into the ER) (Nakano et al. 1989; Fürst and Sandhoff, 1992). The four domains on the precursor, termed saposins A–D by O'Brien et al. (1988), correspond to the mature proteins found in human tissues: A = *sap*–A or saposin A, B = *sap*–B or saposin B or SAP–1 or sulfatide–activator, C = *sap*–C or SAP–2 or saposin C or glucosylceramidase activator protein and D = *sap*–D or saposin D or component C. The positions of cysteine residues are marked by vertical bars and those of the N–glycosylation sites by arrow heads. The positions of the 14 introns and of the known mutations leading to diseases are also given: (a) A1 → T (Met1 → Leu), (Schnabel et al., 1992); (b) C650 → T (Thr217 → Ile) (Rafi et al, 1990; Kretz et al., 1990); (c) 33 bp insertion after G777 (11 additional amino acids after Met259), (Zhang et al., 1990, 1991); (d) G722 → C (Cys241 → Ser), (Holtschmidt et al., 1991); and (e) G1154 → T (Cys385 → Phe), (Schnabel et al., 1991).

tein (Mehl and Jatzkewitz, 1964) several other factors were described but their identity, specificity and function still remain unclear. When sequence data became available it turned out that only two genes code for the five known and putative SAPs (Fürst and Sandhoff, 1992). One gene carries the genetic information for the GM2-activator and the other for the *sap*-precursor which is processed to four homologues proteins including sulfatide activator protein (*sap*-B) and glucosylceramidase activator protein (*sap*-C). The in vivo function of the two other proteins *sap*-A and *sap*-D remains unclear.

Several experimental data (Meier et al., 1991) suggest the mechanism of action of the GM2-activator protein. Hexosaminidase A is a water-soluble enzyme which acts on substrates of the membrane surface only if they extend far enough into the aqueous phase (Fig. 5). Like a razor blade or a lawn-mower the enzyme recognizes and cleaves all substrates (e.g. GD1a-GalNAc) with extended hydrophilic headgroups. However, those GSL-substrates with oligosaccharide headgroups too short to be reached by the water-soluble enzyme, cannot be degraded. Their degradation requires a second component, the GM2-activator, a specialized GSL binding protein, which complexes with the substrate (ganglioside GM2), lifts it from the membrane and presents it to the hexosaminidase A for degradation.

The inherited defect of the GM2-activator results in the fatal AB variant of GM2-gangliosidosis, causing mainly neuronal storage of ganglioside GM2 (for review see Fürst and Sandhoff, 1992). In cell culture, the metabolic defect of cultured fibroblasts from an AB variant patient could be corrected when purified GM2-activator was added to the culture medium. This was achieved with a GM2-activator purified from post mortem human tissues as well as with a purified and carbohydrate free human GM2-activator fusion protein obtained by expressing an appropriate cDNA construct in *E. coli* (Klima et al., 1992). The fusion protein was purified in a single step under a denaturing condition and could be renatured to yield a fully functional GM2-activator.

While GM2-activator and hexosaminidase A represent a selective and precisely tuned machinery for the degradation of only a few structurally similar sphingolipids, *sap*-B stimulates the degradation of many lipids by several enzymes from human, plant and even bacterial origin (Li et al., 1988). Thus *sap*-B seems to act as a kind of physiological detergent with broad specificity and solubilizes glycolipid substrates (Vogel et al., 1991).

Unlike GM2-activator and *sap*-B, *sap*-C has been suggested to form complexes with membrane-associated enzymes and apparently activates them (for review see Fürst and Sandhoff, 1992).

The analysis of sphingolipid storage diseases without detectable hydrolase deficiency resulted in the identification of several point mutations in the GM2-activator gene and in the *sap*-precursor gene (reviewed in Fürst and Sandhoff, 1992) (Fig. 6). Interestingly enough, mutations affecting the *sap*-C domain (Gaucher factor) resulted in a variant form of Gaucher disease, mutations affecting the *sap*-B domain (sulfatide-activator) resulted in variant forms of metachromatic leukodystrophy, whereas a mutation in the initiation codon ATG of the *sap*-precursor resulted in its defect and a defect of several *sap*-proteins. This caused a simultaneous storage of ceramide, glucosylceramide, lactosylceramide, ganglioside GM3 and other GSL with short hydrophilic headgroups in multivesicular storage granules in the patients' tissue (Harzer et al., 1989; Schnabel et al., 1992; Smid et al., 1992).

Acknowledgements

We thank Professor Harun Yusuf (Dhaka, Bangladesh) for reading the manuscript. The work of the authors quoted in this paper was supported by funds of the Deutsche Forschungsgemeinschaft (SFB 284), Fonds der Chemischen Industrie and by Fidia Pharmaforschung GmbH (München, Germany).

References

Basu, S., Ghosh, S., Basu, M., Weng, S., Das, K.K., Hawes, J.W., Li, Z. and Zhang, B. (1990) Effect of hydrophobic moieties of substrates on purified glycosyltransferase activities. *FASEB J.*, 4(7): A2140.

Braun, P.E. and Snell, E.E. (1968) Biosynthesis of sphingolipid bases: II. Keto intermediates in synthesis of sphingosine and dihydrosphingosine by cell-free extracts of Hansenula Cifferri. *J. Biol. Chem.*, 243: 3775–3783.

O'Brien, J.S., Kretz, K.A., Dewji, N., Wenger, D.A., Esch, F. and Fluharty, A.L. (1988) Coding of two sphingolipid activator proteins (SAP-1 and SAP-2) by same genetic locus. *Science* 241: 1098–1101

Brüning, A., Karrenbauer, A., Schnabel, E. and Wieland, F.T. (1992). Brefeldin A-induced increase of sphingomyelin synthesis. *J. Biol. Chem.* 267: 5052–5055.

Carey, D.J. and Hirschberg, C.W. (1981) Topography of sialoglycoproteins and sialyltransferases in mouse and rat liver Golgi. *J. Biol. Chem.* 256: 989–993.

Coleman, P.L., Fishman, P.H., Brady, R.O. and Todaro, G.J. (1975) Altered ganglioside biosynthesis in mouse cell cultures following transformation with chemical carcinogens and x-irradiation. *J. Biol. Chem.* 250: 55–60.

Coste, H., Martel, M.-B., Azzar, G. and Got, R. (1985) UDP glucose-ceramide glycosyltransferase from porcine submaxillary glands is associated with the Golgi apparatus. *Biochim. Biophys. Acta*, 814: 1-7.

Coste, H., Martel, M.-B. and Got, R. (1986) Topology of glucosylceramide synthesis in Golgi membranes from porcine submaxillary glands. *Biochim. Biophys. Acta*, 858: 6–12.

Creek, K.E. and Morré, D.J. (1981) Translocation of cytidine 5'-mono phosphosialic acid across Golgi apparatus membranes. *Biochim. Biophys. Acta*, 643: 292–305.

Daniotti, J.L., Landa, C.A., Rösner, H. and Maccioni, H.J.F. (1991) GD3 prevalence in adult rat retina correlates with the maintenance of a high GD3-/GM2-synthase activity ratio throughout development. *J. Neurochem.*, 57: 2054–2058

Doms, R.W., Russ, G. and Yewdell, J.W. (1989) Brefeldin A restributes resident and itinerant Golgi proteins to the endoplasmic reticulum. *J.Cell. Biol.*, 109: 61–72.

Feizi, T. (1985) Demonstration by monoclonal antibodies that carbohydrate structures of glycoproteins and glycolipids are oncodevelopmental antigens. *Nature*, 314: 53–57.

Fleischer, B. (1981) Orientation of glycoprotein galactosyltransferase and sialyltransferase enzymes in vesicles derived from rat liver Golgi appartus. *J. Cell Biol.*, 89: 246–255.

Fujiwara, T., Oda, K., Yokota, S., Takatsuki, A. and Ikehara, Y. (1988) Brefeldin A causes disassembly of the Golgi complex and accumulation of secretory proteins in the endoplasmic reticulum. *J. Biol. Chem.*, 263: 18545–18552.

Fürst, W. and Sandhoff, K. (1992) Activator proteins and topology of lysosomal sphingolipid catabolism. *Biochim. Biophys. Acta*, 1126: 1–16.

Futerman, A.H. and Pagano, R.E. (1991) Determination of the intracellular sites and topology of glucosylceramide synthesis in rat liver. *Biochem. J.*, 280: 295–302.

Griffiths, G., Hoflack, B., Simons, K., Mellman, I. and Kornfeld, S. (1988) The mannose-6-phosphate receptor and the biogenesis of lysosomes. *Cell*, 52: 329–341.

Gu, X.-B., Gu, T.-J. and Yu, R.K. (1990) Purification to homogeneity of GD3 synthase and partial purification of GM3 synthase from rat brain. *Biochem. Biophys. Res. Commun.*, 166: 387–393.

Hakomori, S.-I. (1984a) Glycosphingolipids as differentiation-dependent, tumor-associated markers and as regulators of cell proliferation. *Trends Biochem. Sci.*, 9: 453–458.

Hakomori, S.-I. (1984b) Tumor-associated carbohydrate antigens. *Annu. Rev. Immunol.*, 2: 103–126.

Harzer, K., Paton, B.C., Poulos, A., Kustermann-Kuhn, B., Roggendorf, W., Grisar, T. and Popp, M. (1989) Sphingolipid activator protein (SAP) deficiency in a 16-week-old atypical Gaucher disease patient and his fetal sibling: biochemical signs of combined sphingolipidoses. *Eur. J. Pediatr.*, 149: 31–39.

Hashimoto, Y., Suzuki, A., Yamakawa, T., Miyashita, N. and Moriwaki, K. (1983) Expression of GM1 and GD1 in mouse liver is linked to the H_2 complex on chromosome 17. *J. Biochem.*, 94: 2043–2048.

Hogan, M.V., Saito, M. and Rosenberg, A. (1988). Influence of monensin on ganglioside anabolism and neurite stability in cultured chick neurons. *J. Neurosci. Res.*, 20: 390–394.

Holleran, W.M., Feingold, K.R., Mao Qiang, M., Gao, W.N., Lee, J.M. and Elias, P.M. (1991). Regulation of epidermal sphingolipid synthesis by permeability barrier function. *J. Lipid Res.*, 32: 1151–1158.

Holtschmidt, H., Sandhoff, K., Kwon, H.Y., Harzer, K., Nakano, T. and Suzuki, K. (1991) Sulfatide activator protein: Alternative splicing generates three messenger RNAs and a newly found mutation responsible for a clinical disease. *J. Biol. Chem.*, 266: 7556–7560.

Honke, K., Gasa, S. and Makita, A. (1988) Purification and some properties of rat liver GM1 synthase. Abstract-Third Rinsho-Ken Int. Conf., Tokyo, p. 126

Hopkins, C.R., Gibson, A., Shipman, M. and Miller, K. (1990) Movement of internalized ligand-receptor complexes along a continuous endosomal reticulum. *Nature*, 346: 335–339.

Iber, H., Kaufmann, R., Pohlentz, G., Schwarzmann, G. and Sandhoff, K. (1989) Identity of GA1-, GM1a- and GD1b synthase in Golgi vesicles from rat liver. *FEBS Lett.*, 248: 18–22.

Iber, H. and Sandhoff, K. (1989) Identity of GD1c, GT1a and GQ1b synthase in Golgi vesicles from rat liver. *FEBS Lett.*, 254: 124–128.

Iber, H., van Echten, G., Klein, R.A. and Sandhoff, K. (1990) pH-Dependent changes of ganglioside biosynthesis in neuronal cell culture. *Eur. J. Cell Biol.*, 52: 236–240.

Iber, H. (1991) Untersuchungen zur Gangliosid-Biosynthese, ihrer Regulation und Lokalisation in Rattenleber-Golgi, Cerebellarzellen und B104-Zellen. Thesis, Institut für Organische Chemie und Biochemie, Universität Bonn.

Iber, H., van Echten, G. and Sandhoff, K. (1991) Substrate specificity of 2->3 sialyltransferases in ganglioside biosynthesis of rat liver Golgi. *Eur. J. Biochem.*, 195: 115–120.

Iber, H., van Echten, G. and Sandhoff, K. (1992a) Fractionation of primary cultured neurons: Distribution of sialyltransferases involved in ganglioside biosynthesis. *J. Neurochem.*, 58: 1533–1537.

Iber, H., Zacharias, C. and Sandhoff, K. (1992b) The c-series gangliosides GT3, GT2 and GP1c are formed in rat liver Golgi by the same set of glycosyltransferase that catalyze the biosynthesis of asialo-, a-, and b-series gangliosides. *Glycobiology*, 2: 137–142.

Jeckel, D., Karrenbauer, A., Burger, K.N.J., van Meer, G. and Wieland, F. (1992) Glycosylceramide is synthesized at the cytosolic surface of various Golgi subfractions. *J. Cell Biol.*, 117: 259–267.

Karlsson, K.-A. (1989) Animal glycosphingolipids as membrane attachment sites for bacteria. *Annu. Rev. Biochem.*, 58: 309–350.

Klima, H., Klein, A., van Echten, G., Schwarzmann, G., Suzuki, K., and Sandhoff, K. (1992) Over-expression of a functionally active human GM2-activator protein in *Escherichia coli*. Biochem. J., 292: 571–576.

Kobayashi, T., Pimplikar, S.W., Parton, R.G., Bhakdi, S. and Simons K. (1992) Sphingolipid transport from the trans-golgi network to the apical surface in permeabilized MDCK cells. *FEBS Lett.*, 300: 227–231.

Kok, J.W., Babia, T. and Hoekstra, D. (1991) Sorting of sphingolipids in the endocytic pathway of HT 29 cells. *J. Cell. Biol.*, 114: 231–239.

Kretz, K.A., Carson, G.S., Morimoto, S., Kishimoto, Y., Fluharty, A.L. and O'Brien J.S. (1990) Characterization of a mutation in a family with saposin B deficiency: A glucosylation site defect. *Proc. Natl. Acad. Sci. USA*, 87: 2541–2544.

Ledeen, R.W. and Yu, R.K. (1982) New strategies for detection and resolution of minor gangliosides as applied to brain fucogangliosides. *Methods Enzymol.*, 83: 139–189.

Li, S.-C., Sonnino, S., Tettamanti, G. and Li, Y.-T. (1988) Characterization of a nonspecific activator protein for the enzymatic hydrolysis of glycolipids. *J. Biol. Chem.*, 263: 6588–6591.

Lippincott-Schwartz, J., Yuan, L.C., Bonifacino, J.S. and Klausner, R.D. (1989) Rapid redistribution of Golgi proteins into the ER in cells treated with BFA: evidence for membrane cycling from Golgi to ER. *Cell*, 56: 801–813.

Lippincott-Schwartz, J., Donaldson, J.G., Schweizer, A., Berger, E.G., Hauri, H.P., Yuan, L.C. and Klausner, R.D. (1990) Microtubuli-dependent retrograde transport of proteins into the ER in the presence of BFA suggests an ER recycling pathway. *Cell*, 60: 821–836.

Lipsky, N.G. and Pagano, R.E. (1985a) Intracellular translocation of fluorescent sphingolipids in cultured fibroblasts: endogenously synthesized sphingomyelin and glucocerebroside analogs pass through the Golgi apparatus en route to the plasma membrane. *J. Cell Biol.*, 100: 27–34.

Lipsky, N.G. and Pagano, R.E. (1985 b) A vital stain for the Golgi apparatus. *Science*, 228: 745–747

Mandon, E.C., van Echten, G., Birk, R., Schmidt, R.R and Sandhoff, K. (1991) Sphingolipid biosynthesis in cultured neurons. Down-regulation of serine palmitoyltransferase by sphingoid bases. *Eur. J. Biochem.*, 198: 667–674.

Mandon, E., Ehses, I, Rother, J., van Echten, G. and Sandhoff, K. (1992) Subcellular localization and membrane topology of serine palmitoyltransferase, 3-dehydrosphinganine reductase and sphinganine N-acyltransferase in mouse liver. *J. Biol. Chem.*, 267: 11144–11148.

Manheller, W. (1990) Sphingosin und seine N- und O-Methylderivate — Synthesen, Metabolismus und intrazellulärer Transport. Thesis, Institut für Organische Chemie und Biochemie, Universität Bonn, Germany.

Matsuura, H., Greene, T. and Hakomori, S.-I. (1989) An α-N-acetylgalactosaminylation at the threonine residue of a defined peptide sequence creates the oncofetal peptide epitope in human fibronectin. *J. Biol. Chem.*, 264: 10472–10476.

Matyas, G.R. and Morré, D.J. (1987) Subcellular distribution and biosynthesis of rat liver gangliosides. *Biochim. Biophys. Acta*, 921: 599–614.

McKanna, J.A., Haigler, H.T. and Cohen, S. (1979) Hormone receptor topology and dynamics: Morphological analysis using ferritin-labeled epidermal growth factor. *Proc. Natl. Acad. Sci. USA*, 76, 5689–5693.

Mehl, E. and Jatzkewitz, H. (1964) Eine Cerebrosidsulfatase aus Schweineniere. *Hoppe-Seyler's Z. Physiol. Chem.*, 339: 260–276.

Meier, E.M., Schwarzmann, G., Fürst, W. and Sandhoff, K. (1991) The human GM2 activator protein: a substrate specific cofactor of hexosaminidase A. *J. Biol. Chem.*, 266: 1879–1887.

Melkerson-Watson, L.J. and Sweely, C.C. (1991) Purification to apparent homogeneity by immunoaffinity, chromatography and partial characterization of the GM3-ganglioside-forming enzyme, CMP-sialic acid: lactosylceramide α2,3-sialyltransferase (SAT-1), from rat-liver Golgi. *J. Biol. Chem.*, 266: 4448–4457.

Merrill, A.H. and Wang, E. (1986) Biosynthesis of long-chain (sphingoid) bases from serine by LM cells. *J. Biol. Chem.*, 261: 3764–3769.

Mesaric, M. and Decker, K. (1990a) Sialyltransferase activities in cultured rat hepatocytes. *Biochem. Biophys. Res. Commun.*, 171: 132–137.

Mesaric, M. and Decker, K. (1990b) Activation of GD3 synthase by sex steroid hormones in cultured rat hepatocytes, *Biochem. Biophys. Res. Commun.*, 171: 1188–1191.

Miller-Prodraza, H. and Fishman, P.H. (1984) Effects of drugs and temperature on biosynthesis and transport of glycosphingolipids in cultured neurotumor cells. *Biochim. Biophys. Acta*, 804: 44–51.

Miller–Prodraza, H., Mansson, J.E. and Svennerholm, L. (1991) Pentasialogangliosides of human brain. *FEBS Lett.*, 288: 212–214.

Moreau, P., Rodriguez, M., Cassagne, M., Morré, D.M. and Morré, D.J. (1991) Trafficking of lipids from the endoplasmic reticulum to the Golgi apparatus in a cell-free system from rat liver. *J. Biol. Chem.*, 266: 4322–4328.

Nagai, Y., Nakaishi, H. and Sanai, Y. (1986) Gene transfer as a novel approach to the gene-controlled mechanism of the cellular expression of glycosphingolipids. *Chem. Phys. Lipids*, 42: 91–103.

Nakaishi, H., Sanai, Y., Shiroki, K. and Nagai, Y. (1988a) Analysis of cellular expession of gangliosides by gene transfection I: GD3 expression in myc-transfected and transformed 3 Y 1 correlates with anchorage-independent growth activity. *Biochem. Biophys. Res. Commun.*, 150: 760–765.

Nakaishi, H., Sanai, Y., Shibuya, M. and Nagai, Y. (1988b) Analysis of cellular expression of gangliosides by gene transfection II: rat 3 Y 1 cells transformed with several DNAs containing onocgenes *(fes, fps, ras & src)* invariably express sialosyl-paragloboside. *Biochem. Biophys. Res. Commun.*, 150: 766–774.

Nakakuma, H., Sanai, Y., Shiroki, K. and Nagai, Y. (1984) Gene-regulated expression of glycolipids: appearance of GD3 ganglioside in rat cells on transfection with transforming gene E 1 of human adenovirus type 12 DNA and its transcriptional subunits. *J. Biochem.*, 96, 1471–1480.

Nakano, T., Sandhoff, K., Stümper, J., Christomanou, H. and Suzuki, K. (1989) Structure of full-length cDNA coding for sulfatide activator, a co-β-glucosidase and two other homologous proteins: Two alternate forms of the sulfatide activator. *J. Biochem. (Tokyo)*, 105:152–154.

Nores, G.A. and Caputto, R. (1984) Inhibition of the UDP-*N*-acetylgalactosamine: GM3, *N*-acetylgalactosaminyl transferase by gangliosides. *J. Neurochem.*, 42: 1205–1211.

Ong, D.E. and Brady, R.N. (1973) In vivo studies on the introduction of the 4-*t*-double bond of the sphingenene moiety of rat brain ceramides. *J. Biol. Chem.*, 248: 3884–3888.

Pagano, R.E. and Sleight, R.G. (1985) Defining lipid transport pathways in animal cells. *Science*, 229: 1051–1057.

Phillips, M.L., Nudelman, E., Gaeta, F.C.A., Pevez, M., Singhal, A.K., Hakomori, S.I. and Paulson, J.C. (1990) ELAM 1 mediates cell adhesion by recognition of a carbohydrate ligand, sialyl-le. *Science*, 250: 1130–1132.

Pohlentz, G., Klein, D., Schwarzmann, G., Schmitz, D. and Sandhoff, K. (1988) Both GA2-, GM2-, and GD2 synthases and GM1b-, GD1a- and GT1b synthases are single enzymes in Golgi vesicles from rat liver. *Proc. Natl. Acad. Sci. USA*, 85: 7044–7048.

Quiroga, S. and Caputto, R. (1988) An inhibitor of the UDP-acetylgalactosamine: GM3, *N*-acetylgalactosaminyl-transferase:purification and properties, and preparation of an antibody to this inhibitor. *J. Neurochem.*, 50: 1695–1700.

Quiroga, S., Panzetta, P. and Caputto, R. (1991) Internalization of the inhibitor of the *N*-acetylgalactosaminyltransferase by chicken embryonic retina cells — Reversibility of the inhibitor effects. *J. Neurosci. Res.*, 30, 414–420.

Rademacher, T.W., Parekh, R.B. and Dwek, R.A. (1988) Glycobiology. *Annu. Rev. Biochem.*, 57, 785–838.

Rafi, M.A., Zhang, X.-L., De Gala, G. and Wenger, D.A. (1990) Detection of a point mutation in sphingolipid activator protein-1 mRNA in patients with a variant form of metachromatic leukodystrophy. *Biochem. Biophys. Res. Commun.*, 166: 1017–1023.

Renfrew, C.A. and Hubbard, A.L. (1991) Degradation of epidermal growth factor receptor in rat liver. Membrane topology through the lysosomal pathway. *J. Biol. Chem.*, 266: 21265–21273.

Richardson, C.L., Keenaan, T.W. and Morré, D.J. (1977) Ganglioside biosynthesis. Characterisation of CMP-*N*-acetylneuraminic acid: Lactosylceramide sialyltransferase in Golgi apparatus from rat liver. *Biochim. Biophys. Acta,* 488: 88–96.

Rother, J., van Echten, G., Schwarzmann, G. and Sandhoff, K. (1992) Biosynthesis of sphingolipids: dihydroceramide and not sphinganine is desaturated by cultured cells. Biochem. Biophys. Res. Commun., 189: 14–20.

Saito, M., Saito, M. and Rosenberg, A. (1984) Action of monensin, a monovalent cationophore, on cultured human fibroblasts: evidence that it induces high cellular accumulation of glucosyl- and lactosylceramide. *Biochemistry,* 23: 1043–1046.

Schnabel, D., Schröder, M. and Sandhoff, K. (1991) Mutation in the sphingolipid activator protein 2 in a patient with a variant of Gaucher disease. *FEBS Lett.,* 284: 57–59

Schnabel, D., Schröder, M., Fürst, W., Klein, A., Hurwitz, R., Zenk, T., Weber, G., Harzer, K., Paton, B., Poulos, A., Suzuki, K. and Sandhoff, K. (1992). Simultaneous deficiency of sphingolipid activator proteins 1 and 2 is caused by a mutation in the initiation codon of their common gene. *J. Biol. Chem.,* 267: 3312–3315.

Schwarzmann, G. and Sandhoff, K. (1990) Metabolism and intracellular transport of glycosphingolipids. "Perspectives in Biochemistry". *Biochemistry,* 29: 10865–10871.

Shukla, G.S. Shukla, A. and Radin, N.S. (1991) Gangliosides inhibit glucosylceramide synthase: a possible role in ganglioside therapy. *J. Neurochem.,* 56: 2125–2132.

Schmid, B., Paton, B.C., Sandhoff, K. and Harzer, K. (1992) Metabolism of GM1 ganglioside in cultured skin fibroblasts: Anomalies in gangliosidoses, sialidoses, and sphingolipid activator protein (SAP, saposin) 1 and prosaposin deficient disorders. *Hum. Genet.,* 89: 573–518.

Stoffel, W., Le Kim, D. and Sticht, G. (1968) Metabolism of sphingosine bases: Biosynthesis of dihydrosphingosine in vitro. *Hoppe-Seyler's Z. Physiol. Chem.,* 349: 664–670.

Stotz, H. (1990) NBD-Sphingolipide — Synthese, Metabolismus und intrazellulärer Transport. Thesis, Institut für Organische Chemie und Biochemie der Universität Bonn, Germany.

Svennerholm L. (1963) Chromatographic separation of human brain gangliosides. *J. Neurochem.,* 10: 613–623.

Svennerholm, L. (1984) Biological significance of gangliosides. In H. Dreyfus, R. Massarelli, L. Freysz and G. Rebel (Eds.), *Cellular and Pathological Aspects of Glycoconjugate Metabolism, Vol. 126,* INSERM, France, pp. 21–44.

Tartakoff, A.M. (1983) Perturbation of vesicular traffic with the carboxyl ionophore monensin. *Cell,* 32: 1026–1028.

Tiemeyer, M., Yasuda, Y. and Schnaar, R.L. (1989) Ganglioside-specific binding protein on rat brain membranes. *J. Biol. Chem.,* 264: 1671–1681.

Trinchera, M. and Ghidoni, R. (1989) The glycosphingolipid sialyltransferases are localized in different sub-Golgi compartments in rat liver. *J. Biol. Chem.,* 264: 15766–15769.

Trinchera, M., Fabbri, M. and Ghidoni, R. (1991) Topography of glycosyltransferases involved in the initial glycosylations of gangliosides. *J. Biol. Chem.,* 266, 20907–20912.

van Echten, G. and Sandhoff, K. (1989) Modulation of ganglioside biosynthesis in primary cultured neurons. *J. Neurochem.,* 52: 207–214.

van Echten, G., Iber, H., Stotz, H., Takatsuki, A. and Sandhoff, K. (1990a) Uncoupling of ganglioside biosynthesis by Brefeldin A. *Eur. J. Cell Biol.,* 51: 135–139.

van Echten, G., Birk, R., Brenner-Weiss, G., Schmidt, R.R. and Sandhoff, K. (1990b) Modulation of sphingolipid biosynthesis in primary cultured neurons by long-chain bases. *J. Biol. Chem.,* 265: 9333–9339.

van Meer, G. (1989) Polarity and polarized transport of membrane lipids in cultured epithelium. *Annu. Rev. Cell Biol.,* 5: 247–275.

Vogel, A., Schwarzmann, G. and Sandhoff, K. (1991) Glycosphingolipid specificity of the human sulfatide activator protein. Eur. *J. Biochem.,* 200: 591–597

Walz, G., Aruffo, A., Kolanus, W., Bevilacqua, M. and Seed, B. (1990) Recognition by ELAM-1 of the sialyl-le determinant on myeloid and tumor cells. *Science,* 250: 1132–1135.

Watanabe, K., Hakomori S., Powell, M.E. and Yokota, M. (1980) The amphiphatic membrane proteins associated with gangliosides: the Paul Bunnell antigen is one of the gangliophilic proteins. *Biochem. Biophys. Res. Commun.,* 92: 638–646.

Wessling-Resnick, M. and Braell, W.A. (1990) The sorting and segregation mechanism of the endocytic pathway is functional in a cell-free system. *J. Biol. Chem.,* 265: 690–699.

Young, W.W., Jr., Lutz, M.S., Mills, S.E. and Lechler–Osborn, S. (1990) Use of Brefeldin A to define sites of glycosphingolipid synthesis GA2/GM2/GD2 synthase is trans to the BFA block. *Proc. Natl. Acad. Sci. USA,* 87: 6838–6842.

Yu, R.K., Gu, T.–J. and Gu, X.–B. (1990) Purification to homogeneity of GD1a synthase from rat brain. *FASEB J.,* 4(7): A 2140.

Yusuf, H.K.M., Pohlentz, G., Schwarzmann, G. and Sandhoff, K. (1983a) Ganglioside biosynthesis in rat liver Golgi apparatus: Stimulation by phosphatidylglycerol and inhibition by tunicamycin. *Eur. J. Biochem.,* 134: 47–54.

Yusuf, H.K.M., Pohlentz, G. and Sandhoff, K. (1983b) Tunicamycin inhibits ganglioside biosynthesis in rat liver Golgi apparatus by blocking sugar nucleotide transport across the membrane vesicles. *Proc. Natl. Acad. Sci. USA,* 80: 7075–7079

Yusuf, H.K.M., Schwarzmann, G., Pohlentz, G. and Sandhoff, K. (1987) Oligosialogangliosides inhibit GM2- and GD3-synthesis in isolated Golgi vesicles from rat liver. *Hoppe-Seyler's Z. Physiol. Chem.,* 368: 455–462.

Zhang, X.-L., Rafi, M.A., De Gala, G. and Wenger, D.A. (1990) Insertion in the mRNA of a metachromatic leukodystrophy patient with sphingolipid activator protein-1 deficiency. *Proc. Natl. Acad. Sci. USA,* 87: 1426–1430.

Zhang, X.-L., Rafi, M.A., De Gala, G. and Wenger, D.A. (1991) The mechanism for a 33-nucleotide insertion in messenger RNA causing sphingolipid activator protein (SAP-1)-deficient metachromatic leukodystrophy. *Hum. Genet.,* 87: 211–215.

L. Svennerholm, A.K. Asbury, R.A. Reisfeld, K. Sandhoff, K. Suzuki, G. Tettamanti and G. Toffano (Eds.)
Progress in Brain Research, Vol. 101
© 1994 Elsevier Science BV. All rights reserved.

CHAPTER 3

Development regulation of ganglioside metabolism

Robert K. Yu

Department of Biochemistry and Molecular Biophysics, Medical College of Virginia, Virginia Commonwealth University, Richmond, VA, U.S.A.

Developmental changes of nervous system gangliosides

A large volume of literature has accumulated showing that the expression of gangliosides in the nervous system is cell-specific and developmentally regulated (see reviews: Ledeen, 1985; Yu and Saito, 1989). Early studies by Svennerholm (1964) and Suzuki (1965) first revealed qualitative and quantitative changes in developing human and rat brains. These initial studies have been followed by intensive investigations of changes in fetal and embryonic nervous tissues (Vanier et al., 1971; Merat and Dickerson, 1973; Yusef et al., 1977; Irwin and Irwin, 1979, 1980; Rosner, 1982; Seyfried et al., 1983, 1984; Kracun et al., 1984; Yu et al., 1988; Bouvier and Seyfried, 1989; Hirabayashi et al., 1989, 1990; Sonnino et al., 1990; Svennerholm et al., 1989, 1991; Rosner et al., 1992) as well as neural cells in tissue culture (Dreyfus et al., 1975, 1980; Yavin and Yavin, 1979; Greis and Rosner, 1990; Thangnipon and Balazs, 1992). By correlating ganglioside changes with such cellular events as neural cell proliferation and migration, axonal and dendritic arborization and synaptogenesis, and myelination, these studies have helped to develop a much better understanding of the cellular localization and the function of gangliosides in neural cell recognition, interaction, and adhesion (Hakomori, 1990; Rahmann and Wiegandt, 1991). These studies have also culminated in the characterization of many developmentally regulated and stage specific gangliosides. Since strict cellular architecture

is required for normal brain function, the expression of gangliosides during brain development must be strictly regulated spatially and temporally. The emerging picture from these studies (Rosner, 1982; Yu et al., 1988; Sonnino et al., 1990; Svennerholm et al., 1989, 1991; Rosner et al., 1992) has revealed the following trends in ganglioside expression during the various developmental stages:

Stage I: Neural tube formation — emergence of GD3.

Stage II: Neuroblast and glioblast proliferation — peak expression and synthesis of GD3, GM3, and O-acetyl GD3, accretion of lacto-series gangliosides 3'-LM1, 3'-isoLM1 and 3',8'-LD1, and emergence of c-series gangliosides.

Stage III: Neurogenesis and neuritogenesis — diminished expression of GD3 and c-series gangliosides, followed by increased synthesis of b-series gangliosides.

Stage IV: Axonal and dendritic arborization, synaptogenesis — increased synthesis of a-series gangliosides, particularly GM1 and GD1a.

Stage V: Myelination — expression and accretion of GM1 and GM4 with the latter more abundant in primate and avian species.

Distortions in ganglioside expression during brain development are known to result in aberrant cellular architecture, leading to mental retardation. This is particularly noteworthy in the various ganglioside catabolic diseases. The importance of regulated ganglioside expression is further underscored by the failed neuronal differentiation in mouse embryos with a T-locus mutation which affects the synthesis of b-series gangliosides (Seyfried, 1987; Novikov and Seyfried, 1991) as well as in transgenic mouse embryos with a defect in the expression of 9-*O*-acetyl GD3, a developmentally regulated ganglioside (Varki et al., 1991). Thus, an understanding of factors that regulate ganglioside expression is critical in understanding their role in nervous system development.

Ganglioside synthesis during development

Ganglioside synthesis at various stages of brain development is necessarily controlled at the genetic and epigenetic levels which are morphoregulated by cell–cell and cell–matrix interactions. In order to elucidate control at these levels, it is important to gain first an understanding of the metabolic basis for ganglioside expression during development. One approach we have undertaken is to correlate the developmental changes in ganglioside composition with enzymatic activities. We found that in rat brain of embryonic day (E) 14, GM3 and GD3 were predominant (Fig. 1 and Table I) (Yu et al., 1988). At E16, b-series gangliosides, such as GD1b, GT1b, and

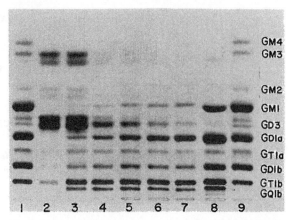

Fig. 1. Thin-layer chromatogram of developing rat brain gangliosides: lane 1 and 9, normal human brain white matter gangliosides; lane 2, E14; lane 3, E16; lane 4, E18; lane 5, E20; lane 6, E22; lane 7, newborn; and lane 8, adult. Each lane contained 3 μg of lipid-bound sialic acid. The developing solvent was chloroform/methanol/aqueous 0.02% CaCl$_2$-H$_2$O (55:45:10 by vol.). Gangliosides were visualized by the resorcinol-HCl reagent (data from Yu et al., 1988).

GQ1b, increased in content. After E18, a-series gangliosides such as GM1a (GM1), GD1a, and GT1a increased in content, and the content of GM3 and

TABLE I

Gangliosides of developing rat embryonic brain

	E14	E16 Day	E18	E20 Day	E22	PO	A
Total S.A.[a]:	42.9±1.4	50.0±1.4	116.2±8.6	160.3	169.9±22.1	286.9±22.1	384.0±59.7
N:	3	3	3	2	3	3	3
				% Distribution[b]			
GM3	19.7	16.8	4.5	3.8	2.9	1.5	1.7
GM2	2.2	1.6	N.D.	N.D.	1.2	1.0	0.4
GM1	N.D.	0.5	1.9	3.1	4.0	5.4	12.9
GD3	44.3	37.6	21.1	12.6	7.9	3.3	0.6
GD1a	1.8	4.9	12.6	19.0	25.1	29.0	27.8
GT1a	2.3	N.D.	N.D.	N.D.	N.D.	N.D.	N.D.
GD2	4.5	5.6	8.3	6.7	6.1	3.4	3.8
GD1b	8.7	8.3	13.3	13.3	12.0	11.0	15.8
GT1b	10.4	15.8	26.8	28.8	29.5	34.4	25.4
GQ1b	3.7	6.9	11.5	12.8	11.1	11.0	11.5

[a] Mean ± S.D. μg/g wet wt.

[b] Values represent average of at least 3 determinations of one representative sample.

GD3 markedly decreased. Because of the nature of these changes in composition, it is evident that one of the means for regulating their expression is by their rates of synthesis. Examination of the biosynthetic pathways (Fig. 2) emanating from the studies of several groups of investigators (Kaufman et al., 1967, 1968; Yip and Dain, 1969; Roseman, 1970; Yu and Lee, 1976; Yu and Ando, 1980; Yohe and Yu, 1980; Yohe et al., 1982; Basu et al., 1987, 1988; Pohlentz et al., 1988) revealed several branching points which constitute unique control points for ganglioside syn-

thesis. Because of the changes in ganglioside composition we observed in embryonic brain, we determined the activities, in brain homogenates, of two key enzymes at one of the branching points: sialyltransferase II (ST-II) for GD3 synthesis and N-acetylgalactosaminyltransferase (GalNAc-T) for GM2 synthesis. The activity of ST-II was constant between E14 and E18 but decreased rapidly from E18 to birth (Fig. 3). In contrast, the activity of GalNAc-T increased between E14 and E18, but was constant from E18 to birth (Fig. 4). These changes in enzyme

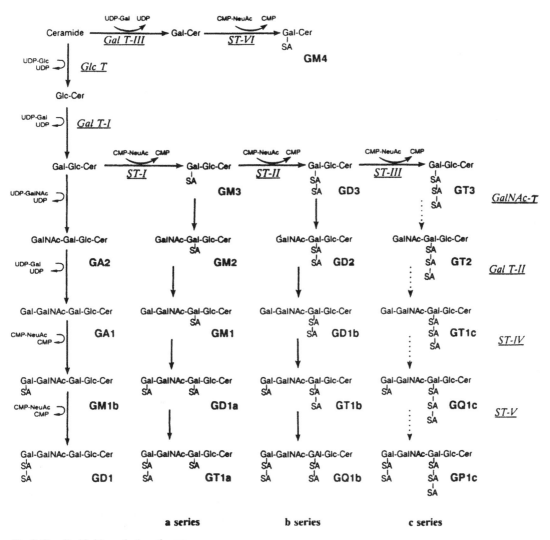

Fig. 2. Ganglioside biosynthetic pathways.

34

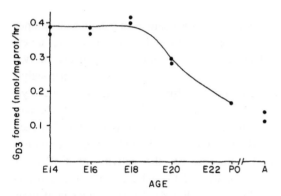

Fig. 3. Profile of GD3 synthesis in developing rat brain. PO, newborn; A, adult (data from Yu et al., 1988).

activities are consistent with the concept that during development there is a shift in synthesis from the c- to the b-, and then to the a-pathway (Yu and Ando, 1980; Rosner 1982; Kasai and Yu, 1983, Rosner et al., 1987). A developmental correlation between the expression and synthesis of the lacto-series gangliosides has also been observed in human fetal brain (Percy et al., 1991). These examples illustrate a remarkable correspondence between the activities of the various glycosyltransferases and the ganglioside expression during development.

Kinetic and substrate specificity studies have indicated that ST-I, ST-II, and ST-III are probably distinct enzyme entities whereas GalNAc-T may be a common enzyme for several pathways (Cumar et al., 1971; Pohlentz et al., 1988; Iber et al., 1991). However, in order to study the mechanisms for the

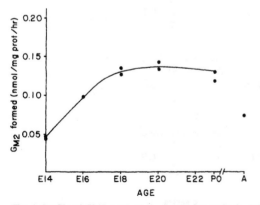

Fig. 4. Profile of GM2 synthesis in developing rat brain. PO, newborn; A, adult (data from Yu et al., 1988).

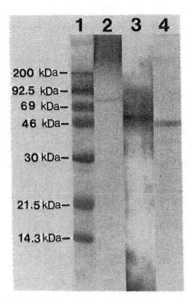

Fig. 5. Sodium-dodecylsulfate polyacrylamide gel electrophoretic pattern of rat brain sialyltransferases. Lane 1, MW standards; lane 2, ST-I; lane 3, ST-II; lane 4, ST-IV. The bands were visualized by silver staining. The percentage of gel was 12%.

regulation of the various pathways, it is necessary to purify the various glycosyltransferases involved (Fig. 2). Progress in this area of enzyme purification has been extremely slow, due probably to the following factors: (a) most of the glycosyltransferases are localized in ER or Golgi membranes and are difficult to solubilize; (b) these enzymes are present in very low amounts and are highly unstable when solubilized; and (c) the activities of the enzymes are relatively difficult to assay in vitro. During the last few years, however, we have witnessed the successful isolation and purification, to homogeneity or apparent homogeneity of several glycosyltransferases in ganglioside synthesis, including ST-IV (Joziassé et al., 1985; Gu, T. et al., 1990), ST-II (Gu, X. et al., 1990), ST-I (Melkerson-Watson and Sweeley, 1991; Preuss et al., 1992), and GalT-II (Chatterjee et al., 1992). Figure 5 shows the SDS-PAGE chromatogram of three sialyltransferases purified from rat brain by successive affinity chromatographies of the detergent solubilized enzymes on CDP-Sepharose and glycolipid-Sepharose. ST-I has an apparent molecular mass of 76 kDa, ST-II 55 kDa, and ST-IV 45 kDa. It is clear

TABLE II

Glycolipid acceptor specificity of rat brain a2-3 Sialyltransferase-IV

Glycolipid	Structure	Relative Activity
Cerebroside	Galβ1-1'Cer	0
Lactosylceramide	Galβ1-4G1cβ1-1'Cer	4
Globoside	GalNAcβ1-3Galα1-4Galβ1-4Glcβ1-1'Cer	0
Gb3	Galα1-4Galβ1-4Glcβ1-1'Cer	0
Forssman antigen	GalNAcα1-3GalNAcβ1-3Galα1-4Galβ1-4Glcβ1-1'Cer	0
GM2	GalNAcβ1-4Galβ1-4-Glcβ1-1'Cer NeuAcα2-3	1
GM3	NeuAcα2-3Galβ1-4Glcβ1-1'Cer	0
GD3	NeuAcα2-8NeuAcα2-3Galβ1-4Glcβ1-1'Cer	0
Asialo-GM1	Galβ1-3GalNAcβ1-4Galβ1-4Glcβ1-1'Cer	54
GM1	Galβ1-3GalNAcβ1-4Galβ1-4Glcβ1-1'Cer NeuAcα2-3	88
GD1ba	Galβ1-3GalNAcβ1-4Galβ1-4Glcβ1-1'Cer NeuAcα2-8NeuAcα2-3	100

that these enzymes are indeed distinct enzyme entities as Pohlentz et al. (1989) have postulated. ST-I is specific for GM3 synthesis from lactosyl ceramide (β-galactoside, α2–3 sialyltransferase) and ST-II is specific for GD3 synthesis from GM3 (αNeuAc ketoside, α2–8 sialyltransferase). However, ST-IV exhibits reactivity toward GA1, GM1 and GD1b in the synthesis of GM1b, GD1a, and GT1b, respectively (Table II). Thus, ST-IV can be considered to be a less specific β-galactoside, a2–3 sialyltransferase. This latter observation provides direct proof that glycosyltransferases in the late Golgi network (trans-Golgi) exhibit less strict substrate specificity (Iber et al., 1991, 1992). Presumably, β-GalNAc-T, whose cDNAs have recently been cloned (Nagata et al, 1992), is also in this same category. The availability of the highly purified enzymes is a prerequisite to understanding the molecular mechanisms by which brain cells coordinate and regulate the expression of various gangliosides during development.

Regulation of ganglioside synthesis

Genetic regulation

Several reports have appeared indicating that the expression of gangliosides may be largely regulated at the genetic level. The expression of b-series gangliosides is associated with the T/t locus located on mouse chromosome 17 (Seyfried, 1987). The gene locus that regulates the expression of GM1 and GD1a in mouse liver has been linked to the H-2 complex on chromosome 17 (Hashimoto et al., 1983). In human meningiomas, the expression of GD3 has been linked with monosomy of chromosome 22 (Fredman et al., 1990). Using gene transfer techniques, it has been shown that cells transfected by DNA- or RNA-tumor viruses express altered and characteristic changes in ganglioside pattern (Hakomori and Kannagi, 1983; Nagai et al., 1986). Nakaishi et al. (1988a,b) further showed that rat 3Y1 fibroblast cells over-expressed GD3 in their nucleus when transfected with c-*myc* DNA (localized intranuclearly), whereas they express 3'-LM1 in the cytoplasm when transfected with extranuclear type oncogenes such as v-*fes*, v-*fps*, v-*ras*, and v-*src*. Presumably, these oncogenes contain the specific transcription factors that activate the appropriate glycosyltransferases. This appears to be the case for β-galactoside, α2–6 sialyltransferase for glycoprotein sialylation whose expression is tissue specific (highest in liver) and is most likely transcriptionally regulated (Paulson et al., 1989; Svensson, et al., 1990, 1992). Evidence from analysis of the gene for this enzyme revealed binding sequences for a number of liver restricted transcription factors such as hepatocyte nuclear factor 1α (HNF-1α), liver specific factors D-binding protein (DBP), and liver enriched transcription activator protein (LAP), as well as the more general transcription factors AP-1 and AP-2. Presumably, the expression of the various glycosyltransferases for ganglioside synthesis may be regulated in an analogous manner.

Intracellular trafficking

Other factors involved in the regulation of ganglioside expression include the proper translocation and sorting of the glycolipid products in multi-glycosyltransferase systems. The concept of a coordinated glycolipid precursor-product relationship in a multi-glyscosyltransferase system was first formulated by Roseman (1970). This was subsequently elaborated by Caputto et al. (1976) who demonstrated the presence of two ganglioside pools, a small pool of highly metabolically active transient precursors for the synthesis of more complex gangliosides and a large pool of end products destined for deposition in the plasma membrane. At present, the localization and trafficking of several glycosyltransferases in the ER, Golgi network and plasma membrane axis are being elucidated (Tettamanti, 1984; Sato et al., 1988; der Meer, 1989; Trinchera and Ghidoni, 1989, 1990a,b; Trinchera et al., 1991; Iber et al., 1992; Young et al., 1992). Disruption of the flow of gangliosides along their biosynthetic pathways can cause alterations in ganglioside expression. Thus, treatment of cultured granule cells with the Golgi uncoupling agent brefeldin A results in the accumulation of LacCer, GM3, and GD3, but not the more complex gangliosides in the a-series (Van Echten et al., 1990). Monensin inhibited the formation of only the more complex glycosphingolipids (Miller-Prodraza and Fishman, 1984; Saito et al., 1984; Hogan et al., 1988), whereas tunicamycin blocks total ganglioside synthesis in the rat liver Golgi apparatus (Yusuf et al., 1983) and in neuronal cells (Guarnaccia et al., 1983).

Metabolic control

A regulatory mechanism for Golgi-localized synthesis is the so-called feedback control in which the activity of ST-II can be inhibited by the end-product GQ1b, and GalNAc-T by end-products GD1a and GT1b (Nores and Caputto, 1984; Yusuf et al., 1987). A similar phenomenon has also been demonstrated for GalNAc-T in which several polysialogangliosides exerted inhibitory effects (Yu et al., 1983). The physiological significance of these in vitro studies is not clear at present as gangliosides themselves are detergent-like substances and the possibility exists that these effects may be attributable to a detergent effect (Yu et al., 1983). Thus, the question arises whether there are natural regulators for the various glycosyltransferases. One example is the discovery of an endogenous inhibitor protein, in the 100,000 g supernatant of rat brain homogenate, of ST-I (Albarracin et al., 1988). Other inhibitors include one for glucosyl- and galactosyltransferase (Costantino-Ceccarini and Suzuki, 1978) and another for galactosaminyltransferase (Quiroga et al., 1984). It is likely that these glycosyltransferase inhibitors may play some role in the regulation of the synthesis of gangliosides in vivo.

Another intriguing mechanism for the regulation of ganglioside synthesis has been described by Iber et al. (1990) who found that in cultured cerebellar cells a shift in the biosynthesis from a- to b-series gangliosides occurred when the pH of the medium was lowered from 7.4 to 6.2, with the effect being reversible. Whether changes in pH occur in vivo to effect the shift in biosynthesis of gangliosides during development remains to be substantiated.

Post-translational modification of glycosyltransferases

Studies from many laboratories have indicated that inducers of differentiation frequently lead to an enhanced glycolipid synthesis (for reviews: Hakamori, 1981, 1990; Burczak et al., 1984). Since these inducers (e. g., sodium butyrate, phorbol esters, various retinoid compounds) are known to alter the phosphorylation state of the cells, the activities of glycosyltransferases may be under phosphorylation-

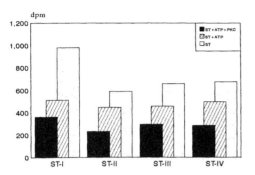

Fig. 6. Effect of PKC and ATP on sialyltransferase activities. (1) The enzyme preparation (70 μg protein) consisted of a Triton extract of the microsomes of rat brain. It was incubated with an appropriate lipid acceptor (40 nmol), cacodylate buffer (25 mM, pH 6.5), MnCl$_2$ (10 mM), and Triton X-100 (0.15%) in a final volume of 100 μl. (2) Sialyltransferase inactivation by ATP. The reaction system consisted of (1) plus the following components: MgCl$_2$ (10 mM), CaCl$_2$ (0.5 mM), phosphatidylserine (2.5 μg), 1,2-dioctanoylglycerol (2.0 μg), and ATP (100 μM).(3) Sialyltransferase inactivation by PKC. The reaction system consisted of (2) plus PKC (50 ng, 0.1 unit). The reaction was carried out at 37°C for 60 min. The glycolipid products were separated by Sephadex G-50 column and their radioactivities counted.

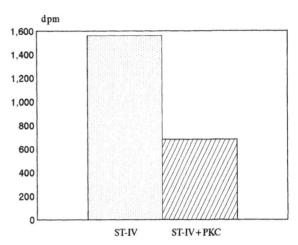

Fig. 7. Inactivation of purified sialyltransferase-IV by PKC. Highly purified rat brain ST-IV was used. For reaction conditions, see Fig. 6.

dephosphorylation control during cellular differentiation. In fact, there is a correlation between ST-I activity and both cAMP-dependent protein kinase (Burczak et al., 1984; Moskal et al., 1987) and protein kinase C (PKC) (Xia et al., 1989) activities. Similarly, GalNAc-T can be activated by cAMP, presumably through the mediation of cAMP-dependent protein kinase (Scheideler and Dawson, 1986). Direct evidence for the above examples, however, is still lacking.

With the availability of highly purified sialyltransferases, this possibility can be examined more directly (Yu, Gu, Gu, Preuss, unpublished data). In initial experiments using a solubilized rat brain sialyltransferase preparation, we found that the addition of 100 μM of ATP resulted in nearly a 50% reduction in the activities of ST-I, -II, -III, and -IV. This is presumably due to their phosphorylation by endogenous protein kinases (Fig. 6). Further addition of a highly purified PKC preparation caused an additional reduc-

tion in the activities of these enzymes. Inhibition of enzyme activity could also be demonstrated using a highly purified rat brain ST-IV preparation (Fig. 7) which could be reversed by treatment with a rat brain membrane bound phosphatase. Analysis of the phosphoamino acids generated from the phosphorylated ST-IV revealed that serine residues were phosphorylated (Fig. 8). Our results thus demonstrate a regula-

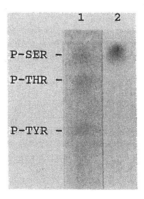

Fig. 8. Analysis of the phosphoamino acids of sialyltransferase IV. ST-IV was phosphorylated by protein kinase C (in the presence of [^{32}P]ATP). After SDS-PAGE, it was transferred to nitrocellulose paper and hydrolyzed in 5.7 N HCl. The hydrolysate was analyzed by electrophoresis at 2000 V. Lane 1 contained standards and was visualized with ninhydrin; lane 2 contained 10 μl of the hydrolysate from ^{32}P-ST-IV sample and was visualized by exposure to an X-ray film.

tory role for the protein kinase systems, which may be activated by external signals, in the regulation of sialyltransferases for ganglioside synthesis. The significance of this regulatory mechanism will undoubtedly be further examined when additional purified glycosyltransferases become available.

Developmental regulation of ganglioside catabolism. Role of myelin-associated neuraminidase

While ganglioside composition can be regulated by synthetic enzymes during development, the degradative enzymes, particularly the neuraminidases (siali-

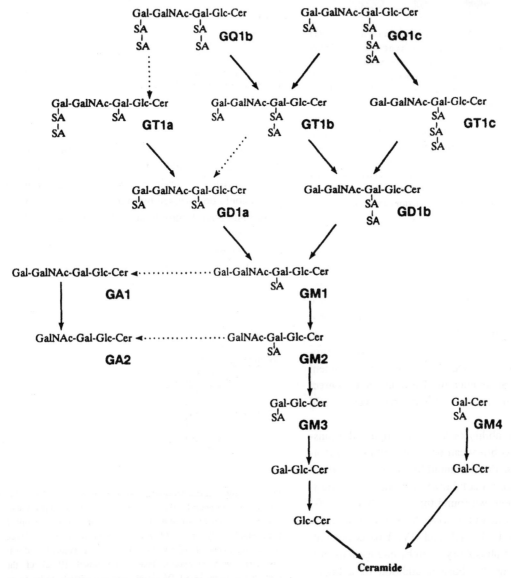

Fig. 9. Degradative pathways of complex gangliosides.

dases) also play an essential role. The degradative pathways for complex gangliosides is shown in Fig. 9. At least five different neuraminidases are known to occur in the brain: two are soluble (Venerando et al., 1975, 1982) and three are membrane bound, located in the synaptosomal (Schengrund and Rosenberg, 1970; Miyagi et al., 1990), lysosomal (Gielen and Harprecht, 1969; Miyagi et al., 1990), and myelin (Yohe et al., 1986) membranes. There is evidence that additional membrane-bound glycosylphosphatidyl inositol anchored species may also exist (Chiarini et al., 1990).

Interest in the myelin-associated neuraminidase stems from the unique ganglioside composition of CNS myelin which is characterized by a high concentration of the monosialoganglioside GM1 and lesser amounts of di- and polysialogangiosides (Suzuki et al., 1967; Ledeen et al., 1973; Yu and Yen, 1975; Yu and Saito, 1989). In primates and avian species, GM4 is another major ganglioside and is expressed specifically in myelin and oligodendroglial cells (Ledeen et al., 1973; Yu and Iqbal, 1979). Furthermore, the ganglioside pattern of CNS myelin is distinct from that of the plasma membrane of the oligodendroglial cell; the latter is more complex with higher contents of GM3, GD3, and polysialogangliosides, such as GT1b and GQ1b (Yu and Iqbal, 1979; Yu et al., 1989). Developmentally, the ganglioside composition of myelin becomes simplified; in rat and mouse CNS myelin, the GM1 content approaches about 90% of the total gangliosides with increasing age (Suzuki et al., 1967; Yu and Yen, 1975). Together, these findings suggest the presence of a specific mechanism for regulation of ganglioside metabolism in the myelin sheath during development.

The existence of a neuraminidase activity intrinsic to myelin was first discovered when we examined interactions between gangliosides and myelin basic protein (MBP) (Yohe et al., 1983). We found that only GM4 specifically interacted with MBP, and became more resistant to the action of *Clostridium perfringens* neuraminidase. The polysialogangliosides neither interacted with MBP, nor were protected from the neuraminidase action. Based on these findings, we proposed that myelin might have a neu-

raminidase activity responsible for its unique ganglioside pattern where GM4 and GM1 are the major ganglioside components. Subsequently we presented evidence for the presence of a neuraminidase activity in purified rat myelin preparations, and later confirmed its existence and further characterized it using intact or delipidated myelin preparations (Yohe et al., 1986; Saito and Yu, 1986). The enzyme could effectively hydrolyze non-ganglioside substrates, *N*-acetylneuraminyl(2-3)lactitol (NL) and fetuin as well as GM3. Interestingly, the enzyme could, though at much slower rates, hydrolyze GM1 and GM2, which are usually resistant to bacterial and viral neuraminidases. On the other hand, GM1 acted as a competitive inhibitor for the hydrolysis of substrates such as GM3 or NL. Based upon time-activity curves on exposure of the enzyme preparation to high temperature or low pH, and K_i values obtained with 2,3-dehydro-2-deoxy-*N*-acetylneuraminic acid, a specific inhibitor for neuraminidase, it was shown that the enzyme activities toward the two substrates, GM3 and NL, were catalyzed by a single enzyme entity.

The role of myelin-associated neuraminidase in ganglioside metabolism was examined using rats ranging from 17 to 97 days of age (Saito and Yu, 1992). The neuraminidase activity directed toward the ganglioside GM3 in the total myelin fraction was high during the period of active myelination and, thereafter, decreased rapidly to the adult level (Fig. 10). This developmental activity pattern correlates well with the unique ganglioside composition which becomes simpler during active myelination with an increasing amount of GM1 and decreasing percentages of di- and polysialogangliosides. In vitro incubation of myelin of young rats under the optimal conditions for neuraminidase action produced a profile of ganglioside changes similar to that observed in in vivo development (Fig. 11). These results strongly suggest that myelin-associated neuraminidase plays a pivotal role in the developmental changes in the ganglioside composition of rat brain myelin.

Having shown that a specific interaction between myelin-associated neuraminidase and GM1 exists (Saito and Yu, 1986), we further characterized this interaction and examined its role in the adhesion of

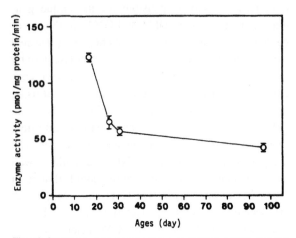

Fig. 10. Developmental changes in neuraminidase activity directed toward GM3 in rat brain myelin. Total myelin fractions were isolated from brains of rats ranging from 17 to 97 days of age. Each point indicates the mean ± S.E. ($n = 3$). (data from Saito and Yu, 1992).

rat oligodendroglial cells to GM1 (Saito and Yu, unpublished). Hydrolysis of N-acetylneuraminyl-lactitol by the enzyme was inhibited by GM1 in a com-

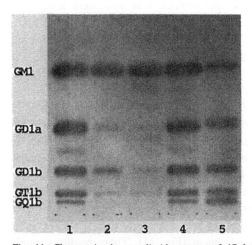

Fig. 11. Changes in the ganglioside pattern of 17-day-old rat myelin during incubation at pH 4.8. After incubation, the composition of gangliosides in myelin membranes was analyzed by the glycolipid-overlay technique with peroxidase-conjugated cholera toxin B subunit. Lane 1, before incubation; lanes 2 and 3, incubated for 2 and 8 h, respectively, without neuraminidase inhibitor (2,3-dehydro-2-deoxy-N-acetylneuraminic acid); lane 4, incubated for 8 h with inhibitor; lane 5, ganglioside standard. (data from Saito and Yu, 1992).

petitive manner, while GM1 itself was not hydrolyzed to any appreciable extent during the incubation. This suggests that GM1 may serve as a competitive inhibitor of the enzyme. Asialo-GM1 showed no inhibitory effect. When a soluble enzyme preparation was applied to a GM1-linked affinity column, the enzyme activity was retained on the column and was recovered from the column only by elution with a buffer containing high concentrations of salt or 5 mM 2,3-dehydro-2-deoxy-N-acetylneuraminic acid (Neu2en5Ac), a competitive inhibitor of neuraminidase. These results indicate that the neuraminidase possesses strong affinity for GM1, but not for asialo-GM1. Additional evidence came from studying the neuraminidase activities in isolated rat brain oligodendroglial perikarya, which are developmentally regulated with a peak at the period of active myelination (Saito et al., 1992). ^{51}Cr-Labeled rat oligodendroglial cells bound preferentially to GM1 developed on a thin-layer plate, but not to other gangliosides such as GM3, GD1a, GD1b, and GT1b. Binding of oligodendroglial cells to GM1 was inhibited by Neu2en5Ac (5 mM). These results indicate that myelin-associated neuraminidase specifically interacts with GM1. The interaction can be characterized as an 'enzyme-substrate' or 'enzyme-inhibitor' interaction, and this interaction may play an important role in the formation and stabilization of the multilamellar structure of the myelin sheath. Thus,

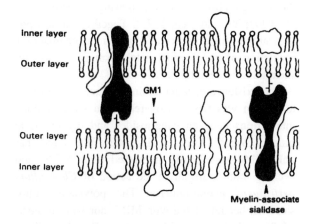

Fig. 12. A schematic diagram showing possible interaction between myelin-associated neuraminidase and GM1.

myelin-associated neuraminidase may degrade poly-sialogangliosides biosynthesized in oligodendroglial cells to GM1 and then binds to GM1 on an adjacent membrane to effect the formation and stabilization of the multilamellar structure of the myelin sheath (Fig. 12) The GM1 ganglioside is then slowly degraded to asialo-GM1 (Saito and Yu, 1986), consistent with its slow turnover rate in myelin (Suzuki, 1970; Ando et al., 1984). A previous study demonstrating the presence of asialo-GM1 in myelin also supports this hypothesis (Kusunoki et al., 1985). Since asialo-GM1 does not bind to neuraminidase, this would allow two adjacent membranes to 'dissociate' from each other, which contributes to the growing process of the multilamellar structure. Further studies are being undertaken to test this dynamic model.

Acknowledgements

The work performed in my laboratory has been done with the collaboration of Drs. Robert Ledeen, Larry Macala, Herbert Yohe, Tian-jue Gu, Xin-bin Gu, Megumi Saito, and Ute Preuss. Financial assistance has been provided by USPHS grants NS-11853, NS-23102, and NS-26994.

References:

Albarracin, I., Lassaga, F.E. and Caputto, R. (1988) Purification and characterization of an endogenous inhibitor of the sialyltransferase CMP-*N*-acetylneuraminate: Lactosylceramide a2,6-*N*-acetylneuraminyltransferase (EC 2.4.99.-). *Biochem. J.*, 254: 559–565.

Ando, S., Tanaka, Y., Ono, Y. and Kon, K. (1984) Incorporation rate of GM1 ganglioside into mouse brain myelin: Effect of aging and modification by hormones and other compounds. *Adv. Exp. Biol. Med.*, 125: 241–248.

Ando, S. and Yu, R.K. (1979) Isolation and characterization of two isomers of brain tetrasialogangliosides. *J. Biol. Chem.*, 254: 12224–12229.

Basu, S., Das, K.K., Schaeper, R.J., Banerjee, P., Daussin, F., Basu, M., Khan, F.A. and Zhang, B.-J. (1988) Biosynthesis in vitro of neuronal and non-neuronal gangliosides. In: R.W. Ledeen, E.L. Hogan, G. Tettamanti, A.J. Yates and R.K. Yu (Eds.), *New Trends in Ganglioside Research: Neurochemical and Neuroregenerative Aspects*, Liviana Press, Padova, pp. 259–273.

Basu, M., De, T., Das, K.K., Kyle, J.W., Chou, H.C., Schaeper,

R.J. and Basu, S. (1987) Glycolipids. *Methods Enzymol.*, 138: 575–607.

Bouvier, J.D. and Seyfried, T.N. (1989) Composition of gangliosides in normal and mutant mouse embryos. *J. Neurochem.*, 52: 460–466.

Burczak, J.D., Soltysiak, R.M. and Sweeley, C.C. (1984) Regulation of membrane-bound enzymes of glycosphingolipid biosynthesis. *J. Lipid Res.*, 25: 1541–1547.

Caputto, R., Maccioni, H.J., Arce, A. and Cumar, F.A. (1976) Biosynthesis of brain gangliosides. *Adv. Exp. Med. Biol.*, 71: 27–44.

Chatterjee, S., Ghosh, N. and Khurana, S. (1992) Purification of uridine diphosphate-galactose: Glucosyl ceramide β1-4 galactosyltransferase from human kidney. *J. Biol. Chem.*, 267: 7148–7153.

Chiarini, A., Fiorilli, A., Siniscalco, C., Tettamanti, G. and Venerando, B. (1990) Subilization of the membrane-bound sialidase from pig brain by treatment' with bacterial phosphatidylinositol phospholipase C. *J. Neurochem.*, 55: 1576–1584.

Costantino-Ceccarini, E. and Suzuki, K. (1978) Isolation and partial characterization of an endogenous inhibitor of ceramide glycosyltransferases from rat brain. *J. Biol. Chem.*, 253: 340–342.

Cumar, F.A., Fishman, P.H. and Brady, R.O. (1971) Analogous reactions for the biosynthesis of monosialo- and disialogangliosides in brain. *J. Biol. Chem.*, 246: 5075–5084.

Daniotti, J.L., Landa, C.A., Gravotta, D. and Maccioni, H.J.F. (1990) GD3 ganglioside is prevalent in fully differentiated neurons from rat retina. *J. Neurosci. Res.*, 26: 436–446.

Daniotti, J.L., Landa, C.A. Rösner, H. and Maccioni, H.J.F. (1991) GD3 prevalence in adult rat retina correlates with the maintenance of a high GD3-/GM2-synthase activity ratio throughout development. *J. Neurochem.*, 57: 2054–2058.

Dreyfus, H., Louis, J.C., Harth, S. and Mandel, P. (1980) Gangliosides in cultured neurons. *Neuroscience*, 5: 1647–1655.

Dreyfus, H., Urban, P.F., Edel-Harth, S. and Mandel, P. (1975) Developmental patterns of gangliosides and of phospholipids in chick retina and brain. *J. Neurochem.*, 25: 245–250.

Fredman, P., Dumanski, J., Davidsson, P., Svennerholm, L. and Collins, V.P. (1990) Expression of the ganglioside GD3 in human meningiomas is associated with monosomy of chromosome 22. *J. Neurochem.*, 55: 1838–1840.

Gries, C. and Rosner, H. (1990) Migration and aggregation of embryonic chicken neurons in vitro: Possible functional implication of polysialogangliosides. *Dev. Brain Res.*, 57: 223–234.

Gu, T.J., Gu, X.-B., Ariga, T. and Yu, R.K. (1990) Purification and characterization of CMP-NeuAc: GM1 (Galβ1-3GalNAc)a2-3 sialyltransferase from rat brain. *FEBS Lett.*, 275: 83–86.

Gu, X.-B., Gu, T.J. and Yu, R.K. (1990) Purification to homogeneity of GD3 synthase and partial purification of GM3 synthase from rat brain. *Biochem. Biophys. Res. Commun.*, 166: 387–393.

Guarnaccia, S.P., Shaper, J.H. and Schnaar, R.L. (1983) Tunicamycin inhibits ganglioside synthesis in neuronal cells. *Proc. Natl. Acad. Sci. USA*, 80: 1551–1555.

Hakomori, S.-i. (1981) Glycosphingolipids in cellular interactions, differentiation and oncogenesis. *Annu. Rev. Biochem.*, 50: 733–764.

Hakomori, S.-i. (1990) Bifunctional role of glycosphingolipids. *J. Biol. Chem.*, 265: 18713–18716.

Hakomori, S.-i. and Kannagi, R. (1983) Glycosphingolipids as tumor-associated and differentiation markers. *J. Natl. Cancer Inst.*, 71: 231–251.

Hashimoto, Y., Suzuki, A., Yamakawa, T., Miyashita, N. and Moriwaki, K. (1983) Expression of GM1 and GD1a in mouse liver is linked to the H-2 complex on chromosome 17. *J. Biochem.*, 94: 2043–2048.

Hirabayashi, Y., Hirota, M., Suzuki, Y., Matsumoto, M., Obata, K. and Ando, S. (1989) Developmentally expressed O-acetyl ganglioside GT3 in fetal rat cerebral cortex. *Neurosci. Lett.*, 106: 193–198.

Hirabayashi, Y., Hyogo, A., Nakao, T., Tsuchiya, K., Suzuki, Y., Matsumoto, M., Kono, K. and Ando, S. (1990) Isolation and characterization of extremely minor gangliosides GM1b and GD1a, in adult bovine brains as developmentally regulated antigens. *J. Biol. Chem.*, 265: 8144–8151.

Hogan, M.V., Saito, M. and Rosenberg, A. (1988) Influence of monensin on ganglioside anabolism and neurite stability in cultured chick neurons. *J. Neurosci. Res.*, 20: 390–394.

Iber, H. and Sandhoff, K. (1989) Identity of GD1c, GT1a and GQ1b synthase in Golgi vesicles from rat liver. *FEBS Lett.*, 254: 124–128.

Iber, H., Van Echten, G., Klein, R.A. and Sandhoff, K. (1990) pH-Dependent changes of ganglioside biosynthesis in neuronal cell culture. *Eur. J. Cell Biol.*, 52: 236–240.

Iber, H., Van Echten, G. and Sandhoff, K. (1991) Substrate specificity of a2-3 sialyltransferases in ganglioside biosynthesis of rat liver Golgi. *Eur. J. Biochem.*, 195: 115–120.

Iber, H., Van Echten, G. and Sandhoff, K. (1992) Fractionation of primary cultured cerebellar neurons. Distribution of sialyltransferases involved in ganglioside Biosynthesis. *J. Neurochem.*, 58: 1533–1537.

Irwin, L.N. and Irwin, C.C. (1979) Changes in ganglioside composition of hippocampus, retina, and optic tectum. *Dev. Neurosci.*, 2: 129–138.

Joziassé, D.H., Bergh, M.L.E., terHart, H.G.T., Koppen, P.L., Hooghwinkel, G.J.M., and Van Den Eijden, D.H. (1985) Purification and enzymatic characterization of CMP-sialic acid: β-galactosyl 1-3-N-acetylgalactosaminide a2-3 sialyltransferase from Human Placenta. *J. Biol. Chem.*, 260: 4941–4951.

Kaufman, B., Basu, S. and Roseman, S. (1967) Studies on the biosynthesis of gangliosides In: A.M. Aronson and B.W.Volk (Eds.), Inborn Disorders of Sphingolipid Metabolism Pergamon Press, N.Y., pp. 193–214.

Kaufman, B., Basu, S. and Roseman, S. (1968) Enzymatic synthesis of disialogangliosides from monosialogangliosides by sialyltransferases from embryonic chicken brain. *J. Biol. Chem.*, 243: 5804–5807.

Kracun, I., Rösner, H., Cosovic, C. and Stavljenic, A. (1984) Topical atlas of the gangliosides of the adult human brain. *J.*

Neurochem., 43: 979–989.

Kusunoki, S., Tsuji, S. and Nagai, Y. (1985) Ganglio-N-tetraosyl-ceramide (asialo-GM1), and antigen common to the brain and immune system: its localization in myelin. *Brain Res.*, 334: 117–124.

Ledeen, R. (1985) Gangliosides of the neurons. *Trends Neurosci.*, 8: 169–174.

Maccioni, H.J.F., Panzetta, P., Arrieta, D. and Caputto, R. (1984a) Ganglioside glycosyltransferase activities in the cerebral hemispheres from developing rat embryos. *Int. J. Dev. Neurosci.*, 2: 13–19.

Maccioni, H.J.F., Panzella, P. and Arrieta, D. (1984b) Some properties of uridine-5'-diphospho-N-acetylgalactosamine: Hematoside N-acetylgalactosaminyltransferase at early and late stages of embryonic development of chicken retina. *Int. J. Dev. Neurosci.*, 2: 259–266.

Melkevson-Watson, L.J. and Sweeley, C.C. (1991) Purification to apparent homogeneity by immunoaffinity chromatography and partial characterization of the GM3 ganglioside forming enzyme, CMP-sialic acid-lactosylceramine a2-3-sialyltransferase (SAT-I), from rat liver Golgi. *J. Biol. Chem.*, 266: 4448–4457.

Merat, A. and Dickerson, J.W.T. (1973) The effect of development on the gangliosides of rat and pig brain. *J. Neurochem.*, 20: 873–880.

Miller-Prodraza, H. and Fishman, P. (1984) Effect of drugs and temperature on biosynthesis and transport of glycosphingolipids in cultured neurotumor cells. *Biochim. Biophys. Acta*, 804: 44–51.

Miyagi, T., Sagawa, J., Konno, K., Handa, S. and Tsuiki, S. (1990) Biochemical and immunological studies on two distinct ganglioside-hydrolyzing sialidases from the particulate fraction of rat brain. *J. Biochem.*, 107: 787–793.

Morre, D.J., Wilkinson, F.E. and Keenan, T.W. (1990) Gangliosides depleted in plasma membrane are directed to internal membranes of rat hepatomas: Evidence for a glycolipid sorting defect in hepato-carcinogenesis. *Biochem. Biophys. Res. Commun.*, 169: 192–197.

Moskal, J.R., Lockney, M.W., Marvel, C.C., Trosko, J.E. and Sweeley, C.C. (1987) Effect of retinoic acid and phorbol-12-myristate-13-acetate on glycosyltransferase activities in normal and transformed cells. *Cancer Res.*, 47: 787–790.

Nagai, Y., Nakanishi, H. and Sanai, Y. (1986) Gene transfer as a novel approach to the gene controlled mechanism of the cellular expression of glycosphingolipids. *Chem. Phys. Lipids*, 42: 91–104.

Nagata, Y., Yamashiro, S., Yodoi, J., Lloyd, K.O., Shiku, H. and Furukawa, K. (1992) Expression cloning of β1,4 N-acetylgalactosaminyltransferase cDNAs that determine the expression of GM2 and GD2 gangliosides. *J. Biol. Chem.*, 267: 12082–12089.

Nakaishi, H., Sanai, S., Shinoki, K. and Nagai, Y. (1988a) Analysis of cellular expression of gangliosides by gene expression I: GD3 expression in *myc* transfected and transformed 3Y1 correlates with anchorage-independent growth activity.

Biochem. Biophys. Res. Commun., 150: 760–765.

Nakaishi, H., Sanai, S., Shibuya, M. and Nagai,Y. (1988b) Analyses of cellular expression of gangliosides by gene transfection II: Rat 3Y1 cells transformed with several DNAs containing oncogenes (*fes, fps, ras,* and *src*) invariably express sialoparagloboside. *Biochem. Biophys. Res. Commun.*, 150: 766–774.

Nores, G.A. and Caputto, R. (1984) Inhibition of the UDP-*N*-Galactosamine: GM3 *N*-galactosaminyltransferase by gangliosides. *J. Neurochem.*, 42: 1205–1211.

Novikov, A.M. and Seyfried, T.N. (1991) Ganglioside GD3 biosynthesis in normal and mutant mouse embryos. *Biochem Genet*, 29: 627–638.

Panzetta, P., Maccioni, H.J.F. and Caputto, T. (1980) Synthesis of retinal gangliosides during chick embryonic development. *J. Neurochem.*, 35: 100–108.

Paulson, J.C., Weinstein, J. and Schauer, A. (1989) Tissue-specific expression of sialyltransferases. *J. Biol. Chem.*, 264: 10931–10934.

Percy, A.K., Gottfries, J., Vilbergsson, G., Mansson, J.-E. and Svennerholm, L. (1991) Glycosphingolipid glycosyltransferases in human fetal brain. *J. Neurochem.*, 56: 1461–1465.

Pohlentz, G., Klein, D., Schwartzmann, G., Schmitz, D. and Sandhoff, K. (1988) Both GA2-, GM2-, and GD2-synthases and GM1b-, GD1a-, and GT1b-synthases are single enzymes in Golgi vesicles from rat liver. *Proc. Natl. Acad. Sci. USA*, 85: 7044–7048.

Preuss, U., Gu, X. and Yu, R.K. (1992) Purification and characterization of CMP-sialic acid: LacCer sialyltransferase (ST-I) from rat brain. *Neurosci. Abs.*, 18: 1328.

Quiroga, S., Caputto, B.L. and Caputto, R. (1984) Inhibition of the chicken retinal UDP-GalNAc: GM3, *N*-acetylgalactosaminyltransferase by blood serum and by pineal gland extracts, *J. Neurosci. Res.*, 12: 269–276.

Rahmann, H. and Wiegandt, H. (1991) Gangliosides and modulation of neuronal functions. In V. Neuhoff and J. Friend,(Eds.), Cell to Cell Signals in Plants and Animals, NATO ASI Series, Vol. H51, Springer-Verlag, Berlin, pp. 212–232.

Roseman, S. (1970) The synthesis of complex carbohydrates by multiglycosyltransferase systems and their potential function in intercellular adhesion. *Chem. Phys. Lipids*, 5: 270–297.

Rosenberg, A., Sauer, A., Nobel, E.P., Gross, H.-J., Chang, R. and Brossmer, R. (1992) Developmental patterns of ganglioside sialosylation coincident with neuritogenesis in cultured embryonic chick brain neurons. *J. Biol. Chem.*, 267: 10607–10612.

Rösner, H. (1982) Ganglioside changes in the chicken optic lobes as biochemical indicators of brain maturation. *Brain Res.*, 236: 49–61.

Rösner, H., Al-Aqtum, M. and Rahmann, H. (1992) Gangliosides and neuronal differentiation. *Neurochem. Int.*, 20: 339–351.

Saito, M. and Yu, R.K. (1986) Further characterization of a myelin-associated Neuraminidase: Properties and substrate specificity. *J. Neurochem.*, 47: 632–641.

Saito, M., Sato-Bigbee, C. and Yu, R.K. (1992) Neuraminidase

activities in oligodendroglial cells of the rat brain. *J. Neurochem.*, 58: 78–82.

Saito, M., Saito, M. and Rosenberg, A. (1984) Action of monensin, a monovalent cationophore, on cultured human fibroblasts: Evidence that it induces high cellular accumulation of glucosyl- and lactosylceramine (gluco- and lactocerebroside), *Biochemistry*, 23: 1043–1046.

Saito, M. and Yu, R.K. (1992) Role of myelin-associated neuraminidase in the ganglioside metabolism of rat brain myelin. *J. Neurochem.*, 58: 83–87.

Sato, C., Black, J.A., and Yu, R.K. (1988) Subcellular distribution of UDP-galactose: Ceramide galactosyltransferase in rat brain oligodendroglia. *J. Neurochem.*, 50: 1887–1893.

Scheideler, M.A. and Dawson, G. (1986) Direct demonstration of the activation of UDP-N-galactosamine: [GM3]*N*-acetylgalactosaminyl transferase by cAMP. *J. Neurochem.*, 46: 1639–1643.

Seyfried, T.N. (1987) Ganglioside abnormalities associated with failed neural differentiation in a T-locus mutant mouse embryos. *Dev. Biol.*, 123: 286–291.

Seyfried, T.N., Bernard, D.J. and Yu, R.K. (1984) Cellular distribution of gangliosides in the developing mouse cerebellum: Analysis using the staggerer Mutant. *J. Neurochem.*, 43: 1152–1162.

Seyfried, T.N., Miyazawa, N. and Yu, R.K. (1983) Cellular localization of gangliosides in the developing mouse cerebellum: Analysis using the weaver mutant. *J. Neurochem.*, 41: 491–505.

Sonnino, S., Bassi, R., Chigorno, V. and Tettamanti, G. (1990) Further studies on the changes of chicken brain gangliosides during prenatal and postnatal life. *J. Neurochem.*, 54: 1653–1660.

Suzuki, K. (1965) The pattern of mammalian brain gangliosides III. Regional and developmental differences. *J. Neurochem.*, 12: 969–979.

Suzuki, K. (1970) Formation and turnover of myelin gangliosides. *J. Neurochem.*, 17: 209–215.

Suzuki, K., Poduslo, S.E. and Norton, W.T. (1967) Ganglioside in the myelin fraction of developing rats. *Biochim. Biophys. Acta*, 24: 604–611.

Svennerholm, L. (1964) The distribution of lipids in the human nervous system. I. Analytical procedure lipids of foetal and newborn brain. *J Neurochem.*, 11: 839–853.

Svennerholm, L., Bostrom, K., Fredman, P., Mansson, J.-E., Rosengren, B. and Rynmark, B.-M. (1989) Human brain gangliosides: Developmental changes from early fetal stages to advanced age. *Biochim. Biophys. Acta*, 1005: 109–117.

Svennerholm, L., Rynmark, B.-M., Vilbergsson, G., Fredman, P., Gottfries, J., Mansson, J.-E. and Percy, A. (1991) Gangliosides in human fetal brain. *J. Neurochem.* 56: 1763–1768.

Svensson, E.C., Soreghan, B. and Paulson, J.C. (1990) Organization of the β-galactoside a2-6 sialyltransferase gene. Evidence for the transcriptional regulation of terminal glycosylation. *J. Biol. Chem.*, 265: 20863–20868.

Svensson, E.C., Conley, P.B. and Paulson, J.C. (1992) Regulated expression of a2-6 sialyltransferase by the liver-enriched transcription factors HNF-1, DNB, and LAP. *J. Biol. Chem.*, 267: 3466–3472.

Thangnipon, W. and Balázs, R. (1992) Developmental changes in gangliosides in cultured cerebellar granule neurons. *Neurochem. Res.*, 17: 45–59.

Trinchera, M. and Ghidoni, R. (1990a) Subcellular biosynthesis and transport of gangliosides from exogenous lactosylceramide in rat liver. *Biochem. J.* 266: 363–369.

Trinchera, M. and Ghidoni, R. (1990b) Precursor-product relationship between GM1 and GD1a biosynthesized from exogenous GM2 ganglioside in rat liver. *J. Biochem.*, 107: 619–623.

Trinchera, M. and Ghidoni (1989) Two glycosphingolipid sialyltransferase are localized in different sub-Golgi compartments in rat liver. *J. Biol. Chem.*, 264: 15766–15769.

Trinchera, M. Pirovano, B., and Ghidoni. R. (1990) Sub-Golgi distribution in rat liver of CMP-NeuAc GM3- and CMP-NeuAc GT1b a2-8 sialyltransferases and comparison with the distribution of the other glycosyltransferase activities involved in ganglioside biosynthesis. *J. Biol. Chem.*, 265: 18242–18247.

Tettamanti, G. (1984) An outline of ganglioside metabolism. *Adv. Exp. Med. Biol.*, 174: 197–211.

Van Echten, G., Iber, H., Stotz, H., Takatsuki, A. and Sandhoff, K. (1990) Uncoupling of ganglioside biosynthesis by brefeldin A. *Eur. J. Cell Biol.*, 51: 135–139.

Vanier, M.T., Holm, M., Ohman, R. and Svennerholm, L. (1971) Developmental profiles of gangliosides in human and rat brain. *J. Neurochem.*, 18: 581–592.

Van Meer, G. (1989) Biosynthetic lipid traffic in animal eukaryotes. *Annu. Rev. Cell Biol.*, 5: 247–275.

Varki, A., Hooshmund, F., Diaz, S., Varki, N.M. and Hedrick, M. (1991) developmental abnormalities in transgenic mice expressing a sialic acid-specific 9-O-acetyltransferase. *Cell*, 65: 65–70.

Xia, X.-J., Gu, X.-B., Sartorelli, A.C. and Yu, R.K. (1989) Effects of inducers of differentiation on protein kinase C and CMP-N-acetylneuraminic acid: Lactosylceramide sialyltransferase activities of HL-60 leukemia cells. *J. Lipid Res.*, 30: 181–188.

Yavin, E. and Yavin, Z. (1979) Ganglioside profile during neural tissue development. Acquisition in prenatal rat brain and cerebral cell cultures. *Dev. Neurosci.*, 2: 25–37.

Yip, M.C.M. and Dain, J.A. (1969) The enzymic synthesis of gangliosides. I. Brain uridinediphosphate-D-galactose: N-acetyl-galactosaminyl-galactosyl-glucosylceramide galactosyltransferase. *Lipids*, 4: 270–277.

Yohe, H.C., Jocobson, R.I. and Yu, R.K. (1983) Ganglioside-basic protein interaction: Protection of ganglioside against neuraminidase action. *J. Neurosci. Res.*, 9: 401–412.

Yohe, H.C., Macala, L.J. and Yu, R.K. (1982) In vitro biosynthesis of a tetrasialoganglioside. *J. Biol. Chem.*, 257: 249–252.

Yohe, H.C., Saito, M., Ledeen, R.W., Kunishita, T., Sclafani, J.R. and Yu, R.K. (1986) Further evidence for an intrinsic neuraminidase in CNS myelin. *J. Neurochem.*, 46: 623–629.

Yohe, H.C. and Yu, R.K. (1980) In vitro biosynthesis of an isomer of trisialo-ganglioside, GT1a. *J. Biol. Chem.*, 255: 608–613.

Young, Jr., W.W., Lutz, M.S. and Blackburn, W.A. (1992) Endogenous glycosphingolipids move to the cell surface at a rate consistent with bulk flow estimates. *J. Biol. Chem.*, 267: 12011–12015.

Yu, R.K. and Ando, S. (1980) Structure of some new complex gangliosides of fish brain. *Adv. Exp. Biol. Med.*, 125: 33–45.

Yu, R.K., Itoh, T., Yohe, H.C. and Macala, L.J. (1983) Characterization of some minor gangliosides of Tay-Sachs brain. *Brain Res.*, 275: 47–52.

Yu, R.K. and Iqbal, K. (1979) Sialosylgalactosylceramide as a specific marker for human myelin and oligodendroglia: Gangliosides of human myelin, oligodendroglia and neurons. *J. Neurochem.*, 32: 293–300.

Yu, R.K. and Lee, S.H. (1976) In vitro biosynthesis of sialosyl-galactoceramide by mouse brain microsomes. *J. Biol. Chem.*, 251: 198–203.

Yu, R.K., Macala, L.J., Farooq, M., Sbaschnig-Aggler, M., Norton, W.T. and Ledeen, R.W. (1989) Ganglioside and lipid composition of bulk-isolated rat and bovine oligodendroglia. *J. Neurosci. Res.*, 23: 136–141.

Yu, R.K., Macala, L.J., Taki, T., Weinfeld, H.M. and Yu, F.S. (1988) Developmental changes in ganglioside composition and synthesis in embryonic rat brain. *J. Neurochem.*, 50: 1824–1829.

Yu, R.K. and Saito, M. (1969) Structure and localization of gangliosides. In: R.U. Margolis and R.K. Margolis, (Eds.),Neurobiology of Glycoconjugates, Plenum, NY pp. 1–42.

Yu, R.K. and Yen, S.I. (1975) Gangliosides in developing mouse brain myelin. *J. Neurochem.*, 25: 223–232.

Yusef, H.K., Merat, A. and Dickerson, J.W.T. (1977) Effect of development on gangliosides of human brain. *J. Neurochem.*, 28: 1299–1304.

Yusef, H.K.M., Pohlentz, G. and Sandhoff (1983) Tunicamycin inhibits ganglioside biosynthesis in rat liver Golgi apparatus by blocking sugar nucleotide transport across the membrane vesicles. *Proc. Natl. Acad. Sci. USA*, 80: 7075–7079.

Yusef, H.K.M., Schwarzmann, G., Pohlentz, G. and Sandhoff, K. (1987) Oligosialogangliosides inhibit GM2 and GD3 synthesis in isolated Golgi vesicles from rat liver. *Hoppe-Seyler's Z. Physiol. Chem.*, 368: 455–462.

L. Svennerholm, A.K. Asbury, R.A. Reisfeld, K. Sandhoff, K. Suzuki, G. Tettamanti and G. Toffano (Eds.)
Progress in Brain Research, Vol. 101

CHAPTER 4

New mass spectral approaches to ganglioside structure determinations

Catherine E. Costello, Peter Juhasz and Hélène Perreault

Mass Spectrometry Resource, Department of Chemistry, Massachusetts Institute of Technology, Cambridge, MA 02139, U.S.A.

Introduction

Over the years, mass spectrometry has been employed quite effectively in the structural determinations of many complex natural products, including gangliosides. The thermal instability and frequent heterogeneity of gangliosides make their analysis a challenging task that tries the limits of any new mass spectral methodology. There is now an exciting array of developments in mass spectrometry; new techniques and instruments offer opportunities for more sensitive and more detailed mass spectral analyses, as well as increased mass range. Liquid secondary ionization (LSI) produces abundant molecular weight-related peaks and fragments that provide structural details. Its effectiveness can be enhanced by utilization of collision-induced decomposition (CID) and a tandem mass spectrometer (MS/MS), and by the preparation of derivatives designed to improve sensitivity and direct fragmentation along structurally informative pathways. Other ionization methods also show great promise for ganglioside analysis: supercritical fluid chromatography/chemical ionization (SFC/CI) and electrospray ionization (ESI), each coupled successfully with both quadrupole and sector mass spectrometers, and matrix-assisted laser desorption ionization (MALDI), usually coupled to a time-of-flight mass spectrometer (TOFMS), but not limited to that analyzer. With these techniques, sensitivity in the pmol range and below can be achieved for molecular weight determinations. SFC/CI and ESI have already been combined with tandem mass spectrometry for ganglioside analysis; MALDI/MS/MS on a tandem magnetic instrument has been demonstrated for peptides and should also be quite feasible for glycolipids. The principles of these methods, their advantages and disadvantages and practical examples of ganglioside derivatizations and mass spectral determinations using each of these approaches are discussed herein, with emphasis on LSIMS/MS and MALDI/TOFMS, the two techniques currently used in our laboratory for ganglioside studies.

Liquid secondary ionization mass spectrometry (LSIMS)

Liquid secondary ionization mass spectrometry (LSIMS) is the general term for the ionization method that generates a secondary ion beam from a sample dissolved in a liquid matrix, upon irradiation with a primary beam of atoms (e.g. Ar^0 or Xe^0: 1–10 keV) or ions (*e.g.*: Cs^+: 1–35 keV). The technique was introduced in the early 1980s by M. Barber et al. (1982), who used an argon neutral beam, and therefore called the method fast atom bombardment (FAB). It later became apparent that charged or neutral primary beams gave very similar results. Use of a 20–35 keV Cs^+ primary beam provides increased sensitivity for molecular weights above 2 kDa and allows operation at low ion source pressure, a factor

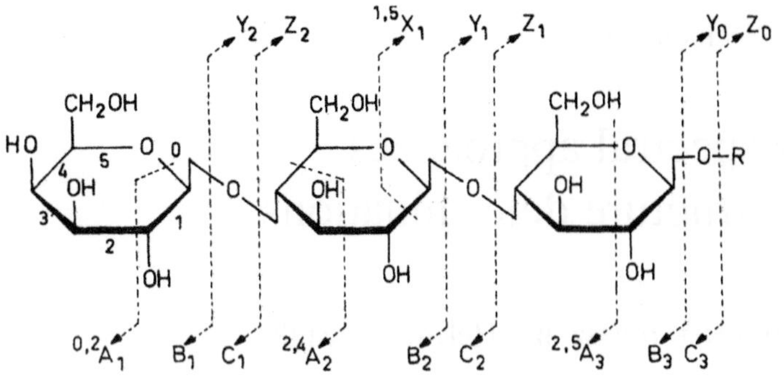

Scheme 1A. Designations of fragment ions in the mass spectra of oligosaccharides and glycoconjugates. Based on Domon and Costello (1988b).

that reduces the likelihood of high voltage arcs in the ion source and makes high resolution determinations easier to perform. LSIMS gives excellent results for glycolipids (Egge and Peter-Katalinic, 1987; Peter-Katalinic and Egge, 1991), providing both molecular weight information and many structural details. For analysis of native gangliosides, the experiment is conducted in the negative-ion mode to capitalize on the sensitivity resulting from the presence of free carboxyl groups in the sialic acid residues. The nomenclature system developed for fragment and CID product ions of oligosaccharides and glycoconjugates (Domon and Costello, 1988a, 1998b; Costello and Vath, 1990) is summarized in Schemes 1A and 1B and is used throughout the text and figures to designate fragment ions. Figure 1a and Scheme 2A show the negative-ion LSIMS spectrum of native GQ1b as an example. Sensitivity is increased and background contribution reduced when continuous-flow is used as the sample introduction method for LSIMS (Chen et al., 1991). LSIMS analysis directly from thin layer plates has also been described (Kushi et al., 1988).

For optimum LSIMS results, samples with larger carbohydrate components and multiple sialic acid residues give improved response after derivatization. Permethylation (often combined with reduction of the amide groups) increases the surface activity of the analyte on the matrix droplet and provides stability to the molecular species. Larson et al. (1987) and Gunnarson (1987) have modified the oligosaccharide permethylation method of Ciucanu and Kerek (1984),

in order to optimize conditions for glycosphingolipid derivatization. The positive-ion LSIMS spectra of the permethylated derivatives have abundant $(M + H)^+$ and/or $(M + Na)^+$ and sequence-related fragment ions, including some fragments that are specific for linkage sites (Peter-Katalinic and Egge, 1991; Salyan et al., 1991; Lemoine et al., 1991). Figure 1b and

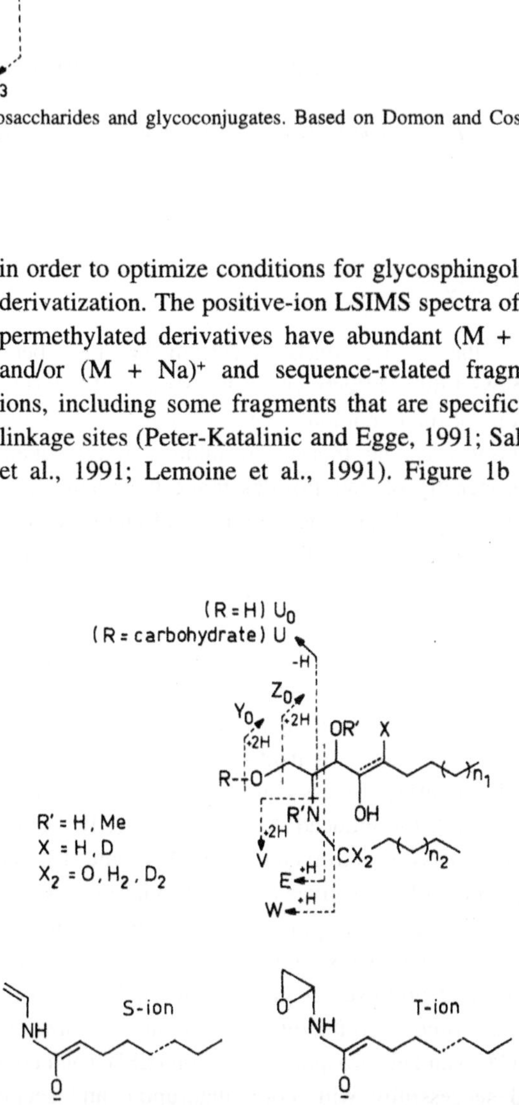

Scheme 1B. Designations of ceramide-related fragment ions in the mass spectra of glycolipids. Reprinted (with permission) from Costello and Vath (1990). In negative-ion spectra, the Y-cleavage does not involve H-migration, the Z-cleavage is accompanied by 2H loss, and the E-cleavage by 1H loss. The S- and T- fragments have the same elemental composition (but opposide charge) in the positive- and negative-ion modes.

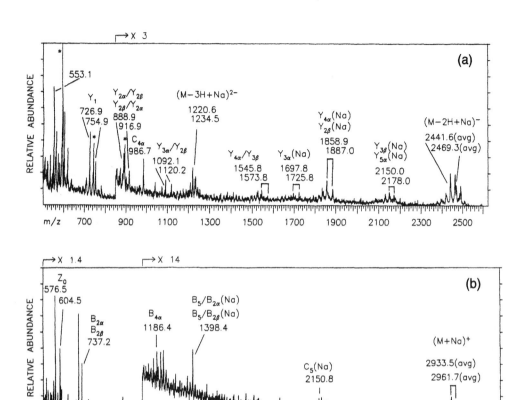

Fig. 1. LSIMS spectra of GQ1b. (a) Negative-ion spectrum in the region m/z 500-2700 of the native ganglioside in CHCl₃/CH₃OH/*N, N, N*-triethanolamine. Asterisks (*) indicate matrix cluster ions. (b) Positive-ion spectrum in the region m/z 500–3200 of the permethylated derivative in CHCl₃/CH₃OH/glycerol/*m*-nitrobenzyl alcohol. Target pretreated with NaCl solution.

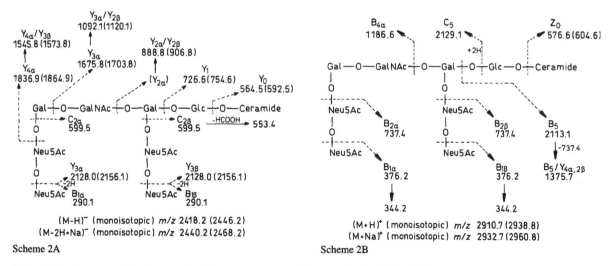

Scheme 2. Assignments of the fragment ions in the mass spectra of the ganglioside GQ1b. (*A*) Negative-ion spectrum of the native compound, shown in Fig. 1a. (*B*) Positive-ion spectrum of the permethylated derivative, shown in Fig. 1b.

Scheme 2B show the LSIMS mass spectrum of permethylated GQ1b, as an example. Permethylated derivatives have been used to ascertain the structures of novel gangliosides, among them trisialosyllactosyl ceramide GT3 from human lung (Månsson et al., 1986) and the pentasialosyl gangliosides GP1b,1c from human brain (Miller-Podroza et al., 1991). Levery et al. (1989, 1990a,b) have used LSIMS and other mass spectral techniques to investigate the cyclization sites in native and derivatized ganglioside lactones. They used crown ethers to scavenge excess sodium ions and reported a resulting improvement in the quality of the mass spectra.

The subsequent reduction may be accomplished by LiAlH$_4$ treatment (Karlsson, 1974; Larson et al., 1987) or by borohydride reduction (Domon et al., 1990, Costello and Vath, 1990). For borohydride reduction, short reduction times affect only the ceramide carbonyl group; with extended reaction time, the N-acetyl groups are also reduced (Costello and Vath, 1990). The MS/MS spectra of these derivatives are discussed below. When the presence of impurities or the solubility characteristics of the glycolipid sample make direct analysis or permethylation of the glycolipid sample problematic, peracetylation (Miller-Podroza et al., 1991, 1992) or pertrimethylsilylation (Merritt et al., 1991) may be employed as an intermediate or final step. The peracetylation procedure results in the formation of lactones when sialic acid is present (McCluer and Evans, 1972; Hansson et al., 1991), and this phenomenon is illustrated in a later section. All of these procedures may be performed on the microscale and are suitable for the analysis of small amounts of material, although efficiencies vary at low levels. The original literature should be consulted for details of optimizing the derivatization conditions.

Tandem mass spectrometry (MS/MS)

Tandem mass spectrometry (MS/MS) offers advantages for specific definition of the individual components, and is particularly appropriate for glycolipid analysis (Domon and Costello, 1988, Costello and Vath, 1990), because these compounds so frequently occur as mixtures of related structures that can vary in the ceramide base and fatty acyl chain lengths and substitution and in the carbohydrate moiety. Collision-induced fragmentation generates further product ions from selected precursors that can be useful in elucidation of the structures of unknown compounds. In the tandem experiment, the molecular or fragment ion of interest is selected in the first stage of separation (MS-1). This precursor ion is then decomposed by collision with a gas, usually helium, and the product ions resulting from its decomposition are separated in a second mass analyzer (MS-2). CID of the $(M-H)^-$ of native gangliosides provides information about the carbohydrate structure and the total ceramide weight(s). The base and fatty acyl weights may be determined by CID of fragments related to the ceramide (Domon and Costello, 1988a; Ladisch et al., 1989; Costello and Vath, 1990). If the resolution for precursor ion selection is not adequate to fully separate the monoisotopic peak of interest, components with molecular weight differences of only a few daltons will not be differentiated. The CID spectrum will contain contributions from multiple species, e.g. both saturated and unsaturated analogs, and the ^{13}C-isotope peaks of the selected peaks. In the examples shown here, a 4-sector JEOL HX110/HX110 instrument with $E_1B_1 - E_2B_2$ (E = electrostatic analyzer, B = magnetic analyzer) configuration was used, with a collision cell located between MS-1 and MS-2. Unit resolution for selection of the precursor ion was always achieved. Most investigations to date have used high energy CID (1–10 keV) on magnetic sector instruments, but it has recently been shown that low energy (200–400 eV) collisions in a triple quadrupole mass spectrometer yield similar results (Kasama and Handa, 1991).

The CID MS/MS spectra of the $(M+H)^+$ ions of permethylated gangliosides contain structurally informative product ions, but the spectra obtained from the permethylated/reduced derivatives provide more complete details of the carbohydrate structure (vide infra). Nevertheless, CID of fragment ions in the spectra of the permethylated compounds does give information about the ceramide structure equivalent to that obtained from the permethylated/reduced

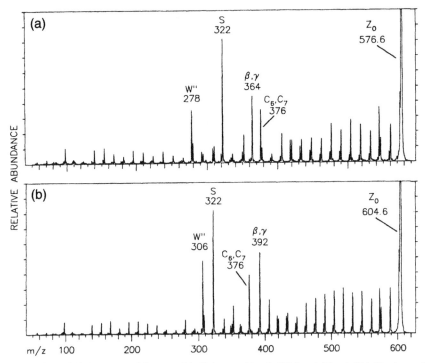

Fig. 2. CID MS/MS spectra of the ceramide Z_0 ions at (a) m/z 576.6 and (b) m/z 604.6 in the positive-ion LSIMS spectra of permethylated GD1b, with 1:1 glycerol/m-nitrobenzyl alcohol matrix. Collision energy 7 keV. For product ion assignments, see Scheme 3 and Table I.

species, with a significant sensitivity increase over the results from analysis of native compounds. Figures 2–4 show the CID mass spectra obtained for the Z_0 fragment in the positive-ion LSIMS spectra of the permethylated derivatives of three gangliosides*:

*Gangliosides (GX_{ny}) are represented with the symbols introduced by Svennerholm (1963), in which X = M,D,T,Q,P indicates the number (1–5) of sialic acid residues, n is the number of the anticipated five uncharged monosaccharides minus the number of those actually found in the respective ganglioside, and y indicates the number of sialic acid residues bound to the inner galactose of the uncharged carbohydrate chain (a = 1, b = 2, c = 3). A general structure for GX_{1y} gangliosides is shown below; R_1 and R_2 are defined in the text:

$$Gal\beta 1 \rightarrow 3GalNAc\beta 1 \rightarrow 4Gal\beta 1 \rightarrow 4Glc\beta 1 \rightarrow 1Gal'Cer$$

```
3                    3
↑        ,           ↑
R₁                   R₂
```

Figure 2, GD1b (R_1, R_2 = Neu5Ac, Neu5Ac = sialic acid); Figure 3, GQ1b [R_1, R_2 = 2Neu5Ac, (Miller-Podroza et al., 1992)]; and Figure 4, GP1b,1c [R_1 = 3Neu5Ac, R_2 = 2Neu5Ac, or the reverse (Miller-Podroza et al., 1991)]. Assignments appear in Table I for the product ions observed in these spectra. The results indicate that the GD1b sample only one isomer contributes to the signal observed at each mass. The m/z 576.6 ion results from the d18:1/18:0 ceramide and the m/z 604.6 ion, from the d20:1/18:0 ceramide. In the cases of GQ1b and GP1b,1c, the lower homolog at m/z 576.6 has d18:1/18:0, but the upper homolog at m/z 604.6 is a mixture of the isomeric ceramides d20:1/18:0 and d18:1/20:0.

Reduction after permethylation yields a product that provides excellent results for electron ionization mass spectrometry of glycosphingolipids (Karlsson, 1974; Hansson et al., 1991) and this derivative is also preferable for CID MS/MS analysis, because it pro-

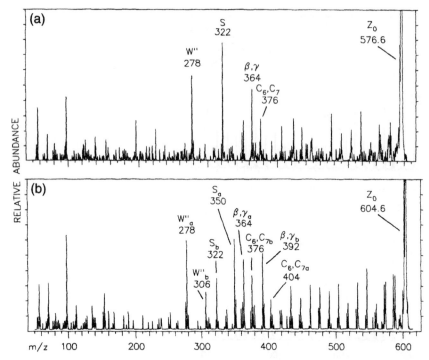

Fig. 3. CID MS/MS spectra of the ceramide Z_0 ions at (a) m/z 576.6 and (b) m/z 604.6 in the positive-ion LSIMS spectra of permethylated GQ1b, with 1:1 glycerol/m-nitrobenzyl alcohol matrix. Collision energy 7 keV. For product ion assignments, see Scheme 3 and Table I.

vides a good site for charge localization and subsequent control of fragmentation (Domon et al., 1990; Costello and Vath, 1990). Unit resolution in precursor ion selection is important for the CID experiment

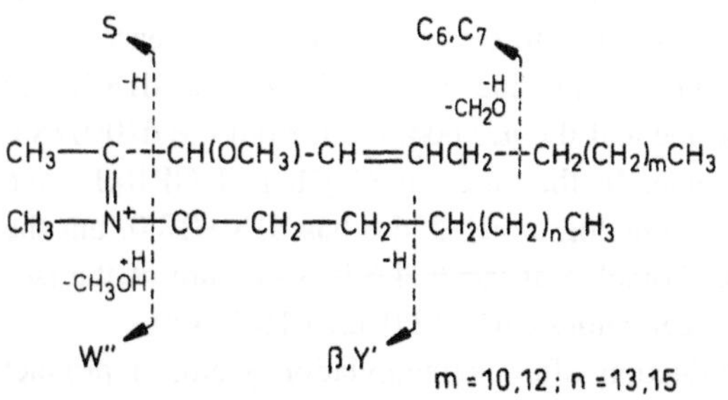

Scheme 3. General structure for the Z_0 fragment ion in the positive-ion LSIMS spectra of permethylated gangliosides, marked to indicate the CID product ions in Fig. 2–4 assigned in Table 1 for the derivatives of GD1b, GQ1b, and GP1b,1c.

on permethylated glycosphingolipids because the molecular ion region may be complicated by the presence of abundant $(M + H - CH_3OH)^+$ ions and the methanol-loss fragment from the upper homolog $(M_2 + H-32)^+$ would fall only 4 units below the $(M_1 + H)^+$ peak of the lower homolog, whose molecular weight is 28 Da below that of M_2. Figure 5 shows the molecular ion region of permethylated, fully-reduced GD1b. Sufficient resolution for unit mass separation can be achieved with a 4-sector tandem mass spectrometer, but is not available during linked magnetic/electric field scans of a 2-sector instrument. Reduction and oxidation at the site of the double bonds shifts components differing only by their degree of unsaturation away from each other in molecular weight; the 2-unit separation in the native compounds is converted to a 16-unit separation upon derivatization. Collisionally-induced dissociation yields product ions that define the carbohydrate sequence and the ceramide structure. Figure 6 shows

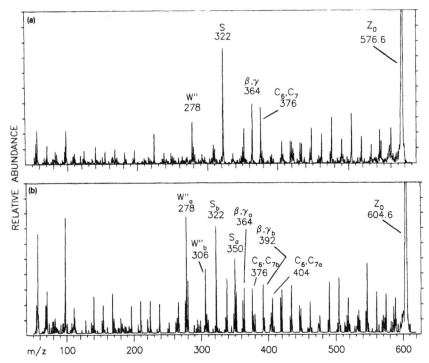

Fig. 4. CID MS/MS spectra of the ceramide Z_0 ions at (a) m/z 576.6 and (b) m/z 604.6 in the positive-ion LSIMS spectra of permethylated GP1b,1c, with 1:1 glycerol/m-nitrobenzyl alcohol matrix. Collision energy 7 keV. For product ion assignments, see Scheme 3 and Table I.

the CID MS/MS spectrum obtained for the $(M + H)^+$ ion, m/z 2178.5, of the higher homolog of the permethylated, reduced derivative of GD1b (Vath et al., 1989). Assignments of the product ions are indicated in Scheme 4. The CID mass spectrum of the $(M+H)^+$ ion of the ceramide-only reduced gangliosides provides information complementary to that of the fully-reduced derivative (Costello and Vath, 1990).

TABLE I

Assignments of product ions in CID spectra of Z_0 ceramide fragments in the positive-ion LSIMS of permethylated gangliosides

Sample	Z_0, m/z	base	FA[a]	Base indicators		Fatty acyl indicators	
				W'' m/z	β,γ[b] m/z	S m/z	C_6,C_7[c] m/z
GD1b	576	d18:1	18:0	278	364	322	376
	604	d20:1	18:0	306	392	322	376
GQ1b and	576	d18:1	18:0	278	364	322	376
GP1b,1c	604	d18:1	20:0	278	364	350	404
	604	d20:1	18:0	306	392	322	376

[a]FA = fatty acyl group.
[b]Cleavage of the C_3 — C_4 bond in the FA group, —H, —CH$_2$O.
[c]Cleavage of the C_6 — C_7 bond in the base, —H, —CH$_2$O.

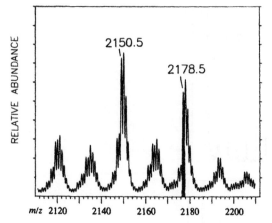

Fig. 5. Molecular ion region in the positive-ion LSIMS spectrum of permethylated, fully reduced GD1b, dissolved in CHCl₃/CH₃OH with glycerol/*m*-nitrobenzyl alcohol matrix. Shaded peak was selected for CID, and the resulting spectrum is shown in Fig. 6.

[252]Cf-plasma desorption mass spectrometry (PDMS)

[252]Cf -plasma desorption mass spectrometry (PDMS) was used by Ohashi et al. (1987) for native glycosphingolipids, but they reported that only the smaller (up to GT) gangliosides gave satisfactory results. Furukawa et al. (1988) employed PDMS for the

analysis of the permethylated derivatives and were successful in determining the structures of *N*-glycolylneuraminic acid-containing gangliosides of cat and sheep erythrocytes. This technique has not been extensively used for ganglioside structural determinations, and later developments in alternative ionization methods for labile compounds have reduced interest in this approach.

Supercritical fluid chromatography combined with chemical ionization mass spectrometry (SFC/CIMS)

Supercritical fluid chromatography combined with chemical ionization mass spectrometry (SFC/CIMS) has been demonstrated to provide very high sensitivity for glycolipids (Kuei et al., 1989) and for ganglioside mixture separation and structural determinations (Merritt et al., 1991). This chromatographic technique takes advantage of the solvent properties of supercritical fluids (Chester et al., 1992) which are used as the mobile phase for the separation. Because the supercritical fluid becomes a gas at atmospheric pressure and upon exposure to the mass spectrometer vacuum system, it is easily handled and removed in the on-line SFC/MS configuration. The sample delivered to the ion source is subjected to chemical ioniza-

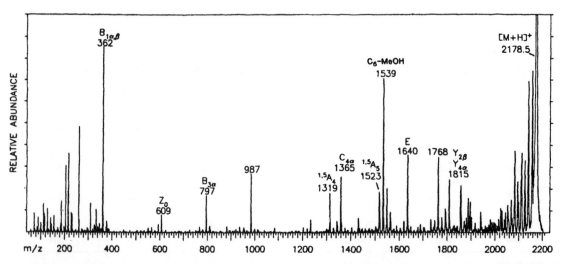

Fig. 6. CID mass spectrum of *m/z* 2178.5, (M + H)⁺ of the higher homolog of permethylated, fully reduced GD1b. Collision energy 7 keV. For product ion assignments, see Scheme 4.

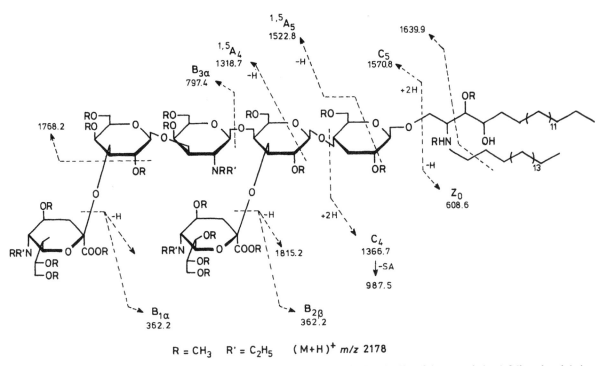

Scheme 4. Assignments of the product ions in the CID mass spectrum of m/z 2178.5, $(M + H)^+$ of the permethylated, fully reduced derivative of the higher homolog of GD1b, Fig. 6.

tion, using either residual mobile phase or added reagent gas for the ionization medium. The choice of derivative and CI reagent gas controls the degree of fragmentation observed in the spectrum. If only a molecular ion is produced, CID may be employed to force fragmentation to yield further structural information. Figure 7 shows the SFC-CIMS spectra of the two homologs of permethylated GM1 (Merritt et al., 1991). Assignments of the fragment ions are given in Scheme 5.

In related studies now being readied for publication, Reinhold and coworkers have amplified the information content available from the SFC-CIMS experiment by introducing the periodate oxidation approach described by Angel et al. (1987) as a degradation step prior to analysis. Periodate oxidation opens the carbohydrate rings at sites between adjacent hydroxyl groups and thus gives indication of the linkage positions. The reaction sites are converted to aldehydes and subsequent borohydride reduction yields alcohols, in a step open to isotopic labeling.

The products are permethylated to improve chromatographic properties. Figure 8 illustrates the use of this approach for a mixture of brain gangliosides. The

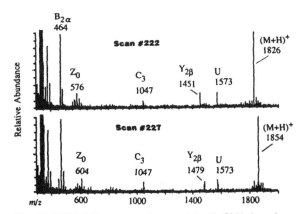

Fig. 7. SFC-CIMS spectra of permethylated GM1 homologs recorded during analysis of a mixture of brain gangliosides (Merritt et al., 1991). Total sample 20 ng injected on the SB-phenyl-5 column. Reagent gas CO_2/CH_3OH. Scans taken at the apex of the chromatographic peaks. Scan 222, d18:1/18:0. Scan 227, d20:1/18:0. Assignments of fragments shown in Scheme 5.

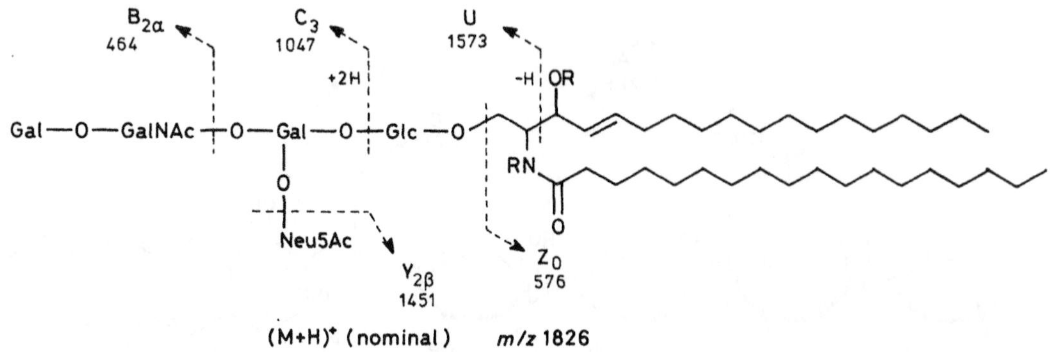

Scheme 5. Assignments of fragment ions in the SFC/CIMS spectrum of the lower homolog (d18:1/18:0) of permethylated GM1, Fig. 7.

mass shifts predicted for the various isomers can be calculated and compared to experimental results to narrow the possible range of isomers, ideally to a single candidate. The mass shifts calculated for the lower homologs of the brain gangliosides in Fig. 8 are: GM1 (m/z 1826 → 1698), GD1a (m/z 2187 → 2013), GD1b (m/z 2187 → 2059) and GT1b (m/z 2548 → 2374). Fragmentation of the derivatives is parallel to that of the native permethylated compounds, with predictable mass shifts, as illustrated in Scheme 6 for GD1a. This method has pmol level sensitivity. Although SFC is not a widely-used technique and still requires special operator skill and experience, its performance for ganglioside analysis is so favorable that it represents a very promising analytical method.

Matrix-assisted laser desorption ionization (MALDI)

Matrix-assisted laser desorption ionization (MALDI) was introduced by Hillenkamp's group in the late 1980s (Karas and Hillenkamp, 1988; Karas et al., 1991), as a method for the determination of molecular weights of proteins and glycoproteins to very high mass, several hundred thousand daltons (kDa). It is also useful for the high-sensitivity determination of proteins and other compounds such as synthetic polymers and oligonucleotides with molecular weights in the 500 Da–170 kDa range. In this ionization method, the sample is mixed with a large excess (10^2–10^3-fold excess in weight) of a matrix which absorbs at the wavelength of the laser being used for irradiation. Ultraviolet, visible and infrared lasers have all been used, but most work reported to date has employed a Nd-YAG laser operated in the frequency-tripled mode at 355 nm or the frequency-quadrupled mode at 266 nm, or the much more economical nitrogen laser operating at 337 nm. Through mechanisms not yet fully understood, some of the energy absorbed by the matrix is transferred to the analyte, and molecular ions of the sample and its dimers, trimers, etc. are observed after mass separation. The method produces a pulsed beam of sample ions and is most easily compatible with time-of-flight mass analysis. TOFMS separation results in high sensitivity because ion collection is very efficient, but this type of analyzer has poor mass resolution. Choice of matrix can affect the sensitivity of this method and analyzer resolution may be improved, e.g. through use of a reflectron TOFMS, or a double-focusing magnetic instrument fitted with an integrating array detector to accommodate the pulsed signal.

Egge et al. (1991) demonstrated that MALDI is useful for the molecular weight determination of native and permethylated neutral glycosphingolipids. We have investigated the application of this technique to ganglioside analysis (Juhasz and Costello, 1992), and have found that the results are dependent both on the ionization mode (positive or negative) and on the choice of matrix and derivatization. For native gangliosides, the negative ion mode generally produces stronger signals that are less complicated by adduct formation and fragmentation. In both positive

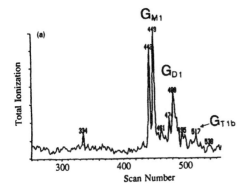

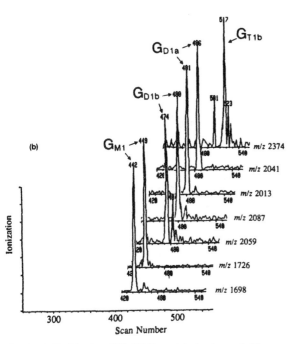

Fig. 8. (a) Positive-ion SFC-CIMS total ionization and (b) mass chromatograms of a mixture of brain gangliosides, after periodate oxidization, sodium borohydride reduction and permethylation. Spectra courtesy of V.N. Reinhold. Molecular weight shifts upon derivatization are discussed in the text. Assignments of fragments for permethylated and oxidized/reduced/permethylated GD1a are shown in Scheme 6.

and negative ion modes, the main fragmentation pathway is sialic acid loss, the primary cleavage occurring at the Neu5Ac-Gal glycosidic bond. The extent of this cleavage is matrix, wavelength and laser power dependent; its occurrence can be used to differentiate isomers, e.g. GD1a and GD1b. The

MALDI/TOFMS method is useful for mixture analysis, as illustrated in Fig. 9, for a mixture of α-galactose (α-fucose)-monosialoganglioside antigens expressed by subsets of rat dorsal root ganglion neurons characterized in an earlier collaborative project (Chou et al., 1988). A VESTEC VT-2000 linear time-of-flight instrument and a nitrogen laser, operating at 337 nm, were used to obtain this spectrum and the others shown. The total sample load applied to the probe in order to obtain this spectrum was about 500 fmol. An earlier LSIMS analysis, although at higher resolution, had used 1 nmol (Chou et al., 1988). The sensitivity increase with MALDI is greater than the fourfold drop in resolution would effect. The structure of the ganglioside is shown in Fig. 9.

MALDI/TOFMS provides a simple and sensitive means for following isolation and derivatization procedures. Fig. 10 shows the tetrasialoganglioside GQ1b as: (A) the native compound; (B) the peracetylated derivative and as the permethylated derivative formed via two routes: (C) through the peracetylated derivative, according to procedures described by Miller-Podroza et al. (1991); and (D) through the pertrimethylsilylated derivative, the perTMS step performed as described by Merritt et al. (1991). It is easily seen that the peracetylation step proceeds to give a product with a narrow molecular weight range, but one which corresponds to the formation of products having internal lactones (Evans and McCluer, 1972; Hansson et al., 1991) rather than to the fully acetylated open derivative: M_r 3523, 3551. The permethylation step reopens the lactones. The expected derivative is formed as the major product by either route, but the sample from the perTMS route has a narrower, lower mass distribution.

Permethylation leads to about a hundredfold sensitivity increase for the gangliosides (Juhasz and Costello, 1992). Choice of matrix is also important, and can affect both the detection limit and the resolution. Fig. 11 shows the molecular ion region of the positive-ion MALDI/TOF mass spectra obtained for 100 fmol and 10 fmol samples of permethylated GT1b with 2-(4-hydroxyphenylazo)benzoic acid matrix, recorded with irradiation from the nitrogen laser, at 337 nm. The resolution in this spectrum is

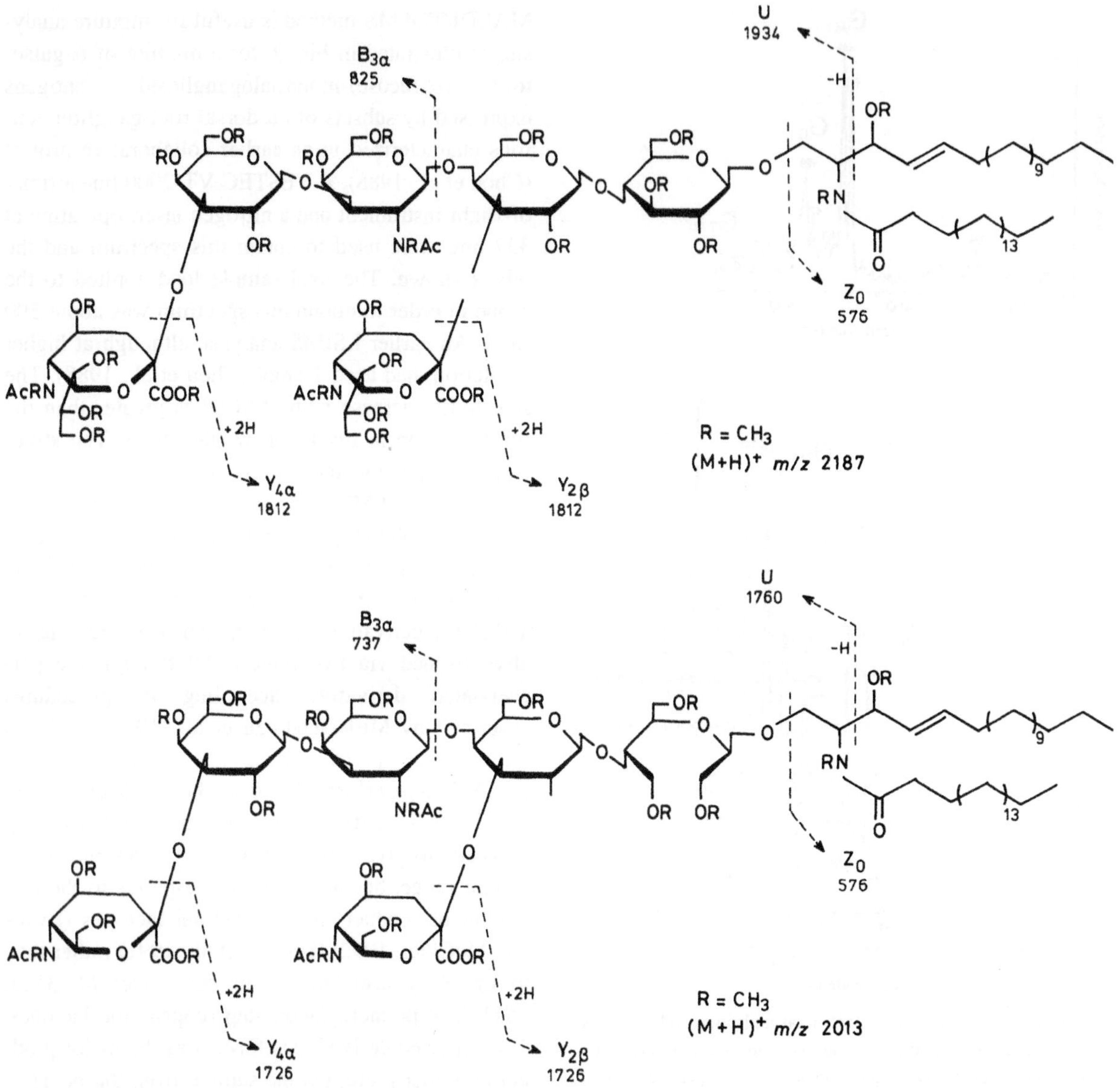

Scheme 6. Assignments of fragment ions in the SFC/CIMS spectra of the lower homolog (d18:1/18:0) of permethylated and oxidized/reduced/permethylated GD1a, Fig. 8.

about 1:400, exceptionally high for this experimental method. Excellent results have also been obtained using 2-thiohydantoin as the matrix and the frequency-quadrupled Nd-YAG laser at 266 nm (Juhasz and Costello, 1992). The selection and evaluation of MALDI matrices is still an active area of research, and even more efficient matrices may well be found, as well as more suitable derivatives. Collision-induced decomposition of MALDI-generated molecular ions from gangliosides and their derivatives should provide MS/MS spectra with a significant sensitivity enhancement over LSIMS/MS.

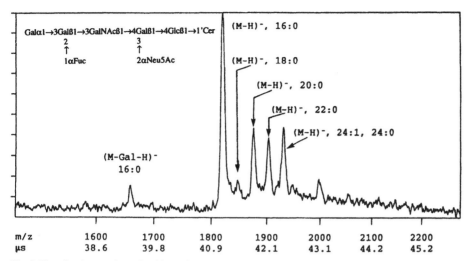

Fig. 9. Negative-ion matrix-assisted laser desorption ionization time-of-flight mass spectrum of the mixture of α-galactose (α-fucose)-mono-sialogangliosides expressed by subsets of rat dorsal root ganglion neurons (Chou et al., 1988), in 2,5-dihydroxybenzoic acid matrix. Total sample size about 500 fmol. All variants have 4E-sphingenine; numbers shown represent fatty acyl substituents. Reprinted, with permission, from Costello and Juhasz, 1992).

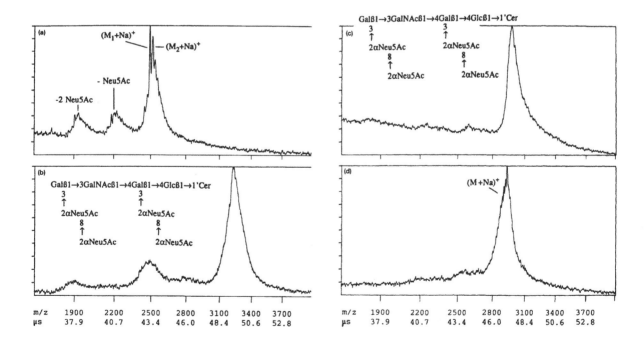

Fig. 10. Positive-ion matrix-assisted laser desorption ionization time-of-flight mass spectra of (a) native and (b) peracetylated tetrasialoganglioside GQ1b. (c) Permethylated derivative, obtained via peracetylated derivative. (d) Permethylated derivative, obtained via per-trimethylsilylated derivative. Matrix 2,5-dihydroxybenzoic acid.

58

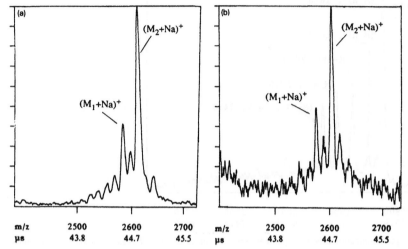

Fig. 11. Positive-ion matrix-assisted laser desorption ionization time-of-flight mass spectra of permethylated GT1b. Matrix 2-(4-hydroxy-phenylazo)benzoic acid. (a) 100 fmol (260 pg). (b) 10 fmol (26 pg).

Electrospray ionization (ESI)

Electrospray ionization (ESI) was first performed more than 20 years ago, but the technique came to prominence recently when Fenn and coworkers described its use for molecular weight determinations of peptides and proteins (Fenn et al., 1989, 1990). In this ionization method the sample solution (the solvent is typically a water-methanol mixture) is intro-duced through a capillary that is maintained at a high voltage with respect to the mass spectrometer ion source. A fine spray consisting of highly charged droplets forms at the tip of the capillary. During the transit to the ion source, analyte ions are released from the droplets via solvent and/or ion evaporation. A distribution of molecular ions that carry a range of charges is observed in the mass spectrum. For proteins, the method has sensitivity in the pmol range

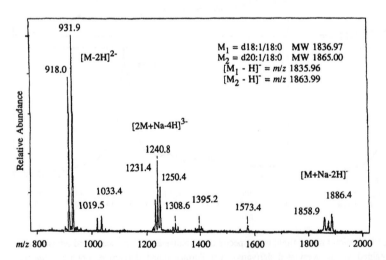

Fig. 12. Negative-ion electrospray mass spectrum of GD1a. Spectrum courtesy of D.A. Gage, S.T. Weintraub and K. Moon.

and has been used to determine molecular weights to over 100 kDa and to observe some types of non-covalently bonded interactions. Because of the sample introduction method, it is particularly amenable to coupling with chromatographic inlet systems. MS/MS spectra may be obtained by two routes: variation of ion source voltages, particularly that of the repeller, and collision-induced decomposition in a triple quadrupole or tandem magnetic sector instrument. ESIMS only now beginning to be applied to the determination of glycoconjugates, including glycoproteins and gangliosides, but seems to have great potential for this field as well. Preliminary results were presented by Huang and Henion(1990), but they have apparently not pursued this area beyond the experiments described at that time. Groups more involved in glycosphingolipid research are now beginning to explore the application of ESIMS to ganglioside analysis. Fig. 12 shows the negative ion ESI mass spectra of 10 pmol of the disialoganglioside GD1a, recorded with the first quadrupole of a Finnigan TSQ 700 instrument by D.A. Gage, S.T. Weintraub and K. Moon. The most abundant peaks observed are the doubly charged monomers and the triply-charged dimers. Loss of sialic acid from the molecular ion is also observed. A small amount of a contaminant that contains an additional galactosamine moiety is clearly seen, although it represents only about 10% of the major component, and thus about 1 pmol of material. ESI/MS/MS has been demonstrated by these investigators, who showed that CID of the $(M-2H)^{2-}$ ion yields a spectrum with singly-charged products. The base peak, arising from the sialic acid residue, was at m/z 290 ($B_{1\alpha}$, $B_{1\beta}$) and the Y_0 ceramide fragment ion was present at m/z 564, albeit at low abundance. These early results are indicative that structurally significant fragments can be formed with high sensitivity, but further experiments are necessary to establish conditions that will produce richer fragmentation patterns.

Conclusions

Recent advances in mass spectral techniques, including improvements in sample purification, derivatization and introduction, and in ionization methods and mass analyzer and signal detection hardware and software offer many exciting possibilities for applications to ganglioside research that will permit structural determinations and mixture profiling at very high sensitivity, in the pmol range and below. Many of these methods have been established for protein and peptide analysis, and the procedures must now be tailored to optimize them for glycolipid studies. The next few years should be eventful and productive as these methods come into more routine use in many laboratories and are incorporated into biochemical and clinical studies.

Acknowledgements

The authors are grateful to K. Biemann for helpful discussions and encouragement, and to B. Domon and J. E. Vath for their contributions to the development of derivatization chemistry and LSIMS/MS methodology for glycosphingolipids. Experimental and data system refinements made by H. Köchling and J. E. Biller facilitated these studies. Highly purified gangliosides GD1b, GT1b, GQ1b and GP1b,1c, and permethylated GD1b and GT1b were kindly provided by J.-E. Månsson and L. Svennerholm (Göteborg University, Sweden). D.K.H. Chou and F.B. Jungalwala (Eunice Kennedy Shriver Center, Waltham, MA) provided the ganglioside mixture from rat dorsal neurons. SFC/MS data was shared by V.N. Reinhold. D.A. Gage, S.T. Weintraub and K. Noon furnished the ESI/MS spectra. The MIT Mass Spectrometry Facility is supported by the NIH Center for Research Resources, Grant No. RR00317 (to K. Biemann). NATO collaborative research grant 900605 assists MALDI/TOFMS studies.

References

Angel, A.-S. Lindh, F. and Nilsson, B. (1987) Determination of binding positions in oligosaccharides and glycosphingolipids by fast-atom-bombardment mass spectrometry. *Carbohydrate Res.*, 168: 15–31.

Barber, M., Bordoli, R.S., Sedgwick, R.D. and Vickerman, J.C. (1982). Fast atom bombardment mass spectrometry. *J. Chem. Soc. Faraday Trans.*, 178: 1291–1296.

Chen, S., Pieraccini, G. and Moneti, G. (1991) Quantitative analysis of the molecular species of monosialogangliosides by continuous-flow fast-atom bombardment mass spectrometry. *Rapid Commun. Mass Spectrom,* 5: 618–621.

Chester, T.L., Pinkston, J.D. and Raynie, D.E. (1992) Supercritical fluid chromatography and extraction. *Anal. Chem.* 64: 153R–170R.

Chou, D.K., Dodd, J., Jessell, T.M., Costello, C.E. and Jungalwala, F.B. (1988) Identification of α-galactose (α-fucose)-asialo-GM1 glycolipid expressed by subsets of rat dorsal root ganglion neurons. *J. Biol Chem.,* 264: 3409–3415.

Ciucanu, I. and Kerek, F. (1984) A simple and rapid method for the permethylation of carbohydrates. *Carbohydrate Res.* 131: 209–217.

Costello, C.E. and Vath, J.E. (1990) Tandem mass spectrometry of glycolipids. In: J.A. McCloskey (Ed.), *Methods in Enzymol, Vol. 193:* Academic Press, Inc., Orlando, FL, pp. 738–768.

Domon, B. and Costello, C.E. (1988a) Structure elucidation of glycosphingolipids and gangliosides using high-performance tandem mass spectrometry. *Biochemistry,* 27: 1534–1543.

Domon, B. and Costello, C.E. (1988b) A systematic nomenclature for carbohydrate fragmentations in FAB-MS/MS spectra of glycoconjugates. *Glycoconjugate J.,* 5: 397–409.

Domon, B., Vath, J.E. and Costello, C.E. (1990) Analysis of derivatized ceramides and neutral glycosphingolipids by high-performance tandem mass spectrometry. *Anal. Biochem,* 184: 151–164.

Egge, H. and Peter-Katalinic, J. (1987) Fast atom bombardment mass spectrometry for structural elucidation of glycoconjugates. *Mass Spectrom., Rev.* 6: 331–393.

Egge, H., Peter-Katalinic, J., Karas, M. and Stahl, B. (1991) The use of fast atom bombardment and laser desorption mass spectrometry in the analysis of complex carbohydrates. *Pure Appl. Chem.,* 63: 491–498.

Fenn, J.B., Mann, M., Meng, C.K., Wong, S. F. and Whitehouse, C. M. (1989) Electrospray ionization for mass spectrometry of large biomolecules. *Science,* 246: 64–71.

Fenn, J.B., Mann, M., Meng, C.K., Wong, S. F. and Whitehouse, C. M. (1990) Electrospray ionization-Principles and practice. *Mass Spectrom. Rev.,* 9: 37–70.

Furukawa, K., Chait, B.T. and Lloyd, K.O. (1988) Identification of *N*-glycolylneuraminic acid-containing gangliosides of cat and sheep erythrocytes. *J. Biol. Chem.,*263: 14939–14947.

Gunnarson, A. (1987) N- and O-alkylation of glycoconjugates and polysaccharides by solid base in dimethyl sulphoxide/alkyl iodide. *Glycoconjugate J.,* 4: 239–245.

Hansson, G.C., Bouhours, J.-F., Karlsson, H. and Carlstedt, I. (1991) Analysis of sialic acid-containing mucin oligosaccharides from porcine small intestine by high-temperature gas chromatography-mass spectrometry of their dimethylamides. *Carbohydrate Res.,* 221: 179–189.

Huang, E.C. and Henion, J.D. (1990) Characterization of glycolipids by atmospheric pressure ionization/tandem mass spectrometry. *Proc. of the 38th ASMS Conf. on Mass Spectrometry and Allied Topics,* Tucson, AZ, June 2–7, 1990, pp. 291–292.

Juhasz, P. and Costello, C.E. (1992) Matrix-assisted laser desorption of underivatized and permethylated gangliosides. *J. Am. Soc. Mass Spectrom.,* 3: 785–796.

Karas, M. and Hillenkamp, F. (1988) Laser desorption ionization of proteins with molecular masses exceeding 10 000 Daltons. *Anal. Chem.,* 60: 2299–2301.

Karas, M., Hillenkamp, F., Beavis, R. and Chait, B. (1991) Matrix-assisted laser desorption/ionization mass spectrometry of biopolymers. *Anal. Chem.,* 63: 1193A–1203A.

Karlsson, K.-A. (1974) Carbohydrate composition and sequence analysis of a derivative of brain disialoganglioside by mass spectrometry, with molecular weight ions at *m/e* 2245. Potential use in the specific microanalysis of cell surface components. *Biochemistry,* 13: 3643–3647.

Kasama, T. and Handa, S. (1991) Structural Studies of gangliosides by fast atom bombardment ionization, low-energy collision-activated dissociation, and tandem mass spectrometry. *Biochemistry,* 30: 5621–5624.

Kuei, J., Her, G.R. and Reinhold, V.N. (1989) Supercritical fluid chromatography of glycosphingolipids. *Anal. Biochem.,* 172: 228–234.

Kushi, Y., Rokukawa, C. and Handa, S. (1988) Direct analysis of glycolipids on thin-layer plates by matrix-assisted secondary ion mass spectrometry: application for glycolipid storage disorders. *Anal. Biochem.,* 175: 167–176.

Ladisch, S., Sweeley, C.C., Becker, H. and Gage, D. (1989) Aberrant fatty acyl α-hydroxylation in human neuroblastoma tumor gangliosides. *J. Biol. Chem.,* 264: 12097–12105.

Larson, G., Karlsson, H., Hansson, G.C. and Pimlott, W. (1987) Application of a simple methylation procedure for the analyses of glycosphingolipids. *Carbohydrate Res.,* 161: 281–290.

Lemoine, J., Strecker, G., Leroy, Y., Fournet, B. and Ricart, G. (1991) Collisional-activation tandem mass spectrometry of sodium adduct ions of methylated oligosaccharides: sequence analysis and discrimination between α-NeuAc-(2 → 3) and α-NeuAc-(2 → 6) linkages. *Carbohydrate Res.,* 221: 209–217.

Levery, S.B., Roberts, C.E., Salyan, M. E. K. and Hakomori, S. (1989) A novel strategy for unambiguous determination of inner esterification site of ganglioside lactones. *Biochem. Biophys. Res. Commun.,* 162: 838–848.

Levery, S.B., Roberts, C.E., Salyan, M.E.K., Bouchon B. and Hakomori, S. (1990a) Strategies for characterization of ganglioside inner esters II – gas chromatography/mass spectrometry. *Biomed. Environ. Mass Spectrom.,* 19: 311–318.

Levery, S.B., Salyan, M.E.K., Roberts, C. E., Bouchon, B. and Hakomori, S. (1990b) Strategies for characterization of ganglioside inner esters I – fast atom bombardment mass spectrometry. *Biomed. Environ. Mass Spectrom.,* 19: 303–310.

Månsson, J.-E., Mo, H., Egge, H., and Svennerholm, L. (1986) Trisialosyllactosylceramide (GT3) is a ganglioside of human lung. *FEBS Lett.,* 196: 259–262.

Merritt, M.V., Sheeley, D.M. and Reinhold, V.N. (1991) Characterization of glycosphingolipids by supercritical fluid chromatography-mass spectrometry. *Anal. Biochem.*, 193: 24–34.

McCluer, R.H., and Evans, J.E. (1972) Ganglioside inner esters. *Adv. Exp. Med. Biol.*, 19: 95.

Miller-Podraza, H., Månsson, J.-E. and Svennerholm, L. (1991) Pentasialogangliosides of human brain. *FEBS Lett.*, 288: 212–214.

Miller-Podraza, H., Månsson, J.-E. and Svennerholm, L. (1992) Isolation of complex gangliosides from human brain. *Biochim. Biophys. Acta.*, 1124: 45–51.

Ohashi, Y., Wang, R., Cotter, R., Fenselau, C. and Nagai, Y. (1987) Californium-252 plasma desorption mass spectrometry applied to saccharides and their conjugates. *Iyo Masu Kenkyukai Koenshu*, 12: 189–192.

Peter-Katalinic, J. and Egge, H. (1991) Desorption mass spectrometry of glycosphingolipids. In: J. A. McCloskey (Ed.), *Methods in Enzymology, Vol. 192*, Academic Press, Inc.; Orlando, FL, pp. 713–733.

Salyan, M.E.K., Stroud, M.R. and Levery, S.B. (1991) Differentiation of type 1 and type 2 chain linkages of native glycosphingolipids by positive-ion fast atom bombardment mass spectrometry with collision-induced dissociation and linked scanning. *Rapid Commun. Mass Spectrom.*, 5: 456–462.

Svennerholm, L. (1963) Chromatographic separation of human brain gangliosides. *J. Neurochem.*, 10: 613–623.

Vath, J.E., Domon, B. and Costello, C.E. (1989) Derivatization of gangliosides and their analysis by high performance tandem mass spectrometry. *Proc. of the 37th ASMS Conf. on Mass Spectrometry and Allied Topics*, Miami Beach, FL., May 21–26, 1989, pp. 770–771.

L. Svennerholm, A.K. Asbury, R.A. Reisfeld, K. Sandhoff, K. Suzuki, G. Tettamanti and G. Toffano (Eds.)
Progress in Brain Research, Vol. 101
© 1994 Elsevier Science BV. All rights reserved.

CHAPTER 5

Principles of glycosphingolipid-oligosaccharide constitution

Herbert Wiegandt

Physiologisch-Chemisches Institut, Philipps-Universitat, Marburg, Germany

Introduction

Throughout cellular evolution of bacteria, fungi, plants and animals, the two hydrocarbon-tailed ceramide has served as a lipophilic anchor for the attachment of carbohydrate to biomembranes. Thereby, mono- or oligosaccharides may be glycosidically linked either directly to C1 of the ceramide sphingoid thus forming the 'classical' glycosphingolipids (GSL), or else to the same position via an inositolphosphodiester as in the glycosylinositolphosphoceramides (GIPC). The majority of the more complex oligosaccharides of the GSL have been detected in direct linkage to ceramide. In contrast, the inositolphosphate-linked carbohydrates of the GIPC have, in several instances, been isolated with lipophilic anchor structures other than the ceramide moiety, e.g. diacylglycerols, alkylacylglycerols and lysoalkylglycerols (Conzelmann et al., 1992).

Depending on their origin, the ceramide as well as the carbohydrate moieties of GSL show a wide range of variability of their chemical structures. However, whereas the double hydrocarbon-chain structure of

the ceramide, except for heterogeneities of sphingoid and fatty acid compositions, have remained basically the same, variations of the carbohydrate chain show up more prominently and account in greater part for differences in physico-chemical behaviour of various GSL components. In addition, most biological properties of GSL observed appear generally to be more influenced by particular structures of their oligosaccharide moieties than ceramide composition.

At present, close to 300 GSL have been reported with oligosaccharides of known chemical constitution (for reviews, see Makita and Taniguchi, 1985; Wiegandt, 1985; Stults et al., 1989). They may be classified into neutral and acidic glycolipids. Acidic GSL components are distinguished by the presence of sulphate, sialic acid (the term sialic acid is used in a broad sense to include all compounds with a 2-*keto*-3-deoxy-nononic acid skeleton), glucuronic acid or pyruvate. In addition. neutral and acidic GSL, as well as inositolphosphoceramide-linked glycolipids may carry zwitterionic 2-aminoethyl-phosphate ($H_2N \cdot CH_2 \cdot CH_2 \cdot O \cdot PO_2^{(-)} \cdot O$—) or 2-aminoethylphosphonate ($H_2N \cdot CH_2 \cdot CH_2 \cdot PO_2^{(-)} \cdot O$—)groups.

Both classes of glycosphingolipids, GSL and GIPC, display core carbohydrate structures. These may carry various additional substituents, and thus form the basis for the great variability in oligosaccharide constitutions. Possibly as a direct consequence of the basic biosynthetic pathways, the core oligosaccharides of glycosylceramides are different from those of glycosylinositolphosphoceramides. And again, for the same reason, the core oligosaccharides of GSL and GIPC are in many cases different from

---

Abbreviations follow the suggestions of the IUPAC-IUB nomenclature committee. Additionally, the following short hand notations have been used: AEP, 2- aminoethylphosphate; AEPn, 2-aminoethylphosphonate; Gangliosides are abbreviated according to Svennerholm's nomenclature (Svennerholm, 1963); GIPC, glycosylinositolphosphoceramide; GPI, glycosylphosphatidylinositol; GSL, glycosphingolipids of the 'classical type', i.e. glycosylceramides; LPPG, lipopeptidophosphoglycan; MAEPn, 2-*N*-methylaminoethylphosphonate; vsg, variant surface glycoprotein

64

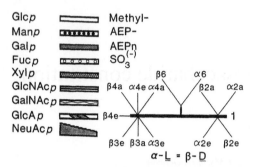

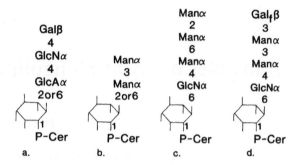

Fig. 1. Scheme of the sketch-type presentation of GSL oligosaccharides. The reducing end position is to the right of the first monosaccharide given. In cases, carbohydrate residues following in sequence are depicted as overlapping previous sugars. All monosaccharides are taken to be in chair conformation.

Fig. 2. Core carbohydrate structures of glycosylinositolphosphoceramides. GIPC from (a) plants (Laine et al., 1980), (b) yeast (Barr et al., 1984), (c) *Trypanosoma* LPPG- and protein anchors (Güther et al., 1992), (d) *Leishmania* lipophosphoglycan-anchor (Thomas et al., 1992.)

those carbohydrates that are glycosidically linked to proteins.

For a closer understanding of the possible significance of the often rather complex ceramide-linked oligosaccharides, an attempt is made here to survey the general principles of their chemical constitutions. And, in order to facilitate an easy one-glance overview allowing for a quick comparison of several related oligosaccharide structures, glycosylpyranosides in Cl- or 1C-chair conformation are represented in some of the figures in a sketch-type way as rectangles filled with various patterns symbolizing the constituent monosaccharide types. Anomeric glycosidic linkage positions are depicted systematically by different positions and angles of the joining monosaccharide rectangles that form oligosaccharide chains (Fig.1).

Glycosyl-inositolphospho-ceramides

The pioneering work of Carter and his coworkers, first reported in 1958. led to the discovery of a new glycolipid class, the glycosylinositol-phosphoceramides from plant seeds (Carter et al., 1958) (Fig. 2).

Later, similar compounds were isolated from plant leaves (for review, see Laine and Renkonen, 1974) (Fig. 2). With a common disaccharide core structure of GlcNα4GlcAα< linked to the 6- or 2-position of inositol-1-phosphate, the latter 'phytoglycosyl'-

ceramides are distinguished from the GIPC isolated from the yeast, *Histoplasma capsulatum* (Barr et al., 1984). The yeast GIPC carbohydrate core structure consists of the disaccharide Manα3Manα< linked to the 2- or 6-position of inositol-1-phosphate (Fig. 2). GIPC may serve in the anchorage of a variety of cell surface glycoconjugates, such as protozoan lipopeptidoglycans or proteins to the lipid bilayer of membranes (for review, see Doering et al., 1990). With the case of the protozoan *Leishmania* lipopeptidophosphoglycan (LPPG) anchor (Turco et al., 1989;

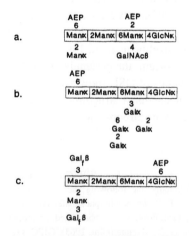

Fig. 3. Comparison of GIPC- and GPI-anchor core structure substitutions. (a) Thy-1 glycoprotein (Homans et al, 1988); (b) *Trypanosoma cruzi* variant surface glycoprotein (Ferguson et al, 1988); (c) *T. cruzi* lipopeptidoglycan (Previato et al., 1992.)

Thomas et al., 1992; Ilg et al., 1992), its GIPC core tetrasaccharide Galfα3Manα3Manα4GlcNα< is distinctly different from that of the protozoan *Trypanosoma* LPPG- or protein-anchoring GIPC Manα2Manα6Manα4GlcNα< (Homans et al., 1988; Güther et al., 1992) and the vertebrate brain Thy-l antigen (Previato et al., 1992) (Figs. 2 and 3). The structural similarity of the *Trypanosoma* GIPC-anchors with that of the Thy-l antigen may indicate the existence of a common evolutionary ancient pathway of biosynthesis for these molecules (Homans et al., 1988) (Fig. 3).

Glycosyl-ceramides of bacteria, plants and protostomians

GSL of bacteria

The discovery of classical GSL in a gram-negative bacterium, *Sphingomonas paucimobilis* (Yamamoto et al., 1978; Kawahara et al.,1991) suggests that the assembly of a ceramide with an oligosaccharide is an ancient structural combination in evolution serving surface protection while permitting controlled permeability (Fig. 4). It can be argued that conservation of similar molecules in vertebrates may be an indication of a common function of glycosphingolipids in higher organisms. Conspicuously, the oligosaccharide of the *S. paucimobilis*-GSL resembles, with its ceramide-linked disaccharide unit GlcNAcα4GlcAα<, the inositolphosphoceramide-linked dihexoside of plants (cf. Figs. 4 and 2). Also, similar to the oligosaccharide structures of the GIPC, where α-anomerically linked monosaccharides far outnumber β-glycosides, all carbohydrate linkages of the *S. paucimobilis*-GSL tetrasaccharide are α-glycosides.

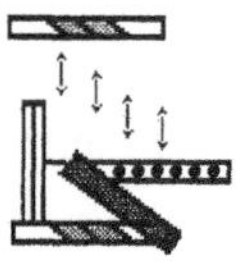

Manα2Galα6GlcNAcα4GlcAα<

Fig. 4. Ceramide mono- and tetrasaccharide from *S. paucimobilis* (Data taken from Yamamoto et al., 1978; Kawahara et al., 1991.)

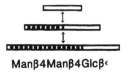

Manβ4Manβ4Glcβ<

Fig. 5. Sito-series GSL oligosaccharides from wheat flour. (Data taken from Laine et al., 1974.)

GSL of Plants

a. Sito-series. One single investigation reports the occurrence of GSL with more extended oligosaccharide moieties in plant material, i.e. three neutral GSL components of obvious biogenic relationship, isolated from wheat flour (Laine et al., 1980). These GSL carry glucosylceramide extended by β4-glycosidically linked mannose residues, thus forming a new carbohydrate series (Fig. 5).

For this monosaccharide sequence, the designation 'sito-series', abbreviated 'St' might be suggested. For naming the intermediary ceramide disaccharide structure Manβ4Glc<, the term 'mactose' has been used emphasizing the structural similarity to lactose, Galβ4Glc<. However, besides the sito-series, other mannose-containing GSL carbohydrate series exist that are also derived from mactose as a metabolic intermediate. The recent generation of an antiserum recognizing the mactose residue of GSL has proved a valuable reagent for the screening of the distribution of this ceramide dihexoside (Itonori et al., 1992). In comparison to the core oligosaccharide of GIPC or the ceramide tetrahexoside of *S paucimobilis* which are compacted by α-glycosides, the 'streched-out' β-linked carbohydrate of the sito-series are reminiscent of fibrillar mannans with similar glycosidic linkages.

GSL of Protostomians

a. Manno-series. Besides its occurrence in GSL of plant material, mannose is also a characteristic constituent of the glycolipids from protostomians. In contrast to the occurrence in phyla of the Protostomia, mannose has not yet been detected in glycosphingolipids of the Deuterostomia. Mannose-containing glycolipids have characteristically been

Manα3Manβ2Manβ‹

Fig. 6. Manno-series GSL oligosaccharides from *H. schlegelii.* (Data taken from Hori et al., 1981b; Itasaka et al., 1983b.)

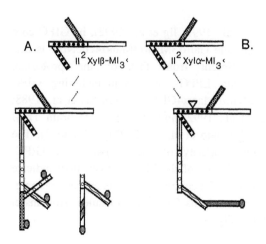

Fig. 8. Xylomollu-series GSL oligosaccharides from *H. schlegelii* (*a*) and *C. sandai* (*b*) (Data taken from (*A*, giving two variant nonreducing ends) and *C. sandai* (*B*) (Data taken from (*A*) Hori et al., 1977a; Hori et al., 1977b; Sugita et al, 1981; Hori et al, 1981a; Itasaka et al, 1983b; Hori et al., 1983, and (*B*) Hori et al., 1968; Sugita et al., 1975; Itasaka et al., 1976; Itasaka and Hori, 1979; Itasaka et al., 1983a)

found in freshwater shellfish (Mollusca: Bivalvia), in the crustaceans, *Euphasia superba* and *Macrobrachium nipponense* and in insects (Insecta: Diptera). A glycolipid-oligosaccharide series consisting solely of mannose residues was described from the mussel *Hyriopsis schlegelii* (Hori et al., 1981b; Itasaka et al., 1983b) (Fig. 6). Since this GSL carbohydrate-series is formed by mannose extensions of mannosylceramide, the designation 'Manno-series', abbreviated 'Ma' is proposed.

b. *Mollu-, isoMollu- and Xylomollu-series.* In addition to manno-series glycolipids, *H. schlegelii* and another freshwater bivalve, *Corbicula sandai*, contain GSL with oligosaccharide structures derived by mannose extensions of glucosylceramide. Similar to the sito-series, mactosylceramide forms the intermediary dihexosylceramide. It is extended by a further α3- or α4-linked mannose residue in forming the mollu- (α3-extension) or isomollu- (α4-extension) series, abbreviated Ml or iMl, respectively (Fig. 7).

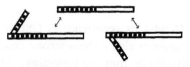

Manα4Manβ4Glc‹ Manα3Manβ4Glc‹
iso-Mollu- (iMl)   Mollu- (Ml)

Fig. 7. Mollu-series GSL oligosaccharides from fresh water bivalves. (Data taken from Itasaka et al., 1976; Sugita et al., 1981.)

Taking galactose for mannose, an analogous principle of sequence isomerisation by α3- and α4-extensions exists in the vertebrate GSL carbohydrates of the globo- and isoglobo-series. Several mollu-series GSL display one structural feature of their oligosaccharides which is reminiscent of what can be observed in some other GSL-series, such as the gastro- and ganglio-series, i.e. a single branching monosaccharide positioned at the second carbohydrate of the extended sugar chain. *H. schlegelii* and *C. sandai* GSL components of the mollu-series carry in such branching 'second sugar substitution' position a β-xylosyl- or α-xylosyl-residue, respectively (Fig. 8).

Additionally, 2-aminoethylphosphate groups may in this case be present as 'second sugar substituents'. As a further structural characteristic, the more extended oligosaccharides of the xylomollu-series GSL are typically enriched in hydrophobic carbohydrate structures. Fucose and frequent terminal methylations may, as in the case of the 4-*O*-methylglucuronic acid constituent, even be combined with an electric charge (Fig. 8). This situation appears comparable to the accumulation of hydrophobic carbohy-

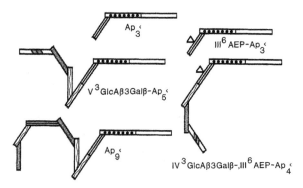

Gal β3GalNAcβ3GlcNAcβ3Galβ3GalNAcα4GalNAcβ4GlcNAcβ3Manβ4Glc‹

Fig. 9. Arthro-series GSL oligosaccharides from Insecta: Diptera. (Data taken from Wiegandt, 1992.)

Fig. 10. Structural comparison of arthro- and neolacto-pentaosyl-ceramide Pl bloodgroup glycolipid: Galα4Galβ4GlcNAcβ3-Galβ4Glc< Arthropentaose: GalNAcα4GalNAcβ4GlcNAcβ3-Manβ4Glc<.

drate residues in the terminal regions of O-specific side-chains of the lipopolysaccharides of gram-negative bacteria.

c. *Arthro-series.* GSL-oligosaccharides that are derived from the mactosyl-residue are not only encountered in plants and molluscs. They are also found in arthropods, amongst them their most highly evolved species, the dipteran insects. GSL from the Calliphoridae, i.e. the blowflies, *Lucilia caesar* and *Calliphora vicina*, mactosylceramide is extended by β-N-acetylglucosamine. This results in the formation of arthrotriaosylceramide, a member of the arthro-series (abbrev. Ap, derived from 'Arthropoda') (for review, see Wiegandt, 1992) (Fig. 9).

Arthro-series oligosaccharides are distinguished by a comparably high content of hexosamines. They may form rather extended carbohydrate chains, and in fact, the longest member so far reported, consists of a straight chain oligosaccharide with nine mono-saccharides that are all in non-repetitive linkage positions (Sugita et al., 1990) (Fig. 9). The arthro-series GSL can be classified into two groups, i.e. the neutral-fraction components and the acidic, glucuronic acid-containing glycolipids. The name 'arthrosides' for the latter group of GSL was suggested with the implied synonymy to the sialic acid-containing 'gangliosides' of the vertebrates. In addition, both groups of arthro-series GSL may carry a zwitterionic 2-aminoethylphosphate substituent in 6-position of

the III-N-acetylglucosamine of the Ap₃Cer-residue (Fig. 9).

The oligosaccharide structures of arthro-series GSL bear conspicuous resemblences to neolacto-series carbohydrates of the vertebrates (Fig. 10).

Interconversion of arthro- and neolacto-series oligosaccharides by substitutions of galactose for mannose and galactose for N-acetylgalactosamine may possibly reflect changes that could have occurred during carbohydrate structural evolution. Such an explanation seems all the more plausible since more recent studies have shown that the substrate specificity of carbohydrate recognizing proteins, such as glycosyltransferases (Yamamoto et al., 1990) and certain toxins (Tyrrell et al., 1992) can indeed be shifted by a few single DNA-base substitutions.

d. *Gastro-series.* Whereas amongst the protostomians investigated, only the GSL of freshwater molluscs, some crustaceans (Itonori et al., 1992) and insects are structurally derived from mactosylceramide, marine gastropods (Mollusca:Gastropoda) have glycolipids based on lactosylceramide. The work of two Japanese groups, Hayashi and coworkers at Osaka and Satake and coworkers at Niigata has shown the presence of two types of 'second sugar substitution', i.e. II²Fucα- and II²Galα- of a GSL-oligosaccharide series common to all the marine gastropods investigated, i.e. the sea hare *Aplysia kurodai* (Araki et al., 1989 (FGp), Abe et al., 1988 (GGp)), the sea snail *Chlorostoma argyrostoma turbinatum* and abalone *Haliotis japonica* (Matsubara and Hayashi, 1982 (FGp)). For this carbohydrate sequence, which is derived by an extension of lactosylceramide with an α-N-acetylgalac-

68

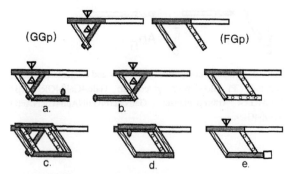

(GGp) (FGp)

a. b.

c. d. e.

a. 3-O-MeGalβ3GalNAcκ3(6'AEP-Galα2)(6'AEP-)Galβ4Glc‹
b. 4-O-MeGlcNAcκ4GalNAcκ3(6'AEP-Galα2)6'AEP-Galβ4Glc‹
c. Fucα2(3-O-Me)Galβ3GalNAcκ3(Fucα2)Galβ3GalNAcκ3(Galα2)Galβ4Glc‹
d. 3-O-MeβGalβ3GalNAcκ3GalβGalNAcκ3(Fucα2)Galβ4Glc‹
e. [3,4-O-(1-carboxyethylidene)]Galβ3GalNAcκ3(Fucα2)6'AEPn-Galβ4Glc‹

Fig. 11. Gastro-series GSL oligosaccharides from marine gastropods (Fucogastro(FGp) and Galagastro- (GGp) subseries).

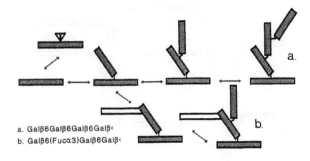

a. Galβ6Galβ6Galβ6Galβ‹
b. Galβ6(Fucα3)Galβ6Galβ‹

Fig. 12. Neogala-series GSL oligosaccharides from marine archeogastropods and Platyhelminthes (Data taken from Matsubara and Hayashi, 1981; Hayashi and Matsubara, 1989: Matsubara and Hayashi, 1986; Persat et al., 1992; Nishimura et al., 1991; Dennis et al,. 1992.)

tosaminyl-residue, the name 'Gastro-series' (abbrev. Gp derived from 'Gastropoda') with 'Fucogastro-' (FGp) and 'Galagastro-' (GGp) subseries is suggested (Fig. 11).

GSL of both of these subseries have been isolated from the same organism, *A. kurodai*. In addition, some of the more complex GSL-oligosaccharides bear structural trisaccharide units carrying characteristic glycosidic linkages of both, the fucogastro- (GalNAcα3(Fucα2)Galβ<) and galagastro- (GalNAcα3(Galα2)Galβ<) subseries (Fig. 11). Whereas the inner region of the gastro-series oligosaccharides may carry multiple substitutions by 2-aminoethylphosphonate or 2-*N*-methylamino-ethylphosphonate, sugar moieties towards the non-reducing end of the oligosaccharide chain are, similar to GSL of the xylomollu-series, frequently substituted by hydrophobic methyl groups (Fig. 10). Recently, the unique pyruvylation of a terminal galagastro-series GSL was reported (Araki et al., 1989).

e. neoGala-series. From marine archeogastropods, i.e. the related sea snails *Turbo cornutus* (Matsubara and Hayashi, 1981; Hayashi and Matsubara, 1989), *Monodonta labio* and *Chlorostoma argyrostoma turbinatum* (Matsubara and Hayashi, 1986), β-galactocerebroside-derived GSL have been isolated. The

carbohydrate moiety of these GSL contain only β-galactopyranosides linked to one another exclusively by β6-linkages. (Fig. 12).

For this monosaccharide sequence and type of linkages, the designation 'neoGala-series' has been used. Surprisingly, GSL of the neogala-series have more recently been encountered in several species of cestode Platyhelminthes, where they putatively contribute to the tegumental-bound glycocalyx (Nishimura et al., 1991; Persat et al., 1992; Dennis et al., 1992). Whereas the marine snail neogala-series GSL-oligosaccharides may carry additional 2-*N*-methylaminoethylphosphonate residues, substitutions with fucose were reported in the case of one tapeworm's glycolipids. A similar functional adaptation to resist environmental constraints may have resulted in the evolution of the neogala-series GSL-oligosaccharide structures in the marine Archeogastropoda as well as in helminth parasites. The assumption of a particularly membrane stabilizing and compacting influence of galactose-containing GSL, whilst permitting specific permeability, is supported by the notion that β-galactocerebroside is found to be concentrated in vertebrate myelin and in membranes exposed to high molecular traffic, such as kidney tubular membranes and those of the intestinal mucosa.

GSL of Deuterostomians

a. Gangliosides of Echinodermata. The dominating ceramide monosaccharide structure in Echinodermata, lower marine animals of the Deuterostomia that have been investigated, is gluco-sylcerebroside (Higuchi et al., 1990). It may be extended by glucose to form ceramide disaccharides with cellobiosyl- (Glcβ4Glcβ<), gentiobiosyl- (Glcβ6Glcβ<), or by galactose to lactosyl- (Galβ4Glcβ<) carbohydrate moieties (Irie et al., 1990; Kawano et al., 1990). Besides these neutral glycolipids, acidic GSL containing sialic acid, i.e. 'gangliosides' have been identified as major components in sea urchins (Echinoidea) and sea stars (Asteroidea). Sialic acid has so far not been detected as a constituent of protostomian GSL. Different from the gangliosides of the more highly evolved Deuterostomia, where sialic acid as a GSL carbohydrate constituent has not been found carrying a glycosidic substituent of another type, the sialoglycolipids of Echinodermata may have sialic acid residues within their oligosaccharide chain (Nagai and Hoshi, 1975; Hoshi and Nagai, 1975; Sugita, 1979a; Sugita, 1979b; Kawano et al., 1990) (Fig. 13).

Substitution by single or multiple sulphate groups in ester linkage appears to be in a similar way a characteristic of deuterostomian GSL as is sialylation. This is exemplified by GSL of the Echinodermata (Kochetkov et al., 1976). In GSL of these lower animals as well as in the higher Deuterostomia, the chordate mammals, sulphate may frequently occupy sub-

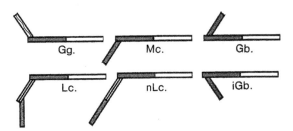

Fig. 14. Definition of GSL oligosaccharides series from vertebrates Gg, GalNAcβ4Galβ4Glc<; Mc, Galβ3Galβ4Glc<; Gb, Galα4Galβ4Glc<; iGb, Galα3Galβ4Glc<; Lc, Galβ3GlcNAcβ3-Galβ4Glc<; nLc, Galβ4GlcNAcβ3Galβ4Glc<.

stitution positions that are identical to those of ganglioside-sialic acid residues. This may indicate that sulphate and sialic acid may to some extent serve similar biological functions.

b. GSL of the Vertebrates. Most GSL of the vertebrates can be classified into five series derived from lactosylceramide by extension with GlcNAcβ3'-, Galα4'-, Galα3'-, Galβ3'- or GalNAcβ3'-residues, i.e. the lacto- (Lc), globo- (Gb), isoglobo- (iGb), muco-(Mc) and ganglio- (Gg) series, respectively (Fig. 14).

From the latter GSL oligosaccharide series, the 'sialoganglio-subseries' (SGg) may be classified separately. This series is formed by a branching 'second sugar substitution' with sialic acid (Fig. 15). Thereby, II^3-galactose substition by a monosialo- (II^3NeuAcα<), disialo- (II^3NeuAcα8NeuAcα<) and trisialo- (II^3NeuAcα8NeuAcα8NeuAcα<) residue distinguishes the designated sialoganglio-subseries a, b and c, respectively (Fig. 15).

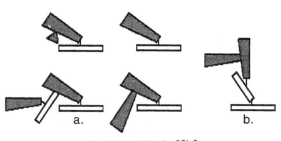

a. NeuAcα6Glcβ8NeuAcα6Glcβ<
b. NeuAcα8NeuAcα6Glcβ6Glcβ<

Fig. 13. Sialo-glycosphingolipids of the sea urchin. (Data taken from Nagai and Hoshi, 1975; Hoshi and Nagai, 1975.)

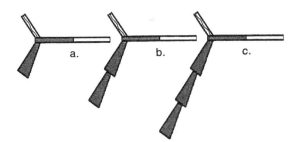

Fig. 15. Sialo-glycosphingolipids of the vertebrates: sialoganglio-subseries a, b and c. (*a*) Ganglioside GM2a (II^3NeuAc-Gg$_3$Cer); (*b*) Ganglioside GD2b (II^3[NeuAc]$_2$-Gg$_3$ Cer); (*c*) Ganglioside GT2c (II^3[NeuAc]$_3$-Gg$_3$Cer).

A third sugar substitution by Galβ4' instead of Galβ3' separates neolacto-series oligosaccharides from the higher lacto-series (Fig. 14). Very extensive structural variations of the sugars belonging to these oligosaccharide series are introduced by carbohydrate chain elongation and/or branchings, as well as by single or multiple substitutions with branching or terminal fucose. Different from fucose in GSL of the Protostomia, in deuterostomian glycolipids this desoxysugar is not positioned within a linear oligosaccharide chain. One or several negative electric charges may be introduced in branching or terminal positions to the GSL oligosaccharides by sulphate- and mono- or oligosialic acid-, as well as glucuronic acid residues (for reviews, see Makita and Taniguchi, 1985; Wiegandt, 1985).

In recent years, a considerable number of GSL have been characterized from cell culture and normal tissues that show mixed oligosaccharide series characteristics. Such hybrids of the globo-, lacto- and ganglio-series may even combine glycosidic linkage types of all three series (Fig. 16).

It is at present not yet clearly understood, which functional properties of cells might decisively depend on the expression of GSL of particular oligosaccharide series. Glycolipids of several series, including sulphatide and gangliosides, are already formed in chick blastoderm cells, i.e. at a very early stage of ontogenetic development (Felding-Habermann et al.,

1986). In contrast, as it appears, fully matured and differentiated cells more or less express GSL of only a single oligosaccharide series. In adult human tissues, some coincidence appears to exist between GSL distribution and the germ layer derivation of cells. Thereby, GSL of the globo-, lacto- and sialoganglio-series are frequently localized in tissues of a meso-, endo- or (neuro)ecto-dermal cellular origin, respectively. However, GSL carbohydrate repertoires are animal-specific, thus adding to the difficulty in ascribing any biological significance to GSL oligosaccharide series. This is illustrated by the GSL of mammalian erythrocytes which contain gycolipids of different oligosaccharide series, i.e. humans, horse, dog, sheep and pig have Gb-, cattle and rabbit nLc-, guinea pig Gg- and rat SGg-series. In contrast, in the human system, bone-marrow stem cell-derived granulocytes carry lacto-series GSL (Macher et al., 1981). In humans, type-I chain GSL oligosaccharides of the lacto-series glycolipids are more prominent in foetal-embryo than in adult tissues where type-II neolacto-series GSL predominate. Interestingly, in mother's milk, free oligosaccharides of type-I chain structures are more concentrated than type-II neolacto-series sugars.

One remarkable exception to the animal-species dependency of GSL carbohydrate types is the distribution shown by the GSL-components of the vertebrate central nervous system. Except for the galacto-cerebroside and sulphatide of oligodendritic cells, the glycolipids of the neuroectodermal brain cells of all vertebrates are derived from the sialoganglio-series (for review, see Wiegandt, 1985). Indeed, this fact may be taken as one indication of a special role played by these particular gangliosides for the vertebrate brain, possibly involving neuronal functions and the maintenance of the nervous system.

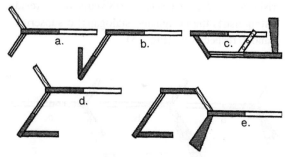

a. Gg/Lc-, b. nLc/Gb-, c. iGb/nLc-, d. Gg/nLc/iGb-, e. SGg/nLc/iGb-series

Fig. 16. Hybrid type GSL components with glycosidic linkages characteristic of several oligosaccharide series (a) Gg/Lc-series, Kannagi et al., 1983; (b) nLc/Gb-series, Kannagi et al., 1984; (c) iGb/nLc-series, Naiki et al., 1975; (d) Gg/nLc/iGb-series, Niimura et al., 1988; (e) nLc/SGg-series, Gillard et al., 1988; (f) SGg/nLc/iGb-series, Nohara et al., 1990).

References

Abe, S., Watanabe, Y., Araki, S., Kumanishi, T., Satake, M. (1988) Immunochemical and histochemical studies on a phosphonoglycosphingolipid, SGL-II, isolated from the sea Gastropod *Aplysia kurodai. J. Biochem.*, 104: 220–226.

Araki, S., Abe, S., Ando, S., Kon, K., Fujiwara, N., Satake, M.

(1989) Structure of phosphonoglycosphingolipid containing pyruvylated galactose in nerve fibers of *Aplysia kurodai. J. Biol. Chem.*, 264: 19922–19927.

Barr, K., Laine, R.A., and Lester, R.L. (1984) Carbohydrate structures of three novel phosphoinositol-containing sphingolipids from the yeast *Histoplasma capsulatum. Biochemistry,* 23: 5589–5596.

Carter, H.E., Celmer, W.D., Galanos, D.S., Gigg, R.H., Lands, W.E., Law, J.H., Muller, K.L., Nakayama, T., Tomizawa, H.H., and Weber, E. (1958) Biochemistry of the sphingolipids. X. Phytoglycolipid, a complex phytosphingosine-containing lipid from plant seeds. *J. Am. Oil Chem. Soc.*, 35: 335–343.

Conzelmann, A., Puoti, A., Lester, L. and Desponds, C. (1992) Two different types of lipid moieties are present in glycophosphoinositol-anchored membrane proteins of *Saccharomyces cerevisiae. EMBO J.*, 11: 457–466.

Dennis, R.D., Baumeister, S., Geyer, R., Peter-Katalinic, J., Hartmann, R., Egge, H., Geyer, E., and Wiegandt, H. (1992) Glycosphingolipids in cestodes. Chemical structures of ceramide monosaccharide, disaccharide, trisaccharide and tetrasaccharide from metacestodes of the fox tapeworm, *Taenia crassiceps* (Cestode: Cyclophyllidea). *Eur. J. Biochem.*, 207: 1053–1062.

Doering, T.L., Masterson, W.J., Hart, G.W. and Englund, P.T. (1990) Biosynthesis of glycosyl phosphatidylinositol membrane anchors. *J. Biol. Chem.*, 265: 611–614.

Felding-Habermann, B., Jennemann, R., Schmitt, J. and Wiegandt, H. (1986) Glycolipid biosynthesis in early chick embryos. *Eur. J. Biochem.*, 166: 651–658.

Ferguson, A.J., Homans, S.W., Dwek, R.A. and Rademacher, T.W. (1988) Glycosylphosphatidylinositol moiety that anchors *Trypanosoma cruzi* variant surface glycoprotein to the membrane. *Science*, 239: 753–759.

Gillard, B.K., Blanchard, D., Bouhours, J.-F., Carton, J.-P., van Kuik, J.A., Kamerling, J.P., Vliegenthart, J.F.G. and Marcus, D.M. (1988) Structure of a ganglioside with Cad blood group antigen activity. *Biochemistry*, 27: 4601–4606.

Güther, L.S., Cardoso de Almeida, M.L., Yoshida, N. and Ferguson, M.A.J. (1992) Structural studies on the glycophosphatidylinositol membrane anchor of *Trypanosoma cruzi* lG7-antigen. *J. Biol. Chem.*, 267: 6820–6828.

Hayashi, A. and Matubara, T. (1989) A new homologue of phosphonoglycosphingolipid, *N*-methylaminoethylphosphonyltrigalactosylceramide. *Biochim. Biophys. Acta*, 1006: 89–96.

Higuchi, R., Natori. and T., Komori, T. (1990) Isolation and characterization of Acanthacerebroside B and structure elucidation of related, nearly homogeneous cerebrosides. *Liebigs Ann. Chem.*, 51–55.

Homans, S.W., Ferguson,. A.J., Dweg, R.A., Rademacher, T.W., Anand, R. and Williams, A.F. (1988) Complete structure of the glycosyl phosphatidylinositol membrane anchor of rat brain Thy-l glycoprotein. *Nature*, 333: 269–272.

Hori, T., Itasaka, O. and Kamimura, M. (1968) Isolation of sphingoethanolamine from pupae of the Green-Bottle Fly, *Lucilia caesar. J. Biochem.*, 64: 125–128.

Hori, T., Sugita, M., Kanbayashi, J. and Itasaka, O. (1977a) Studies on glycosphingolipids of fresh-water bivalves. III. Isolation and characterization of a novel globoside containing mannose from spermatozoa of the fresh-water bivalve, *Hyriopsis schlegelii. J. Biochem.*, 81: 107–114.

Hori, T., Takeda, H., Sugita, M. and Itasaka, O. (1977b) Studies on glycosphingolipids of fresh-water bivalves. IV. Structure of a branched globoside containing mannose from spermatozoa of the fresh-water bivalve *Hyriopsis schlegelii. J. Biochem.*, 82: 1281–1285.

Hori, T., Sugita, M., Ando, S., Kiwahara, M., Kumauchi, K., Sugie, E. and Itasaka,O. (1981a) Characterization of a novel glycosphingolipid, ceramide nonahexoside, isolated from spermatozoa of the fresh-water bivalve, *Hyriopsis schlegelii. J. Biol. Chem.*, 256: 10979–10985.

Hori, T., Sugita, M. and Shimizu, H. (1981b) Identification of β-D-mannosylceramide in hepatopancreas of the fresh-water bivalve *Hyriopsis schlegelii. Biochim. Biophys Acta*, 665: 170–173.

Hori, T., Sugita, M., Ando, S., Tsukada, K., Shiota, K., Tsuzuki, M. and Itasaka, O. (1983) Isolation and characterization of a 4-O-methylglucuronic acid-containing glycosphingolipid from spermatozoa of a fresh water bivalve, *Hyriopsis schlegelii. J. Biol. Chem.*, 258: 2239–2245.

Hoshi, M. and Nagai, Y. (1975) Novel sialosphingolipids from spermatozoa of the sea urchin *Anthocidaris crassispina. Biochim. Biophys. Acta*, 388: 151–162.

Ilg, T., Etges, R., Overath, P., McConville, J., Thomas-Oates, J., Thomas, J., Homans, S.W. and Ferguson, MA.J. (1992) Structure of *Leishmania mexicana* lipophosphoglycan. *J. Biol. Chem.*, 267: 6834–6840.

Irie, A., Kubo, H., Inagaki, F. and Hoshi, M. (1990) Ceramide dihexosides from the spermatozoa of the starfish, *Asterias amurensis,* consists of gentiobiosyl-, cellobiosyl-, and lactosylceramide. *J. Biochem.*, 108: 531–536.

Itasaka, 0., Sugita, M., Yoshizaka, H. and Hori, T. (1976) Determination of the anomeric configurations of *Corbicula* ceramide di- and trihexoside by chromium trioxide oxidation. *J. Biochem.*, 80: 935–936.

Itasaka, O. and Hori, T. (1979) Studies on glycosphingolipids of fresh-water bivalves. V. The structure of a novel ceramide octasaccharide containing mannose-l-phosphate found in the bivalve *Corbicula sandai. J. Biochem.*, 85: 1469–1481.

Itasaka, O., Kosuge, M., Okayama, M. and Hori, T. (1983a) Characterization of a novel ceramide octasaccharide isolated from whole tissue of a fresh-water bivalve, *Corbicula sandai. Biochim. Biophys. Acta*, 750: 440–446.

Itasaka, O., Sugita, M., Kataoka, H. and Hori, T. (1983b) Di- and trimannosylceramides in hepatopancreas of a fresh-water bi-valve, *Hyriopsis schlegelii. Biochim. Biophys. Acta*, 751: 8–13.

Itonori, S., Hiratsuka. M., Sonku, N., Tsuji, H., Itasaka, O., Hori, T. and Sugita, M. (1992 Immunogenic properties of mannose-containing ceramide disaccharide and immunochemical detection of its hapten in the two kins of crustacean, *Euphausia superba* and *Macrobrachium nipponense. Biochim. Biophys. Acta*, 1123: 263–268.

Kannagi, R., Levery, S.B. and Hakomori, S. (1983) Sequential change of carbohydrate antigen associated with differentiation of murine leukemia cells: iI-antigenic conversion and shifting of glycolipid synthesis. *Proc. Natl. Acad. Sci. USA*, 80: 2844–2848.

Kannagi, R., Levery, S.B. and Hakomori, S. (1984) Hybrid type glycolipids (lactoganglio series) with a novel branched structure. *J. Biol. Chem.*, 259: 8444–8451.

Kawahara, K., Seydel, U., Matsuura, M., Danbara, H., Rietschel, E.T. and Zähringer, U. (1991) Chemical structure of glycosphingolipids isolated from *Sphingomonas paucimobilis. FEBS Lett.*, 292: 107–110.

Kawano, Y., Higuchi, R. and Komori, T. (1990) Isolation and structure of five new gangliosides. *Liebigs Ann. Chem.*, 43–50.

Kochetkov, N.K., Smirnova, G.P. and Chekareva, N.V. (1976) Isolation and structural studies of a sulfated sialosphingolipid from the sea urchin *Echinocardium cordatum. Biochim. Biophys. Acta*, 424: 274–283.

Laine, R.A. and Renkonen, O. (1974) Ceramide di- and trihexosides of wheat flour. *Biochemistry*, 13: 2837–2843.

Laine, R.A., Hsieh, T.C.-Y. and Lester, R.L. (1980) Glycophosphoceramides from plants. In: C.C. Sweeley (Ed.), *ACS, Symposium Series*, Washington, American Chemical Society, pp. 65–78.

Macher, BA., Klock, J.C., Fukuda, M.V. and Fukuda, M. (1981) Isolation and structural characterization of human lymphocyte and granulocyte gangliosides. *J. Biol. Chem.*, 256: 1968–1974.

Makita, A. and Taniguchi, N. (1985) Glycosphingolipids. In: H. Wiegandt (Ed.), Glycolipids. *New Comprehensive Biochemistry, Vol. 10*, Elsevier, Amsterdam, pp. 1–100.

Matsubara, T. and Hayashi, A. (1981) Structural studies on glycolipid of shellfish glycolipids from *Turbo cornutus. J. Biochem.*, 89: 645–650.

Matsubara, T. and Hayashi, A (1982) Structural studies on glycolipids of shellfish. V. A novel pentaglycosylceramide from abalone, *Haliotis japonica. Biochim. Biophys. Acta*, 711: 551–553.

Matsubara, T. and Hayashi, A. (1986) Structural studies on glycolipids of shellfish. V. Gala-6 series glycosphingolipids of the marine snail, *Chlorostoma argyrostoma turbinatum. J. Biochem.*, 99: 1401–1408.

Nagai, Y. and Hoshi, M. (1975) Sialosphingolipids of sea urchin eggs and spermatozoa showing a characteristic composition for species and gamete. *Biochim. Biophys. Acta*, 388: 146–151.

Naiki, M., Fong, J., Ledeen, R. and Marcus, D.M. (1975) Structure of the human blood group P1 glycosphingolipid. *Biochemistry*, 14: 4831–4836.

Niimura, Y., Tomori, M., Toida, T. and Ishizuka, I. (1988) Unique gangliosides in transporting organs of salmon. *Rinsho-ken Conf. 3rd.* 92–93.

Nishimura, K., Suzuki, A. and Kino, H. (1991) Sphingolipids of a cestode *Metroliasthes cotornix. Biochim. Biophys. Acta*, 1986: 141–150.

Nohara, K., Suzuki, M., Inagaki, F., Ito, H. and Kaya, K. (1990) Identification of novel gangliosides containing lactosylaminyl-GM1 structure from rat spleen. *J. Biol. Chem.*, 265: 14335–14339.

Persat, F., Bouhours, J.F., Mojon, M. and Petavy, A.F. (1992) Glycosphingolipids with Galβ1-6Gal sequences in metacestodes of the parasite *Echinococcus multilocularis. J. Biol. Chem.*, 267: 8764–8769.

Previato, J.O., Gorin, P.A.J., Mazurek, M., Xavier, M.T., Fourtnet, B., Wieruszesk, J.M. and Mendonca-Previato, L. (1992) Primary structure of the oligosaccharide chain of lipopeptidophosphoglycan of epimastigote forms of *Trypanosoma cruzi. J. Biol. Chem.*, 267: 2518–2526.

Stults, C.L.M., Sweeley, C.C. and Macher, B.A. (1989) Glycosphingolipids: structure, biological source, and properties. *Methods Enzymol.*, 179: 167–214.

Sugita, M., Shirai, S., Itasaka, O. and Hori, T. (1975) Neutral glycosphingolipids containing mannose from the bivalve *Corbicula sandai. J. Biochem.*, 77: 125–130.

Sugita, M. (1979a) Studies on the glycosphingolipids of the starfish, *Asterina pectinifera* II. Isolation and characterization of a novel ganglioside with an internal sialic acid residue. *J. Biochem.*, 86: 289–300.

Sugita, M. (1979b) Studies on the glycosphingolipids of the starfish, *Asterina pectinifera* III. Isolation and structural studies of two novel gangliosides containing internal sialic acid residues. *J. Biochem.*, 86: 765–772.

Sugita, M., Yamamoto, A., Masuda, S., Itasaka, O. and Hori, T. (1981) Studies on glycosphingolipids of fresh-water bivalves. VI. Isolation and chemical characterization of neutral glycosphingolipids from spermatozoa of the fresh-water bivalve *Hyriopsis schlegelii. J. Biochem.*, 90: 1519–1535.

Sugita, M., Inagaki, F., Naito, H. and Hori, T. (1990) Studies on glycosphingolipids in larvae of the green-bottle fly, *Lucilia caesar*: Two neutral glycosphingolipids having large straight oligosaccharide chains with eight and nine sugars. *J. Biochem.*, 107: 899–903.

Svennerholm, L. (1963) Chromatographic separation of human brain gangliosides. *J. Neurochem.*, 10: 612–623.

Thomas, J.R., McConville, M.J., Thomas-Oates, J.E., Homans, S.W., Ferguson, A.J., Gorin, P.A.J., Greis, K.D. and Turco, S.J. (1992) Refined structure of the lipophosphoglycan of *Leishmania donovani. J. Biol. Chem.*, 267: 6829–6833.

Turco, S.J., Orlandi, P.A., jr., Homans, St. W., Ferguson, M.A.J., Dwek, RA. and Rademacher, Th.W. (1989) Structure of the phosphosaccharide-inositol core of the *Leishmania donovani* lipophosphoglycan. *J. Biol. Chem.*, 264: 6711–6715.

Tyrrell, G.C., Ramotar, K., Toye, B., Boyd, B., Lingwood, C.A. and Branton, J.C. (1992) Alteration of the carbohydrate binding specificity of verotoxins from Galαl-4Gal to GalNAcβ1-3Galαl-4Gal and vice versa by site directed mutagenesis of the binding subunit. *Proc. Natl. Acad. Sci. USA*, 89: 524–528.

Wiegandt. H. (1985) Gangliosides. In: H. Wiegandt (Ed.), *Glycolipids. New Comprehensive Biochemistry, Vol. 10*, Elsevier, Amsterdam, pp. 199–260.

Wiegandt, H. (1992) Insect glycolipids. *Biochim. Biophys. Acta*, 1123: 117–126.

Yamamoto, A., Yano, I., Masui, M. and Yabuuchi, E. (1978) Isolation of a novel sphingoglycolipid containing glucuronic acid and 2-hydroxy fatty acid from *Flavobacterium devorans* ATCC. *J. Biochem.*, 83: 1213–1216.

Yamamoto, F., Clausen, H., White, T., Marken, J. and Hakomori, S. (1990) Molecular genetic basis of histo-blood group ABO system. *Nature*, 345: 229–233.

SECTION II

Gangliosides in Neuronal Differentiation and Development

L. Svennerholm, A.K. Asbury, R.A. Reisfeld, K. Sandhoff, K. Suzuki, G. Tettamanti and G. Toffano (Eds.)
Progress in Brain Research, Vol. 101

CHAPTER 6

Gangliosides turnover and neural cells function: a new perspective

Guido Tettamanti and Laura Riboni

Department of Medical Chemistry and Biochemistry, The Medical School, University of Milan, Via Saldini 50, Milan, 20133, Italy.

Introduction

Gangliosides, sialic acid containing glycosphin-golipids, are typical components of the plasma membrane of vertebrate cells, and are asymmetrically located in the outer leaflet, with the oligosaccharide portion exposed on the cell surface, and the ceramide portion inserted into the lipid layer (Svennerholm, 1985; Wiegandt, 1985). Although prevailing at the membrane level, gangliosides are located also intra-cellularly, linked partly to the organelles responsible for their intracellular traffic and metabolism, partly to soluble protein carriers (Sonnino et al., 1979).

Gangliosides, with their oligosaccharide portion, constitute recognition sites at the cell surface, and are well suited for interacting with a variety of extracellular substances. This is the basis for their implication in receptor function, and cell-cell and cell-substratum recognition processes (Hakomori, 1981). Moreover, through their ceramide portions, they share the dynamic flow of the lipid components of the membrane, and interact with membrane-bound functional proteins, modulating their activity, and influencing the membrane mediated transfer of information (Fishman, 1988; Hakomori, 1990). Finally, metabolites (sphingosine, ceramide, etc.) of membrane-bound sphingolipids have been recently found to regulate a number of enzymatic events (Hannun and Bell, 1989; Merrill and Jones, 1990). Thus, the possibility exists that ganglioside turnover play a role in the formation of intracellular metabolic regulators. The evidence supporting this possibility, with particular reference to neural cells function, is focused in the present report.

Functional biochemistry of gangliosides in the nervous system

Basic chemical and conformational properties of gangliosides

Gangliosides are contained in all the cell types occurring in the nervous system, and are highly concentrated in neurons where they constitute about 1/10 of the total membrane bound lipids (Wiegandt, 1985). Although present on the whole neuron surface they are more concentrated in the synaptic region (Hansson et al., 1977).

Gangliosides are heterogenous in both their oligosaccharide and ceramide portions, and different gangliosides (sometimes very many) occur in the same cell. The characteristic component of gangliosides is sialic acid, with a number of sialosyl residues per ganglioside molecule varying from 1 to 7 in the

Abbreviations: ganglioside nomenclature is according to Svennerholm (1980). Glc,glucose; Gal, galactose; GalNAc, *N*-acetylgalactosamine; GlcNAc, *N*-acetylglucosamine; NeuAc, *N*-acetylneuraminic acid; Fuc, fucose; EGF, epidermal growth factor; PDGF, plateled derived growth factor; n-CAM, neural cell adhesion molecule.

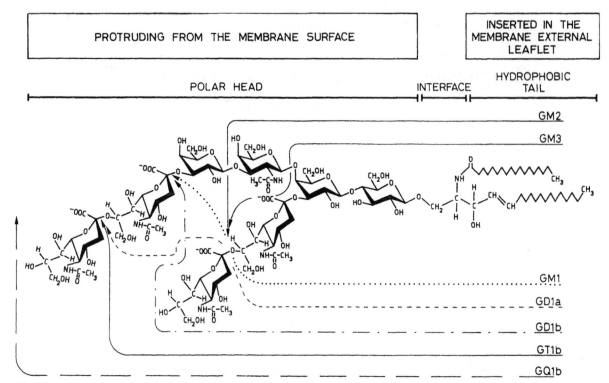

Fig. 1. Structures and structural similarities of the main gangliosides occurring in the vertebrate nervous system.

brain gangliosides of most vertebrates (Yu and Saito, 1989). The sialic acid residue(s) is (are) α-glycosidically attached to a neutral oligosaccharide core which may contain glucose, galactose, N-acetylgalactosamine, N-acetylglucosamine and fucose. On the basis of the neutral oligosaccharide core gangliosides are classified in different series (Yu and Saito, 1989). The gangliosides that are predominant in the nervous tissue (see their structures in Fig. 1) belong to the ganglioseries. In brain gangliosides the most abundant sialic acid is N-acetylneuraminic acid, followed by N-glycolylneuraminic acid, and 4-, or 9-O-acetyl-N-acetylneuraminic acid; inner esters of sialic acid are also present (Riboni et al., 1986). The sialic acid containing oligosaccharide is β-glycosidically linked to ceramide, formed by a long chain fatty acid (commonly stearic acid) and a long chain base (C18 or C20), mostly in the unsaturated form (sphingosine).

The content and composition of brain gangliosides are under genetic and epigenetic control. A genetic control correlates the ganglioside composition to animal species, regions or areas of the nervous system, development, and aging (Suzuki and Yamakawa, 1991). Among epigenetic influences functional (Caputto, 1988) and environmental (Straedel-Flaig et al., 1987) factors have been recognized.

The conformational features of the ganglioside molecules are becoming known. The ceramide portion of gangliosides adopts a rigid conformation, with the two hydrocarbon chains parallelly oriented, and closely packed together (Pascher, 1976; Harris and Thornton, 1978; Pascher et al., 1992). The oligosaccharide portion of gangliosides is conformationally dependent on the high stability of sialic acid, the number of sialic acid residues, and the type of sugars vicinal to sialic acid (Czarniecki and Thornton, 1977). In ganglioside GM1 the plane of inner sialic acid is perpendicular to that of the neighbouring N-acetylgalactosamine (Sillerud et al., 1982), and the region involving inner galactose, N-acetylgalac-

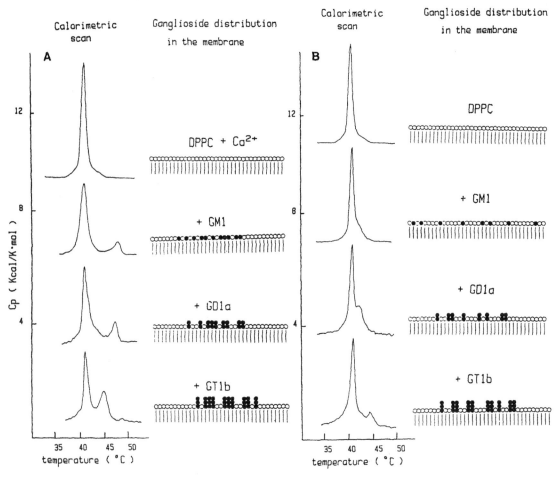

Fig. 2. Calorimetric evidence for lateral phase separation of gangliosides present (<5% on a molar basis) on the surface of dipalmitoyl-phosphatidylcholine (DPPC) unilamellar vesicles in the presence (A) or absence (B) of Ca^{2+} ions. The peaks displayed on the calorimetric scan correspond to the heat required for gel-liquid transition of the phospholipid molecules at the transition temperature. The main peak corresponds to the typical transition of DPPC; the second minor peak, when present, corresponds to a family of DPPC molecules secluded in a membrane domain different from the rest of the membrane and due to an enriched presence of ganglioside(s) (ganglioside enriched domain). The presence of this domain is an expression of the distribution of ganglioside(s) partly as molecular dispersions partly as 'clusters'. Lateral phase separation appears to be favoured by the presence of Ca^{2+} ions and by an increasing number of sialosyl residues per ganglioside molecule.

tosamine and sialic acid, defines an oxygen-rich surface suitable for interaction with cations (Koerner et al., 1983). Moreover, the glycosidic linkages of the branched trisaccharide core segment are more rigid than the linkages of the disaccharides Gal-GalNac, and Gal-Glc (Acquotti et al., 1990). Finally, the intramolecular hydrogen bond between the NH group of long chain base and the vicinal glycosidic oxygen

tends to impose a shovel position between the oligosaccharide chain and ceramide (Abrahamsson et al., 1977). However, forces arising from the ganglioside interactions with the other components of the membrane may prevent this bend, providing a projected conformation (McDaniel et al., 1984; Wynn and Robson, 1986). Chemical composition and conformation are the two parameters defining the geometrical constraint of the molecule, which in turn governs both the binding and aggregation properties of gangliosides.

Physicochemical and binding properties of gangliosides

Gangliosides have amphiphilic properties and undergo micellization above the critical micellar concentration, comprised in the 10^{-7}–10^{-9} M range (Corti et al., 1987). Differently from phospholipids, gangliosides maintain a micellar structure in an extremely wide range of concentrations and generally do not form bilayer structures (Curatolo, 1987). It is well established that gangliosides decrease membrane fluidity (Bertoli et al., 1981). In addition, they contribute to the electrostatic potential of the membrane (Langner et al., 1988; Thompson and Brown, 1988) and influence the thermodynamic and geometrical features of the membrane (Maggio, 1985).

Under established conditions gangliosides undergo lateral phase separation with formation of enriched microdomains ('clusters') (Myers et al., 1984; Masserini and Freire, 1986) (Fig. 2). These clusters can be distinguished from molecularly dispersed gangliosides by specific ligands like enzymes (Masserini et al., 1988; Palestini et al., 1991). This indicates that, besides chemical diversity, also aggregation diversity may play a role in governing ganglioside-ligand interaction.

Gangliosides possess a high binding potential. A number of agents interacts primarily with the oligosaccharide portion of gangliosides. Among them are toxins of bacterial or non-bacterial origin, viruses, serotonin, fibronectin, antibodies (for a review see Tettamanti and Masserini, 1987). The specificity of these interactions is seemingly due to particular carbohydrate sequences that can be shared by different gangliosides. The most specific interaction is that of cholera toxin with GM1 (Holmgren et al., 1974) and Fuc-GM1 (Masserini et al., 1992).

An example of interactions involving the ganglioside micelle is the binding of GM1 with bovine serum albumin (Tomasi et al., 1980), with formation of two well-defined complexes (a monomer and a dimer) characterized by a stoichiometry of one GM1

micelle/ one albumin molecule. Similar complexes were described to occur between micellar gangliosides and enzymes, like the neural cytosolic sialidase and α-L-fucosidase, and result in enzyme inactivation (Venerando et al., 1985; Masserini et al., 1985). Dissociation of complexes restores, at least partially, the enzyme activity (Venerando et al., 1987). Since micelle-like assemblies of gangliosides (clusters) are present at the membrane level, it is possible that under physiological conditions some ganglioside-protein interactions mimic those of micellar gangliosides.

Seemingly specific interactions occur between gangliosides and membrane proteins, yet being unclear whether molecularly dispersed or aggregated forms of gangliosides are involved in the process. The occurrence in brain of plasma membrane proteins that specifically bind to gangliosides, namely GM1 and GT1b, has been demonstrated (Yasuda et al., 1988; Tiemeyer et al., 1990; Fueshko and Schengrund, 1990; Sonnino et al., 1992). These proteins may be involved in the process of ganglioside internalization via endocytosis.

Effects of gangliosides on cell or animal systems

Gangliosides are implicated in important neurobiological events like neurodifferentiation, neuritogenesis, synaptogenesis, and neuronal survival after injury.

Neuronal, as well as glial, differentiation is coupled to changes in the ganglioside content and pattern (Ledeen, 1989). The total ganglioside content increases till completion of differentiation and peculiar qualitative changes occur, in relation to the particular stage of development. Some gangliosides can be considered 'markers' of differentiation stages (Rahmann, 1992). These changes were observed in full brains (or brain areas) and in primary cultures of neurons, with a remarkable parallelism between the two systems (Fig. 3). Support to the concept that gangliosides are actively involved in neuronal differentiation was provided by the evidence that exogenously administered gangliosides have neuritogenic and synaptogenic effects (Obata et al., 1977; Ledeen, 1984). Cultured cells of neural nature (neurotumoral

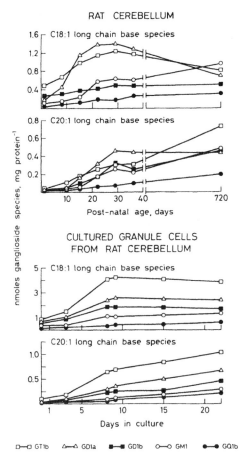

RAT CEREBELLUM

CULTURED GRANULE CELLS
FROM RAT CEREBELLUM

□—□ GT1b   △—△ GD1a   ■—■ GD1b   ○—○ GM1   ●—● GQ1b

Fig. 3. Effect of differentiation and aging on the specific concentration of the main gangliosides (GM1, GD1a, GD1b, GT1b, GQ1b) in rat cerebellum and in cultured granule cells obtained from 8-day-old rat cerebellum. Each ganglioside was separated into the two molecular species containing C18:1 and C20:1 long chain base. Note: (a) the parallel behaviour of the different ganglioside species in the two models; (b) the marked increase of both C18:1 and C20:1 long chain base species during differentiation (up to 30 days in rats; up to 7–8 days in culture in cerebellar granule cells); and (c) the decrease of C18:1 long chain base species with aging after differentiation, in contrast with the continuous increase with aging of the C20:1 long chain base species.

cells, like neuroblastoma and pheochromocytoma cells, and primary cultures of neurons) respond to the presence of gangliosides in the medium with outgrowth of processes, and formation, under favourable conditions, of synaptic-like contacts (Dreyfus et al., 1980; Spoerri, 1983; Ledeen et al., 1990). Cases are

known where the neuritogenic effect of gangliosides is exhibited by cells cultivated in serum free, although hormone supplemented, media (Durand et al., 1986; Nakajima et al., 1986). In other cases the neuro-differentiation effect of gangliosides is observed only in the presence of fetal calf serum or nerve growth factor (Ferrari et al., 1983; Facci et al., 1984) indicating the presence in the medium of factors that are primarily responsible for the effect. It should be reminded that the absence of added differentiating or trophic factors in the medium does not exclude that such substances are formed by the same cells during incubation. In fact, it was reported (Leon et al., 1988) that dopaminergic neurons from embryonic mouse mesencephalon, cultured in a serum free medium, undergo a ganglioside-stimulated differentiation that becomes evident only after a certain density of cells is reached in culture. As well, some clones of neuroblastoma cells are strongly induced by exogenous gangliosides to produce a microtubule associated protein, MAP-2 (that regulates tubulin polymerization and microtubule stabilization), dependently on the cell density in culture (Ferreira et al., 1990). Therefore, the hypothesis (Varon et al., 1988) can be considered that gangliosides influence neuro-differentiation by modulating, or activating, receptors of extracellular (including exogenously added) differentiation factors.

It is generally accepted that survival of adult neurons depends on the presence of neuronotrophic factors that can be of neural or extra-neural origin (Berg, 1984). In order to respond to these factors neurons are assumed to possess the corresponding receptors. The interplay between functioning neurons and neuronotrophic factors is the basis for the potential ability of damaged neurons to undergo repair after injury. Gangliosides appear to be involved in the response of neurons to neuronotrophic factors (Varon et al., 1988; Cuello, 1990). In fact, the life-span of neural cells cultivated in vitro can be increased by addition of gangliosides (Leon et al., 1988; Juurlink et al., 1991; Rosner et al., 1992). Moreover, the functional recovery of central nervous system neurons, after injuries of various origin (traumatic, ischemic, toxic) can be substantially enhanced by administration of gangliosides (Cuello et al., 1989; Hadjiconstantinou and Neff, 1990; Schneider et al., 1992). In these conditions exogenous gangliosides can either sensitize damaged neurons to locally released neuronotrophic factors and/or protect neurons against neuronotoxic factors (Skaper et al., 1988; Seren et al., 1990). The latter ones include glutamate excess, a condition that is produced also in ischemic brain. In the case of cultured cerebellar granule cells that die after transient exposure to high doses (50 μM) of glutamate, gangliosides protect from glutamate excitotoxicity by inhibiting protein kinase C (preventing its translocation to the plasma membrane), and normalizing distorted intracellular free Ca^{2+} dynamics (Favaron et al., 1988; De Erausquin et al., 1990; Manev et al., 1990).

It should also be remembered that gangliosides are able to modulate the expression of mRNA encoding cytoskeletal proteins in neurohybrid and pheochromocytoma cells, and of tubulin mRNA in substantia nigra and striatum of rats after unilateral hemitransection (Yavin et al., 1988). This suggests the possibility that some of the above neurobiological effects derive from the ability of gangliosides to influence the gene expression machinery.

Effects of gangliosides at the molecular level

Many papers ascertained the capability of gangliosides to affect the activity of a number of functional proteins (see Table I). Both the plasma-membrane linked pumps and channels of Na^+ and Ca^{2+} appear to be modulated by gangliosides, particularly GM1 (Leon et al., 1981; Spiegel et al., 1986; Nagata et al., 1987; Spiegel, 1988; Slenzka et al., 1990; Hilbush and Levine, 1992), suggesting an important role of gangliosides in the regulation of cation fluxes through the membrane. The interactions between Ca^{2+} and gangliosides were the object of accurate investigations (reviewed by Rahmann, 1992) and led to suggest the hypothesis (Thomas and Brewer 1990; Rahmann, 1992) that calcium-ganglioside interactions act as modulators of neuronal function, particularly synaptic transmission and long-term adaptation. The processes of protein phosphorylation-dephosphorylation are modulated by gangliosides, through

TABLE I

Functional proteins modulated by gangliosides

1. Proteins involved in the ion fluxes through the plasma membrane

  — Na^+, K^+ ATP-ase
  — Na^+ channel(s)
  — Ca^{2+} ATP-ase
  — Ca^{2+} channel(s)

2. Enzymes involved in protein phosphorylation-dephosphorylation

  — Endo- and ecto-protein kinases
  — Protein phosphatase
  — Ca^{2+}-calmodulin

3. Enzymes involved in the formation and removal of metabolic second messengers.

  — Adenylate cyclase
  — Cyclic nucleotide phosphodiesterase
  — Phosphoinositide specific phospholipase C
  — Mono- and di-acylglycerol lipases
  — Arachidonic acid lipooxygenase and cyclooxygenase

4. Receptors

  — EGF receptor
  — PDGF receptor
  — Insulin receptor
  — Serotonin receptor
  — Receptors for neurotrophic and neuritogenic factors (?)
  — Receptors for adhesion molecules (n-CAM, vitronectin, fibronectin)
  — High affinity choline uptake system

either their influence (activation/inactivation) on protein kinases (Ledeen, 1989) protein phosphatase (Yates et al., 1989), cyclic neucleotide phosphodiesterase (Yates et al., 1989; Higashi and Yamagata 1992) and their binding to Ca^{2+}-calmodulin (Higashi et al., 1992). In the cases of the tyrosine protein kinases associated with receptors of specific growth factors (EGF, PDGF, insulin) the ganglioside action seems to be specific, since an individual ganglioside is involved in the process (Igarashi et al., 1989). In all other cases several gangliosides appear to be effective, but at different optimal concentrations (Nagai et al., 1986; Kreutter et al., 1987; Fukunaga et al., 1990). Therefore at the concentration where the most potent effector is active, the other effectors are inactive or very poorly active. In the case of protein kinases gangliosides interact directly with the enzyme, with the only exception of the kinases phosphorylating the different components of myelin basic protein, where the ganglioside effect is at the substrate level (Chan, 1989). A relevant problem is whether gangliosides, under physiological conditions, are capable of affecting the different enzymes involved in protein phosphorylation-dephosphorylation. In some cases, like the Ca^{2+}-calmodulin-dependent protein kinase II, which is localized at synaptic junctions (Quinet et al., 1984), and the ecto-protein kinase of GOTO neuroblastoma cells, which is located on the outer leaflet of the plasma membrane (Tsuji et al., 1985), the possibility exists of direct enzyme-ganglioside interactions. With regard to the intracellularly located enzymes it cannot be excluded that they interact with the gangliosides occurring in the cytosol (Sonnino et al., 1979) or with ganglioside metabolites.

The implication of gangliosides in receptor function has a large experimental support. Gangliosides themselves were suggested to act as receptors, after the demonstration that GM1 specifically interacts with the B-subunit of cholera toxin (Holmgren et al., 1974) and is instrumental to the penetration of the active A subunit into the cell membrane (Brady and Fishman, 1979). Gangliosides exert a modulatory role on EGF, PDGF and insulin receptors (Hokomori, 1990; Weis and Davis, 1990), an individual ganglioside or a ganglioside metabolite being involved. The receptors for neuronotrophic, and neuritogenic factors (Varon et al., 1988), as well as the serotonin receptor (Berry-Kravis and Dawson, 1985) and the high affinity choline uptake system (Maysinger et al., 1992) are potently modulated by gangliosides. Also, receptors for adhesion molecules, like vitronectin and n-CAM, are influenced by gangliosides (or ganglioside metabolites) in a likely specific manner (Cheresh et al., 1986; 1988; Doherty et al., 1992). A recent line

of investigation concerns the presence in brain of proteins (or glycoproteins) that act as membrane receptors for gangliosides (Yasuda et al., 1988; Tiemeyer et al., 1990; Fueshko and Schengrund 1990; Sonnino et al., 1992). These interactions seem to be specific, suggesting that different receptorial proteins are present for different gangliosides.

Evidence is accumulating showing that gangliosides influence the formation of metabolic second messengers. Both brain adenylate cyclase (Partington and Daly, 1979; Claro et al., 1991) and cyclic nucleotide phosphodiesterase (Davis and Daly, 1980; Yates et al., 1989) are markedly activated by gangliosides, provided that proper conditions and critical concentrations are employed. The occurrence of a substantial, although delayed, increase of phosphoinositide breakdown was described in several neural cell systems (Ferret et al., 1987; Vaswani et al., 1990; Claro et al., 1991), suggesting a regulatory action of gangliosides on the activation of phosphoinositide specific phospholipase C. Also, the mono-, and diacylglycerol lipases of primary cultures of chicken neurons are activated by gangliosides (Freysz et al., 1991), indicating a possible role of gangliosides in the control of diacylglycerol concentration. Finally, there is indirect, but definite, indication that the formation of eicosanoids in brain, especially after ischemic injury and reperfusion, is influenced by gangliosides (Petroni et al., 1989). Most, if not all, of the effects of gangliosides in the production of metabolic second messengers, were observed in cellular systems. Therefore, it cannot be excluded that these effects are elicited not by gangliosides but by ganglioside metabolic derivatives.

Mechanism(s) of ganglioside action: the multivalent role of gangliosides

At the present time a precise and definite model for the mechanism of action of gangliosides cannot be proposed. The levels of ganglioside implications are so wide and diverse that a unitary concept for a causal relationship between ganglioside and a single expression of biological function is difficult to understand (Tettamanti, 1988) (Table II). The oligosaccharide portions of gangliosides, possess a remarkable

TABLE II

The multivalent role of gangliosides

Gangliosides can serve as:

(1) Molecular tools of carbohydrate nature for appropriate interactions between the plasma membrane and extracellular substances (or other cells).

(2) Modulators of membrane-bound or intracellular functional proteins (receptors; enzymes; ion channels; ion carriers; etc.), implicated in transmembrane signalling processes.

(3) Precursors of intracellular metabolic regulators (sphingosine and derivatives; ceramide and derivatives; etc.)

potential of (specific?) interactions with a variety of external legands of different nature. On this basis the concept that gangliosides can simply serve as molec-ular tools for appropriate interactions between the cell plasma membrane and extracellular substances (or other cells) is valid. However, it does not explain all the observed functional implications of gangliosides. The fluid character of the membrane into which ganglioside are embedded allows gangliosides to flow on the membrane and possibly to undergo reversible processes of lateral phase separation. This facilitates ganglioside interaction with, and modulation of, functional membrane proteins (receptors; enzymes; ion pumps; ion channels; G-proteins; etc.). Similar interactions with functional proteins can be envisaged to occur also intracellularly, provided that ganglioside soluble carriers are present and/or transport systems are operating in order for gangliosides to be transferred to the side of the membrane facing the ligands. By means of these interactions gangliosides may modulate events like transmembrane signalling, metabolic second messenger production, and

TABLE III

Recognized regulatory effects of sphingosine, sphingosine derivatives and ceramide on individual enzymes

Affected enzyme	Effector	Effect
Protein kinase C	Sphingosine	Inhibition
	N, monomethyl-sphingosine	Inhibition
	N,N,dimethyl-sphingosine	Inhibition
	N,N,N,trimethyl-sphingosine	Inhibition
src Protein tyrosine kinase	Sphingosine	Inhibition
	N,monomethyl-sphingosine	Inhibition
	N,N,dimethyl-sphingosine	Activation
EGF receptor associated tyrosine kinase	Sphingosine	Activation
	N,N,dimethyl-sphingosine	Activation
Ca^{2+}-calmodulin-dependent protein kinase	Sphingosine	Inhibition
Ca^{2+}-calmodulin-dependent phosphodiesterase	Sphingosine	Inhibition
Phosphatidic acid phosphohydrolase	Sphingosine	Inhibition
Phospholipase D	Sphingosine	Activation
Phospholipase C	Sphingosine	Activation
Phospholipase A_2	Sphingosine	Inhibition
CTP:phosphocholine cytidyltransferase	Sphingosine	Inhibition
Diacylglycerol kinase	Sphingosine	Activation
Na^+,K^+-ATPase	Sphingosine	Inhibition
Cytosolic protein phosphatase	Ceramide	Activation
Membrane-bound protein kinase	Ceramide	Activation
Diacylglycerol kinase	Ceramide	Inhibition

protein phosphorylation-dephosphorylation. This concept does not exclude that ganglioside metabolites — not gangliosides per se — are responsible for the effects.

A further mechanism of action may consist in the production of metabolic second messengers of sphingoid nature. This is a new issue arising from the growing body of evidence that sphingosine, or sphingosine derivatives (*N*-monomethylsphingosine, *N,N*-dimethylsphingosine, *N,N,N*-trimethyl-sphingosine, sphingosine-1-phosphate), ceramide and ceramide-l-phosphate, act as potent regulators of protein kinase C, other protein kinases and enzymes involved in the regulation of signal transduction (see Table III) (Hannum and Bell, 1989; Merrill and Jones, 1990; Liscovitch and Lavie, 1990; Merrill, 1991; Kolesnick 1990, 1992; Younes et al., 1992). The view arising from this evidence is that particular extracellular substances, after binding to the proper plasma membrane linked receptor, trigger the degradation of membrane-bound sphingolipids, with formation of sphingosine or ceramide which become available in the cell interior and regulate the target enzymes, either directly or as derivatives (Fig. 4). Particularly interesting are the findings (reviewed by Kolesnick, 1992) that different agents inducing monocyte differentiation of HL-60 cells (such as tumor necrosis factor-, interferon, vitamin D-3, retinoic acid), stimulate the neutral plasma membrane-sphingomyelinase with formation of ceramide ('sphingomyelin pathway'). The same pathway seems to mediate interleukin-1 effects in EL4 thymoma cells (Mathias et al., 1993). Formed ceramide activates a membrane bound protein kinase with subsequent phosphorylation of proteins that promote monocytic differentiation. The rapid kinetics of activation of the sphingomyelin pathway, the ability of cell permeable ceramide analogs to bypass receptor activation and mimic the effect of monocytic differentiation factors, and the possibility to reconstitute the sequence of events in cell-free systems, are considered (Kolesnick, 1992) good criteria to support the notion that the sphingomyelin pathway may serve as a signalling system (Fig. 4). An alternative procedure for the sphingomyelin pathway, suggested by Merrill (1991), postulates sphingomyelin internalization and

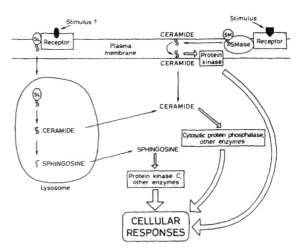

Fig. 4. Formation of metabolic regulators of sphingoid nature (sphingosine, ceramide) from the degradation of sphingomyelin (SM) or sphingolipids (SL) in general. The 'sphingomyelin pathway' illustrated in the right hand part of the scheme is based on sphingomyelin degradation catalyzed by the neutral plasma membrane-linked sphingomyelinase (SMase). This enzyme is activated by the interaction of some external stimuli with the corresponding receptors. The sphingolipid pathway (that may include also sphingomyelin) illustrated in the left hand part of the scheme, is based on intralysosomal degradation of sphingolipids, following their endocytosis and sorting to lysosomes. The possibility is suggested that endocytosis and/or sorting to lysosomes is influenced by interactions of external stimuli with receptors associated to sphingolipids.

intralysosomal degradation with formation of ceramide (Fig. 4). This scheme has been extended to sphingolipids in general (Merrill, 1991) and, therefore, is applicable in principle to gangliosides too.

Ganglioside turnover and formation of sphingoid metabolic regulators

Ganglioside metabolism occurs essentially in the cell body with the participation of different subcellular organelles. This implicates the occurrence of efficient mechanisms for sorting and routing of gangliosides between the plasma membrane and the sites of biosynthesis and biodegradation. In the case of neuronal cells the machinery for ganglioside biosynthesis and biodegradation is located in the perikaryon.

Ganglioside metabolism consists in: de novo biosynthesis; biodegradation; salvage processes; and direct glycosylation following internalization.

De novo biosynthesis

The precursor for ganglioside biosynthesis is ceramide, which results from the sequential action of the following enzymes operating in the endoplasmic reticulum: serine palmitoyltransferase, producing 3-dehydrosphinganine; 3-dehydrosphinganine NADPH oxidoreductase, producing sphinganine; sphinganine N-acyltransferase, producing dihydroceramide; dihydroceramide desaturase producing ceramide (Walter et al., 1983; Radin, 1984; Rother et al., 1992; Mandom et al., 1992). Gangliosides are then formed by sequential addition of monosaccharide units to ceramide. The process takes place in the Golgi apparatus with the involvement of membrane-bound glycosyltransferases and the corresponding sugar-nucleotides. The sequence of glycosylations is based on the specificity, compartmentation and topology of the involved enzymes. Initiation of the three distinct lines (a,b,c) of the ganglioseries gangliosides is dependent on the strict specificity of sialyltransferase-1 (SAT-1) sialyltransferase-2 (SAT-2), and sialyltransferase-3 (SAT-3), which act on lactosylceramide, GM3 and GD3, and produce gangliosides GM3, GD3 and GT3, respectively (Basu et al., 1987; Schwarzmann and Sandhoff, 1990). Further glycosylations are catalyzed by enzymes having broader specificity. Particularly, a single enzyme, N-acetyl-galactosamine transferase I (GalNAc T-1), converts lactosylceramide to GA2 (the asialoderivative of GM2), GM3 to GM2, and GD3 to GD2, and a single sialyltransferase (SAT-4) catalyzes the transformation of GA1 (the asialoderivative of GM1) to GM1b, of GM1 to GD1a, and of GD1b to GT1b (this enzyme is also capable of transforming GM3 to GD3) (Pohlentz et al., 1988). Similarly, a single galactosyltransferase (Gal T-2) converts GM2 to GM1, GD2 to GD1b and GT2 to GT1c, and one sialytransferase (SAT-5) catalyzes the conversion of GM1b to GD1c, of GD1a to GT1a, of GT1b to GQ1b and of GQ1c to GP1c (SAT-5 also converts GD3 to GT3) (Iber and Sandhoff, 1989; Iber et al., 1992a).

The glucosyltransferase forming glucosylceramide from ceramide, as well as the galactosyltransferase forming lactosylceramide, show an aspecific distribution along the different Golgi cisternae (Futerman and Pagano, 1991; Trinchera et al., 1991a; Jeckel et al., 1992). The glycosyltransferases producing the different gangliosides from lactosylceramide show a gradient distribution on the Golgi system. Earlier sialosylation prevails in the cis/medial Golgi, and later glycosylations in the trans Golgi/trans Golgi network (Trinchera et al., 1989; Iber et al., 1992b). This implies that the growing glycolipid moves from the endoplasmic reticulum through the Golgi cysternae, presumably via a vesicle flow (Wattenberg, 1990), which is also assumed to translocate mature gangliosides to the plasma membrane. Of course, the gangliosides residing at the plasma membrane level possess oligosaccharide chains of different complexity. This indicates that the different gangliosides undergo different glycosylation steps; particularly, that some of them escape further glycosylations possibly leaving the Golgi apparatus before the sites where these glycosylations occur. Newly synthesized gangliosides reach axons and nerve terminal membranes by fast axonal transport (Ledeen, 1989).

An intriguing aspect of ganglioside biosynthesis regards the membrane topology of the active site of the involved enzymes. All the enzymes implicated in the biosynthesis of ceramide as well as the glucosyltransferase forming glucosylceramide and the galactosyltransferase forming lactosylceramide have the active site oriented toward the cytosol (Coste et al., 1986; Futerman and Pagano, 1991; Trinchera et al., 1991b; Jeckel et al., 1992; Mandon et al., 1992). Instead, the glycosyltransferases involved in further glycosylations show a lumenal orientation (Schwarzmann and Sandhoff, 1990). Therefore lactosylceramide should be translocated from the cytosolic to the lumenal side of the Golgi membrane in order to be further glycosylated.

Biodegradation

Ganglioside biodegradation consists in the sequential removal of individual sugar residues, catalyzed by exo-glycohydrolases, ultimately leading to the formation of ceramide (Sandhoff and Christomanou 1979), that is split by ceramidase into sphingosine and fatty acid (Spence et al., 1986). The glycohydrolases involved in this process reside in the lysosomes. Infact, gangliosides accumulate in the cells after administration of chloroquine, the lysosomotropic drug that blocks lysosomal enzymes (Klinghardt et al., 1981; Riboni et al., 1991), and in inborn lysosomal diseases characterized by a defect of enzymes involved in ganglioside biodegradation (O'Brien, 1989; Sandhoff et al., 1989). Moreover, highly purified preparations of lysosomes were proven to carry enzymes affecting ganglioside breakdown (Fiorilli et al., 1989). The lysosomal degradation of gangliosides, and related neutral glycosphingolipids, requires the accessibility of the different saccharide units to the lysosomal enzymes, mostly present in the lysosol. It is suggested (Fürst and Sandhoff, 1992) that parts of the endosomal membranes — possibly those enriched in components derived from the plasma membrane — budd off into the endosomal lumen and form intra-endosomal vesicles carrying the glycolipid moieties outside. These vesicles, after endosome fusion with lysosome, become intralysosomally located with their gangliosidic components facing the lysosol. In many cases a physiologically high rate of hydrolysis is warranted by the presence of small non-enzymic cofactors, called sphingolipid activator proteins, some of them behaving as sphingolipid binding proteins (see the review by Fürst and Sandhoff, 1992).

Salvage metabolic pathways

The existence of recycling processes, where fragments from ganglioside biodegradation are used for biosynthetic purposes, was demonstrated by supplying animals or cultured cells with gangliosides, carrying radioactivity in different portions of the molecule, and following the metabolic fate of radioactivity. The rationale of this approach relied on the observation that exogenous gangliosides, after being taken up by cells, enter the pool of endogenous gangliosides and undergo regular metabolic processing (Schwarzmann et al., 1986; Huang and Dietsch, 1991). It was demonstrated in liver (in the case of animal experi-

ments) (Ghidoni et al., 1983; 1986; 1987; 1989; Trinchera et al., 1990) and in cultured cerebellar granule cells (Ghidoni et al., 1989; Riboni and Tettamanti 1991; Van Echten et al., 1990a and 1990b; Riboni et al., 1990; 1991; 1992). that gangliosides liberate, upon degradation, galactose, N-acetylgalactosamine, sialic acid, fatty acid and sphingosine, that are re-used for the biosynthesis of novel gangliosides, glycoproteins, glycerophospholipids, and sphingomyelin (Fig. 5). Also bigger metabolites, like glucosylceramide and ceramide, seem to be submitted to metabolic salvage.

In cultured cerebellar granule cells, compounds of catabolic origin started being detectable at 10–15 min of pulse, and compounds of biosynthesis from recycled catabolites after 15–30 min of pulse (Riboni et al., 1992). In the same cells, fed with GM1 carrying the radioactivity in the sphingosine moiety, it was shown that most of the sphingosine produced by ganglioside breakdown was metabolically recycled, whereas only a small percentage underwent complete degradation (Riboni et al., 1990). Similar results (unpublished observations) were obtained using exogenous gangliosides carrying the radioactivity in different sugar moieties. Moreover, the recycling process appeared to be blocked by inhibiting lysosome function by chloroquine or preventing endocytosis by low temperature treatment (Riboni and Tettamanti 1991; Riboni et al., 1992). All this means that recycling of catabolic fragments: (a) follows endocytosis and lysosomal breakdown; (b) is a rapid process; and (c) displays a high degree of metabolic salvage. Of course, recycling of metabolites formed in lysosomes implies their exit from lysosomes, possibly by the action of specific carriers (Pisoni et al., 1991).

Direct glycosylation processes

These processes consist in the direct glycosylation of membrane-bound gangliosides, internalized into the cell, with formation of more complex gangliosides. The occurrence of this process was demonstrated by the use of exogenously administered gangliosides, carrying a proper label (Fig. 6). For example, administration to rats (Trinchera et al., 1990) or cultured neurons (Ghidoni et al., 1989) of GM1 radiolabelled in the terminal galactose moiety resulted in the formation of GD1a; as well, administration of GM2, radiolabelled in the hexosamine moiety, in the formation of GM1 and GD1a. In cultured neurons, inhibition of endocytosis (Riboni et al., 1992) blocked direct glycosylation, implying that the exogenously added ganglioside had to be internalized to reach the intracellular site where glycosylations occur (presumably the Golgi apparatus). Therefore, it is necessary to assume the existence of a specific sorting of the ganglioside-carrying vesicle from the early endosomal compartment to the Golgi apparatus, parallelly and independently from the sorting to lysosomes.

This kind of lipid traffic is strongly supported by the evidence that in cultured fibroblasts (Schwarzmann et al., 1986), cultured kidney cells (Kok et al., 1989) and rat liver (Trinchera et al., 1990) a pulse of administered biotinylated, fluorescent, or radiolabeled glycolipid, respectively, is followed by the appearance of the same compound in the Golgi apparatus.

Role of the plasma membrane in ganglioside turnover

In addition to the overall turnover that has been described in the previous section, gangliosides might undergo a partial turnover at the level of the plasma membrane. This process affects some terminal sugars of gangliosides, particularly sialic acid. In fact, both

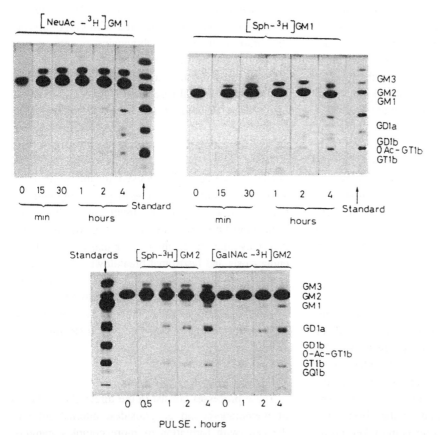

Fig. 5. Time course of the metabolic processing of different exogenous gangliosides (GM2, GM1), labeled in different parts of the molecule (sialic acid, [NeuAc-^{3}H]GM1; sphingosine, [Sph-^{3}H]GM1; *N*-acetylgalactosamine, [GalNAc-^{3}H]GM2; sphingosine, [Sph-^{3}H]GM2) by rat cerebellar granule cells differentiated in vitro. Note the formation of: (a) gangliosides simpler than the additioned one (GM3 from GM2; GM2 and GM3 from GM1), that are degradation products; and (b) gangliosides more complex than the additioned one that derive from recycling of degradation fragments (sialic acid, sphingosine, *N*-acetyl-galactosamine) (salvage processes), or direct glycosylation.

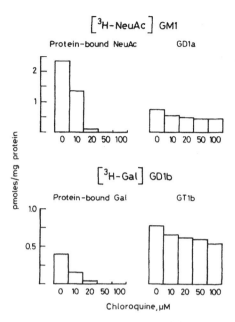

Fig 6. Effect of increasing amounts of chloroquine on salvage and direct glycosylation processes in rat cerebellar granule cells differentiated in vitro fed with GM1 labeled at the sialic acid moiety ([NeuAc-^{3}H]GM1), or GD1b labeled at the galactose moiety ([Gal-^{3}H]GD1b). Incorporation of labeled NeuAc and Gal into protein is an index only of salvage processes (recycling of saccharide units, liberated during ganglioside degradation); incorporation of radioactivity into GD1a (from GM1) and GT1b (from GD1b) is an expression of direct glycosylation plus salvage processes.

sialidase and sialyl-transferase activities are present in the plasma membranes of many cells, particularly neural cells. Brain membrane-bound sialidase was shown to be linked to synaptosomal plasma membranes (Tettamanti et al., 1972; Schengrund and Nelson 1975) and to face the extracellular environment (Scheel et al., 1982). The same enzyme appears to be linked to the membranes by a glycosylphosphatidylinositol anchor (Chiarini et al., 1990). On the other hand, sialyltransferases were reported to occur in the outer neural surface (Schengrund and Nelson 1975; Preti et al., 1980; Durrie et al., 1987) and to have a different specificity than the Golgi-linked sialyltransferases (Durrie et al., 1987).

The proof that desialosylation of gangliosides by membrane-bound sialidase is operative in living cells has been recently provided (Riboni et al., 1991), using primary cultures of cerebellar neurons. It was

observed that GD1a and GD1b, inserted into the cell plasma membrane, are degraded to GM1 under conditions of complete block of lysosomal function by chloroquine, or of endocytosis by low temperature. Parallelly, studies performed on brain slices in the presence of CMP-NeuAc demonstrated the local labeling of membrane-bound gangliosides as the result of the action of the plasma membrane-bound sialyltransferase (Durrie et al., 1987). Altogether these findings support the hypothesis that plasma membranes, particularly neural plasma membranes, possess a sialosylation-desialosylation system that may modulate locally the degree of ganglioside sialosylation (Tettamanti et al., 1980). This system may be instrumental to some functional cell performances and/or to trigger ganglioside endocytosis.

Role of endocytosis in ganglioside turnover

Under culture conditions cells are estimated to internalize via endocytosis about half their plasma membrane per hour (Steinman et al., 1983). This event, which is quite rapid, is followed by equally rapid processes of re-cycling and re-synthesis of the membrane components in order to maintain dynamically cell surface integrity. In the overall turnover of plasma membrane a key role is played by the sorting processes that from the early endosomes convey vesicles: (a) directly back to the plasma membrane (retroendocytosis); (b) to the Golgi apparatus; and (c) to late endosomes and further to lysosomes (Pagano, 1990) (Fig. 7). These sorting routes have different consequences in the metabolic turnover of the individual protein and lipid components of the membrane. In route: (a) no molecular modifications occur; in route (b) molecular remodelling is expected to occur; and in route (c) biodegradation takes place (only partially to terminal waste products) followed by re-use of fragments for biosynthetic purposes, and concurrent refilling by ex novo biosynthesis. Moreover, it cannot be excluded that during the processes of endocytotic vesicle fusion, sorting and routing, some of the membrane components are extruded and conveyed (with possible involvement of carrier proteins in the case of lipids) to the plasma membrane (Pagano, 1990).

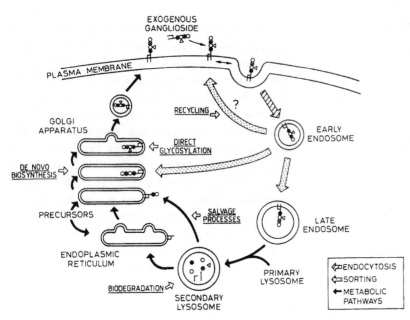

Fig. 7. Interconnection between intracellular flow and turnover of plasma membrane bound gangliosides.

Gangliosides, as well as glycosphingolipids in general, appear to undergo endocytosis (Schwarzmann et al., 1987; Van Meer, 1989; Schwarzmann and Sandhoff, 1990). Particularly, electron microscopic studies in cultured fibroblasts (Schwarzmann et al., 1987) employing biotinyl-GM1 showed that the biotin label was present on the cell surface, coated pitlike structures, endocytotic vesicles, lysosomes and Golgi cisternae. Indirect evidence supporting ganglioside endocytosis was also provided by studies with cells cultivated in vitro, and fed with radiolabelled gangliosides (Riboni et al., 1991; Riboni and Tettamanti, 1991). A definite proof that endocytosis of membrane-bound gangliosides follows a receptor-mediated endocytic pathway is not available. However, the observation that in both cultured fibroblasts and cerebellar granule cells (Sonnino et al., 1989; 1992) exogenous GM1 (labeled with a photoreactive probe) rapidly binds to one (or few) membrane-bound protein(s) before being internalized and metabolized, strongly suggests this hypothesis. Intracellular destination of endocytosed gangliosides is only partly understood. Thus far no approaches have been devised to inspect the occurrence of the

ganglioside direct return to the plasma membrane via the early endosome intracellular flow (retro-endocytosis) or via transfer proteins. Noteworthy, the occurrence of the latter process in artificial model systems (Ledeen et al., 1990) and the direct recycling of a fluorescent analog of glucosylceramide from early endosomes to the plasma membrane (Kok et al., 1992) were demonstrated. However, no suggestions can be given regarding the impact of these events in ganglioside turnover. Instead, convincing evidence was provided for vesicle transport of gangliosides from the early endosome compartment to lysosomes (Schwarzmann et al., 1987; Riboni et al., 1991; Riboni and Tettamanti, 1991), and the Golgi apparatus (Schwarzmann et al., 1986; Tettamanti et al., 1988; Trinchera et al., 1990). The percent distribution of endocytosed gangliosides to: (a) lysosomes; and (b) Golgi apparatus is presently difficult to evaluate and likely different from cell to cell. However, in cultured fibroblasts and cerebellar granule cells the major portion seems to be targeted to lysosomes (Schwarzmann and Sandhoff, 1990).

Internalization of gangliosides by endocytosis is expected to be a rapid event. Infact, and as already mentioned, lysosomal breakdown and salvage pathways of gangliosides take place in minutes in the cell systems employed so far (Riboni et al., 1991). Since in these systems previous insertion of exogenous gangliosides into the cell membrane is required, and both lysosomal biodegradation and salvage pathways are consequent to endocytosis, it should be inferred that ganglioside endocytosis occurs in terms of minutes, if not less. Therefore it can be suggested that the gangliosides present in endocytosed membrane are rapidly turned over. This conclusion, added to the observation that the rapid salvage pathways have a particular importance in ganglioside metabolism, contrasts with the turnover rates of gangliosides determined on tissues or cultured cells, with the use of radioactive simple precursors (glucose, N-acetylmannosamine, serine, fatty acid). Half-lives ranging from a few days to several weeks were reported (Ledeen, 1989). In order to provide an acceptable explanation of this seeming contradiction, more knowledge has to be gained especially on: (a) the ability of gangliosides to be recycled back to the membrane via retro-endocytosis, or carrier-protein transport; (b) the precise quantification of salvage pathways; and (c) the possibility for ganglioside molecules to be internalized at a different frequency than that of other membrane components.

Gangliosides as precursors of sphingosine and ceramide

As previously mentioned the hypothesis has been launched that sphingosine and ceramide serve as metabolic second messengers and that sphingolipids may produce them under particular conditions of cell stimulation by external substances. With regards to gangliosides two questions appear to be crucial in assessing the validity of this hypothesis: (a) by which pathway are sphingosine and ceramide formed from gangliosides; and (b) which external stimuli are able to modify the rate of production of sphingosine and ceramide from gangliosides.

Sphingosine and ceramide are present in free form in cultured cells and tissues (Merrill et al., 1988; Van Veldhoven et al., 1989; Dressler et al., 1990; Golkorn et al., 1991). Both molecules are assumed to be formed during sphingolipid degradation (Merrill,

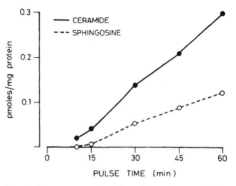

Fig. 8. Time course of sphingosine and ceramide formation from exogenous GM1 by rat cerebellar granule cells fed with 10^{-5}M GM1 labeled at the sphingosine level. Cells were used after 8 days in culture, when they were completely differentiated.

1991). Quite recently, it was demonstrated that ceramide and sphingosine are formed in cultured cerebellar granule cells during degradation of administered GM1 (radiolabeled at the sphingosine moiety) (Riboni et al., 1992) (Fig. 8). Both substances appeared to be produced very rapidly (10 and 15 min

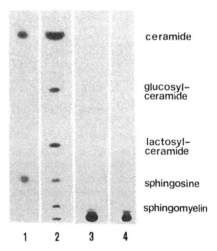

Fig. 9. Block of sphingosine and ceramide formation from exogenous GM1 by rat cerebellar granule cells upon treatment with 50 μM chloroquine (resulting in a block of intralysosomal degradation) or incubation at 4°C (that blocks endocytosis). 1: standard sphingosine and ceramide; 2: metabolic products from exogenously administered [Sph-^{3}H]GM1 (2×10^{-6} M), after 2-h pulse followed by 4-h chase, under control incubation conditions; 3: as 2 in the presence of 50 μM chloroquine; 4: as 2 at 4°C. Cerebellar granule cells were used at 8 days in culture.

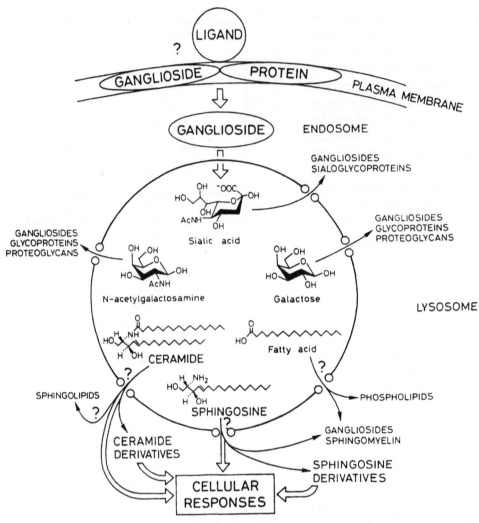

Fig. 10. A hypothetical scheme for the formation of metabolic regulators of sphingoid nature (sphingosine and its derivatives, ceramide and its derivatives) through the ganglioside pathway. This pathway consists in: (a) external ligand-receptor-ganglioside interaction at the cell surface capable of modifying ganglioside endocytosis; (b) ganglioside endocytosis; (c) intralysosomal degradation of gangliosides with formation of catabolic fragments including sphingosine and ceramide; (d) exit of sphingosine and ceramide from lysosomes; and (e) availability of metabolic regulators of sphingoid nature at the intracellular sites proper for eliciting cellular responses. Steps with a question mark (?) are still to be demonstrated.

after pulse for ceramide and sphingosine, respectively). Current experiments on the same cells showed (Fig. 9) that the formation of sphingosine and ceramide was completely abolished when endocytosis was prevented by low temperature (4°C) and the lysosomal apparatus blocked by chloroquine. Moreover, it was markedly reduced in the presence of Brefeldin A, a substance known to impair some steps

of the intracellular vesicle flow, including the route from late endosomes to lysosomes (Lippincott-Schwartz et al., 1991). All this supports the view that the formation of sphingosine and ceramide from membrane-bound gangliosides follows ganglioside internalization by endocytosis, sorting to lysosomes, and breakdown in the lysosomal apparatus (Fig. 10). How sphingosine and ceramide leave the lysosomal apparatus in order to be available at the subcellular sites where they exert regulatory roles is not known. Sphingosine surely leaves the lysosomal apparatus since it is recycled for the Golgi apparatus-assisted biosynthesis of both gangliosides and sphingomyelin (Tettamanti et al., 1988; Riboni et al., 1991, 1992). Data on the ceramide exit from lysosomes are not available. However, it has been established (Pagano, 1990) that ceramide can easily undergo transmembrane movement and be rapidly translocated from one to another intracellular site.

An open question is to what extent ganglioside endocytosis and biodegradation contribute to produce physiologically active concentrations of sphingosine and ceramide (or their derivatives). In a study, where cerebellar granule cells were submitted to a 1-h pulse to 4-h chase with 2×10^{-6} M exogenous ganglioside, it was calculated that 2 pmol and 20 pmol/10^6 cells of sphingosine and ceramide were produced, respectively (Riboni et al., 1992). Considering that the used cells were under basal conditions and assuming that the levels of free sphingosine and ceramide in cerebellar granule cells are in the range of those deter-

mined in other cells, it should be inferred that gangliosides do contribute to the maintenance of near-basal levels of free sphingosine and ceramide in cultured cerebellar granule cells. Recent findings (Rother et al., 1992), demonstrating that in cerebellar granule cells free sphingosine is not an intermediate of de novo sphingolipid biosynthesis, give further relevance to the possible ganglioside origin of this sphingoid molecule.

A second question is whether extra-, and/or intracellular stimuli might influence the sequence of events producing sphingosine and ceramide from gangliosides. In the case of cultured cerebellar granule cells it has been already established that higher amounts of both compounds are produced by differ-

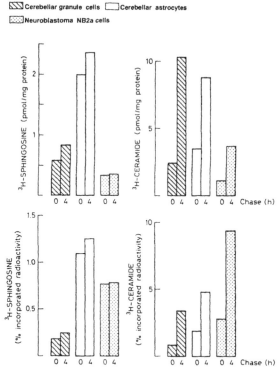

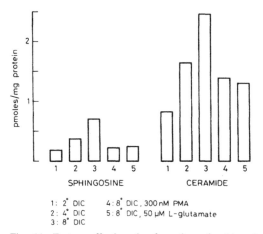

Fig. 11. Factors affecting the formation of sphingosine and ceramide from exogenous GM1 by rat cerebellar granule cells differentiated in vitro. Cells were submitted to a 2-h pulse with 2×10^{-6} M GM1, followed by a 4-h chase. 1,2,3: cells at the 2nd, 4th and 8th day in culture; 4: same as 3, in the presence of 300 nM phorbol-12-myristate, 13-acetate (PMA). 5: same as 3 with a 15-min exposure to 50 μM L-glutamate before ganglioside treatment.

Fig. 12. Different behaviour of different cells of neural origin, cultivated in vitro, in the formation of sphingosine and ceramide from exogenous GM1. Cells were submitted to a 2-h pulse with 2×10^{-6} M [Sph-[³H]GM1] followed, or not, by a chase period of 4 h. Formed radiolabeled sphingosine and ceramide are expressed as pmol, mg protein⁻¹, or as percentage of total incorporated radioactivity.

entiated than by undifferentiated cells (Riboni et al., 1992) (Fig. 11). In the same cells, made deficient in protein kinase C by long-term exposure to phorbol 12-myristate 13-acetate (PMA) or treated with excitotoxic amounts (50 μM) of glutamate, the formation of sphingosine and ceramide from ganglioside GM1 was markedly lower than that of control cells (Fig. 11). Finally, under seemingly similar culture conditions and in the presence of the same amounts of radioactive exogenous ganglioside (GM1), different neural cells produce different amounts of sphingosine and ceramide (Fig. 12). When compared to cerebellar granule cells (that produce much more ceramide than sphingosine), cerebellar astrocytes produce a markedly higher proportion of sphingosine; both cerebellar astrocytes and murine neuroblastoma Neuro 2a cells incorporate into sphingosine and ceramide a percentage of total radioactivity higher than the corresponding values from cerebellar granule cells. All these pieces of evidence, although fragmentary, encourage to suggest the possible existence of a relationship between formation of metabolic sphingoid-regulators from gangliosides and some functional states of the cell.

Perspectives

Gangliosides appear to be involved in many nerve cell functions and the concept of a multifunctional role for these molecules seems the most flexible and suitable way to explain different functional behaviours (Hakomori, 1990). This wider view of interpreting ganglioside function includes the challenging hypothesis that gangliosides contribute to generate a new family of metabolic second messengers, comprising sphingosine, ceramide and some of their derivatives (Fig. 10). On the basis of the present knowledge the only conceivable link between ganglioside metabolism and formation of compounds of a sphingoid nature is intra-lysosomal breakdown of membrane-bound gangliosides subsequent to endocytosis. Therefore the suggestion is that external ligand binding to ganglioside affects the rate of ganglioside endocytosis and/or their sorting from endosomes to lysosomes. By consequence their metabolic processing and the subsequent formation of second messengers of sphingoid nature turns out to be modified. The final effect is a modulation of cellular responses. The various aspects of this new perspective, as well as assessment of its applicability to definite aspects of nerve cell function, must be precisely defined and require the development of new experimental designs and ideas.

Acknowledgements

The experimental part of this work, accomplished in the Authors' laboratory, was partially supported by grants from the Consiglio Nazionale delle Ricerche, C.N.R., Rome, Italy (grant no. 91.01246.PF 70, Targeted Project 'Biotechnology and Bioinstrumentation'; and grant no. 91.00052.PF 99, Targeted Project 'Genetic Engineering').

References

Abrahamsson, S., Dahlen, B., Lofgren, H., Pascher, J. and Sandell, S. (1977) Structure of biological membranes. In: S.Abrahamsson and J. Pascher (Eds), *Structure of biological membranes*, Plenum Press, New York, pp. 1–23.

Acquotti, D., Poppe, L., Dabrowski, J., Von der Lieth, C.W., Sonnino, S. and Tettamanti, G. (1990) Three dimensional structure of the oligosaccharide chain of GM1 ganglioside revealed by a distance — mapping procedure: a rotating and laboratory frame nuclear over — hauser enhancement investigation of native glycolipid in dimethylsulfoxide and in water — Dodecylphosphocholine solution. *J. Am. Chem. Soc.*, 112: 7772–7776.

Basu, M., De, T., Das, K.K., Kyle, J.W., Chon, H-C., Schaeper, R.J. and Basu, S. (1987), Glycolipids. *Methods Enzymol.*, 138: 575–607.

Berg, D.K. (1984) New neuronal growth factors. *Annu. Rev. Neurosci.*, 7: 149–170.

Berry-Kravis, E. and Dawson, G.G. (1985) Possible role of gangliosides in regulating an adenylate cyclase-linked 5-hydroxytryptamine (5-HT1) receptor. *J. Neurochem.*, 45: 1739–1747.

Bertoli, E., Masserini, M., Sonnino, S., Ghidoni, R., Cestaro, B. and Tettamanti, G. (1981) Electron paramagnetic resonance studies on the fluidity and surface dynamics of egg phosphatidylcholine vesicles containing ganglioside. *Biochim. Biophys. Acta*, 467: 196–202.

Brady, R.O. and Fishman, P.H. (1979) Biotransducers of membrane mediated information. *Adv. Enzymol.*, 50: 303–324.

Caputto, R. (1988) Aspects of the biosynthesis of gangliosides and its regulation in relation to sensorial stimulation. In: R.W.

Ledeen, E.L. Hogan, G. Tettamanti, A.J. Yates and R.K. Yu (Eds), *New Trends in Ganglioside Research: Neurochemical and Neurogenerative Aspects, Fidia Research Series, vol. 14*, Liviana Press/Springer Verlag, Padova/Berlin, pp. 3–14.

Chan, K.F.J. (1989) Effects of gangliosides on protein phosphorylation in rat brain myelin. *Neurosci. Res. Commun.*, 5: 95–104.

Cheresh, D.A., Pierschbacher, M.D., Herzig, M.A. and Mujoo, K. (1986) Disialogangliosides GD2 and GD3 are involved in the attachment of human melanoma and neuroblastoma cells, to extracellular matrix proteins. *J. Cell. Biol.*, 102: 688–696.

Cheresh, D.A., Pytela, R., Pierschbacher, M.D., Ruoslahti, E. and Reisfeld, R.A. (1988) An arg-gly-asf-directed adhesion receptor on human melanoma cell exists in a calcium-dependent functional complex with the disialoganglioside GD2, In: R.W. Ledeen, E.L. Hogan, G.Tettamanti, A.J. Yates and R.K. Yu (Eds), *New Trends in Ganglioside Research: Neurochemical and Neurogenerative Aspects Fidia Research Series, Vol.14*, Liviana Press/Springer Verlag, Padova/Berlin, pp. 203–217.

Chiarini, A., Fiorilli, A., Siniscalco, C., Tettamanti, G. and Venerando, B. (1990) Solubilization of the membrane-bound sialidase from pig brain by treatment with bacterial phosphatidylinositol phospholipase C. *J. Neurochem.*, 55: 1576–1584.

Claro, E., Wallace, M.A., Fain, J.N., Nair, B.G., Patel, T.B., Shanker, G. and Baker, H.J. (1991) Altered phosphoinositide-specific phospholipase C and adenylyl cyclase in brain cortical membranes of cats with GM1 and GM2 gangliosidois. *Mol. Brain. Res.*, 11: 265–271.

Corti, M., Cantù, L. and Sonnino S. (1987) Fundamentals of physico-chemical properties of gangliosides in solution. In: H. Rahmann (Ed.), *Gangliosides and Modulation of Neuronal Functions, NATO ASI Series, Series H.: Cell Biology, Vol. 7*, Springer/Verlag, Berlin, pp. 101–118.

Coste, H., Martel, M.B. and Got, R. (1986) Topology of glucosylceramide synthesis in Golgi membranes from porcine submaxillary glands. *Biochim. Biophys. Acta*, 858: 6–12.

Cuello, A.C. (1990) Glycosphingolipids that can regulate nerve growth and repair. *Adv. Pharmacol.*, 21: 1–50.

Cuello, A.C., Garofalo, L., Kenisberg, R.L. and Maysinger, D. (1989) Gangliosides potentiate in vivo and in vitro effects of nerve growth factor on central cholinergic neurons. *Proc. Natl. Acad. Sci., U.S.A.*, 86: 2056–2060.

Curatolo, W. (1987) The physical properties of glycolipids. *Biochim. Biophys. Acta*, 906: 111–136.

Czarniecki, M.F. and Thornton, E.R. (1977) Carbon-13 nuclear magnetic resonance spin-lattice relaxation in the *N*-acylneuraminic acids. Probes for internal dynamics and conformational analysis. *J. Am. Chem. Soc.*, 99: 8273–8278.

Davis, C.W. and Daly, J.W. (1980) Activation of rat cerebral cortical 3′, 5′-cyclic nucleotide phosphodiesterase activity by gangliosides. *Mol. Pharmacol.*, 17: 206–211.

De Erausquin, G.A., Manev, H., Guidotti, A., Costa, E. and Brooker, G. (1990) Gangliosides normalize distorted single-cell intracellular free Ca^{2+} dynamics after toxic doses of glutamate in cerebellar granule cells. *Proc. Natl. Acad. Sci., U.S.A.*, 87: 8017–8021.

Doherty, P., Ashton, S.V., Skaper, S.D., Leon, A. and Walsh, F.S. (1992) Ganglioside modulation of neural cell adhesion molecule and *N*-cadherin-dependent neurite outgrowth. *J. Cell. Biol.*, 117: 1093–1099.

Dressler, K.A. and Kolesnick, R.N. (1990) Ceramide 1-phosphate a novel phospholipid in human leukemia (HL-60) cells. Synthesis via ceramide from sphingomyelin. *J. Biol. Chem.*, 265: 14917–14929.

Dreyfus, H., Louis, J.C., Harth, S. and Mandel, P. (1980) Gangliosides in cultured neurons. *Neuroscience*, 5: 1647–1655.

Durand, M., Guerold, B., Lombard-Golly, D. and Dreyfus, H. (1986) Evidence for the effects of gangliosides on the development of neurons in primary cultures, In: G. Tettamanti, R.W. Ledeen, K. Sandhoff, Y. Nagai and G. Toffano (Eds), *Gangliosides and neuronal plasticity, Fidia Research Series, Vol. 6*, Liviana Press/Springer Verlag, Padova/Berlin. pp. 295–308.

Durrie, R., Saito, M. and Rosenberg, A. (1987) Glycolipid sialosyltransferase activity in synaptosomes exhibits a product specificity for (2-8)disialosyl lactosyl ceramide (ganglioside GD3). *J. Neurosci. Res.*, 18: 456–465.

Facci, L., Leon, A., Toffano, G., Sonnino, S., Ghidoni, R. and Tettamanti, G. (1984) Promotion of neuritogenesis in mouse neuroblastoma cells by exogenous gangliosides. Relationship between the effect and the cell association of ganglioside GM1. *J. Neurochem.*, 42: 299–305.

Favaron, M., Manev, H., Alho, H., Bertolino, M., Ferret, B., Guidotti, A. and Costa, E. (1988) Gangliosides prevent glutamate and kainate neurotoxicity in primary neuronal cultures of neonatal rat cerebellum and cortex. *Proc. Natl. Acad. Sci., U.S.A.*, 85: 7351–7355.

Ferrari, G., Fabris, M. and Gorio, A. (1983) Gangliosides enhance neurite outgrowth in PC12 cells. *Dev. Brain. Res.*, 8: 215–222.

Ferreira, A., Busciglio, J., Landa, C. and Caceras, A. (1990) Ganglioside-enhanced neurite growth: evidence for a selective induction of high-molecular-weight MAP-2. *J. Neurosci.*, 10: 293–302.

Ferret, B., Massarelli, R., Freysz, L. and Dreyfus, H. (1987) Effect of exogenous gangliosides on the metabolism of inositol compounds in chick neurons in culture. *C.R. Acad. Sci.*, 304: 97–99.

Fiorilli, A., Venerando, B., Siniscalco, C., Monti, E., Bresciani, R., Caimi, L., Preti, A. and Tettamanti, G. (1989) Occurrence in brain lysosomes of a sialidase active on gangliosides. *J. Neurochem.*, 53: 672–680.

Fishman, P.H. (1988) Gangliosides as cell surface receptors and transducers of biological signals. In: R.W. Ledeen, E.L. Hogan, G. Tettamanti, A.J. Yates and R.K. Yu (Eds), *New Trends in Ganglioside Research: Neurochemical and Neuroregenerative Aspects, Fidia Research series, Vol. 14*, Liviana Press/Springer Verlag, Padova/Berlin, pp. 183–201.

Freysz, L., Farooqui, A.A., Horrocks, L.A., Massarelli, R. and Dreyfus, H. (1991) Stimulation of mono- and diacylglycerol lipase activities by gangliosides in chicken neuronal cultures. *Neurochem. Res.*, 16: 1241–1244.

Fueshko, S.M. and Schengrund, C.L. (1990) Murine neuroblastoma cells express ganglioside binding sites on their cell surface. *J. Neurochem.*, 54: 1791–1797.

Fürst, W. and Sandhoff, K. (1992) Activator proteins and topology of lysosomal sphingolipid catabolism. *Biochim. Biophys. Acta*, 1126: 1–16.

Futerman, A.H. and Pagano, R.E. (1991) Determination of the intracellular sites and topology of glucosylceramide synthesis in rat liver. *Biochem. J.*, 280: 295–302.

Fukunaga, K., Miyamoto, E. and Soderling, T.R. (1990) Regulation of Ca^{2+}/calmodulin-dependent protein kinase II by brain gangliosides. *J. Neurochem.*, 54: 102–109.

Ghidoni, R., Sonnino, S., Chigorno, V., Venerando, B. and Tettamanti, G. (1983) Differences in liver ganglioside patterns in various inbred strains of mice. *Biochem. J.*, 209: 885–888.

Ghidoni, R., Trinchera, M., Venerando, B., Fiorilli, A., Sonnino, S. and Tettamanti, G. (1986) Incorporation and metabolism of exogenous GM1 ganglioside in rat liver. *Biochem. J.*, 237: 147–155.

Ghidoni, R., Trinchera, M., Sonnino, S., Chigorno, V. and Tettamanti, G. (1987) The sialic acid of exogenous GM1 ganglioside is recycled for biosynthesis of sialoglycoconjugates in rat liver. *Biochem. J.*, 247: 157–164.

Ghidoni, R., Riboni, L. and Tettamanti, G. (1989) Metabolism of exogenous gangliosides in cerebellar granule cells, differentiated in culture. *J. Neurochem.*, 53: 1567–1574.

Goldkorn, T., Dressler, K.A., Muindi, J., Radin, N.S., Mendelsohn, J., Menaldino, D., Liotta, D. and Kolesnick, R.N. (1991) Ceramide stimulates epidermal growth factor receptor phosphorylation in A431 human epidermoid carcinoma cells. *J. Biol. Chem.*, 266: 16092–16097.

Hadjiconstantinou, M. and Neff, N.H. (1990) Treatment with GM1 ganglioside reverses dopamine D-2 receptor, *Eur. J. Pharmacol.*, 181: 137–139.

Hakomori, S.I. (1981) Glycosphingolipids in cellular interaction, differentiation and oncogenesis. *Annu. Rev. Biochem.*, 50: 733–764.

Hakomori, S.I. (1990) Bifunctional role of glycosphingolipids. Modulators for transmembrane signaling and mediators for cellular interactions. *J. Biol. Chem.*, 265: 18713–18716.

Hannun, Y.A. and Bell, R.M. (1989) Function of sphingolipids and sphingolipid breakdown products in cellular regulation. *Science*, 243: 500–507.

Hansson, H.A., Holmgren, J. and Svennerholm, L. (1977) Ultrastructural localization of cell membrane GM1 ganglioside by cholera toxin. *Proc. Natl. Acad. Sci, U.S.A.*, 74: 3782–3786.

Harris, P.L. and Thornton, E.R. (1978) Carbon-13 and proton nuclear magnetic resonance studies of gangliosides. *J. Am. Chem. Soc.*, 100: 6738–6745.

Higashi, H. and Yamagata, T. (1992) Mechanism for ganglioside-mediated modulation of a calmodulin-dependent enzyme. *J. Biol. Chem.*, 267: 9839–9843.

Higashi, H., Omori, A. and Yamagata, T. (1992) Calmodulin a ganglioside-binding protein. *J. Biol. Chem.*, 267: 9831–9838.

Hilbush, B.S. and Levine, S.M. (1992) Modulation of a Ca^{++} signaling pathway by GM1 ganglioside in PC12 cells, *J. Biol. Chem.*, 267: 24789–24795.

Holmgren, J., Mansson, J.E. and Svennerholm, L. (1974) Tissue receptor for cholera enterotoxin: structural requirements of GM1 ganglioside in toxin binding and inactivation., *Med. Biol.*, 52: 224–233.

Huang, R.T.C. and Dietsch, E. (1991) Cellular incorporation and localization of fluorescent derivatives of gangliosides, cerebroside and sphingomyelin. *FEBS Lett.*, 281: 39–42.

Iber, H. and Sandhoff, K. (1989) Identity of GD1c, GT1a and GQ1b synthase in Golgi vesicles from rat liver. *FEBS Lett.*, 254: 124–128.

Iber, H., Zacharias, C. and Sandhoff, K. (1992a) The c-series gangliosides GT3, GT2 and GP1c are formed in rat liver Golgi by the same set of glycosyltransferase that catalyse the biosynthesis of asialo-, a-, and b-series gangliosides. *Glycobiology*, 2: 137–142.

Iber, H., Van Echten, G. and Sandhoff, K. (1992b) Fractionation of primary cultured cerebellar neurons: distribution of sialyltransferases involved in ganglioside biosynthesis. *J. Neurochem.*, 58: 1533–1537.

Igarashi, Y., Nojiri, H., Hanai, N. and Hakomori, S.I. (1989) Gangliosides that modulate membrane protein function. *Methods Enzymol.*, 179: 521–541.

Jeckel, D., Karrenbauer, A., Burger, K.N.J., Van, Meer G. and Wieland, F. (1992) Glucosylceramide is synthetized at the cytosolic surface of various Golgi subfractions. *J. Cell. Biol.*, 117: 259–267.

Juurlink, B.H.J., Munoz, D.G. and Ang, L.C. (1991) Motoneuron survival in vitro: effects of pyruvate, α-ketoglutarate, gangliosides and potassium. *Neurosci. Lett.*, 133: 25–28.

Klinghardt, G.W., Fredman, P. and Svennerholm, L. (1981) Chloroquine intoxication induces ganglioside storage in nervous tissue: a chemical and histopathological study of brain, spinal cord, dorsal root ganglia, and retina in the miniature pig. *J. Neurochem.*, 37: 897–908.

Koerner, T.A.W., Prestegard, J.H., Demou, P.C. and Yu, R.K. (1983) High resolution proton NMR studies of gangliosides 1. use of homonuclear two-dimensional spin-echo J-correlated spectroscopy for determination of residue composition and anomeric configurations. *Biochemistry*, 22: 2676–2687.

Kok, J.W., Eskelinen, S., Hoekstra, K. and Hoekstra, D. (1989) Salvage of glucosylceramide by recycling after internalization along the pathway of receptor-mediated endocytosis. *Proc. Natl. Acad. Sci., U.S.A.*, 86: 9896–9900.

Kok, J.W., Hoekstra, K., Eskelinen, S. and Hoekstra, D. (1992) Recycling pathways of glucosylceramide in BHK cells: distinct involvement of early and late endosomes, *J. Cell Sci.*, 103: 1139–1152.

Kolesnick, R.N. (1990) Ceramide 1-phosphate a novel phospholipid in human leukemia (HL-60) cells. Synthesis via ceramide from sphingomyelin. *J. Biol. Chem.*, 265: 14917–14929.

Kolesnick, R. (1992) Ceramide: a novel second messenger. Trends *Cell Biol.*, 2: 232–236.

Kreutter, D., Kim, J.Y.H., Goldenring, J.R., Rasmussen, H., Ukomadu, Y., DeLorenzo, R.J. and Yu, R.K. (1987) Regulation of protein kinase C activity by gangliosides. *J. Biol. Chem.*, 262: 1633–1637.

Langner, M., Winiski, A., Eisenberg, M. and McLaughlin, S. (1988) The electrostatic potential adjacent to bilayer membranes containing either charged phospholipids or gangliosides. In: R.W. Ledeen, E.L. Hogan, Y. Nagai, G. Tettamanti, A.J. Yates and R.K. Yu (Eds), *New Trends in Ganglioside Research: Neurochemical and Neuroregenerative Aspects, Fidia Research Series, Vol.14*, Liviana Press/Springer Verlag, Padova/Berlin, pp. 121–131.

Ledeen, R.W. (1984) Biology of gangliosides: neuritogenic and neurono-trophic properties. *J. Neurosci. Res.*, 12: 147–159.

Ledeen, R.W. (1989) Biosynthesis, metabolism, and biological effects of gangliosides. In: R.U. Margolis and R.K. Margolis (Eds), *Neurobiology of glycoconjugates*. Plenum Press, New York, pp. 43–83.

Ledeen, R.W., Wu, G., Vaswani, K.K. and Cannella, M.S. (1990) Comparison of exogenous and endogenous gangliosides as modulators of neuronal differentiation. In: L.A. Horrocks, N.H. Neff, A.J. Yates and M. Hadjicostantinou (Eds), *Trophic Factors and the Nervous System, Fidia Research Foundation Symposium Series, Vol.3*, Raven Press, New York, pp. 17–33.

Leon, A., Facci, L., Toffano, G., Sonnino, S. and Tettamanti, G. (1981) Activation of (Na$^+$, K$^+$)-ATPase by nanomolar concentrations of GM1 ganglioside. *J. Neurochem.*, 37: 350–357.

Leon, A., Dal Toso, R., Presti, D., Benvegni, D., Facci, L., Kirschner, G., Tettamanti, G. and Toffano, G. (1988) Development and survival of neurons in dissociated fetal mesencephalic serum-free cell cultures: II. Modulatory effects of gangliosides. *J. Neurosci.*, 8: 746–753.

Lippincott-Schwartz, J., Yuan, L., Tipper, C., Amherdt, M., Orci, L., and Klausner, R.D. (1991) Brefeldin A's effects on endosomes, lysosomes, and the TGN suggest a general mechanism for regulating organelle structure and membrane traffic. *Cell*, 67: 601–616.

Liscovitch, M. and Lavie, Y. (1990) Sphingoid bases: sphingolipid derived modulators of signal transduction. *TIGG*, 2: 470–484.

Maggio, B. (1985) Geometric and thermodynamic restrictions for the self-assembly of glycosphingolipid-phospholipid systems. *Biochim. Biophys. Acta*, 815: 245–258.

Mandom, E.C., Ehses, I., Rother, J., Van Echten, G. and Sandhoff, K. (1992) Subcellular localization and membrane topology of serine palmitoyltransferase, 3-dehydrosphinganine reductase and sphinganine *N*-acyltransferase in mouse liver. *J. Biol. Chem.*, 267: 11144–11148.

Manev, H., Favaron, M., Vicini, S. and Guidotti, A. (1990) Ganglioside-mediated protection from glutamate-induced neuronal death. *Acta Neurobiol. Exp.*, 50: 475–488.

Masserini, M. and Freire, E. (1986) Termotropic characterization of phosphatidylcholine vesicles containing GM1 ganglioside with homogeneous ceramide chain length composition. *Biochemistry*, 25: 1043–1049.

Masserini, M., Giuliani, A., Venerando, B., Fiorilli, A., D'Aniello, A. and Tettamanti, G. (1985) alfa-Fucosidase-ganglioside interactions. Action of alfa-L-fucosidase from the hepatopancreas of Octopus vulgaris on a fucose-containing ganglioside (Fuc-GM1). *Biochem. J.*, 229: 595–603.

Masserini, M., Palestini, P., Venerando, B., Fiorilli, A., Acquotti, D. and Tettamanti, G. (1988) Interactions of proteins with ganglioside-enriched microdomains on the membrane; the lateral phase separation of molecular species of GD1a ganglioside having homogeneous long chain base composition is recognized by vibrio-cholerae sialidase. *Biochemistry*, 27: 7973–7978.

Masserini, M., Freire, E., Palestini, P., Calappi, E. and Tettamanti, G. (1992) Fuc-GM1 ganglioside mimics the receptor function of GM1 for cholera toxin. *Biochemistry*, 31: 2422–2426.

Mathias, S., Younes, A., Kan, C.-C., Orlow, I., Joseph, C. and Kolesnick, R.N. (1993) Activation of the sphingomyelin signaling pathways in intact EL4 cells and in a cell-free system by IL-1β. *Science*, 259: 519–522.

Maysinger, D., Leavitt, B.R., Zorc, B., Butula, I., Fernandes, L.G. and Rebeiro-da-Silva, A. (1992) Inhibition of high affinity choline uptake in the rat brain by neurotoxins: effect of monosialoganglioside GM1. *Neurochem. Int.*, 20: 289–297.

McDaniel, R.V., McLaughlin, A., Winiski, A.P., Eisenberg, M. and McLaughlin, S. (1984) Bilayer membranes containing the ganglioside GM1: models for electrostatic potentials adjacent to biological membranes. *Biochemistry*, 23: 4618–4624.

Merrill, A.H. Jr. (1991) Cell regulation by sphingosine and more complex sphingolipids. *J. Bioenerg. Biomembr.*, 23: 83–104.

Merrill, A.H. Jr. and Jones, D.D. (1990) An update of the enzymology and regulation of sphingomyelin metabolism. *Biochim. Biophys. Acta*, 1044: 1–12.

Merrill, A.H. Jr., Wang, E., Mullins, R.E., Jamison, W.C.L., Nimkar, S. and Liotta, D.C. (1988) Quantitation of free sphingosine in liver by high-performance liquid chromatography. *Anal. Biochem.*, 171: 373–381.

Myers, M., Wortman, C. and Freire, E. (1984) Modulation of neuraminidase activity by the physical state of phospholipid bilayers containing gangliosides GD1a and GT1b. *Biochemistry*, 23: 1442–1448.

Nagai, Y., Tsuji, S. and Nakajima, J. (1986) Bioactive gangliosides: gangliosides in signal transduction. In: S.Tucek (Ed.), *Metabolism and development of the nervous system*, Academia Praha, J.Wiley and Sons, Chichester, pp. 113–118.

Nagata, Y., Ando, M., Iwata, H., Hara, A. and Taketomi, T. (1987) Effect of exogenous gangliosides on amino acid uptake and Na$^+$, K$^+$ -ATPase activity in superior cervical and nodose ganglia of rats. *J. Neurochem.*, 49: 201–207.

Nakajima, J., Tsuji, S. and Nagai, Y. (1986) Bioactive gangliosides: analysis of functional structures of the tetrasialoganglioside GQ1b which promotes neurite outgrowth. *Biochim. Biophys. Acta*, 876: 65–71.

Obata, K., Oide, M., Handa, S. (1977) Effects of glycolipids on in vitro development of neuromuscular junction. *Nature*, 266: 369–371.

O'Brien, J.S. (1989) β-Galactosidase deficiency (GM1 gangliosidosis, galactosidosis and Morquio syndrome type B); ganglioside sialidase deficiency (mucolipidosis IV). In: C.R. Scriver, A.L. Beaudet, W.S. Sly and D.Valle (Eds), *The Metabolic Bases of Inherited Disease, 6th edn*, McGraw-Hill, New York, pp. 1797–1806.

Pagano, R.E. (1990) Lipid traffic in enkaryotic cells: mechanism for intracellular transport and organelle-specific enrichment of lipids. *Curr. Opinion Cell Biol.*, 2: 652–663.

Palestini, P., Masserini, M., Fiorilli, A., Calappi, E. and Tettamanti, G. (1991) Evidence of nonrandom distribution of GD1a ganglioside in rabbit brain microsomal membranes. *J. Neurochem.*, 57: 748–753.

Partington, C.R. and Daly, J.W. (1979) Effect of gangliosides on adenylate cyclase activity in rat cerebral cortical membranes. *Mol. Pharmacol.*, 15: 484–491.

Pascher, I. (1976) Molecular arrangements in sphingolipids. Conformation and hydrogen bonding of ceramide and their implication on membrane stability and permeability. *Biochim. Biophys. Acta*, 455: 433–451.

Pascher, I., Lundmark, M., Nyholm, P.-G. and Sundell, S. (1992) Crystal structures of membrane lipids. *Biochim. Biophys. Acta*, 1113: 339–373.

Petroni, A., Bertazzo, A., Sarti, S. and Galli, C. (1989) Accumulation of arachidonic acid cyclo-and lipoxygenase products in rat brain during ischemia and reperfusion: effects of treatment with GM1-lactone. *J. Neurochem.*, 53: 747–752.

Pisoni, R.L. and Thoene, J.G. (1991) The transport systems of mammalian lysosomes. *Biochim. Biophys. Acta*, 1071: 351–373.

Pohlentz, G., Klein, D., Schwarzmann, G., Schmitz, D. and Sandhoff, K. (1988) Both GA2, GM2, and GD2 synthases and GM1b, GD1a, and GT1b synthases are single enzymes in Golgi vesicles from rat liver. *Proc. Natl. Acad. Sci, U.S.A.*, 85: 7044–7048.

Preti, A.S., Fiorilli, A., Lombardo, A., Caimi, L. and Tettamanti, G. (1980) Occurrence of sialyltransferase activity in the synaptosomal membranes prepared from calf brain cortex. *J. Neurochem.*, 35: 281–296.

Quinet, C.C., McGuiness, T.L. and Greengard, P. (1984) Immunocytochemical localization of calcium/calmodulin-dependent protein kinase II in rat brain. *Proc. Natl. Acad. Sci. U.S.A.*, 81: 5604–5608.

Radin, N.S. (1984) Biosynthesis of the sphingoid bases: a provocation. *J. Lipid Res.*, 25: 1536–1540.

Rahmann, H. (1992) Calcium-ganglioside interactions and modulation of neuronal functions, In: N.N. Osborne (Ed.), *Current aspects of the Neurosciences, Vol. 4*, Mc Millan Press, London, pp. 87–125.

Riboni, L. and Tettamanti, G. (1991) Rapid internalization and intracellular metabolic processing of exogenous ganglioside by cerebellar granule cells differentiated in culture. *J. Neurochem.*, 57: 1931–1939.

Riboni, L., Sonnino, S., Acquotti, D., Malesci, A., Ghidoni, R., Egge, H., Mingrino, S. and Tettamanti, G. (1986) Natural occurrence of ganglioside lactones. Isolation and characterization of GD1b inner ester from adult human brain. *J. Biol. Chem.*, 261: 8514–8519.

Riboni, L., Prinetti, A., Pitto, M. and Tettamanti, G. (1990) Patterns of endogenous gangliosides and metabolic processing of exogenous gangliosides in cerebellar granule cells during differentiation in culture. *Neurochem. Res.*, 15: 1175–1183.

Riboni, L., Prinetti, A., Bassi, R. and Tettamanti, G. (1991) Cerebellar granule cells in culture exhibit a ganglioside-sialidase presumably linked to the plasma membrane. *FEBS Lett.*, 287: 42–46.

Riboni, L., Bassi, R., Sonnino, S. and Tettamanti, G. (1992) Formation of free sphingosine and ceramide from exogenous ganglioside GM1 by cerebellar granule cells in culture. *FEBS Lett.*, 300: 188–192.

Rosner, H., Al-Aqtum, M., Sonnentag, U., Wurster, A. and Rahmann, H. (1992) Gangliosides and neuronal differentiation. *Neurochem. Int.*, 20: 409–419.

Rother, J., Van Echten, O., Schwarzmann, G. and Sandhoff, K. (1992) Biosynthesis of sphingolipids: dihydroceramides and not sphinganine is desaturated by cultured cells. *Biochim. Biophys. Res. Commun.*, 189: 14–20.

Sandhoff, K. and Christomanou, H. (1979) Biochemistry and genetics of gangliosidoses. *Hum. Genet.*, 50: 107–143.

Sandhoff, K., Conzelmann, E., Neufeld, E.F., Kaback, M.M. and Suzuki, K. (1989) The GM2 gangliosidoses. In: C.R. Scriver, A.L. Beudet, W.S. Sly and D. Valle (Eds), *The metabolic basis of inherited disease, 6th edn*, McGraw-Hill, New York, pp. 1807–1839.

Scheel, G., Acevedo, E., Conzelmann, E., Nehrkorn, H. and Sandhoff, K. (1982) Model for the interaction of membrane-bound substrates and enzymes. Hydrolysis of ganglioside GD1a by sialidase of neuronal membranes isolated from calf brain. *Eur. J. Biochem.*, 127: 245–253.

Schengrund, C.-L. and Nelson, J.T. (1975) Influence of cation concentration on the sialidase activity of neuronal synaptic membranes. *Biochem. Biophys. Res. Commun.*, 63: 217–223.

Schneider, J.S., Pope, A., Simpson, K., Taggart, J., Smith, M.G. and Di Stefano, L. (1992) Recovery from experimental parkinsonism in primates with GM1 ganglioside treatment. *Science*, 256: 843–846.

Schwarzmann, G. and Sandhoff, K. (1990) Metabolism and intracellular transport of glycosphingolipids. *Biochemistry*, 29: 10865–10871.

Schwarzmann, G., Hinrichs, U., Sonderfeld, S., Marsh, D. and Sandhoff, K. (1986) Metabolism of exogenous gangliosides in cultured fibroblasts and in cerebellar cells. In: L. Freysz, H. Dreyfus, R. Massarelli, and S. Gatt (Eds), *Enzymes of lipid metabolism II*, Plenum Press, New York, pp. 553–562.

Schwarzmann, G., Marsh, D., Herzag, V. and Sandhoff, K. (1987) In vitro incorporation and metabolism of gangliosides. In: H.

Rahmann (Ed.) *Gangliosides and Modulation of Neuronal Functions, NATO ASI Series, Series H: Cell Biology, vol. 7*, Springer Verlag, Berlin, pp. 217–229

Seren, M.S., Lipartiti, M., Lazzaro, A., Rubini, R., Mazzari, S., Facci, L., Vantini, G., Zanoni, R., Zanotti, A., Bonvento, G. and Leon, A. (1990) Monosialoganglioside effects following cerebral ischemia: relationship with anti-neuronotoxic and pro-neuronotrophic properties, In: L.A. Horrocks, N.H. Neff, A.J. Yates and M. Hadjiconstantinou (Eds), *Trophic Factors in the Nervous System, Fidia Research Foundation Symposium Series, Vol. 3*, Raven Press, New York, pp. 339–348.

Sillerud, L.O., Yu, R.K. and Schafer, D.E. (1982) Assignement of the carbon-13 nuclear magnetic resonance spectra of gangliosides GM4, GM3, GM2, GM1, GD1a, GD1b and GT1b. *Biochemistry*, 21: 1260–1271.

Skaper, S.D., Favaron, M., Facci, L., Vantini, G., Fusco, M., Ferrari, G. and Leon, A. (1988) Ganglioside involvement in neuronotrophic interactions, In: R.W. Ledeen, E.L. Hogan, G. Tettamanti, A.J. Yates and R.K. Yu, (Eds) *New trends in ganglioside research: neurochemical and neuroregenerative aspects, Fidia Research Series, Vol. 14*, Liviana Press/Springer Verlag, Padova/Berlin, pp. 351–360.

Slenzka, K., Appel, R. and Rahmann, H. (1990) Influence of exogenous gangliosides (GM1, GD1a, GMix) on a Ca^{2+}-activated Mg^{2+}- dependent ATPase in cellular and subcellular brain fractions of the djungarian dwarf hamster (*phodopus sungorus*). *Neurochem. Int.*, 17: 609–614.

Sonnino, S., Ghidoni, R., Marchesini, S. and Tettamanti, G. (1979) Cytosolic gangliosides: occurrence in calf brain as ganglioside-protein complexes. *J. Neurochem.*, 33: 117–121.

Sonnino, S., Chigorno, V., Acquotti, D., Pitto, M., Kirschner, G. and Tettamanti, G. (1989) A photoreactive derivative of radiolabelled GM1 ganglioside: preparation and use to establish the involvement of specific proteins in GM1 uptake by human fibroblasts in culture. *Biochemistry*, 28: 77–84.

Sonnino, S., Chigorno, V., Valsecchi, M., Pitto, M. and Tettamanti, G. (1992) Specific ganglioside-cell protein interactions: a study performed with GM1 ganglioside derivative containing photoactivable azide and rat cerebellar granule cells in culture. *Neurochem. Int.*, 20: 315–321.

Spence, M.W., Reed, S. and Cook, H.W. (1986) Acid and alkaline ceramidases of rat tissues. *Biochem. Cell Biol.*, 64: 400–404.

Spiegel, S. (1988) Gangliosides are biomodulators of cell growth, In: R.W. Ledeen, E.L. Hogan, G. Tettamanti, A.J. Yates and R.K. Yu, (Eds), *New trends in ganglioside research: neurochemical and neuroregenerative aspects, Fidia Research Series, Vol. 14*, Liviana Press/Springer Verlag, Padova/Berlin, pp. 405–421.

Spiegel, S., Handler, J.S. and Fishmann, P.H. (1986) Gangliosides modulate sodium transport in cultured toad kidney epithelia. *J. Biol. Chem.*, 261: 15755–15760.

Spoerri, P.E. (1983) Effects of gangliosides on the in vitro development of neuroblastoma cells: an ultrastructural study. *Int. J. Rev. Neurosci.*, 1: 383–391.

Straedel-Flaig, C., Beck, J.P., Gabelle, M. and Rebel, G. (1987) Adaptation of Zajdela ascitic hepatoma cells to monolayer growth: change in the cell ganglioside pattern. *Cancer Biochem. Biophys.*, 9: 233–244.

Steinman, R.M., Mellman, I.S., Muller, W.A. and Cohn, Z.A. (1983) Endocytosis and the recycling of plasma membrane. *J. Cell Biol.*, 96: 1–27.

Suzuki, A. and Yamakawa, T. (1991) Gangliosides. In: R. Dulbecco (Ed.), *Encyclopedia of Human Biology, Vol. 3*, Academic Press, San Diego, pp. 725–735.

Svennerholm, L. (1980) Ganglioside designation. In: L. Svennerholm, P. Mandel, H. Dreyfus, P.F. Urban (Eds), *Structure and Function of Gangliosides, Adv. Exp. Med. Biol., Vol. 125*, Plenum Press, New York, p. 11.

Svennerholm, L. (1985) Biological significance of gangliosides, In: *Cellular and pathological aspects of glycoconjugate metabolism.* Colloque INSERM/CNRS, Les Editions INSERM, Paris, pp. 21–44.

Tettamanti, G. (1988) Towards the understanding of the physiological role of gangliosides. In: R.W. Ledeen, E.L. Hogan, G. Tettamanti, A.J. Yates and R.K. Yu (Eds), *New Trends in Ganglioside Research: Neurochemical and Neuroregenerative Aspects, Fidia Research Series, Vol. 14*, Liviana Press, Springer/Verlag, Padova/Berlin, pp. 625–646.

Tettamanti, G. and Masserini, M. (1987) Gangliosides and the transfer of information through the plasma membrane. In: E. Bertoli, D. Chapman, A. Cambria and U. Scapagnini (Eds), *Biomembrane and Receptor Mechanism, Fidia Research Series, Vol. 7*, Liviana Press/Springer Verlag, Padova/Berlin, pp. 223–260.

Tettamanti, G., Morgan, I.G., Gombos, G., Vincendon, G. and Mandel, P. (1972) Sub-synaptosomal location of brain particulate neuraminidase. *Brain Res.*, 47: 515–518.

Tettamanti, G., Preti, A., Cestaro, B., Masserini, M., Sonnino, S. and Ghidoni, R. (1980) Gangliosides and associated enzymes at the nerve ending membranes. In: C. Sweely (Ed.), *Cell surface glycolipids. Am. Chem. Soc. Symp., 1 Series 128*, pp. 321–343.

Tettamanti, G., Ghidoni, R. and Trinchera, M. (1988) Recent advances in ganglioside metabolism. *Indian J. Biochem. Biophys.* 25: 106–111.

Thomas, P.P. and Brewer, G.J. (1990) Gangliosides and synaptic transmission. *Biochim. Biophys. Acta*, 1031: 277–289.

Thompson, T.E. and Brown, R.E. (1988) Biophysical properties of gangliosides. In: R.W. Ledeen, E.L. Hogan, Y. Nagai, G. Tettamanti, A.J. Yates and R.K. Yu, (Eds), *New trends in Ganglioside Research: Neurochemical and Neuroregenerative Aspects, Fidia Research Series, Vol. 14*, Liviana Press/Springer Verlag, Padova/Berlin, pp. 65–78.

Tiemeyer, M., Yasuda, Y. and Schnaar, R.L. (1990) Ganglioside-specific binding protein on rat brain membranes. *J. Biol. Chem.*, 264: 1671–1681.

Tomasi, M., Roda, C., Ausiello, C., D'Agnolo, G., Venerando, B., Ghidoni, R., Sonnino, S. and Tettamanti, G. (1980) Interaction of GM1 ganglioside with bovine serum albumin. Formation and isolation of multiple complexes. *Eur. J. Biochem.*, 111: 315–324.

Trinchera, M. and Ghidoni, R. (1989) The glycosphingolipid sialyltransferases are localized in different sub-Golgi compartments in rat liver. *J. Biol. Chem.*, 264: 15766–15769.

Trinchera, M., Ghidoni, R., Greggia, L. and Tettamanti, G. (1990) The *N*-acetylgalactosamine residue of exogenous GM2 ganglioside is recycled for glycoconjugate biosynthesis in rat liver. *Biochem. J.*, 266: 103–106.

Trinchera, M., Fiorilli, A. and Ghidoni, R. (1991a) Localization in the Golgi apparatus of rat liver UDP-Gal: glucosylceramide β1–4galactosyltransferase. *Biochemistry*, 30: 2719–2724.

Trinchera, M., Carrettoni, D. and Ghidoni, R. (1991b) Topography of glycosyltransferases involved in the initial glycosylations of gangliosides. *J. Biol. Chem.*, 266; 20907–20912.

Tsuji, S., Nakajima, J., Sasaki, T. and Nagai, Y. (1985) Bioactive Gangliosides. IV. Ganglioside GQ1b/Ca²⁺ dependent protein kinase activity exists in the plasma membrane fraction of neuroblastoma cell line, GOTO. *J. Biochem.*, 97: 969–972.

van Echten, G., Birk, R., Brenner-Weiss, G., Schmidt, R.R. and Sandhoff, K. (1990a) Modulation of sphingolipid biosynthesis in primary cultured neurons by long chain bases. *J. Biol. Chem.*, 265: 9333–9339.

van Echten, G., Iber, H., Stotz, H., Takatsuki, A. and Sandhoff, K. (1990b) Uncoupling of ganglioside biosynthesis by brefeldin A. *Eur. J. Cell. Biol.*, 51: 135–139.

van Meer, G. (1989) Lipid traffic in animal cells. *Annu. Rev. Cell Biol.*, 5; 247–275.

van Veldhoven, P.P., Bishop, W.R. and Bell, R.M. (1989) Enzymatic guantification of sphingosine in the picomole range in cultured cells. *Anal. Biochem.*, 183: 177–189.

Varon, S., Pettmann, B., Manthorpe, M. (1988) Extrinsic regulations of neuronal maintenance and repair, In: R.W. Ledeen, E.L. Hogan, G. Tettamanti, A.J. Yates, and R.K. Yu, (Eds), *New Trends in Ganglioside Research: Neurochemical and Neurogenerative Aspects, Fidia Research Series, Vol. 14*, Liviana Press/Springer Verlag, Padova/Berlin, pp. 607-623.

Vaswani, K.K., Wu, G. and Ledeen, R.W. (1990) Exogenous gangliosides stimulate breakdown of neuro-2A phosphoinositides in a manner unrelated to neurite outgrowth. *J. Neurochem.*, 55: 492–499.

Venerando, B., Fiorilli, A., Masserini, M., Giuliani, A. and Tettamanti, G. (1985) Interactions of pig brain cytosolic sialidase with gangliosides. Formation of catalytically inactive enzyme-ganglioside complexes. *Biochim. Biophys. Acta*, 833: 82–92.

Venerando, B., Fiorilli, A., Caimi, L. and Tettamanti, G. (1987) Interactions of pig brain cytosolic sialidase with gangliosides. The formation of catalytically inactive enzyme-ganglioside complexes requires homogeneous ganglioside micelles and is a reversible phenomenon. *J. Biochem.*, 102: 1167–1176.

Walter, V.P., Sweeney, K. and Morrè, D.J. (1983) Neutral lipid precursors for gangliosides are not formed by rat liver homogenates or by purified cell fractions. *Biochim. Biophys. Acta*, 750: 346–352.

Wattenberg, B.W. (1990) Glycolipid and glycoprotein transport through the Golgi complex are similar biochemically and kinetically. Reconstitution of glycolipid transport in a cell free system. *J. Cell. Biol.*, 111: 421–428.

Weis, F.M.B. and Davis, R.J. (1990) Regulation of epidermal growth factor receptor signal transduction role of gangliosides. *J. Biol. Chem.*, 265: 12059–12066.

Wiegandt, H. (1985) Gangliosides. In: H. Wiegandt (Ed.), *New Comprehensive Biochemistry; Glycolipids, Vol. 10*, Elsevier, Amsterdam, pp. 199–260.

Wynn, C.G. and Robson, B. (1986) Calculation of the conformation of glycosphingolipids. 2. GM1 and GM2 gangliosides. *J. Theor. Biol.*, 123: 221–230.

Yasuda, Y., Tiemeyer, M., Blackburn, C.C. and Schnaar, R. (1988) Neuronal recognition of gangliosides: evidence for a brain ganglioside receptor. In: R.W. Ledeen, E.L. Hogan, G. Tettamanti, A.J. Yates and R.K. Yu (Eds), *New Trends in Ganglioside Research: Neurochemical and Neuroregenerative Aspects, Fidia Research Series, Vol. 14*, Liviana Press/Springer Verlag, Padova/Berlin, pp. 229–243.

Yates, A.J., Walters, J.D., Wood, C.L. and Johnson, J.D. (1989) Ganglioside modulation of cyclic AMP-dependent protein kinase and cyclic nucleotide phosphodiesterase in vitro. *J. Neurochem.*, 53: 162–167.

Yavin, E., Consolazione, A., Gil, S., Ginzburg, I., Leon, A., Del Toso, R. and Rybak, S. (1988) Gangliosides mediate cytoskeletal gene expression during growth and after injury of the nervous system, In: R.W. Ledeen, E.L. Hogan, G. Tettamanti, A.J. Yates, R.K. Yu (Eds), *New Trends in Ganglioside Research Neurochemical and Neurogenerative Aspects, Fidia Research Series, Vol. 14*, Liviana Press/Springer Verlag, Padova/Berlin, pp. 579–593.

Younes, A., Kahn, D.W., Besterman, J.M., Bittman, R., Byun, H. and Kolesnick, R.N. (1992) Ceramide is a competitive inhibitor of diacylglycerol kinase in vitro and in intact human leukemia (HL-60) Cells. *J. Biol. Chem.*, 267: 842–847.

Yu, R.K. and Saito, M. (1989) Structure and localization of gangliosides, In: R.U. Margolis and R.K. Margolis (Eds.), *Neurobiology or glycoconjugates*. Plenum Press, New York, pp. 1–42.

L. Svennerholm, A.K. Asbury, R.A. Reisfeld, K. Sandhoff, K. Suzuki, G. Tettamanti and G. Toffano (Eds.)
Progress in Brain Research, Vol. 101
© 1994 Elsevier Science BV. All rights reserved.

CHAPTER 7

Gangliosides as modulators of neuronal calcium

Gusheng Wu and Robert W. Ledeen

Departments of Neurosciences and Physiology, New Jersey Medical School - UMDNJ, 185 South Orange Avenue, Newark, N J 07103-2757, U.S.A.

Introduction

The historical connection between gangliosides and neurons goes back to the 1930s when Ernst Klenk first discovered these substances in gray matter of normal and pathological brains and named them for their apparent localization in neurons. Although they are now recognized as ubiquitous components of virtually all vertebrate cells, their unusually high concentration in the neuron, together with their distinctive structural and developmental patterns, has given rise to the idea of a special role(s) in neuronal functioning. This widely held view has been reinforced by many experiments over the past decade which have demonstrated significant trophic effects of gangliosides expressed in vitro upon addition to neurons and neuroblastoma cells in culture and in vivo upon administration to animals suffering lesions to the central or peripheral nervous system. These 'neuronotrophic' effects have turned out in some cases to depend on the ability of gangliosides to modulate intracellular calcium levels, these modulatory effects being manifested by both exogenous ganglioside as well as the natural endogenous component of the cell. Examples of such phenomena will be recounted in this presentation.

In keeping with the theme of this meeting on the biological functions of gangliosides it is appropriate

to recall the many seminal contributions of Professor Lars Svennerholm, in whose honor this meeting is being held. The 'modern' era of ganglioside research, what might be called the 'structure-function' period, began in the early 1960s, and this author (RWL) recalls his own entry into the field coinciding with the appearance of a paper by Professor Svennerholm (1962) on the sugar sequence of Tay-Sachs and normal brain gangliosides. This represented one of the first successful efforts to depict a structural outline of gangliosides from the nervous system. There followed, approximately a year later (Svennerholm, 1963), one of the most frequently cited papers in the ganglioside literature which provided not only a chromatographic separation technique but also a nomenclature system that became part of the discipline's vocabulary. To him we are indebted for several other methodological techniques that facilitated progress in the field (e.g., Svennerholm, 1957; Svennerholm and Fredman, 1980). There is hardly an area of ganglioside research that has not benefited from the contributions of Professor Svennerholm and the high standards which he helped to establish in our field.

Returning to the theme of this meeting, biological function, we would like to describe some recent experiments of our laboratory that attempted to explore the connection between neuronal gangliosides and calcium flux. We believe these and related findings in other laboratories can explain at least some of the neuronotrophic properties of these molecules, including neuritogenic as well as neuroprotective effects. These studies entailed spectrofluorimetric

---

Abbreviations: BBG, bovine brain ganglioside mixture; cholera B, cholera toxin B subunit; DMEM, Dulbecco's modified Eagle medium; FBS, fetal bovine serum; N'ase, neuraminidase.

measurement of $[Ca^{2+}]_i$ during manipulation of endogenous or exogenous gangliosides as well as observation of resulting morphological changes.

Methods

Cell culture

Neuroblastoma cells were obtained from the following sources: Neuro-2a from the American Type Culture Collection (CCL 131; Rockville, MD); N18 as a gift from Dr. Marshall Nirenberg; NG108-15 as a gift from Dr. Joseph Moskal. These were routinely cultured in Dulbecco's modified Eagle medium (DMEM) supplemented with 10% fetal bovine serum (FBS), in an atmosphere of 5% CO_2/95% humidified air at 37°C. Treatment with neuraminidase (N'ase) and B-subunit of cholera toxin (cholera B) were as previously described (Wu and Ledeen, 1991). Cholera B (purchased from List Biological Labs, Campbell CA) was heat inactivated at 56°C for 30 min, a procedure shown to destroy any residual cholera toxin A activity (Shen and Crain, 1990). Calcium ionophore treatment, together with quantification of survival and neurite stability, were carried out as before (Nakamura et al., 1992). In experiments involving Amiloride (Sigma Chemical Co., St. Louis, MO), this was dissolved in dimethylsulfoxide at a concentration of 100 mM and this stock solution subsequently diluted with DMEM-5% FBS to 2–20 μM; the resulting dimethylsulfoxide did not affect cell survival or differentiation.

Measurement of adsorbed ganglioside

To quantify the amounts of ganglioside entering the three pools associated with the cell surface (Radsak et al., 1982; Facci et al., 1984), we employed in separate runs [³H]GM1 (8810 DPM/nmol) and [³H]GD1a (801 DPM/nmol), each at a concentration of 100 μM. Neuro-2a cells growing in DMEM/10% FBS in 6-well tissue culture plates were washed twice with warm DMEM and then incubated with 1 ml of the above ganglioside-containing medium for 2 h at 37°C. Following removal of this medium, the cells were incubated first with 1 ml DMEM/10% FBS ('serum removable' fraction) and then with 1 ml

trypsin (0.1% in DMEM), each for 30 min at 37°C. After each of the above steps the cells were carefully washed 2–3 times with Dulbecco's PBS without Ca^{2+} and Mg^{2+}, and these washes were combined with the related initial medium for counting. Cell residues were dissolved in a mixture of 5% SDS and 0.5 N NaOH for protein quantification and counting. Cell protein in each well was 0.40–0.45 mg. To determine the effect of serum on ganglioside absorption, parallel experiments were carried out in which Neuro-2a cells were incubated with ganglioside as above, but with 10% FBS added to the DMEM. In these trials there was no 'serum-removable' fraction.

Measurement of intracellular free calcium $[Ca^{2+}]_i$

Intracellular free calcium in cells treated with N'ase, cholera B or exogenous ganglioside was determined with the fluorescent Ca^{2+}-sensitive indicator, Fura-2 (Molecular Probes, Eugene, OR) (Grynkiewicz et al., 1985). Neuro-2a and N18 cells, grown as monolayers in DMEM with 10% FBS, were incubated with Fura-2/AM (5 μM) and sulphinpyrazone (250 μM) for 0.5 h at 37°C; the latter is an inhibitor of organic anion transport and was used to prevent leakage of the acid form of Fura-2 from the cells (Di Virgilio et al., 1988). Cells were detached by gently triturating with a Pasteur pipet and transferred with medium to tubes. The cells were then treated with N'ase (1 unit/ml) and/or cholera B (5 μg/ml). Aliquots containing 0.5×10^6 suspended cells were centrifuged at the designated time points and the supernatant containing unabsorbed dye was discarded. The cell pellet was resuspended in 5 ml of HEPES buffer (20 mM, pH 7.2) containing 140 mM NaCl, 5 mM KCl, 1 mM $MgSO_4$, 10 mM glucose, 0.1% bovine serum albumin, and 1.8 mM $CaCl_2$, and transferred to a quartz cuvette. Readings were taken in a SPEX Fluomax fluorescence spectrometer (Spex Industries, Inc., Edison, NJ) equipped with a thermostatically controlled holder and a magnetic stirrer. Excitation wavelengths were set at 340 and 380 nm and emission wavelength at 505 nm. Cells were equilibrated 1–2 min in the buffer until the signal stabilized, and measurements were then carried out for 1 min. For calcium calibration R_{max} and R_{min} were determined on similar preparations by exposing the cells to 0.1 mM digitonin followed by 15 mM EGTA.

To determine the effect of exogenous gangliosides on intracellular calcium, Neuro-2a cells were preincubated with 100 μM bovine brain ganglioside mixture (BBG) (gift from Fidia Research Laboratories, with composition 21% GM1, 39.7% GD1a, 16% GD1b, 19% GT1b). This incubation was carried out at 37°C for 1 h in the presence of 10% FBS prior to dye loading. Gangliosides remaining in the medium were removed together with dye by centrifugation. After the signal stabilized the base level of $[Ca^{2+}]_i$ was scanned for 150 sec. Cells were then exposed to the calcium ionophore ionomycin (1 μM) in the solution previously described for suspending the cell pellet (without added bovine serum albumin, which absorbed ionomycin) and measurement continued to 550 sec.

Results

Adsorption of ganglioside by Neuro-2a cells

Exogenous gangliosides added to cultured Neuro-2a cells associated with the membrane in three distinguishable modes (Table I): (a) a loosely attached pool susceptible to removal by serum ('serum removable'); (b) a somewhat more tightly associated pool that was released by trypsin ('trypsin removable'); (c) a serum- and trypsin-stable component representing the 'membrane inserted' pool. When GM1 alone was added the amount of total associated ganglioside was only slightly below that previously reported for association of this ganglioside with Neuro-2a cells in serum-free medium (Facci et al., 1984) while the percentage distribution was roughly similar. In the present study we have shown that GD1a associates in a similar manner, although the 'serum removable' pool appears to increase somewhat at the expense of the 'trypsin removable' pool when compared to GM1. The amounts of GM1 entering the two more stable pools decreased significantly in the presence of serum while the amounts of absorbed GD1a were essentially unchanged. The study of Facci et al. (1984) demonstrated that the amount of absorbed GM1 did not change appreciably between 1 and 2 h.

Modulatory effects of exogenous gangliosides

It was recently demonstrated that exogenous gangliosides exerted a neuroprotective effect toward Neuro-2a neuroblastoma cells that were subjected to

TABLE I

Association of exogenous gangliosides with neuro-2a cells

Medium for incubation	Radiolabeled ganglioside	Total nmol/mg protein	Serum removable nmol/mg protein	Trypsin removable nmol/mg protein	Membrane nmol/mg protein
DMEM	GM1	4.65 ± 1.33	2.63 ± 0.73 (56.6%)	1.02 ± 0.28 (21.9%)	1.01 ± 0.34 (21.7%)
	GD1a	4.74 ± 1.33	3.32 ± 0.38 (70.0%)	0.50 ± 0.12 (10.6%)	0.92 ± 0.15 (19.4%)
DMEM 10%FBS	GM1	0.74 ± 0.08		0.25 ± 0.04 (33.8%)	0.49 ± 0.07 (66.2%)
	GD1a	1.54 ± 0.13		0.66 ± 0.05 (42.9%)	0.88 ± 0.12 (57.1%)

Cells growing in 6-well tissue culture plates were washed twice with warm DMEM, and then incubated with 1 ml ganglioside-containing medium as described in text. In the lower set of data the medium contained FBS and in the upper it did not. These results represent one of three typical experiments. Each data point is the average ± S.D of 6 wells, each well containing 0.40–0.45 mg of protein.

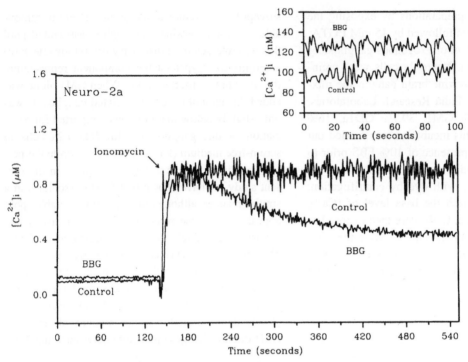

Fig. 1. Effect of BBG on calcium ionophore ionomycin elevated [Ca²⁺]ᵢ in Neuro-2a cells in a representative measurement. [Ca²⁺]ᵢ was determined by using Fura-2 as described in text. Cells were preincubated with BBG (100 μM) for 1 h before dye loading. Both BBG and dye were removed by centrifugation. Ionomycin (1 μM) was added at 150 sec. Cells preincubated with BBG exhibited reduced [Ca²⁺]ᵢ with partial recovery from calcium overloading caused by ionomycin. Inset panel shows base level of [Ca²⁺]ᵢ before ionomycin stimulation; cells preincubated with BBG possessed higher [Ca²⁺]ᵢ. Result shown here was repeated twice.

calcium ionophore cytotoxicity (Nakamura et al., 1992). This effect, manifested in relation to both neurite stability and cell survival, was presumed to reflect the ability of gangliosides to modulate calcium flux. In this model calcium efflux was thought to be enhanced in a process that partially reduced the calci-

um overload caused by the ionophore. Support for this interpretation comes from the present experiment in which use of the fluorescent calcium indicator Fura-2 revealed that 100 μM bovine brain ganglioside mixture (BBG) caused significant reduction in [Ca²⁺]ᵢ following the sharp increase caused by iono-

TABLE II

Effect of ganglioside on ionomycin-elevated intracellular calcium in neuro-2a cells

	n	Base level (μM)	Peak (μM)	Plateau (μM)
Control	3	0.102 ± 0.006	0.931 ± 0.065	0.872 ± 0.041
BBG	3	0.125 ± 0.003	0.860 ± 0.055	0.497 ± 0.065
		P < 0.01		*P < 0.001*

BBG-treated cells (1 h preincubation) were loaded with Fura-2/AM and both dye and BBG were removed by centrifugation. The cells were suspended in HEPES-buffered saline containing 1.8 mM Ca²⁺ and [Ca²⁺]ᵢ was measured by spectrofluorometry. Ionomycin was added after 150 sec and readings continued to 550 sec.

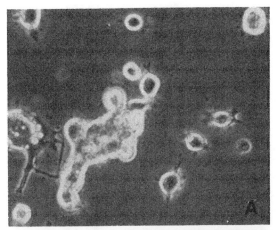

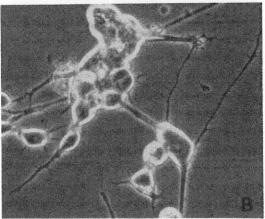

Fig. 2. Morphological comparison of NG108-15 cells treated with calcium ionophore A23187 (1 μM) alone (*A*), and A23187 plus bovine brain ganglioside mixture (100 μM) (*B*). Cells were exposed for a period of 48 h. Magnification: 200×.

mycin (Fig. 1, Table II). The fact that preincubation with BBG was only partially successful in restoring $[Ca^{2+}]_i$ to its original level may account for the observation that the neuroprotective effect was only partial (Nakamura et al., 1992).

The same experiment demonstrated a modest but statistically significant increase in $[Ca^{2+}]_i$ due to ganglioside alone (Fig. 1, inset; Table II), similar in magnitude to that previously seen with $^{45}Ca^{2+}$ (Wu et al., 1990). In the latter study this rise was postulated as the cause of the ganglioside-induced neuritogenesis observed in Neuro-2a (and possibly other neuroblastoma) cells, since removal of exogenous calcium negated the trophic effect.

To determine whether the neuroprotective effect of exogenous gangliosides toward Neuro-2a represents a general phenomenon applicable to other cell types, we tried a similar experiment with the NG108-15 hybrid cell line. During 48 h exposure to varying concentrations of the calcium ionophore A23187, both survival and neurite stability of these cells were considerably enhanced in the presence of 100 μM BBG (Figs. 2 and 3).

Modulatory effects of endogenous GM1

The presence of neuraminidase (N'ase) in the culture medium was shown to produce a powerful neuritogenic effect on Neuro-2a (Wu and Ledeen, 1991) and NG108-15 cells (Ledeen and Wu, 1992). The

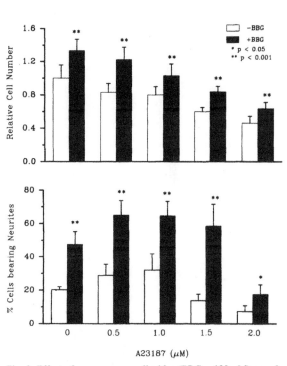

Fig. 3. Effect of exogenous gangliosides (BBG = 100 μM) on calcium ionophore A23187-treated NG108-15 cells. Cells were grown for 48 h in DMEM-10% FBS containing various amounts of A23187 alone (blank bars), and A23187 plus BBG (solid bars). Cell numbers (upper panel) were monitored with MTT assay; all values (*n*=12) represent comparison with control in which neither A23187 nor BBG was present. For the percentage of cells bearing neurites (lower panel), the data (*n*=4) represent direct counts of more than 300 cells/well. Error bars indicate S.D.

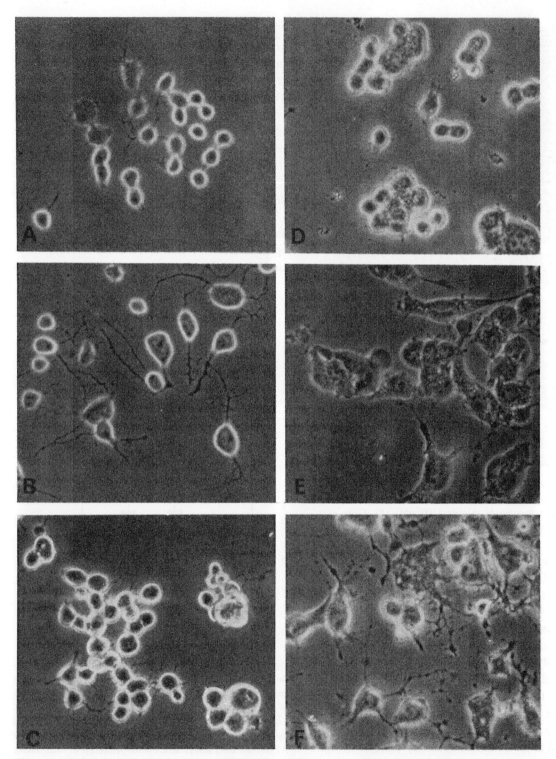

Fig. 4. Morphological comparison of Neuro-2a and N18 cells treated with neuraminidase and cholera B. Neuro-2a cells (*A–C*) were treated for 48 h. *A*. control, medium alone (DMEM-5% FBS); *B*. neuraminidase (1 unit/ml); *C*. neuraminidase plus cholera B (2.5 μg/ml). N18 cells were treated for 72 h, (*D–F*). *D*. control (DMEM-5% FBS); *E*. neuraminidase (0.5 unit/ml); *F*. neuraminidase plus cholera B (2.5 μg/ml). Note that cholera B reversed neuraminidase-stimulated neuritogenesis in Neuro-2a, while cholera B elevated neurite formation in neuraminidase-treated N18 cells. Magnification: 200×.

fact that the effect was blocked by the B subunit of cholera toxin (Cholera B) suggested the neurite outgrowth was related to the increase of GM1 on the surface of the cells rather than the other changes caused by N'ase. A subsequent report (Masco et al., 1991) employing N18 neuroblastoma cells reported the opposite effect, i.e. enhancement of neurite outgrowth by cholera B. In the present study we obtained similar results with N18 cells; Fig. 4 depicts the disparate morphological changes undergone by N18 and Neuro-2a cells in the presence of cholera B. Measurement of $[Ca^{2+}]_i$ with Fura-2 also revealed opposite responses of these two cell types which paralleled the morphological changes (Fig. 5): whereas cholera B eliminated the N'ase-induced rise in $[Ca^{2+}]_i$ of Neuro-2a cells, the combination of cholera B plus N'ase caused $[Ca^{2+}]_i$ to rise in N18 cells. As seen in Fig. 5, the latter cells did not respond measurably to N'ase alone, while Neuro-2a experienced no calcium change with cholera B alone (not shown).

Tentative identification of calcium channel

In the previous study (Wu and Ledeen, 1991) N'ase-enhanced neurite outgrowth was inhibited by reduction of extracellular calcium or addition of LaCl₃ (a general inhibitor of calcium channels), suggesting possible modulation of calcium channels by the increased membrane GM1. We have attempted to identify the type of calcium channel that may be involved by employing various pharmacologica

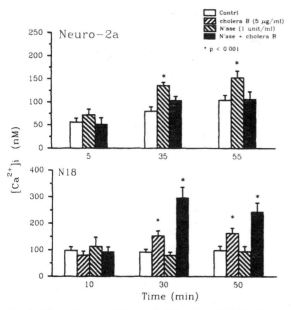

Fig. 5. Effect of neuraminidase and cholera B on $[Ca^{2+}]_i$ in Neuro-2a and N18 cells. $[Ca^{2+}]_i$ was monitored with Fura-2 dye as described in text. Note that neuraminidase increased $[Ca^{2+}]_i$ in Neuro-2a cells which could be reversed by cholera B. In N18 cells, cholera B dramatically elevated $[Ca^{2+}]_i$ in the neuraminidase-treated cells.

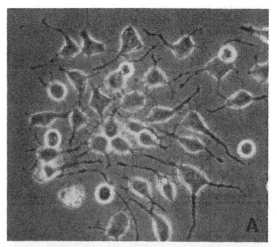

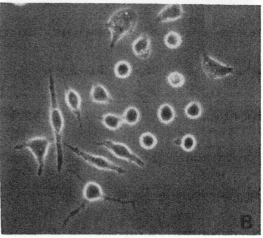

Fig. 6. Morphological effect of amiloride on neuraminidase-stimulated neurite outgrowth. Neuro-2a cells were grown for 48 h in DMEM-5% FBS supplemented with neuraminidase (1 unit/ml) alone (A), and neuraminidase plus amiloride (20 μM) (B). Amiloride, a T-type calcium channel blocker, inhibited neurite formation induced with neuraminidase. Magnification: 200×.

calcium channel blockers. The results, though not definitive, suggest T-channel involvement, based for example on blockade of N'ase-induced neuritogenesis with amiloride at 20 µM (Fig. 6); even lower concentrations were effective (not shown). This agent was described as a specific T-channel blocker at low concentration (Tang et al., 1988). Similar inhibition was seen with nickel (not shown), also reported to show preferential blockade of T-type calcium channels (Narahashi et al., 1987). Finally, verapamil, a specific inhibitor of L-type channels (Miller, 1987), was found to have no effect.

Discussion

This study adds to the mounting evidence that cell surface (endogenous) GM1 has a role in modulation of calcium flux, although apparently not by a single mechanism since the GM1-binding ligand cholera B can cause either reduction or enhancement of $[Ca^{2+}]_i$ depending on cell type and conditions. Our previous results (Wu and Ledeen, 1991) demonstrated that not only Neuro-2a but also the B104 and B50 neuroblastoma cell lines responded to N'ase with enhanced neuritogenesis accompanied by increased $^{45}Ca^{2+}$ influx, while the present study further illustrated calcium modulation by use of the fluorescent indicator dye Fura-2. In our experiments the N18 cell line treated with N'ase plus cholera B showed neurite outgrowth and elevation of $[Ca^{2+}]_i$, opposite to the above three lines. This finding is in agreement with the recent report on N18 cells by Masco et al. (1991).

We have also provided more data illustrating the calcium-modulatory role of exogenously applied gangliosides, a small portion of which enters the plasma membrane bilayer. This study has confirmed a previous report (Facci et al., 1984) on the nature of GM1 absorption by Neuro-2a cells and we have shown that the level of absorbed GM1 is reduced in the presence of serum (FBS). Ganglioside GD1a, on the other hand, whose absorption resembled that of GM1 in serum-free medium, did not show diminished absorption with serum (Table I). Other members of the BBG mixture were not examined here, but it may be

noted that GD1b behaved similarly to GM1 when associating with C1-1-D cells (Callies et al., 1977). The amount of GM1 we observed to be inserted into the Neuro-2a membrane in the presence of FBS (490 pmol/mg protein) exceeds the endogenous GM1 level (150 pmol/mg protein) approximately 3-fold (Wu et al., 1991a). Assuming the four major components of BBG show somewhat similar absorption behavior, it would appear that incubation of neuroblastoma cells with this ganglioside mixture elevates the membrane content of gangliotetraose ganglioside well above the level of such species in untreated cells and increases the total ganglioside level perhaps 50% or more. The predominant ganglioside in this cell is GM3 (Wu et al., 1991a), a structure shown to be ineffective in protecting Neuro-2a cells against calcium toxicity (Nakamura et al., 1992). We have not yet determined which of the molecular species in the BBG mixture is the effective agent in modulating calcium flux in the present paradigm, or whether the resulting activity depends on a mixture. It is often assumed in such studies that it is the inserted ganglioside pool which is responsible for whatever biological/biochemical changes are observed, and while that is frequently a reasonable assumption the possibility of contributions by one or both of the other pools cannot be arbitrarily discounted.

Other studies have pointed to a similar modulatory role for GM1. One example is the report on anti-GM1 antibody-enhancement of depolarization-induced release of GABA from rat brain slices, postulated to result from GM1 participation in calcium channel function (Frieder and Rapport, 1987). Preincubation of hippocampus slices with GM1, as well as N'ase treatment, increased potentiation of synaptic response recorded from pyramidal neurons following high frequency stimulation of Schaffer collateral-commissural fibers (Wieraszko and Seifert, 1985, 1986). In this model of long-term potentiation, involving increase of calcium in the pre- and post-synaptic compartments, the results pointed to GM1 as critical in maintaining calcium at the prescribed levels. Cholera B and anti-GM1 antibody abolished the potential, as did Arthrobacter ureafaciens N'ase (which hydrolyzes GM1 to asialo GM1). This paradigm was believed to have monitored the efficiency of glutamatergic pathways in the hippocampus. In another primary culture cholera B applied to cerebellar granule neurons caused $[Ca^{2+}]_i$ to rise, resulting in a trophic effect which enhanced cell survival (Wu et al., 1991b).

Similar phenomena have been reported in non-neural systems as well, as seen in studies with thymocytes in which cholera B caused pronounced increase of $[Ca^{2+}]_i$ concomitant with mitosis (Dixon et al., 1987). The same was observed with quiescent 3T3 cells (Spiegel and Panagiotopoulos, 1988), but cholera B inhibited mitosis in the same cells that had been transformed or were in a state of rapid growth (Spiegel and Fishman, 1987). It was concluded that endogenous GM1 can function as a bimodal regulator of positive or negative signals, depending on the metabolic state of the cell. Cardiac myocytes provide another example, these cells responding to N'ase treatment with enhanced calcium influx (Langer et al., 1976; Nathan et al., 1980; Yee et al., 1989). Although the authors of those studies interpreted their results primarily in terms of sialic acid removal from glycoproteins, it is equally plausible that increase in GM1 content was the operative change. It is interesting to note that the authors postulated the enhanced influx to be mediated by T-type calcium channels, similar to our current proposal. While the data thus far obtained in our study for Neuro-2a cells are consistent with this channel type, more definite evidence is clearly needed.

With regard to the effects of exogenous GM1 (and other gangliosides), this study has confirmed previous observations by our group (Wu et al., 1990) and others (Spoeri et al., 1990; Guerold et al., 1992) that calcium modulation is an important, possibly key factor in the trophic effects observed. Gangliosides added to the cell culture medium can cause either influx or efflux of calcium depending on cell type and culture condition. The enhanced influx described in the earlier report (Wu et al., 1990), and confirmed in the present study with Fura-2, is believed responsible for the neurite outgrowth that characterizes that system. This neuritogenic effect was previously shown to be indiscriminate in regard to sialolipid structure (Byrne et al., 1983; Cannella et al., 1988a; 1988b). On the other hand, the neuroprotective effect toward Neuro-2a cells subjected to calcium ionophore neurotoxicity (Nakamura et al., 1992) can, on the basis of the present study, be attributed to the ability of exogenous gangliosides to facilitate removal of some of the excess $[Ca^{2+}]_i$. As mentioned, this effect has shown some specificity in regard to ganglioside structure. We previously found that the protective effect was largely neutralized by dichlorobenzamil, inhibitor (though not entirely selective) of the Na^+-Ca^{2+} exchanger. Potentiation of that system is thus one possible mechanism, although this does not rule out other possible mechanisms (e.g. activation of Ca^{2+}-ATPase) whereby gangliosides might serve to promote calcium homeostasis.

Gangliosides in the above paradigm did not offer full or long-lasting protection to the Neuro-2a cells, nor to the NG108-15 cells employed in the present study, and for that and other reasons it is apparent that the protective mechanism differs fundamentally from that observed with the excitotoxicity model for primary neuronal cultures (Vaccarino et al., 1987; Favaron et al., 1988; Manev et al., 1990). In the latter case gangliosides were shown to reduce $^{45}Ca^{2+}$ accumulation after glutamate removal from the cells, thereby reducing the late increase in $[Ca^{2+}]_i$ believed to be responsible for the neurotoxic effect of glutamate. A significant aspect of those studies was the finding that gangliosides inhibited glutamate-induced translocation of protein kinase C from cytosol to the membrane fraction of granule neurons. GM1 and a few of its analogs that were protective prevented the protracted increase in $[Ca^{2+}]_i$ but failed to control the immediate rise of calcium elicited by glutamate, whereas glutamate receptor antagonists did the opposite and were not protective (De Erausquin et al., 1990). The latter study suggested that gangliosides assert their protective effect through inhibition of protein kinase C-catayzed phosphorylation of membrane components participating in calcium flux modulation, this inhibition resulting from reduced kinase translocation. An inhibitory effect of gangliosides on protein kinase C from brain has been reported by Kreutter et al. (1987).

In terms of potential clinical application it is worth noting that rats subjected to cortical focal ischemia showed significant reduction in tissue calcium when treated with GM1 (Karpiak et al., 1991). A role for ganglioside-calcium complexes in synaptic transmission, based on ideas of charge interaction and binding affinity, was originally proposed that involves facilitated release of neurotransmitters and long-term functional adaptation of the neuronal membrane (Rahmann, 1983). Considering that the ganglioside-calcium interaction at the membrane has been characterized as only weakly attractive (Langer et al., 1988), other modes of interaction at the nerve ending (and elsewhere) can be considered. Arguments against localization of gangliosides at the synapse, and in favor of a more widespread distribution over the neuronal surface, have been presented (Ledeen, 1978, 1989; Rösner et al., 1992).

In attempting to correlate the findings of many laboratories the principal generality that emerges in regard to ganglioside function is that of protein modulation, encompassing a variety of enzymes, receptors, ion channels, and cell adhesion molecules residing in the plasma membrane. Additionally, the possibility of interaction with intracellular constituents must be considered in view of the occurrence of gangliosides and other glycosphingolipids in certain intracellular organelles (Ledeen et al., 1976; Gravotta and Maccioni, 1985; Symington et al., 1987; Ledeen et al., 1988). The protein modulation concept has the advantage of rationalizing the wide diversity of ganglioside structures occurring in nature, with the postulate that each structure is designed for optimal interaction with a particular protein. It remains to be elucidated in more precise molecular terms how the proteins involved in calcium flux and the maintenance of calcium homeostasis interact with particular gangliosides, of which GM1 appears to have a prominent role.

Acknowledgments

Supported by U.S. Public Health Service Grant NS04834 and a grant from the Fidia Research Laboratories. We are indebted to Drs. Abraham Aviv and Jeff Gardner for use of their spectrofluorimeter and helpful discussions on methodology and interpretation of calcium measurements.

References

Byrne, M.C., Ledeen, R.W., Roisen, F.J., Yorke, G. and Sclafani, J.R. (1983) Ganglioside-induced neuritogenesis: Verification that gangliosides are the active agents, and comparison of molecular species. *J. Neurochem.*, 41: 1214–1222.

Callies, R., Schwarzmann, G., Radsak, K., Siegert, R. and Wiegandt, H. (1977) Characterization of the cellular binding of exogenous gangliosides. *Eur. J. Biochem.*, 80: 425–432.

Cannella, M.S., Roisen, F.J., Ogawa, T., Sugimoto, M. and Ledeen, R.W. (1988a) Comparison of epi-GM3 with GM3 and GM1 as stimulators of neurite outgrowth. *Dev. Brain Res.*, 39: 137–143.

Cannella, M.S., Acher, A.J. and Ledeen, R.W. (1988b) Stimulation of neurite outgrowth in vitro by a glycero-ganglioside. *Int. J. Dev. Neurosci.*, 6: 319–326.

De Erausquin, G.A., Manev, H., Guidotti, A., Costa, E. and Brooker, G. (1990) Gangliosides normalize distorted single-cell intracellular free Ca^{2+} dynamics after toxic doses of glutamate in cerebellar granule cells. *Proc. Natl. Acad. Sci. USA*, 87: 8017–8021.

Di Virgilio, F., Fasolato, C. and Steinberg, T.H. (1988) Inhibitions of membrane transport system for organic anions block fura-2 excretion from PC12 and N2A cells. *Biochem. J.*, 256: 959–963.

Dixon, S.J., Stewart, D., Grinstein, S. and Spiegel, S. (1987) Transmembrane signaling by the B subunit of cholera toxin: increased cytoplastomic free calcium in rat lymphocytes. *J. Cell Biol.*, 105: 1153–1161.

Facci, L., Leon, A., Toffano, G., Sonnino, S., Ghidoni, R. and Tettamanti, G. (1984) Promotion of neuritogenesis in mouse neuroblastoma cells by exogenous gangliosides. Relationship between the effect and the cell association of ganglioside GM1. *J. Neurochem.*, 42: 299–305.

Favaron, M., Manev, H., Alho, H., Bertolino, M., Ferret, B., Guidotti, A. and Costa, E. (1988) Gangliosides prevent glutamate and kainate neurocytoxicity in primary neuronal cultures of neonatal rat cerebellum and cortex. *Proc. Natl. Acad. Sci. U.S.A.*, 85: 7351–7355.

Frieder, B and Rapport, M.M. (1987) The effect of antibodies to gangliosides on Ca^{2+} channel-linked release of γ-aminobutyric acid in rat brain slices. *J. Neurochem.*, 48: 1048–1052.

Gravotta, D. and Maccioni, H.J.F. (1985) Gangliosides and sialosylglycoproteins in coated vesicles from bovine brain. *Biochem. J.*, 225: 713–725.

Grynkiewicz, G., Poenie, M. and Tsien, R.Y. (1985) A new generation of Ca^{2+} indicators with greatly improved fluorescence properties. *J. Biol. Chem.*, 260: 3440–3450.

Guerold, B., Massarelli, R., Forster, V., Freysz, L. and Dreyfus, H. (1992) Exogenous gangliosides modulate calcium fluxes in cultured neuronal cells. *J. Neurosci. Res.*, 32: 110–115.

Karpiak, S.K., Wakade, A., Tagliavia, A. and Mahadik, S.P. (1991) Temporal changes in edema, Na+, K+ and Ca++, in focal cortical stroke: GM1 ganglioside reduces ischemic injury. *J. Neurosci. Res.*, 30: 512–520.

Kreutter, D., Kim, J.Y.H., Goldenring, J.R., Rasmussen, H., Ukomadu, C., DeLorenzo, R.J. and Yu, R.K. (1987) Regulation of protein kinase C activity by gangliosides. *J. Biol. Chem.*, 262: 1633–1637.

Langer, G.A., Frank, J.S., Nudd, L.M. and Seraydarian, K. (1976) Sialic acid: effect of removal on calcium exchangeability of cultured heart cells. *Science*, 193: 1013–1015.

Langer, M., Winiski, A., Eisenberg, M., McLaughlin, A. and McLaughlin, S. (1988) The electrostatic potential adjacent to bilayer membranes containing either charged phospholipids or gangliosides. In: R.W. Ledeen, E.L. Hogan, G. Tettamanti, A.J. Yates and R.K. Yu (Eds), *New Trends in Ganglioside Research: Neurochemical and Neuroregenerative Aspects*, Liviana Press, Padova, pp. 121–131.

Ledeen, R.W. (1978) Ganglioside structures and distribution: are they localized at the nerve ending? *J. Supramolec. Struct.*, 8: 1–17.

Ledeen, R.W. (1989) Biosynthesis, metabolism, and biological effects of gangliosides. In: R.U. Margolis and R.K. Margolis (Eds), *Neurobiology of Glycoconjugates*, Plenum Publishing Corp., pp. 43–83.

Ledeen, R.W. and Wu, G. (1992) Ganglioside function in the neuron. *TIGG*, 4: 174–187.

Ledeen, R.W., Parsons, S.M., Diebler, M.F., Sbaschnig-Agler, M. and Lazereg, S. (1988) Gangliosides composition of synaptic vesicles from *Torpedo* electric organ. *J. Neurochem.*, 51: 1465–1469.

Ledeen, R.W., Skrivanek, J.A., Tirri, L.J., Margolis, R.K. & Margolis, R.U. (1976) Gangliosides of the neuron: Localization and origin. *Adv. Exp. Med. Biol.*, 71:83–104.

Manev, H., Favaron, M., Vicini, S., Guidotti, A. and Costa, E. (1990) Glutamate-induced neuronal death in primary cultures of cerebellar granule cells: Protection by synthetic derivatives of endogenous sphingolipids. *J. Pharmacol. Exp. Therap.*, 252: 419–427.

Masco, D., Van de Walle, M. and Spiegel, S. (1991) Interaction of ganglioside GM1 with the B subunit of cholera toxin modulates growth and differentiation of neuroblastoma N18 cells. *J. Neurosci.*, 11: 2443–2452.

Miller, R.J. (1987) Multiple calcium channels and neuronal function. *Science*, 235: 46–52.

Nakamura, K., Wu, G. and Ledeen, R.W. (1992) Protection of Neuro-2a cells against calcium ionophore cytotoxicity by gangliosides. *J. Neurosci. Res.*, 31: 245–253.

Narahashi, T., Tsunoo, A. and Yushii, M. (1987) Characterization of two types of calcium channels in mouse neuroblastoma cells. *J. Physiol.*, 383: 231–249.

Nathan, R.D., Fung, S.J., Stocco, D.M., Barron, E.A. and Markwald, R.R. (1980) Sialic acid: regulation of electrogenesis in cultured heart cells. *Am. J. Physiol.*, 239: C197–C207.

Radsak, K., Schwarzmann, G. and Wiegandt, H. (1982) Studies on the cell association of exogenously added sialo-glycolipids. *Hoppe Seylers Z. Physiol. Chem.*, 363: 263–272.

Rahmann, H. (1983) Functional implication of gangliosides in synaptic transmission. *Neurochem. Int.*, 5: 539–547.

Rösner, H., Al-Aqtum, M., Sonnentag, U., Wurster, A. and Rahmann, H. (1992) Cell surface distribution of endogenous and effects of exogenous gangliosides on neuronal survival, cell shape and growth in vitro. *Neurochem. Int.*, 20: 409–419.

Shen, K.-F. and Crain, S.M. (1990) Cholera toxin-B subunit blocks excitatory effects of opioids on sensory neuron action potentials indicating that GM1 ganglioside may regulate G_s-linked opioid receptor functions. *Brain Res.*, 531: 1–7.

Spiegel, S. and Fishman, P.H. (1987) Gangliosides as bimodal regulators of cell growth. *Proc. Natl. Acad. Sci. U.S.A.*, 84: 141–145.

Spiegel, S. and Panagiotopoulos, C. (1988) Mitogenesis of 3T3 fibroblasts induced by endogenous gangliosides is not mediated by cAMP, protein kinase C, or phosphoinositides turnover. *Exp. Cell Res.*, 177: 414–427.

Spoerri, P.E., Dozier, A.K. and Roisen, F.J. (1990) Calcium regulation of neuronal differentiation: the role of calcium in GM1-mediated neuritogenesis. *Dev. Brain Res.*, 56: 177–188.

Svennerholm, L. (1957) Quantitative estimation of sialic acids. *Biochim. Biophys. Acta*, 24: 604–611.

Svennerholm, L. (1962) The chemical structure of normal human brain and Tay-Sachs gangliosides. *Biochem. Biophys. Res. Commun.*, 9: 436–414.

Svennerholm, L. (1963) Chromatographic separation of human brain gangliosides. *J. Neurochem.*, 10: 613–623.

Svennerholm, L. and Fredman, P. (1980) A procedure for the quantitative isolation of brain gangliosides. *Biochim. Biophys. Acta*, 617: 97–109.

Symington, F.W., Murray, W.A., Bearman, S.I. and Hakomori, S. (1987) Intracellular localization of lactosylceramide, the major human neurophil glycosphingolipid. *J. Biol. Chem.*, 262: 11356–11363.

Tang, Ch.-M., Presser, F. and Morad, M. (1988) Amiloride selectively blocks the low threshold (T) calcium channel. *Science*, 240: 213–215.

Vaccarino, F., Guidotti, A. and Costa, E. (1987) Ganglioside inhibition of glutamate-mediated protein kinase C translocation in primary cultures of cerebellar neurons. *Proc. Natl. Acad. Sci. U.S.A.*, 84: 8707–8711.

Wieraszko, A. and Seifert, W. (1985) The role of monosialoganglioside GM1 in the synaptic plasticity: in vitro study on rat hippocampal slices. *Brain Res.*, 345: 159–164.

Wieraszko, A. and Seifert, W. (1986) Evidence for the functional role of monosialioganglioside GM1 in synaptic transmission in the rat hippocampus. *Brain Res.*, 371: 305–313.

Wu, G. and Ledeen, R.W. (1991) Stimulation of Neurite outgrowth in neuroblastoma cells by neuraminidase: putative role of GM1 ganglioside in differentiation. *J. Neurochem.*, 56: 95–104.

Wu, G., Vaswani, K.K., Lu, Z.-H. and Ledeen, R.W. (1990) Gangliosides stimulate calcium flux in Neuro-2A cells and require exogenous calcium for neuritogenesis. *J. Neurochem.*, 55: 484–491.

Wu, G., Lu, Z.-H. and Ledeen, R.W. (1991a) Correlation of gangliotetraose gangliosides with neurite forming potential of neu-

roblastoma cells. *Dev. Brain Res.*, 61: 217–228.

Wu, G., Nakamura, K. and Ledeen, R.W. (1991b) Influence of gangliosides on calcium flux and neuronal differentiation. *Trans. Am. Soc. Neurochem.*, 22: 173.

Yee, H.F., Weiss, J.N. and Langer, G.A. (1989) Neuraminidase selectively enhances transient Ca^{2+} current in cardiac myocytes. *Am. J. Physiol.*, 256: C1267–C1272.

L. Svennerholm, A.K. Asbury, R.A. Reisfeld, K. Sandhoff, K. Suzuki, G. Tettamanti and G. Toffano (Eds.)
Progress in Brain Research, Vol. 101

CHAPTER 8

Cell adhesion molecule (NCAM and N-cadherin)-dependent neurite outgrowth is modulated by gangliosides

Frank S. Walsh[1], Stephen D. Skaper[2] and Patrick Doherty[1]

[1]*Department of Experimental Pathology, UMDS, Guy's Hospital, London Bridge, London SE1 9RT, U.K.*
and [2]Fidia Research Laboratories Abano Terme, 35031, Italy

Introduction

During development axons extend in highly stereospecific manners to innervate target tissues. In recent years a large number of environmental cues have been implicated in these processes. These include soluble factors secreted by intermediate and final target tissues, and a variety of membrane associated components. There are at least four well defined receptor systems in the neuronal membrane that mediate neurite outgrowth over a wide variety of neuronal substrata such as glial cells and skeletal muscle. These are calcium-independent adhesion molecules N-CAM (Walsh and Doherty, 1991) and L1 (Seilheimer and Schachner, 1988; Williams et al., 1992), the calcium dependent cell adhesion molecule (CAM) called N-cadherin (Takeichi, 1991) and the integrins (Reichardt and Tomaselli, 1991). The CAMs, NCAM, L1 and N-cadherin are potent inducers of neurite outgrowth and do so via a homophilic (self) binding mechanism. However, it is becoming increasingly clear that some of these CAMs can also be involved in heterophilic interactions. In certain cases the heterophilic partner is known. Thus L1 can interact with the axonin-1 (TAG-1) glycoprotein to mediate neurite outgrowth. (Kuhn et al., 1992). N-cadherin has also been found to interact with R-cadherin although it is not known if this interaction is involved in axonal growth (Inuzuka et al., 1991). N-CAM mediated axonal growth has

been found to be homophilic (Doherty et al., 1990) but is remarkably dynamic in other respects. From a temporal point of view neurons appear to lose responsiveness to NCAM with increasing age. We have shown that in three distinct populations of neurons there is a down-regulation in responsiveness to NCAM. Retinal ganglion neurons, cerebellar neurons and hippocampal neurons have each been shown to exhibit this phenomenon (Doherty et al., 1990, 1991a, 1992a). A number of features of the NCAM protein appear to be involved in this process including regulation of the level of polysialic acid (PSA) on neuronal NCAM and the pattern of alternative splicing of the NCAM gene. Polysialic acid which is made up of long chains of $\alpha2$-8 linked sialic acid is found on NCAM during early embryogenesis while at later stages it is present in very much lower amounts. This transition can be mimicked experimentally using the enzyme endoneuraminidase-*N* with removal of PSA leading to a dramatic decrease in the ability of neurons to respond to the NCAM in the substrate. NCAM is also found in a large number of individual isoforms due to the process of alternative splicing. One exon called VASE has been found to insert a small sequence of only ten amino acids into the fourth immunoglobulin domain in the NCAM protein. Developmentally there are dramatic changes in the usage of this particular exon. Neurons early in development do not appear to have high amounts of VASE exon usage in NCAM while older neurons can have up to 50% of the total NCAM transcripts containing VASE (Small and Akeson, 1990; Walsh et al., 1992). Associated with this we have also shown a remarkable loss of neuronal responsiveness when responsive neurons are plated on a substrate of cells expressing NCAM containing the VASE exon. The detailed mechanism by which VASE reduces NCAMs ability to stimulate axonal growth are currently being investigated. However, it is likely that it is associated with the ability of NCAM to participate in intracellular signalling events. We have shown (Doherty et al., 1991b; Doherty and Walsh 1992; Williams et al., 1992) that NCAM, L1 and N-cadherin appear to be operating in axonal growth processes via activation of specific second messenger pathways. A number of components of this pathway which we refer to as the CAM pathway have been identified. In particular the opening of neuronal N- and L-type calcium channels is a key step in this pathway. Upstream of the calcium channels a pertussis toxin sensitive G-protein has been identified. No involvement of protein kinase A or C has been found in the CAM pathway. It is likely that the ability of CAMs to activate axonal growth through the CAM pathway can be modulated pharmacologically. One class of agents that could possibly act via this pathway are gangliosides. It has been known for some years that gangliosides added exogenously to neurons in cell culture can significantly enhance neurite outgrowth from both cell lines and primary cultures (Facci et al., 1984; Doherty et al., 1985a,b). Although the mechanism of action of gangliosides in this type of model is not known considerable evidence suggests that extracellular calcium appears to be important for ganglioside responses (Spoerri et al., 1990; Wu et al., 1990). Also gangliosides have been shown to be able to alter the activity of a Ca^{2+} dependent protein kinase in PC12 cells with removal of extracellular calcium or blockade of L-type calcium channels inhibiting the response (Hilbush and Levine, 1991). As our studies with CAMs and other studies on gangliosides suggest calcium influx to be a common site of action we have attempted to determine whether the two systems are related.

Ganglioside GM1 can potentiate CAM-dependent neurite outgrowth

We have used simple co-culture models to analyse the effects of CAMs on neurite outgrowth. Full length cDNAs for a variety of CAMs have been placed in a number of expression vectors and transfected into 3T3 fibroblasts. Clones of 3T3 cells that stably express high levels of the CAMs (NCAM, L1 and N-cadherin) at the cell surface can then be selected for study. Various populations of neurons, both primary neurons and PC12 cells are plated on the transfected monolayers and the pattern of neurite outgrowth measured after about 20–40 h. Using this type of bioassay it has been possible to show that NCAM, L1 and N-cadherin are potent inducers of axon growth. To study ganglioside effects on neurite outgrowth we have restricted our analyses to N-CAM and N-cadherin as examples of the immunoglobulin superfamily and the cadherin family of the CAMs respectively. Also these studies have used PC12 cells as the probe 'neurons' in order to analyse the role of the NGF signalling pathway in ganglioside action.

In a first set of experiments PC12 cells were grown on monolayers of 3T3 fibroblasts or the same cells expressing transfected NCAM or N-cadherin. In addition sister cultures were set up with 50 μg/ml GM1 to determine whether it had any effects in this system. Full details of the experiments can be found in Doherty et al. (1992b) with a summary presented here. Figure 1 shows the results of this experiment for N-cadherin. PC12 cells do not extend neurites on 3T3 fibroblasts and this is unaffected by addition of 50 μg/ml GM1. However, a different picture emerges wherever PC12 cells are grown on monolayers of 3T3 cells expressing transfected NCAM or N-cadherin. The CAMs induce neurite outgrowth in the absence of added NGF and this is unaffected by antibodies to NGF (Doherty et al., 1991b). Addition of 50 μg/ml GM1 leads to a potentiation of the CAM response. We have quantitated these effects by measuring the length of the longest neurite on each PC12 cell (Table I). N-cadherin expression in the monolayer led to a 254% increase in the length of the longest neurite; this figure increased to 424% in the presence

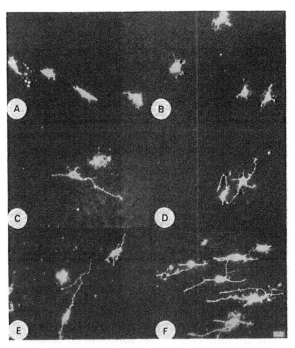

Fig. 1. GM1 effects on PC12 cell morphology on control and N-cadherin expressing cell monolayers. PC12 cells, specifically identified by immunostaining with a mouse anti-rat Thy-1 MAb, were grown for 40 h on a substratum of confluent 3T3 cells expressing transfected N-cadherin (c,d,e,f) in control media (a,c,d,) or media supplemented with 50 μg/ml GM1 (b,e,f,) Bar = 50 μm.

μg/ml GM1 (for details see Doherty et al., 1992b). A number of gangliosides (GT1b, GM3 and GD1a) were found to have similar effects to GM1, however asialo GM1 and the oligosaccharide portion of GM1 were without effect.

GM1 effects are due to cell association

One question to arise from these studies is whether the effects were due to stable association of GM1 with the PC12 cells. To determine the answer PC12 cells were grown in culture media (Sato) for 16 h in the presence of differing concentrations of GM1 from 6.25 to 100 μg/ml. At that point the cells were sub-cultured and re-plated on control 3T3 and N-cadherin expressing monolayers. At concentrations of GM1 from 6.25 to 100 μg/ml there was a significant enhancement of neurite outgrowth only on the N-cadherin monolayers (Table II). Thus it is clearly the GM1 which is stably incorporated in the cell membrane that is inducing the enhanced neurite outgrowth on the N-cadherin monolayer (see Doherty et al., 1992b). It has also been possible to measure the level of incorporated GM1 in the cell membrane by binding cholera toxin to PC12 cells were treated with GM1 at 100 μg/ml for 90 min. Significant increases in B-cholera toxin binding (the levels were about twofold greater than the endogenous levels) were found at up to 20 h following pretreatment with GM1 and this was not decreased by treatment with trypsin (Doherty et al., 1992b). Thus a short pre-treatment of PC12 cells with GM1 can lead to an increase in the

of GM1 at 50 μg/ml. A dose response curve for GM1 showed concentrations between 0 and 100 μg/ml to have no effect on the growth of PC12 cells over 3T3 monolayers. However for both NCAM and N-cadherin expressing monolayers significant effects were detected at 25 μg/ml which were maximal at 100

TABLE I

GM1 potentiates NCAM and N-Cadherin-dependent neurite outgrowth

	3T3 monolayer	NCAM transfectant	N-cadherin transfectant
Control	100 ± 8.6 (98)	211 ± 16.8 (120)	254 ± 27.8 (96)
GM1 50 μg/ml	121 ± 11.9 (89)[n.s.]	292 ± 21.0 (109)***	424 ± 27.2 (110)***

PC12 cells were cultured for 40 h in SATO media (control) or SATO supplemented with 50 g/ml GM1 on monolayers of control or transfected 3T3 cells as indicated. Co-cultures were fixed and the length of the longest neurite on Thy-1 stained PC12 cells was determined. The results show mean neurite length per cell (±1 S.E.M.) for 90–120 PC12 cells sampled in replicate cultures. The 100% control value was 13.4 m. n.s. = not significant; ***$P < 0.0025$.

116

TABLE II

N-Cadherin response from control and GM1 Pre-treated PC12 Cells

GM1 (µg/ml)	Mean neurite length, % control
0	140 ± 9
6	202 ± 12
25	212 ± 13
100	187 ± 10

PC12 cells were incubated overnight at 37°C in SATO media supplemented with 0, 6, 25 or 100 µg/ml GM1. The cultures were washed and dissociated as normal and cultured on confluent monolayers of 3T3 cells or N-cadherin expressing 3T3 cells. After 20 h the co-cultures were fixed and the mean length of the longest neurite per cell determined. Mean neurite lengths are shown for growth on N-cadherin monolayers and these values are the mean ± S.E.M. for 150–180 PC12 cells per data point. The values are expressed as a percentage of the growth measured on control 3T3 cell monolayers (measured as 12.3 ± 1.0 n = 153).

cell membrane content of GM1, and this is associated with an enhancement of neurite outgrowth at up to 48 h, which is the longest time so far analysed.

GM1 effects are antagonised by pertussis toxin

Downstream of the initial CAM recognition event there is activation of a G-protein, upstream of the calcium influx that is required for CAM dependent neurite outgrowth. We have used pertussis toxin which ribosylates and consequently inactivates G-proteins to analyse this system further. Pertussis toxin blocks N-cadherin dependent neurite outgrowth (Doherty et al., 1991b) and under these circumstances GM1 addition at 100 µg/ml has no effect on neurite outgrowth. Thus signalling through the second messenger pathway is required for gangliosides to exert their effects.

Conclusions

A large number of studies have reported, over many years, that exogenously added gangliosides can have positive effects on neurite outgrowth. The molecular mechanism underlying these positive effects, howev-

er, remains unclear and it has often been difficult to determine whether gangliosides mimic the full range of trophic effects of neurotrophic factors such as NGF or whether they have more limited effects. We have shown (Doherty et al., 1985a,b) that in dorsal root ganglion neurons, which are NGF responsive, GM1 could enhance some NGF effects but that it could not substitute for NGF by directly stimulating neuronal survival. It therefore seems likely that gangliosides act by modulating signals that lead to neurite outgrowth and do not by themselves possess full neurotrophic activity.

Recently the NGF pathway of signal transduction has been significantly dissected with the identification of the NGF-receptor as a tyrosine kinase (Bothwell, 1991). It is clear that NGF has many effects on cells such as PC12 cells with downstream events after NGF action being associated with rapid changes in the patterns of tyrosine phosphorylation. There are then longer term changes in the levels of a variety of cellular components including NCAM and N-cadherin (Doherty et al., 1988, 1991a). However, in our hands gangliosides were not found to affect any of these late transcription dependent NGF responses suggesting that gangliosides do not directly modulate NGF signalling in this system. In the PC12 system NGF is actively promoting a switch from an adrenal to a neuronal phenotype and is not acting as a survival factor. The actual extension of neurites is equally dependent on a permissive substratum and this requires integrin or CAM receptor function in neurons. Thus it is equally likely that GM1 modulates integrin or CAM signals rather than NGF signals per se.

In order to analyse the possible interactions of gangliosides with CAM signalling pathways we have set up a series of co-cultures of PC12 cells with monolayers of 3T3 fibroblasts or the same monolayer cells expressing transfected NCAM or N-cadherin. We have shown that a direct incorporation of GM1 into the cell membrane of PC12 cells, in a trypsin insensitive manner, is associated with a dramatic enhancement of NCAM and N-cadherin-dependent neurite outgrowth. In contrast no effects were found for PC12 cells cultured on parental 3T3 cells. There was also a degree of specificity in the response as asialo-GM1 and the oligosaccharide portion of GM1 were without effect.

The mechanism for NCAM, N-cadherin and integrin dependent neurite outgrowth has been studied by us. NCAM and N-cadherin appear to operate in neurite outgrowth by activation of a second messenger pathway that culminates in an influx of calcium via the neuronal L- and N- type calcium channels (Doherty et al., 1991b). Protein kinase A and C do not seem to be a part of this signalling pathway, however, the activity of a pertussis toxin sensitive G-protein is required. The site of GM1 action is not clear but some possibilities can be immediately discounted. GM1 does not have any effects on PC12 cells when they are in the presence of 3T3 fibroblasts. Thus GM1 on its own cannot activate the CAM specific second messenger pathway, at least not in PC12 cells. It is likely that GM1 is acting either to enhance flux through the calcium channels or some other calcium dependent target within the cell. It is, however, also possible that GM1 acts upstream of the pathway, for instance by directly interacting with the CAMs by modulating their binding activity.

Complementary data to the present study have been provided by Hilbush and Levine (1991). They showed in PC12 cells that the phosphorylation of a specific peptide in tyrosine hydroxylase was greatly increased in the presence of GM1 but only after NGF addition or membrane depolarisation by potassium. It seems likely that in this study a ganglioside potentiation of a calcium-dependent signalling pathway has been identified. Further evidence for this postulate has recently been obtained (B.S. Hilbush and J. Levine pers. comm.). Potassium depolarisation of PC12 cells leads to an increase in intracellular calcium and GM1 can potentiate this response.

A number of other studies have shown that exogenous gangliosides can modulate the function of calcium channels. B-cholera toxin which binds exclusively to GM1 can stimulate an increase in free calcium in lymphocytes (Dixon et al., 1987). Also antibodies to GM1 can enhance the release of GABA following potassium depolarisation (Frieder and Rapport, 1987). There is therefore widespread evidence that exogenous ganglioside may modulate the function of calcium channels. Whether these effects are direct or indirect will have to be determined experimentally. However, the systems discussed here are in place to analyse these phenomena further. Gangliosides have recently been shown to have therapeutic potential in a number of pathologies including human spinal cord injury (Geisler et al., 1992) and stroke (Skaper and Leon, 1992). In order to design logical interventive strategies some idea of mechanism of action is required. The ability of gangliosides to modulate CAM function associated with axonal growth in co-culture systems provides a model system for further analysis. The assays have the possibilities for defining structure-function relationships among different naturally occurring or synthetic gangliosides. There is also the possibility for designing drug therapies based on the interactions of different classes of drugs that may be active in the CAM pathway of axonal growth.

References

Bothwell, M. (1991) Keeping track of neurotrophin receptors. *Cell*, 65: 915–918.

Dixon, S.J., Stewart, D., Grinstein, G. and Spiegel, S. (1987) Transmembrane signalling by the B subunit of cholera toxin: increased cytoplasmic free calcium in rat lymphocytes. *J. Cell Biol.* 105: 1153–1161.

Doherty, P., Dickson J.G., Flanigan, T.P. and Walsh, F.S. (1985a) Ganglioside GM1 does not initiate, but enhances neurite regeneration of nerve growth factor-dependent sensory neurons. *J. Neurochem.*, 44: 1259–1265.

Doherty, P., Dickson, J.G., Flanigan,T.P., Leon, A., Toffano, G. and Walsh, F.S. (1985b) Molecular specificity of ganglioside effects on neurite regeneration of sensory neurons *in vitro*. *Neurosci. Lett.*, 62: 193–198.

Doherty, P., Mann, D.A. and Walsh, F.S. (1988) Comparison of the effects of NGF, activators of protein kinase C, and a calcium ionophore on the expression of Thy-1 and NCAM in PC12 cell cultures. *J. Cell Biol.*, 107: 333–340.

Doherty, P., Ashton, S.V., Moore, S.E., Mann, D.A. and Walsh, F.S. (1991a) Neurite outgrowth in response to transfected NCAM and N-cadherin reveals fundamental differences in neuronal responsiveness to CAMs. *Neuron*, 6: 247–258.

Doherty, P., Ashton, S.V., Moore, S.E. and Walsh, F.S. (1991b) Morphoregulatory activities of NCAM and N-cadherin can be accounted for by G protein-dependent activation of L- and N-type neuronal Ca^{2+} channels. *Cell*, 67: 21–33.

Doherty, P. and Walsh, F.S. (1992) Cell adhesion molecules, sec-

118

ond messengers and axonal growth. *Curr. Opinion Neurobiol.,* and 2: 595–601.

Doherty, P. Skaper, S.D., Moore, S.E., Leon, A. and Walsh, F.S. (1992a) A developmentally regulated switch in neuronal responsiveness to NCAM and N-cadherin in the rat hippocampus. *Development,* 115: 885–892.

Doherty, P., Ashton, S., Skaper, S.D., Leon, A. and Walsh, F.S. (1992b) Ganglioside modulation of neural cell adhesion molecule and N-cadherin dependent neurite outgrowth. *J. Cell Biol.* 117: 1093–1099.

Facci, L., Leon, A., Toffano, G., Sonnino, S., Ghidoni, R. and Tettamanti, G. (1984) Promotion of neuritogenesis in mouse neuroblastoma cells by exogenous gangliosides. Relationship between the effect and the cell association of ganglioside GM1. *J. Neurochem.,* 42: 229–305.

Frieder, B. and Rapport, M.M. (1987) The effect of antibodies to gangliosides on Ca^{2+} channel-linked release of γ-aminobutyric acid in rat brain slices. *J. Neurochem.,* 48: 1048–1052.

Geisler, F.H., Dorsey, F.C. and Coleman, W.P. (1992) GM-1 ganglioside in human spinal cord injury. *J. Neurotrauma,* 9 (Suppl. 2): 517–530.

Hilbush, B.S. and Levine, J.M. (1991) Stimulation of Ca^{2+} dependent protein kinase by GM1 ganglioside in nerve growth factor-treated PC12 cells. *Procl. Natl. Acad. Sci. U.S.A.,* 88: 5616–5620.

Inuzuka, H., Miyatani, S. and Takeichi, M. (1991) R-cadherin: a novel Ca^{2+} dependent cell-cell adhesion molecule expressed in the retina. *Neuron,* 7: 69–79.

Kuhn, T.B., Stoeckli, E.T., Condrau, M.A., Rathjen, F.G. and Sonderegger, P.(1991) Neurite outgrowth on immobilized axonin-1 is mediated by heterophilic interaction with L1 (G4). *J. Cell Biol.,* 115: 1113–1126.

Reichardt, L.F. and Tomaselli, K.J. (1991) Extracellular matrix molecules and their receptor: functions in neural development. *Annu. Rev. Neurosci.,* 14: 531–570.

Seilheimer, B. and Schachner, M. (1988) Studies of adhesion molecules mediating interactions between cells of peripheral nervous system indicate a major role for L1 in mediating sensory neurons growth on Schwann cells in culture. *J. Cell Biol.,* 107: 341–351.

Skaper, S.D. and Leon, A. (1992) Monosialogangliosides, neuroprotection, and neuronal repair processes. *J. Neurotrauma.,* 9(Suppl. 2): 507–516.

Small, S.J., Akeson, R. (1990) Expression of the unique NCAM VASE exon is independently regulated in distinct tissues during development. *J. Cell Biol.,* 111: 2089–2096.

Spoerri, P.E., Sozier, A.K. and Rosien, F.J. (1990) Calcium regulation of neuronal differentiation: the role of calcium in GM1-mediated neuritogenesis. *Dev. Brain. Res.,* 56: 177–188.

Takeichi, M. (1991) Cadherin cell adhesion receptors as a morphogenetic regulator. *Science, (Wash. DC),* 251: 1451–1455.

Walsh, F.S. and Doherty, P. (1991) Structure and function of the gene for neural cell adhesion molecule. *Semin. Neurosci.,* 3: 271–284.

Walsh, F.S., Furness, J., Moore, S.E., Ashton, S.V. and Doherty, P. (1992) Use of the NCAM-VASE exon by neurons is associated with a specific downregulation of NCAM dependent neurite outgrowth in the developing cerebellum and hippocampus. *J. Neurochem.,* 59: 1959–1962.

Williams, E., Doherty, P., Turner, G., Reid, R.A., Hemperley, J.J. and Walsh, F.S. (1992) Calcium influx into neurons can soley account for cell contact dependent neurite outgrowth stimulated by transfected L1. *J. Cell Biol.* (In press)

Wu, G., Vaswani, K.K., Lu, Z. and Ledeen, R.W. (1990) Gangliosides stimulate calcium flux in Neuro-2A cells and require exogenous calcium for neuritogenesis. *J. Neurochem.,* 55: 484–491.

L. Svennerholm, A.K. Asbury, R.A. Reisfeld, K. Sandhoff, K. Suzuki, G. Tettamanti and G. Toffano (Eds.)
Progress in Brain Research, Vol. 101
119

CHAPTER 9

Significance of ganglioside-mediated glycosignal transduction in neuronal differentiation and development

Yoshitaka Nagai[1] and Shuichi Tsuji[2]

[1]Tokyo Metropolitan Institute of Medical Science, Bunkyo-ku, Tokyo 113, Japan, Department of Neurobiology, Brain Research Institute, Niigata University, Niigata City, Niigata 951, Japan. and Glycobiology Research Group of the Frontier Research Program, The Institute of Physical and Chemical Research (RIKEN), Wako City, Saitama, 351-01, Japan and [2]Department of Glyco Molecular Biology, Glycobiology Research Group of the Frontier Research Program, The Institute of Physical and Chemical Research (REKEN), Wako City, Saitama, 351-01, Japan

Introduction

A great molecular diversity of complex carbohydrate chains has been utilized in cell biology and biotechnology practice for detecting or sorting particular cells. Along this line of study of glycoconjugates as cell surface markers or antigens, it has been presumed that cells also utilize this diversity for cell to cell recognition, including adhesion of microorganisms to animal and plant cells. Such a system of cell to cell recognition and responses is highly devel- oped in the nervous tissues which embody the highly intricate, interacting systems of neuron to neuron, as well as neuron to glial cell. In fact these tissues are extremely rich in glycoconjugates, particularly gangliosides, strongly suggesting important involvement of carbohydrate recognition in the nervous system. In this regard it is of particular interest to note recent findings that certain gangliosides, when exogenously added to cultured cells, elicit various biological responses which seem to be closely associated with signal transduction pathway of the cells. Four types of such ganglioside-mediated responses are presently recognized (Table I).

GM3-Type responses (type 1, or low responder type)

Ganglioside GM3 modulates growth factor-depen- dent (EGF, PDGF, insulin) cell growth and associat- ed receptor-mediated tyrosine phosphorylation (Bremer et al., 1983; Bremer et al., 1984; Bremer et al., 1986; Nojiri et al., 1991). Gangliosides have a potential to inhibit cell growth and to guide differen- tiation of human myelocytic leukemic cells, HL-60, either into monocytic, macrophage-like cells, or into mature granulocytes, depending upon the type of exogenous gangliosides (Nojiri et al., 1986, 1988). Several derivatives of GM3 and sphingosine and its

TABLE I

Four modes of carbohydrate (glycolipid)-mediated signal trans- duction

1. *Functional modulation of receptor in micro domain (Type I)*
   Alteration of receptor function by annular (boundary) lipid molecules; e.g., GM3 ganglioside and its derivatives
2. *Glycomessage-glycobinder interaction (Type II)*
   Lectin/antibody/toxin-membrane glycoconjugate interaction
3. *Glycomessage-glycorecepter interaction (Type III)*
   (Recognizing, transducing, responding)
   Intercellular (Endo) type signal transduction
   Cell surface (Ecto) type signal transduction
4. *Direct transfer to cell nuclei (genes) of glycomessage (Type IV)*
   Carbohydrate-mediated recognition (interaction) to nucleo- protein supramolecular structure or regulation of gene activa- tion by specific glycosylation.

derivatives are also uniquely involved in signal trans-duction (Hannun and Bell, 1989; Igarashi et al., 1989; Igarashi and Hakomori, 1991). Effective doses of these bioactive gangliosides and their derivatives are usually in a range of μM concentrations, in contrast to other cases which elicit activity in nanomolar range. Along this line we once proposed to name this group of responses 'M-type' as compared to 'N-type' where nanomolar concentrations are sufficient to induce activity (Nagai and Tsuji, 1988).

Considering the necessity of high concentrations of exogenous gangliosides, it is likely that the activity may develop as a consequence of the formation of annular or boundary lipid (gangliosides) structure with either the receptor protein molecule or protein components of the signal transduction pathway in the membrane. This presumably results in a conforma-tional change or in the polymerization state of respec-tive protein components and their subunits. Such boundary (annular) lipid micro-domain structure was proposed to explain the active state of membrane pro-tein components in association with glycolipids (Yamakawa and Nagai, 1978; Nagai and Iwamori, 1980) (Fig.1). Such annular lipid is distinct from other lipids in the bulk matrix phase in that it can no longer show thermodynamically highly cooperative, first-order transition and thereby a latent heat of tran-sition. The annular structure is presumed to be formed by at least 35 lipid molecules surrounding protein molecules, as in the case of $[Ca^{2+}\text{-}Mg^{2+}]$-ATPase from sarcoplasmic reticulum (Lee, 1977). However, this thermodynamic model does not wholly explain the necessity of a particular ganglioside for this activity.

Lectin/antibody/toxin-type responses (Type II)

Lectin is known to affect growth and differentiation of cells, as seen from the blastogenic effect of concanavalin A on lymphocytes. A bacterial toxin lectin, for example, cholera toxin B subunit that prin-cipally recognizes GM1 and GD1b gangliosides, regulates cultured cells bimodally and interferes with signal transduction in the cells (Spiegel et al., 1985; Spiegel and Fishman, 1987; Spiegel, 1988; Spiegel, 1989). Certain polyclonal antibodies such as anti-GM3 and anti-GM1 also regulate growth of certain cultured cells (Kanfer and Hakomori, 1983). We recently observed that monoclonal anti-body to GD3 ganglioside specifically and reversibly inhibits adenovirus EIA oneogene DNA-transfected, transformed cells which express GD3 on the cell surface, but does not inhibit parent cells, 3YI, which-contain little GD3 (Sanai and Nagai, unpublished).

All these phenomena again indicate the occurrence of another type of carbohydrate-mediated response

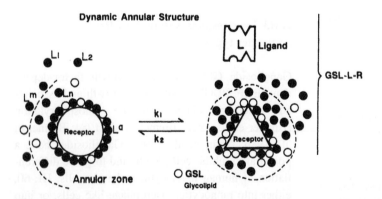

Fig. 1. Dynamic annular structure in protein-lipid interaction in membrane. Annular (boundary) lipid (L^a) is in equilibrium with matrix (bulk phase) lipids ($L_1L_2, \ldots . L_m,L_n$) on lateral diffusion. When annular lipid molecules attain to a sufficient and definite number around receptor protein molecule (R), the receptor molecule changes its conformation into an activated form (GSL-L-R complex) to receive ligand.

of the cells, although the underlying molecular mechanism and signal transduction pathway remain unclear.

Glycoreceptor-type responses (Type III)

There are two types of protein receptors that receive carbohydrate signals. One serves merely for specific binding of carbohydrate ligands, while the other not only binds to the ligand but induces physiological responses of the receptor-bearing cells. The former type are the lectins and the latter should be called glycoreceptors.

We reported in 1983 that GQ1b has a strong potency to promote neurite outgrowth in an oligopolar fashion of two human neuroblastoma cell lines, GOTO and NB-1 (Tsuji et al., 1983).

This biological response is understandable on the basis of a glycoreceptor which specifically recognizes the carbohydrate structure of gangliosides

TABLE II

Evidence for glycoreceptor to GQ1b ganglioside in human neuroblastoma cell line, GOTO

---

1. Necessity of ganglioside for neurite outgrowth is strict and requires only GQ1b but not other homologous gangliosides like GD1a, GD1b, GT1a, GT1b and GQ1c and ceramide.
2. A few nM GQ1b (5 ng/ml) are sufficient to promote neurite outgrowth.
3. The promotion is suppressed with micromolar order of GQ1b oligosaccharide in a concentration dependent manner.

(Nakajima et al., 1986) (Table II, Fig, 2). The activity is mediated by initial, sugar chain-dependent, recognitive interaction of GQ1b with the glycoreceptor on the cell surface membrane. Several studies suggest that the glycoreceptor and GQ1b interaction is directly coupled with subsequent phosphorylation at cell surface loci of particular membrane proteins (54, 60

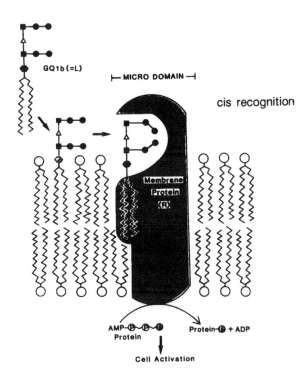

Fig. 2. Glycoreceptor that recognizes and interacts with GQ1b ganglioside.

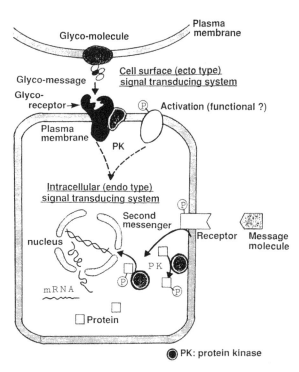

Fig. 3. Two major types of signal transducing systems (ecto type and endo type).

TABLE III

Characterization of GQ1b-dependent ecto-protein phosphorylation

1. With a few exceptions ATP does not appear to permeate across surface membrane of the cells.
2. $[\gamma^{-32}P]$ATP added to the culture medium very shortly results in many radiolabeled protein bands, all of which are susceptible to mild trypsin treatment.
3. Phosphorylation does not occur with *ortho*-$[^{32}P]$-phosphoric acid (10 μCi/well) for 30 mins, whereas with $[\gamma^{-32}P]$ATP (1 μM; 10 μCi/well) phosphorylation occurs by this time. On longer incubation after the addition of *ortho*-$[^{32}P]$phosphoric acid to the medium, phosphorylation does occur, but with a different pattern from that of the short term phosphorylations.
4. Exogenous GQ1b and $[\gamma^{-32}P]$ATP specifically stimulate ^{32}P-labeling of three proteins (64, 60 and 54 kDa). The phosphorylation occurs at specific amino acids, Ser and Thr.
5. Mild trypsin (0.01%) treatment of the intact cells causes the three phosphorylated bands (64, 60 and 54 kDa) disappear almost immediately. During the treatment the amount of the total SDS-solubilized protein remains almost unchanged. No change in attachment to the plates and in cell shape is observed.

TABLE IV

Evidence for coupling of GQ1b-dependent promotion of neurite outgrowth with cell ecto-protein phosphorylation

1. Both ganglioside-dependent neurite outgrowth and ecto-protein phosphorylation require specifically GQ1b and proceed at nM concentrations of the ganglioside in a time- and dose-dependent manner.
2. Both events are suppressed with μM order of GQ1b oligosaccharide in a concentration-dependent manner.
3. Photoaffinity labeling of the cell surface with 8-azido-ATP results in complete inhibition of the ecto-protein phoshporylation and also abolishes the GQ1b-dependent promotion of neurite outgrowth.
4. K-252b, a specific, cell impermeable and non-cytotoxic inhibitor of protein phosphorylation and neurite outgrowth promotion in living cells.

and 64 kDa), in the presence of extracellular ATP (Fig.3, Table II, Table III) (Tsuji et al., 1985; Tsuji et al.,1988a; Nagai and Tsuji, 1989; Tsuji et al., 1992).

K-252b (Fig.4), a hydrophylic alkaloid which was isolated as an inhibitor of protein kinase C from cul-

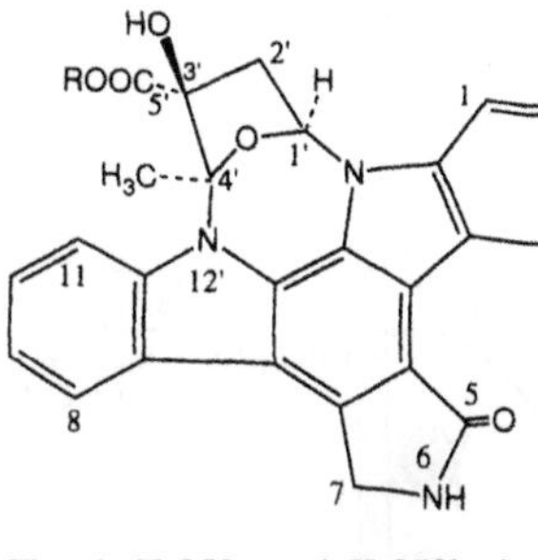

Fig. 4. K-252a and K-252b, isolated from the culture broth of *Nocardiopsis* sp. as protein kinase inhibitor (Kase et al., 1986). K-252a (R:CH₃), carboxylic acid methyl ester derivative of K-252b is permeable into cells and potently inhibits several protein kinases in vitro, including protein kinase C and cyclic nucleotide-dependent kinases (Kase et al., 1987). K-252b (R:H) is a cell-impermeable compound showing a potent inhibitory activity rather selectively to protein kinase C in vitro.

ture broth of *Nacardiopsis* sp. by Kase et al. (1986, 1987), was later found to be a most potent, selective inhibitor of ecto-protein kinases, ecto-protein phosphorylation (Tsuji et al., 1992), and neuritogenesis (Nagashima et al., 1991; Tsuji et al., 1992). This strongly fluorescent, indole-carbazole type of compound is non-cytotoxic and its fate in the cell can be easily traced (Nagashima et al., 1991). The compound, so far tested is totally cell impermeable in cultured cell lines, in contrast to its structurally related compound, K-252a, which was isolated from the same microbial culture and is also a potent, non-selective and cell permeable inhibitor of several protein kinases (Fig.4) (Kase et al., 1987). K-252b has also been reported to inhibit nerve growth factor (NGE)-induced neurite outgrowth of PC12 cells (Nagashima et al., 1991), and to inhibit selective NGF antagonistic effects on cholinergic neurons of basal forebrain without showing any cytotoxicity (Kniisel and Hefti, 1991).

The ecto-phosphorylation found in GOTO cells was similarly found in other cultured neuronal cells like PC12h, Neuro 2a and NIE-115, and also in glial cells such as RCR-1 (astrocyte), C-6 (astroglioma) and 354-A (peripheral glioma-like schwannoma) (unpublished). In the latter cases, phosphorylated

TABLE V

Profile of GQ1b-mediated neuritogenesis and cell surface (ecto) type protein phosphorylation

	Neuritogenesis	Ecto-protein phosphorylaton
Specificity for requirement GQ1b (5 nm) + oligo-GQ1b	GQ1b, nM range (3–5 pmol/ml) inhibitory (conc. dependent 0.1–1. 0 µM)	GQ1b, nM range (3–5 pmol/ml) inhibitory (conc. dependent 0.025–1. 0 µM)
ATP addition	not necessary	necessary
K–252b (1 µM) (cell membrane impermeable inhibitor of protein kinases)	inhibitory	inhibitory

bands on SDS-PAGE are very few in number as compared to those of neuronal cells showing a greater number of the bands. It is very likely that the ecto-phosphorylation system may mostly develop in neurons.

Muramoto et al. (1988) recently demonstrated successfully that K-252b inhibits functional synapse formation between cultured neurons of rat cerebral cortex. Meanwhile, K-252b inhibits phorbol ester-induced long-term potentiation (LTP) as a possible neural basis for learning and memory in rat hippocampal slices (Reymann et al., 1988). Moreover, perfusion of GQ1b in guinea pig hippocampal slices induces the characteristic post-tetanic, long-term increase (LTP) in the magnitude of CA1-evoked responses (Ando, S. and Kato, H., pers. commun.). Thus, glycoreceptor-mediated signaling pathway should be of predominant importance for synapse formation and neuronal network activity in the central nervous system. Yada et al. (1991) recently reported that exogenous GQ1b but not other gangliosides, specifically induces terminal differentiation in cultured mouse keratinocytes by remarkably increasing the mass content of inositol 1,4,5-triphosphate (IP_3) and the intracellular Ca^{2+} levels, and that GQ1b alone promotes the translocation of protein kinase C from cytosol to membrane. Whether a similar ecto-phosphorylation system might be involved in this case is not certain. The source of extracellular ATP for the ecto-phosphorylation remains unclear (Table V). It is

likely that synaptic vesicles known to be extremely enriched in ATP (0.1–0.2 M) together with neurotransmitters may be the source of phosphate donor for ecto-phosphorylation in neuronal cells.

In the pursuit of the molecular mechanism of ecto-protein phosphorylation, the identification, isolation and molecular cloning of ecto-protein kinases, as well as their substrates, are most urgent. There have been many reports that gangliosides affect protein kinases (Nagai and Tsuji, 1988).

We recently reported that GQ1b (80 nM) significantly and specifically affects phosphorylation and also dephosphorylation of the 72 kDa protein in rat neuronal membrane fraction for a very short time, but does not noticeably affect ATPase in the same fraction (Nakaoka et al., 1992). The results suggest that the transphosphorylation of the 72 kDa protein is distinctly affected by the interaction of GQ1b with either the responsible enzymes or the 72 kDa protein as a substrate.

Glycosignals, cell nuclei targeting and regulation of gene activity (Type IV)

Following the above-mentioned three types of responses, there is an entirely different type of carbohydrate reaction which occurs within cell nuclei. Sialyl cholesterol (SC), a synthetic amphipathic sialyl compound (Fig. 5), has a potency to initiate and pro-

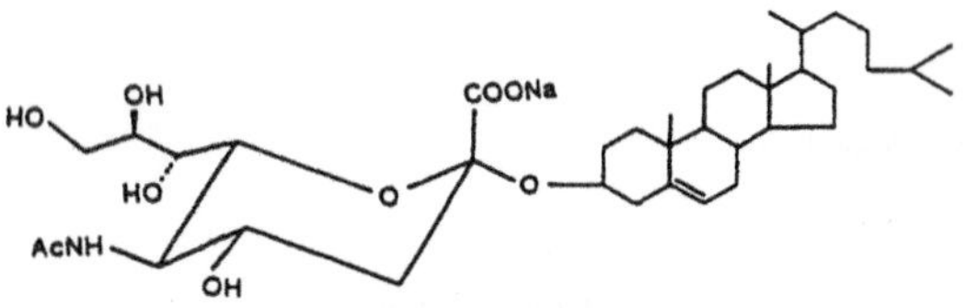

124

Fig. 5. α-*N*-acetylneuraminyl cholesterol.

mote neurite outgrowth in mouse neuroblastoma cells, Neuro 2a, irrespective of its α or β form (Tsuji et al., 1988b). Interestingly, almost 40% of the compound added to the cells are promptly transported into the cell nucleus (Yamashita et al., 1991). Moreover, the chromatin fraction prepared from the SC-treated cells shows distinct enhancement of transcriptional activity (Table VI). Such migration is not observed with the GM1 ganglioside under the same conditions. SC remains intact in the cell nucleus, indicating that the nuclear SC affects de novo RNA synthesis. The result suggests that the biological effect of SC may be mediated at the transcription level. Neither the intracellular level of total Ca^{2+} nor the levels of inositol 1,4,5,-triphosphate (IP_3) changes, though SC increases the rate of both Ca^{2+} influx and efflux. An excess amount of W-7, an inhibitor of Ca^{2+}/CaM kinases, does not interfere with

TABLE VI

De nova RNA synthetic activity of chromatin fractions obtained from α-sialyl cholesterol (α-SC), β-sialyl cholesterol (β-SC), or GM1-treated mouse neuroblastoma cells, Neuro 2a (Yamashita et al., 1991)

Treatment	UMP incorporated[a] (dpm/mg DNA/10 min)	Ratio treated/non-treated
α-SC (1 mM)	2800 ± 180	2.33
β-SC (1 mM)	2000 ± 130	1.67
GM1 (1 mM)	1100 ± 100	0.92
Non-treated cells	1200 ± 60	1.00

[a] The figures are the mean values obtained for five different experiments. The transcription activities (UMP incorporation) of chromatin fractions which were obtained from treated or non-treated cells were evaluated by the incorporation of [³H]UTP into the acid-insoluble fraction.

the SC-dependent neuritogenesis. Thus, the results are in accord with the possibility that SC in the nucleus plays a key role in neuritogenesis without using several of the more common second messengers and that SC may directly modulate gene expression.

This may represent the fourth type of the carbohydrate-mediated biological response and a novel glycosignal transduction pathway. Hart et al. (1989) and Jackson and Tjian (1988) also demonstrated that O-GlcNAc which is glycosidically linked to the hydroxyl of serine or threonine occurs on important nuclear pore glycoproteins (cytoskeletal proteins) as well as on many chromatin proteins, including factors that regulate gene transcription. They also demonstrated an astonishing abundance of O-GlcNAc-bearing proteins associated with polytene chromosomes of *Drosophila* embryos using a new assay method for glycosylation sites with bovine milk galactosyl transferase and UDP-[³H]-galactose, and also with fluorescent WGA lectin (Hart et al., 1989).

Acknowledgements

We would like to express our thanks to Dr. J. Nakajima, Dr. T. Nakaoka, Mr. T. Yamashita, and Mr. H. Yamamoto for their contribution throughout this study. This study was supported by the following grants: Grant-in-Aids for Scientific Research on Priority Area (Nos. 04250106 and 04268216) and Grant-in-Aids for Scientific Research (Nos. 03454157 and 03680139) from the Ministry of Education, Science and Culture, Japan; The research Grant(3A-2) for Nervous and Mental Disorders from the Ministry of Health and Welfare, Japan.

References

Bremer, E.G. and Hakomori, S.I. (1982) GM3 ganglioside induces hamster fibroblast growth inhibition in chemically-defined medium: Ganglioside may regulate growth factor receptor function. *Biochem. Biophys. Res. Commun.*, 106: 711–718.

Bremer, E.G., Hakomori, S.I., Bowen-Pope, D.F., Raines, E. and Ross, R. (1984) Ganglioside-mediated modulation of cell growth, growth factor binding and receptor phosphorylation. *J. Biol. Chem.*, 259: 6818–6825.

Bremer, E.G., Schlessinger, J. and Hakomori, S.I. (1986)

Ganglioside-mediated modulation of cell growth: Specific effects of GM3 on tyrosine phosphorylation of the epidermal growth factor receptor. *J. Biol. Chem.*, 216: 2434–2440.

Hannun, Y.A. and Bell, R.M. (1989) Functions of sphingolipids and sphingolipid breakdown products in cellular regulation. *Science*, 243: 500–507.

Hart, G.W., Haltiwanger, R.S., Gordon, D.H. and Kelly, W.G. (1989) Nucleocytoplasmic and cytoplasmic glycoproteins. In: *Carbohydrate Recognition in Cellular Function, Ciba Foundation Symposium, Vol. 145,* John Willey and Sons, New York, pp. 102–118.

Igarashi, Y. and Hakomori, S.I. (1991) Modulatory effect of gangliosides and their derivatives on transmembrane signal transduction. In: *Gangliosides: The Pharmacology of Neuronal Plasticity,* Fidia Res. Foundation Symposium (Abstracts), Italy, pp. 12–18.

Igarashi, Y., Hakomori, S., Toyokuni, T., Dean, B., Fujita, S., Sugimoto, M., Ogawa, T., El-Ghendy, K. and Racker, E. (1989) Effect of chemically well-defined sphingosine and its *N*-methyl derivatives on protein kinase C and *src* kinase activities. *Biochemistry*, 28: 6796–6800.

Jackson, S.P. and Tjian, R. (1988) O-Glycosylation of eukaryotic transcription factors: Implication for mechanisms of transcriptional regulation. *Cell*, 55: 125–133.

Kanfer, J.N. and Hakomori, S.I. (1983) *Sphingolipid Biochemistry, Handbook of Lipid Research, Vol. 3,* Plenum Press, New York, p. 355.

Kase, H., Iwahashi, K. and Matsuda, Y. (1986) K-252a, a potent inhibitor of protein kinase C from microbial origin. *J. Antibiotics*, 39: 1059–1065.

Kase, H., Iwahashi, K., Nakanishi, S., Matsuda, Y., Yamada, K., Takahashi, M., Murakata, C., Sato, A. and Kaneko, M. (1987) K-252 compounds, novel and potent inhibitors of protein kinase C and cyclic nucleotide-dependent protein kinases. *Biochem. Biophys. Res. Commun.*, 142: 436–440.

Kniisel, B. and Hefti, F. (1991) K-252b is a selective and non-toxic inhibitor of nerve growth factor action on cultured brain neurons. *J. Neurochem.*, 57: 955–962.

Lee, A.G. (1977) Annular events: lipid-protein interactions. *TIBS* 2: 231–233.

Muramoto, K., Kobayashi, K., Nakanishi, S., Matsuda, Y. and Kuroda, Y. (1988) Functional synapse formation between cultured neurons of rat cerebral cortex: Block by a protein kinase inhibitor which does not permeate the cell membrane. *Proc. Jpn. Acad. Ser. B.*, 64: 319–322.

Nagai, Y. and Iwamori, M. (1980) Brain and thymus gangliosides: their molecular diversity and its biological implications and a dynamic annular model for their function in cell surface membranes. *Mol. Cell. Biochem.*, 29: 81–90.

Nagai, Y. and Tsuji, S. (1988) Cell biological significance of gangliosides in neural differentiation and development. In: R.W. Ledeen, E.L. Hogan, G. Tettamanti, A.J. Yates and K.K. Yu (Eds), *New Trends in Ganglioside Research: Neurochemical and Neuroregenerative Aspects, Fidia Res. Series, Vol. 14,* Liviana Press, Padova and Springer Verlag, Berlin, pp. 329–350.

Nagai, Y. and Tsuji, S. (1989) Bioactive ganglioside-mediated carbohydrate recognition in coupling with ecto-protein phosphorylation. In: *Carbohydrate Recognition in Cellular Function, Ciba Foundation Symposium, Vol. 145,* John Wiley and Sons, New York, pp. 119–134.

Nagashima, K., Nakanishi, S. and Matsuda, Y. (1991) Inhibition of nerve growth factor-induced neurite outgrowth of PC12 cells by a protein kinase inhibitor which does not permeate the cell membrane. FEBS letts., 293: 119–123.

Nakajima, J., Tsuji, S. and Nagai, Y. (1986) Bioactive gangliosides: analysis of functional structures of tetrasialo-ganglioside GQ1b which promotes neurite outgrowth. *Biochim. Biophys. Acta*, 876: 65–71

Nakaoka, T., Tsuji, S. and Nagai, Y. (1992) Bimodal regulation of protein phosphorylation by a ganglioside in rat brain membrane. *J. Neurosci. Res.*, 31: 724–730.

Nojiri, H., Takaku, F., Terui, Y., Miura, Y. and Saito, M. (1986) Gangliosides GM_3: An acidic membrane component that increases during macrophage-like cell differentiation can induce monocytic differentiation of human myeloid and monocytoid leukemic cell lines HL-60 and U937. *Proc. Natl. Acad. Sci. USA*, 83; 782–786.

Nojiri, H., Kitagawa, S., Nakamura, M., Kirito, K., Enomoto, Y. and Saito, M. (1988) Neulacto-series gangliosides induces granulo-cytic differentiation of human promyelocytic leukemia cell line HL-60. *J. Biol. Chem.*, 263: 7443–7446.

Nojiri, H., Stoud, M. and Hakomori, S.I. (1991) A specific type of ganglioside as a modulator of insulin-dependent cell growth and insulin receptor tyrosine kinase activity. *J. Biol. Chem.*, 266: 4531–4537.

Reymann, K.G., Brödemann, R., Kase, H. and Matties, H. (1988) Inhibitors of calmodulin and protein kinase C block different phases of hippocampal long-term potentiation. *Brain Res.*, 461: 388-392.

Spiegel, S. (1988) Insertion of ganglioside GM1 into rat glioma C6 cells renders them susceptible to growth inhibition by the B subunit of cholera toxin. *Biochim. Biophys. Acta*, 969: 249–256.

Spiegel, S. (1989) Inhibition of protein kinase C-dependent cellular proliferation by interaction of endogenous ganglioside GM1 with the B subunit of cholera toxin. *J. Biol. Chem.*, 264: 16512–16517.

Spiegel, S. and Fishman, P. (1987) Gangliosides as bimodal regulators of cell growth. *Proc. Natl. Acad. Sci. USA*, 84: 141–145.

Spiegel, S., Fishman, P.H. and Weber, R.J. (1985) Direct evidence that endogenous GM1 ganglioside can mediate thymocyte proliferation. *Science*, 230: 1285–1289.

Tsuji, S., Arita, M. and Nagai, Y. (1983) GQ1b, a bioactive ganglioside that exhibits novel nerve growth factor (NGF)-like activities in the two neuroblastonia cell lines. *J. Biochem.*, 94: 303–306.

Tsuji, S., Nakajima, J., Sasaki, T. and Nagai, Y. (1985) Bioactive gangliosides. IV. Ganglioside GQ1b/Ca^{2+} dependent protein kinase activity exists in the plasma membrane fraction of neu-

126

roblastoma cell line, GOTO. *J. Biochem.*, 97: 969–972.

Tsuji, S., Yamashita, T. and Nagai, Y. (1988a) A novel, carbohydrate signal-mediated cell surface protein phosphorylation: ganglioside GQ1b stimulates ecto-protein kinase activity on the cell surface of a human neuroblastoma cell line, GOTO. *J. Biochem.*, 104: 498–503.

Tsuji, S., Yamashita, T., Tanaka, M. and Nagai, Y. (1988b) Synthetic sialyl compounds as well as natural gangliosides induce neuritogenesis in a mouse neuroblastoma cell lines (Neuro 2a). *J. Neurochem.*, 50: 414–423.

Tsuji, S., Yamashita, T., Matsuda, Y. and Nagai, Y. (1992) A novel glycosignaling system: GQ1b-dependent neuritogenesis of human neuroblastoma cell line, GOTO, is closely associated with GQ1b-dependent ecto-type protein phosphorylation. *Neurochem Int.*, in press.

Yada, Y., Okano, Y. and Nozawa, Y. (1991) Ganglioside GQ1b-induced terminal differentiation in cultured mouse keratinocytes. *Biochem. J.*, 279: 665–670.

Yamakawa, T. and Nagai, Y. (1978) Glycolipids at the cell surface and their biological functions. *TIBS*, 3: 128–131.

Yamashita, T., Tsuji, S. and Nagai, Y. (1991) Sialyl cholesterol is translocated into cell nuclei and it promotes neurite outgrowth in a mouse neuroblastoma cell line. *Glycobiology*, 1: 149–154.

L. Svennerholm, A.K. Asbury, R.A. Reisfeld, K. Sandhoff, K. Suzuki, G. Tettamanti and G. Toffano (Eds.)
Progress in Brain Research, Vol. 101

CHAPTER 10

Ca²⁺-Ganglioside-interaction in neuronal differentiation and development

Hinrich Rahmann, Harald Rösner, Karl-Heinz Körtje, Heinz Beitinger and V. Seybold

Zoological Institute, University of Stuttgart-Hohenheim, D-70593 Stuttgart 70 (Hohenheim), Germany

Introduction

A controlled exchange of calcium between the extra-cellular space (mM Ca^{2+}) and the neuroplasm (μM Ca^{2+}) is considered to be an essential prerequisite for almost every stage of neuronal activity, especially for electroresponsiveness, synaptic transmission of information, and for long-term adaptive events like neuronal differentiation, regeneration and, last but not least, for storage of information. It is generally accepted that this transmembraneous calcium-exchange can be modulated in a complex way by external factors, such as physical parameters as lateral surface pressure and temperature or chemical parameters as ion milieu, enzyme activities, hormones and drugs. On this background during the last decade lipid research is focused more and more on those compounds, which due to their physico-chemical properties, together with membrane proteins, might be involved in neuronal calcium-regulation.

With regard to this gangliosides (= amphiphilic sialic acid containing glycosphingolipids) which are highly accumulated in a complex composition in the outer leaflet of neuronal, particularly in synaptic membranes, and which show very peculiar binding properties towards calcium and peptides, became favourite candidates as neuromodulators to be essentially involved in the process of cell differentiation, particularly of synaptic transmission and long-lasting neuronal adaptation (Rahmann et al., 1976; Thomas

and Brewer, 1990; Rahmann, 1987a,b, 1992; all reviews).

Together with neutral glycolipids, glycoproteins and glycosaminoglycans gangliosides comprise the glycocalyx of the cell surface of deuterostomian animals (echinoderms and chordates, including vertebrates; Hilbig and Rahmann, 1987). The precise functional role of ganglioside is still poorly understood. Nevertheless gangliosides were shown to be implicated in a variety of cellular events, such as cell recognition, cell-to-cell contact formation, receptor binding and modulation, immunological properties and biosignal transduction. Significant correlations have been found particularly between cell growth, and differentiation, oncogenic transformation and tumor progression on the one side, and profound changes in ganglioside composition and biosynthesis on the other (Nakamura and Saito, 1990; Spiegel and Buckley, 1990; Taki and Honda, 1990; Ledeen and Wu, 1992; all reviews).

Since, however, all biochemical reactions within an organism have to be traced back to basic physico-chemical properties of the involved molecules, the interest, especially of ganglioside research, has become more and more focused on fundamental physico-chemical properties of these glycosphingolipids which were supposed to fulfill modulatory functions for membrane receptors, ion pumps and ion channels (Hakomori, 1987; Kaczmarek, 1988; Varon et al., 1988; Yates et al., 1988; Slenzka et al., 1990). During the last few years surprising data were obtained concerning ganglioside effects on the conformation (Koerner et al., 1987), the membrane potential, and the electrostatic potential adjacent to the membrane surface (McDaniel and McLaughlin, 1984; Beitinger et al., 1987; Schifferer et al., 1988, 1991; Thompson and Brown, 1988; Maggio et al., 1990). In summary, these data indicate that gangliosides significantly affect both, the topological and geometrical features of the membrane and the adjacent environment. In particular, gangliosides modify the intermolecular packing, curvature, interfacial micropolarity, asymmetry, electrostatics and free energy depending on the molecular composition and conformation, and by this the phase state within the membrane (Montich et al., 1988; Beitinger et al., 1989b; Maggio et al., 1990; Fidelio et al., 1991; Rahmann et al., 1992).

On the basis of the well-known functional properties of calcium for cellular homeostasis and of the peculiar physico-chemical properties of gangliosides in general, the aim of the present paper is to summarize investigations of basic interactive properties of gangliosides with respect to calcium and with functional membrane-bound proteins as fundamentals for the understanding of the role of gangliosides in differentiation and development. With regard to this, experimental data will be presented concerning: (a) the ultrastructural localization of calcium, a calcium pump (high-affinity Ca^{2+}-ATPase) and of gangliosides in nerve fiber terminals; (b) the effect of exogenous gangliosides on Ca^{2+}-deprived cultured neurons; and (c) calcium-ganglioside-peptide interactions in artificial mono- and bilayer systems.

Experimental procedures

Electronmicroscopical experiments

Ultrastructural analyses of synaptic terminals were performed with the brain of larval and adult cichlid fish (*Oreochromis mossambicus*) according to common electron microscopical procedures. Calcium localization was demonstrated with a cytochemical precipitation reaction using phosphate-buffered glutaraldehyde as a primary and a $K_2Cr_2O_7/OsO_4$ mix-

ture as a secondary fixative. Proof for the Ca^{2+}-specificity in electron dense deposits was established by means of electronspectroscopic imaging (ESI), respectively electron energy loss spectroscopy (EELS) using a Zeiss CEM 902 (Körtje et al., 1990a). Ca^{2+}-ATPase reaction products were localized as cerium phosphate precipitates (Körtje et al., 1990b). In order to localize c-series gangliosides the monoclonal mouse antibody Q211, followed by an anti-mouse IgM antibody, coupled to colloidal gold sols as an electron dense marker was used (Seybold et al., 1989).

Cell culture

Cell suspensions were prepared from optic lobes of 7-days-incubated chicken embryos (Greis and Rösner, 1990) and plated in modified Eagle's Medium (MEM 1% Ultroser, antibiotics) with a Ca^{2+}-concentration of < 20 μM. In some cases this was increased to 0.6 mM by the addition of sterile $CaCl_2$. In case of exogenous ganglioside application 10–100 μM GM1 or GD1b and GT1b (Fidia, Abano Terme, Italy) was added immediately after plating. Likewise cultures were treated with phosphatidylcholine (PC) or -serine (PS). Neuron viability, especially neurite growth, was scored by determination of the proportion of viable cells with processes, the length of which was at least twice the cell diameter (Rösner et al., 1992). In addition, scanning-electronmicroscopic pictures were prepared from GM1-treated and control cultures following standard procedures.

Monolayer experiments

The lipid monolayers were prepared at the air–water interface using a teflon trough and a dynal movable barrier. The surface pressure was measured by a filter paper Wilhelmy balance, and the surface potential was registered by a vibrating plate capacitor according to procedures as described previously (Probst et al., 1984; Schifferer et al., 1988, 1991; McLaughlin, 1989).

The surface pressure/area and surface potential/area isotherms were registered from monolayers prepared from the following compounds: GM1-mono, GD1a-di-, and GT1b-trisialoganglioside

(Fidia Research Laboratories, Abano Terme, Italy), sulfatide, ceramide (Sigma, Germany), dioleolylphosphatidylcholine (DOPC; Avanti Polar Lipids Inc., Birmingham, U.K.) and valinomycin (Boehringer, Mannheim, Germany). All compounds were spread as 1×10^{-3} M chloroform/methanol (2:1) solutions; their monofilms were compressed at a constant rate of 20 cm²/min on a subphase of TEA/HCl (5 mM) buffer, pH 7.4 (triethanolamine, Merck, Darmstadt, Germany), sometimes containing 0.01 mM Ca^{2+}. All experiments were performed at least 2–5 times at a temperature of $20 \pm 0.5°C$.

Planar bilayer experiments

Optically black lipid membranes were formed, according to methods described previously (Rahmann et al., 1992), on a hole in the septum of a teflon cell. In addition to the gangliosides mentioned before, the brain ganglioside mixture (GMix = Cronassial; Fidia), dioleoyl phosphatidylcholine (PC) and -serine (PS; both from Avanti Polar Lipids Inc.) and gramicidin A (kindly provided by Dr. H. J. Apell, Konstanz, Germany) were used. The membrane forming solutions contained 1–2.5% (w/v) of the pure lipid or lipid mixture, respectively, in *n*-decane. The electrolyte solutions contained 0.5 up to 1000 mM CsCl, 5 mM TEA/HCl buffer, and for some experiments also 0.01 or 0.5 mM $CaCl_2$, which in other experiments had been replaced by 0.5 mM EDTA.

The mean single channel conductance, Λ, and the mean life-time τ, of gramicidin A were measured as described by Bamberg et al., 1976. The records of conductance fluctuations were analyzed by counting the number of steps within a given conductance interval. Thus it was possible to check the data with respect to the conductance and life-time probability distribution of the single channel events.

Results

Subcellular localization of calcium, Ca^{2+}-ATPases and of gangliosides in synaptic terminals

The aim of a first set of experiments was to determine the localization of calcium, a high-affinity cal-

cium pump (Ca^{2+}-ATPase) and of gangliosides in synaptic nerve fiber terminals during differentiation, because the exact knowledge of the subcellular location of these compounds according to our introductory remarks is of utmost importance for the interpretation of all physiological, biochemical and, last but not least, physico-chemical indications according to which calcium-ganglioside interactions might act as modulators for synaptic functional proteins.

Localization of calcium. Our data obtained so far have revealed, for the first time, high extracellular accumulations of endogenous calcium deposits within the synaptic cleft of nerve fiber terminals, especially at the very local zone of synaptic contact (Figs. 1a,b) in addition to the well-known intracellular Ca^{2+}-storage sites (endoplasmic reticulum, mitochondria, storage vesicles). Direct proof of calcium specificity has been established by electron energy loss spectroscopy (EELS; Fig. 1c) or indirectly by pretreatment of the ultrathin section with the calcium-specific chelator EGTA.

The density of endogenous calcium within the synaptic clefts of fish brain was found to be much more pronounced than in those of mammals (data not shown). This correlates with the fact that extracellular calcium concentration in the CNS of bony fish is about 3 mM versus 1 mM in mammals (Veh and Sander, 1981), and to the degree of polysialylation of brain gangliosides of cold-blooded vertebrates in ecophysiological adaptation to their habitat temperature (Rahmann, 1983). Concerning neuronal differentiation, a development-depending accumulation of calcium in the synaptic cleft can be paralleled with the level of structural organization of the synaptic substructures (Fig. 2).

Localization of a high-affinity Ca^{2+}-ATPase. The high concentration of calcium in the synaptic cleft in differentiated neurons indicates that in this area a large number of calcium binding structures and most probably also of calcium pumps must be available. Calcium-dependent ATPases which are assumed to act as calcium pumps, use the energy released by the hydrolyzation of ATP to transport Ca^{2+}-ions out of

130

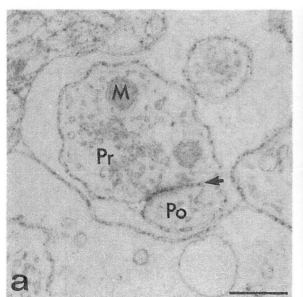

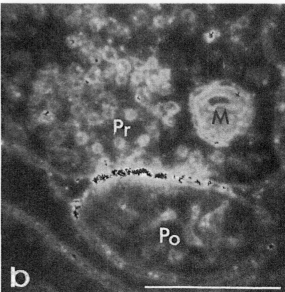

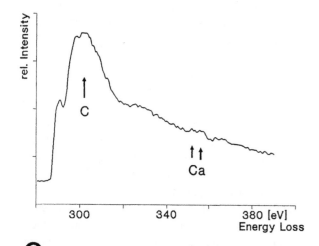

Fig. 1. Transmission electronmicroscopic (a) and electronspectroscopic imaging aspect (b) of a synapse in the optic tectum of a cichlid fish for the demonstration of calcium localization (arrow) mainly in the synaptic cleft. Bar = 0.5 μm; Pr = presynapse with synaptic vesicles; po = postsynapse. (c) Electron energy loss spectrum of the synaptic cleft in a brain slice section containing endogenous calcium. C = carbon edge; Ca = calcium edge.

the synaptoplasm back into the extracellular space against the high concentration gradient of about four orders of magnitude.

The ultrastructural localization of a high-affinity

Ca^{2+}-ATPase within the synaptic terminals of fish brain was demonstrated using cerium phosphate as an indicator. Reaction products were found, especially on the inner sides, both of the presynaptic as well as

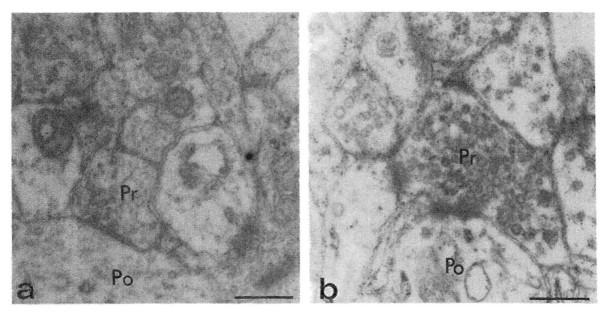

Fig. 2. Increase of calcium within the synaptic cleft of developing neurons from the optic tectum of a cichlid fish. (a) not yet stabilized synapse; (b) stabilized synapse. Bar = 0.5 μm; Pr = presynapse; po = postsynapse.

of the postsynaptic membranes, with special preference for the local zone of synaptic contact (Figs. 3a–d). Up to now, it is still an open question whether the distinct localization of reaction products either on the presynaptic, or postsynaptic membrane or both depends upon the state of neuronal differentiation during ontogenetical development or upon the level of functional excitation. The cerium specificity of the precipitates was demonstrated by means of EELS. The distinct localization of a high-affinity Ca^{2+}-ATPase at the inner side of the synaptic membranes is in full agreement with the high extracellular Ca^{2+}-concentration within the synaptic cleft, indicating that these Ca^{2+}-ATPase deposits reflect the location of the calcium pump being responsible for an active efflux of calcium out of the synaptoplasm back into the extracellular space following depolarization.

Differentiation-dependent distribution of polysialogangliosides on neuronal cell surfaces. On the basis of the distinct localization of calcium and of a high-affinity Ca^{2+}-ATPase (calcium pump) in the very local zone of synaptic contact it was of special inter-

est to investigate the subcellular location of gangliosides, which due to biochemical analyses were reported to be highly accumulated and specifically composed in synaptosomal fractions (Ledeen, 1984). For this purpose the immunocytochemical technique was used to depict ganglioside epitopes by means of the monoclonal antibody Q211 which recognizes polysialogangliosides from the c-pathway (Rösner et al., 1988). Since the antibody-gold complex was found to be too large for penetration into the intercellular spaces of brain tissue, neuronal primary cell cultures of fish brain which were shown to contain high amounts of polysialogangliosides (Seybold and Rahmann, 1985), were chosen for better accessibility. A precise localization of polysialoganglioside epitopes in a clustered organization was found on the surface of perikarya (Fig. 4b), of sprouting neurons (Fig. 4c) and of synaptic contact zones (Fig. 4a). The expression of polysialoganglioside epitopes occurred dependent on the developmental stage of neurogenesis: in early postmitotic cells both the membranes of cell bodies and outgrowing fiber processes were labelled. With further developmental progress, how-

132

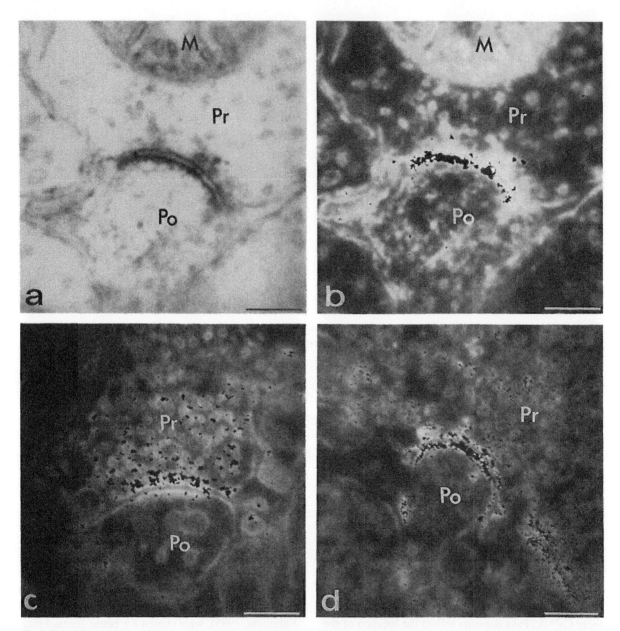

Fig. 3. Transmission electronmicroscopic (a) and electronspectroscopic imaging aspects (b to d) of a synapses in the optic tectum of a cich-lid fish for the demonstration of the localization of a high-affinity Ca^{2+}-ATPase (calcium pump) labelled by cerium phosphate mainly at the inner sides of the postsynaptic (a + b), presynaptic (c) or both (d) membranes. Bar = 0.2 μm. M = mitochondrion; Pr = presynapse; po = postsynapse; V = synaptic vesicles.

ever, ganglioside epitopes were recognized merely on the surface of the outgrowing fibers and especially of synaptic terminals (Fig. 4d). Neuronal primary cell cultures of adult fish did not express polysialogan-glioside epitopes any more. However, after pre-treat-ment of cells with proteinase (trypsin) a re-activation of the epitopes occurred (Fig. 4d), thus indicating that with proceeding neurogenesis formerly surface-locat-

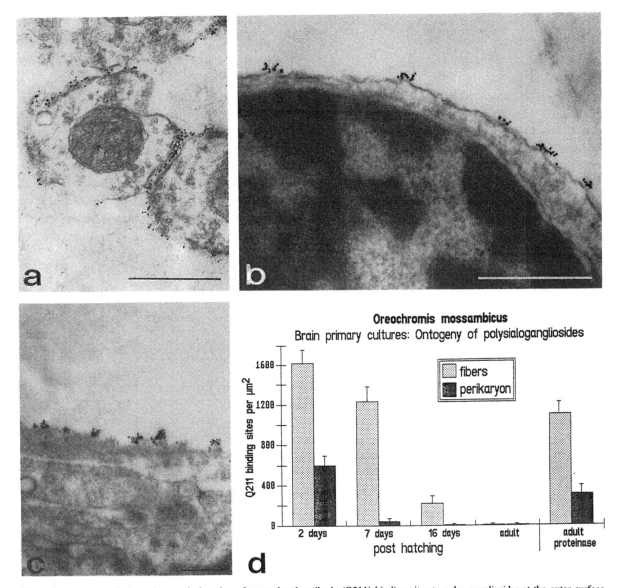

Fig. 4. Transmission electronmicroscopic imaging of monoclonal antibody (Q211) binding sites to polar gangliosides at the outer surface only of cell cultured fish brain neurons using immuno-gold. Enlargement demonstrating the accumulation of binding sites at synaptic terminal (a), neuronal cell body (b) and nerve fiber (c). Bar = 0.5 μm. (d) Development-depending expression of anti-polysialoganglioside epitopes and their re-activation by proteinase-treatment.

ed epitopes of these glycosphingolipids become covered with proteins.

This developmental and distinct enrichment of polysialoganglioside epitopes at synaptic contact zones suggests, together with previous electron microscopical detections of GM1-ganglioside with cholera toxin at the outer surface of synaptic terminals (Hansson et al., 1977), an involvement of

134

gangliosides in the formation and functional maintenance of synaptic contacts.

Effect of exogenous gangliosides on shape and neurite growth of embryonic chicken neurons cultured at low Ca²⁺-concentration

On the basis of the meanwhile well-known correlation between the development-dependent expression of individual ganglioside species and the degree of neuronal differentiation (neurogenesis; Fig. 5, comp. Rösner et al., 1992), it was of special interest to investigate whether or not calcium might be involved in these interactions. Therefore, in a series of exten-

sive in vitro studies the influence of ganglioside application on outgrowing embryonic calcium-deprived chicken brain neurons was investigated.

Under reduced Ca²⁺-conditions (< 20 μM) embryonic optic neurons from chickens (E7) adhered quite well to the polylysine-coated bottoms of plastic dishes and survived for several days (Greis and Rösner, 1990). As shown by means of scanning-electron microscopy and immunostaining with mAb Q211 against polysialogangliosides, the normal shape of the cells, however, was remarkably changed (Fig. 6a): the Ca²⁺-deprived neurons had a very flat shape and developed extended, often many, short processes.

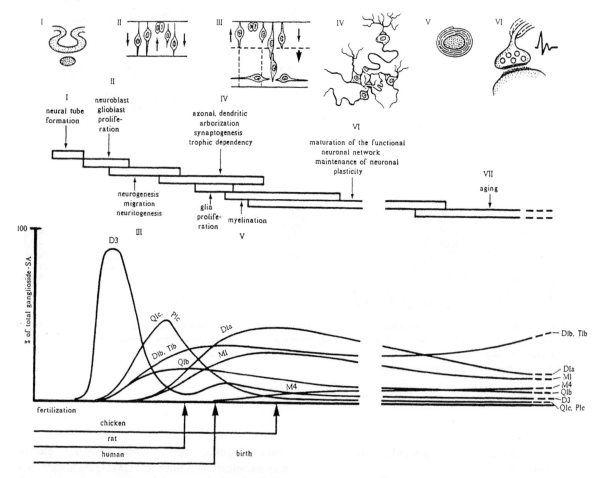

Fig. 5. Schematic presentation of the correlation between the relative developmental and aging profiles of brain gangliosides (relative composition of individual fractions) and the status of neuronal differentiation in higher vertebrates.

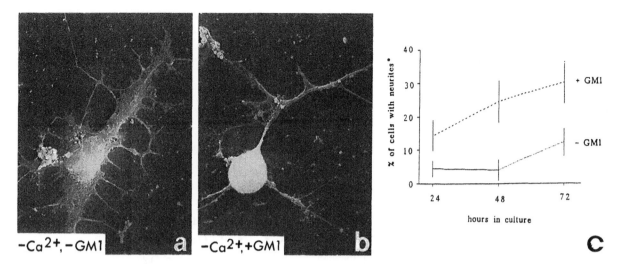

Fig. 6. Primary neurons with neurites of more than twice cell diameter from optic lobes of 7-day incubated chicken embryos cultured for 24 h in MEM with less than 20 μM Ca²⁺ in the absence (a) and presence of 50 μg/ml exogenous ganglioside GM1 (b). Note the development of 'normal' neuronal cell shape in the presence of GM1. Quantitative data (c) from 15 determinations by counting 5 areas of 3 dishes per experiment and culture time.

In the presence of exogenous gangliosides (GM1, GD1a; up to 100 μM), however, the cells developed almost normal shape with rounded cell bodies and thin, often branched, processes and growth cones (Fig. 6b). Furthermore, these cells had a much denser organization of the actin-cytoskeleton, especially within their processes and growth cones, as can be followed after actin-staining with rhodamine-labelled phalloidine (data not shown). Last but not least, the neurite growth was remarkably increased in the GM1-treated cultures: within the first 48 h in low Ca²⁺-culture only 4% of control neurons developed neurites although the cells survived well. In the presence of GM1, however, 23% of cells had at this time already neurites with length of more than twice cell body diameter (Fig. 6c). Control experiments with PC and PS revealed no comparable effects of these phospholipids (data not shown).

In summary these data speak in favour of the assumption that exogenous gangliosides have a protective effect in Ca²⁺-deprived neurons by supporting their regulation of Ca²⁺-homeostasis, possibly by modulation of membrane Ca²⁺-ATPases. They obviously improve the organization of the actin cytoskeleton and most likely by this the neuronal cell shape

and growth. This interpretation of the presented data is supported by increasing experimental evidence suggesting a modulatory role of gangliosides in the phenomenon of Ca²⁺ cycling across the plasma membrane (for review see Ledeen and Wu, 1992).

Effect of Ca²⁺ on surface properties of gangliosides in pure and mixed monolayers
Pure lipid monolayer studies. The aim of this set of experiments was to determine the influence of calcium on the molecular space requirement or packing behavior and the surface potential of the water/lipid interface for gangliosides and different lipids, like sulfatide, ceramide and PC (Fig. 7) that are common compounds in biological membranes. For the better understanding of the possible role of gangliosides in the formation and function of biological membranes it is important to know how the addition of Ca²⁺ might modify the surface pressure and surface potential behavior with respect to other lipids.

The surface pressure and surface potential isotherms on 5 mM TEA-HCl (pH 7.4) subphase, for a phospholipid (DOPC), sulfatide, ceramide, and 3 gangliosides of increasing complexity are shown in Figs 8a–f.

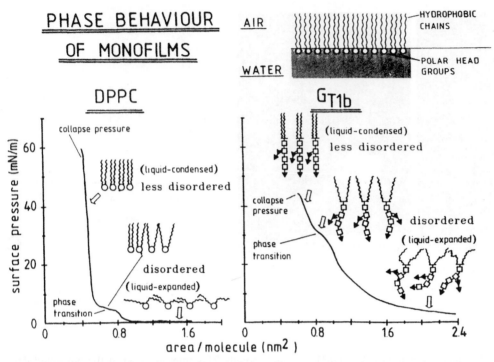

Fig. 7. Phase behavior of phosphatidylcholine (DPPC) and ganglioside (GT1b) in a monolayer film at the air–water interface. The molecular space requirement of the molecules decreases with increasing surface pressure including phase transition from expanded to condensed liquid state and collapse of film. Pressure conditions in biological membranes are in the range between 20 and 30 mN/m.

The surface pressure/area curve of DOPC is of the form typically shown by lipids in the liquid-expanded state. Unlike the glycosphingolipids, the molecular packing of DOPC is determined by the two unsaturated fatty acid chains in the hydrophobic portion and a relatively small and zwitterionic polar head group. During monolayer compression, both surface pressure and surface potential of DOPC-films increase for areas smaller than 1.14 nm²/molecule. This can be correlated with the beginning of the removal ('lift off') of the hydrocarbon chains from the water surface. The collapse of the film occurs at a limiting area less than 0.6 nm²/molecule at about 45 mN/m. The maximum in surface potential (ΔV = +440 mV) is obtained at the surface pressure of film collapse.

During compression of ganglioside monolayers, the surface pressure/area isotherms exhibit a shoulder, which can be interpreted as a transition between one disordered state and another possibly less disordered state (Probst et al., 1984; Beitinger et al., 1989b). The pressure of additive charges or uncharged sugar residues in the larger polar head group of the different glycosphingolipids determines and modifies the molecular conformation, molecular packing, and electric potential at the air/water interface. The mean molecular areas are significantly increased due to the electrostatic repulsions by one or more negative charges.

The surface potential (ΔV) of all gangliosides investigated give rise to negative values with only slight influences of the increasing number of sialic acids. With decreasing surface area of the ganglioside monolayers, the surface potential slightly increases before the transition shoulder is reached. Thereafter, the surface potential shows a decrease up to the collapse of films. Surface potential-values (ΔV) of gan-

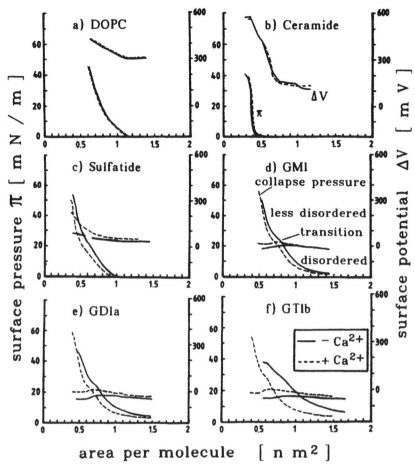

Fig. 8. Different influences of Ca²⁺ (0.01 mM) on the molecular area (surface pressure π) and surface potential (ΔV) isotherms of monolayers derived from different gangliosides (GM1, GD1a, GT1b), phosphatidylcholine (DOPC), ceramide and sulfatide on TEA-HCl subphase; buffer pH 7.4; $20 \pm 0.5^{\circ}$C; accuracy 10 mV.

glioside monolayers are quite similar. This indicates that variations in the molecular structure of the different gangliosides, like the number of negative charges per ganglioside which leads to appreciable changes in the molecular packing, do not cause larger changes in the surface potential.

The addition of 0.01 mM Ca²⁺ to the subphase has no effect on DOPC films, but a pronounced effect on the properties of the glycosphingolipids, especially of the ganglioside monolayers (Fig. 8, interrupted lines). These data confirm that bivalent cations, especially Ca²⁺, do not exhibit a significant interaction with neutral phospholipids but substantial interactions with negatively charged glycosphingolipids. The major features of the addition of Ca²⁺ to the ganglioside monolayers are:

1. condensation of the monolayers as indicated by a decrease of the surface pressure at definite molecular areas;
2. higher film stability as indicated by a significant rise in collapse pressure;
3. shift of the phase transition shoulder to lower surface pressure values; and
4. slight increase of surface potential in direction to more positive values.

The data obtained so far give evidence for strong

molecular rearrangements, especially of gangliosides caused by calcium, which were quite different from those of phospholipids.

Since it has been previously shown, that the dipole potential contribution of the terminal CH_3-groups in closely packed lipid monolayers (surface pressure π > 20 mN/m) with a value of +0.35 D is constant (Vogel and Möbius, 1988), it was of interest to investigate the influence of calcium (0.01 and 1 mM) on the effective total dipole moment ($\Delta \perp \mu$ (D)) and the effective local dipole moment ($\Delta \perp \mu_\alpha(D)$) and the effective local surface potential ($\Delta V_\alpha(mV)$) of the hydrated polar head groups of close-packed ($\pi = 30$ mN/m) monolayers of the investigated lipids at the water/monolayer interface. Figure 9 shows that, in addition to the data discussed already with regard to the mean molecular area (a) and surface potential (b), the effective local surface potential (ΔV_α) of the polar head group region in case of the gangliosides remained almost uninfluenced when calcium was added. The effective total dipole moment ($\Delta \perp \mu_\alpha$) and the effective local dipole moment ($\Delta \perp \mu_\alpha$) of the polar head group region of close-packed ($\pi = 30$ mN/m) ganglioside and phosphatidylserine monolayers, however, were significantly affected by calcium. With regard to all these molecular parameters investigated, calcium exhibits absolutely no effects on DOPC (lecithin) surface properties.

Mixed lipid-peptide monolayer studies. In a further set of experiments, the effect of the negatively charged monosialoganglioside GM1 on the molecular organization and the physico-chemical properties of lipid/peptide (ion carrier valinomycin) systems, and its influenceability by Ca^{2+} was investigated in monolayers at the air/water interface (Beitinger et al., 1987; Schifferer et al., 1988, 1991). During monolayer compression the different isotherms exhibit a two-dimensional transition from the liquid-expanded to the liquid-condensed state with increasing surface pressures, a decrease in mean molecular area, and a shift of collapse pressure to higher values, thus indicating an increasing stabilization effect of the peptide in the ganglioside film. The mean surface potential of the complex was increased. The addition of 0.01 mM

Ca^{2+} to the subphase induced phase separation and/or aggregate formation between the ganglioside and the peptide (Fig. 10). This demixing effect is probably due to a formation of small immiscible patches or complete phase separation.

These data strongly support the assumption, that the addition or removal of Ca^{2+} can immediately lead to local membrane variations. In biological membranes, where gangliosides were found to be arranged in an ordered or clustered form (cf. Fig. 4) this could have an effect on the interaction between gangliosides and functional proteins through the formation of more or less fluid ranges and/or the activation or inhibition of membrane proteins (cf. Fig. 13).

Effect of Ca^{2+} on single channel conductance and mean channel life-time of gramicidin A in mixed DOPC/ganglioside bilayers. Based on the results from the monolayer experiments it was of particular interest to investigate the effect of gangliosides in membraneous bilayer systems with defined charged lipid composition on the single channel characteristics of the monovalent ion channel gramicidin A and its influenceability by Ca^{2+} (Rahmann et al., 1992).

At first the single channel characteristics of gramicidin A were determined in pure DOPC membranes as well as in mixed membranes of negatively charged gangliosides (or DOPS) and neutral lecithin. When analyzing several hundred single channel events with respect to their current increment the corresponding histograms (Figs. 11a, b) reveal a significant increase of conductance (Λ) and of mean life time (opentime τ) of gramicidin A channels with increasing mole fraction of the negatively charged lipids and increasing polarity and complexity of the lipids.

In final experiments the addition of Ca^{2+} (0.01 mM) to the bath solution of the black membrane chamber containing 150 mM CsCl, 5 mM TEA/HCl buffer (pH 7.4) was investigated, revealing in the case of DOPC/GMix bilayers (8:2), only minor reductions of the efficiency of the single channel conductance (Fig. 11c). However, under these conditions the proportion of 'low-conductance' channel populations ('minis') had been significantly affected by the calcium application.

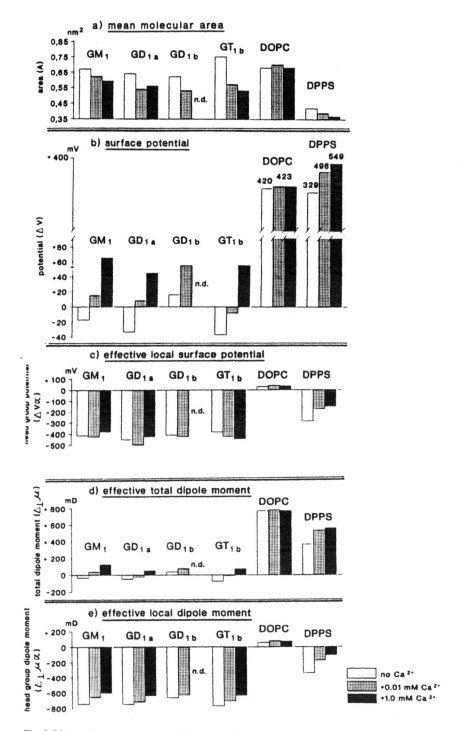

Fig. 9. Mean molecular area Λ (a), surface potential ΔV (b), effective local surface potential ΔV_α of the polar head group (c), effective total dipole moment $\Delta \perp \mu$ (D) and effective local dipole moment $\Delta \perp \mu_\alpha$ (e) of the polar head group region of close-packed ($\pi = 30$ mN/m) ganglioside (GM1, GD1a, GD1b, GT1b) and phospholipid monolayers. (The polar headgroup dipole moments were calculated by substracting the dipole moment of two closely packed hydrocarbon chains [$\mu^{CH}_3 = +0.35$ D; 36] from the molecular dipole moment ($\Delta \perp \mu$).

140

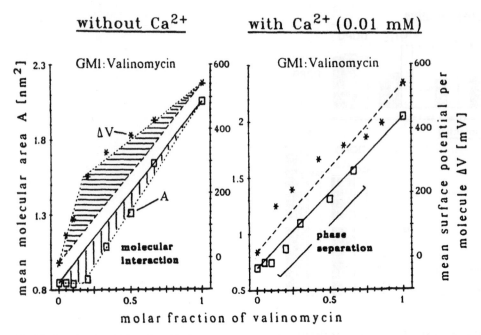

Fig. 10. Ganglioside (GM1) and peptide (ion carrier valinomycin) monolayers at the air–water interface in the presence or absence of Ca^{2+} (0.01 mM). Mean molecular area (A) and surface potential (ΔV) of mixed monolayers at surface pressure of 20 mN/m. Line of ideality represents values calculated according to the additivity rule (Gaines); i.e. the components are immiscible (phase separation or aggregate formation). Deviations from this line indicate miscibility of both components.

Summing up

Our ultrastructural data concerning the localization of calcium, of a high-affinity calcium ATPase (calcium pump) and of poly-sialoganglioside epitopes in close vicinity of the very local zone of contact within synaptic terminals are in good agreement with our physico-chemical results and recent biochemical data dealing with activatory effects of neuronal protein phosphorylation and ATPases (Rahmann et al., 1992; for review) by exogenous gangliosides. Taken together, these findings support the hypothesis of a modulatory function of gangliosides in synaptic transmission, which had been advanced by us about 15 years ago (Rahmann et al., 1976; Rahmann, 1987a,b, 1992).

According to our model Ca^{2+}-ganglioside-complexes are assumed to be localized in a clustered, annulus-like arrangement in the outer leaflet both of

the pre- and postsynaptic membranes around functional proteins (e.g. ion carriers, ion channels, receptors; Fig. 12). In case of an ion channel, for example, during the resting state of a synaptic terminal, phase separation between the Ca^{2+}-ganglioside-complex and the adjacent protein exists (detail Fig. 12a). The membrane in this local zone is kept rigid or 'tightened' for substance penetration. For this assumption evidence for a calcium-induced phase separation between gangliosides and peptides was obtained from our physico-chemical studies (cf. Fig. 10; Beitinger et al., 1987; Schifferer et al., 1988, 1991; Rahmann et al., 1992).

Following first functional activation of the synapse (= transmission) local changes in ion concentration and in electrical field strength at the outer surface of the membrane cause an opening of the ion channel (Fig. 12b) and a release of Ca^{2+} from the negative binding sites at the gangliosides (Fig. 12c). As a

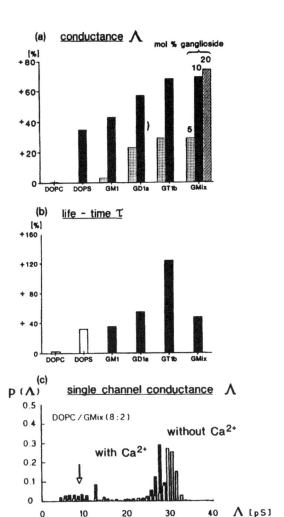

Fig. 11. Increase of conductance Λ (a; in %) and of mean life-time (open time) τ* (b; in %) of gramicidin A in different mixed membranes related to pure phosphatidylcholin (DOPC) membranes. (c) Effect of 0.5 mM Ca²⁺ on single channel conductance of gramicidin A in DOPC- and mixed DOPC/GMix-membranes (8:2) indicating a proportional increase of low-conductance channel populations ('minis').

result of this, conformational changes in the calcium-ganglioside complexes will occur. Evidence for this was obtained by means of the polarography and liposome experiments with gangliosides-Ca²⁺-interactions where it had been shown that an addition of monovalent cations or neurotransmitters (ACh) to the complex induced a release of the Ca²⁺ from its former

ganglioside binding sites (Probst et al., 1979; Wörner et al., 1988). The ion-concentration- or voltage-dependent influx of calcium into the synapse fulfills the task of a primary messenger for the internalization of the electrically encoded information:

By release of calcium from the gangliosides, the very local zone of synaptic membrane now becomes more fluid, thus causing an alteration of the membrane's permeability (Fig. 12c). The intrasynaptic calcium induces second messenger-cascades, thus triggering the transmitter release, and, in turn the activation of the postsynaptic membrane.

During the repolarization phase the resting potential is restored by the re-organization of the ion-balance by means of activation of ion pumps, enzymatic degradation and/or re-uptake of parts of the transmitter. Calcium is pumped out of the neuroplasm by means of ganglioside-modulated, membrane-bound Ca²⁺-ATPase. Calcium then re-binds to the ganglioside, which induces phase separation between the gangliosides and the ion-channel proteins (Fig. 12a), thus causing a re-tightening of the membrane for a new transmission cycle.

Last but not least, the different equipment of neuronal membranes with differently composed ganglioside 'cocktails' among the warm- versus the cold-blooded vertebrates (extreme: antarctic icefish) obviously is correlated with large differences in the concentration of extracellular calcium in the CNS (Rahmann, 1987a,b, 1992; Veh and Sander, 1981). This speaks in favour of the strategy of nature to keep the very local zone of synaptic contacts within the nervous system in homeoviscosity for an optimal transmission process although the physical environment (e.g. temperature) of the membrane has changed drastically (Fig. 12; Rahmann, 1987a, 1992; Rahmann et al., 1992).

On this background the ganglioside molecules might function as neuromodulators controlling the efficiency of the channel-protein to ensure the depolarization of the membrane. (A comparable mechanism had been detected already by Dyer and Benjamins (1990) with galactocerebrosides, compounds closely related with gangliosides, which were found to participate in the opening of Ca²⁺-channels

142

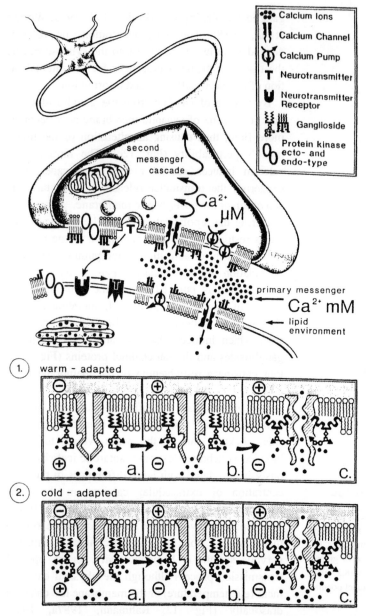

Fig. 12. Summarizing model of a synaptic terminal demonstrating the localization of calcium, functional membrane-bound proteins (e.g. Ca^{2+}-ATPase) and ganglioside clusters being arranged in an annulus-like way around proteins. Details a–c: Model of the possible role of calcium–ganglioside interactions as modulators for a functional channel protein under warm-adaptation versus cold-adaptation. a to b: opening of ion channel by electrogenic (ionic) alterations induces phase separation between ganglioside and protein; c: interaction between ganglioside and protein following electrogenically induced Ca^{2+}-release modulates the ion influx through the channel (a to c: status of membrane changes from more rigid to more fluid phase).

Involvement of Gangliosides in Molecular Facilitation in Synapses

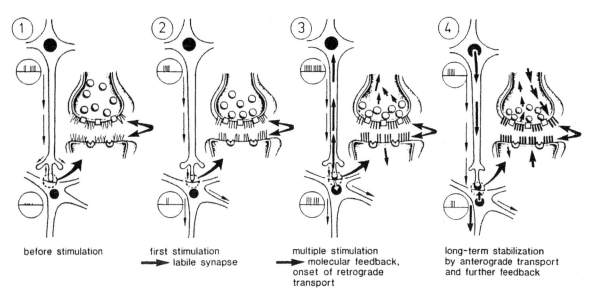

before stimulation	first stimulation	multiple stimulation	long-term stabilization

before stimulation

first stimulation
➤ labile synapse

multiple stimulation
➤ molecular feedback,
onset of retrograde
transport

long-term stabilization
by anterograde transport
and further feedback

Fig. 13. Functional scheme of the model of the involvement of gangliosides in molecular facilitation in synapses during neuronal differentiation.

in oligodendroglial cells.) In the case of long-term adaptive events the 'cocktail', i.e. the composition of synaptic gangliosides, is being organized and maintained by means of the retrograde and anterograde neuronal transport system between the synaptic terminal and the perikaryon (Fig. 13; Rahmann, 1992; Rahmann et al., 1992).

References

Bamberg, E., Noda, K., Gross, E. and Läuger, P. (1976) Single-channel parameters of gramicidin A, B and C. *Biochim. Biophys. Acta.*, 419: 223–228.

Beitinger, H., Probst, W., Möbius, D. and Rahmann, H. (1987) Influence of Ca^{2+} and temperature on the interaction of gangliosides with valinomycin in mixed monolayers at the air/water interface. *J. Biochem.*, 102: 963–966.

Beitinger, H., Vogel, V., Möbius, D. and Rahmann, H. (1989b) Surface potentials and electric dipole moments of ganglioside and phospholipid monolayers: contribution of the polar head-group at the water/lipid interface. *Biochim. Biophys. Acta.*, 984: 293–300.

Dyer, C.A. and Benjamins, J.A. (1990) Glycolipids and transmembrane signaling: Antibodies to galactocerebroside cause an

influce of calcium in oligodendrocytes. *J. Cell. Biol.*, 111: 625–633.

Fidelio, G.D., Ariga, T. and Maggio, B. (1991) Molecular parameters of gangliosides in monolayers: Comparative evaluation of suitable purification procedures. *J. Biochem.*, 110: 12–16.

Greis, C. and Rösner, H. (1990) C-pathway polysialogangliosides in the nervous tissue of vertebrates, reacting with the monoclonal antibody Q211. *Brain Res.*, 517: 105–110.

Hakomori, S. (1987) Ganglioside-mediated modulation of growth factor receptor function and cell adhesion. In: H. Rahmann (Ed.), *Gangliosides and Modualtion of Neuronal Functions, NATO ASI Series H7*, Springer, Berlin, Heidelberg, New York, pp. 465–479.

Hansson, H.A., Holmgren, J. and Svennerholm, L. (1977) Ultrastructural localization of cell membrane GM1 ganglioside by cholera toxin. *Proc. Natl. Acad. Sci. U.S.A.*, 74: 3782–3786.

Hilbig, R. and Rahmann, H. (1987) Phylogeny of vertebrate brain gangliosides. In: H. Rahmann (Ed.), *Gangliosides and Modulation of Neuronal Functions, NATO ASI Series H7*, Springer Verlag, Berlin, pp. 333–350.

Kaczmarek, L.K. (1988) The role of protein kinase C in the regulation of ion channels and neurotransmitter release. *TINS*, 10: 30–37.

Koerner, T.A.W., Prestgard, J.H. and Yu, R.K. (1987) Oligosaccharide structure by two dimensional proton nuclear magnetic resonancy spectroscopy. In: E. Ginsburg (Ed.),

Methods in Enzymology, Complex Carbohydrates, Vol. 138, Part E, pp. 38–59.

Körtje, K.H., Freihöfer, D. and Rahmann, H. (1990a) Cytochemical localization of calcium in the central system of vertebrates. *Ultramicroscopy*, 32: 12–17.

Körtje, K.H., Freihöfer, D. and Rahmann, H. (1990b) Cytochemical localization of high-affinity Ca^{2+}-ATPase activity in synaptic terminals. *J. Histochem. Cytochem.*, 38: 895–900.

Ledeen, R.W. (1984) Biology of gangliosides: neuritogenic and neuronotrophic properties. *J. Neurosci. Res.*, 12: 147–159.

Ledeen, R.W. and Wu, G. (1992) Ganglioside function in the neuron. *Trends Glycosci. Glycotechn.*, 4: 174–187.

Maggio, B., Ariga, T. and Yu, R.K. (1990) Ganglioside GD3 lactones: Polar head group mediated control of the intermolecular organization. *Biochemistry*, 29: 8729–8734.

McDaniel, R. and McLaughlin, S. (1984) The interaction of calcium with gangliosides in bilayer membranes. *Biochim. Biophys. Acta*, 819: 153–160.

McLaughlin, S. (1989) The electrostatic properties of membranes. *A. Rev. Biophys. Chem.*, 18: 113–136.

Montich, G.G., Cosa, J.J. and Maggio, B. (1988) Interaction of l-anilinonaphthalene 8-sulfonic acid with interfaces conating cerebrosides, sulfatides and gangliosides. *Chem. Phys. Lipids.*, 49: 111–117.

Nakamura, M. and Saito, M. (1990) Molecular mechanism of cell differentiation by ganglioside. *Trends Glycosci. Glycotechn.*, 2: 33–39.

Probst, W., Möbius, D. and Rahmann, H. (1984) Modulatory effects of different temperature and Ca^{2+} concentrations on ganglioside and phospholipids in monolayers at air/water interfaces and their possible functional role. *Cell Mol. Neurobiol.*, 4: 157–174.

Probst, W., Rösner, H., Wiegandt, H. and Rahmann, H. (1979) Das Komplexationsvermögen von Gangliosiden für Ca^{2+}. I. Einfluß mono- und divalenter Kationen sowie von Acetylcholin. *Hoppe-Seylers Z. Physiol. Chem.*, 360: 979–986.

Rahmann, H. (1983) Critique: Functional implication of gangliosides in synaptic transmission. *Neurochem. Int.*, 5: 539–547.

Rahmann, H. (1987a) Brain Gangliosides: neuro-modulators for environmental adaptation. In: P. Dejours, L. Bolis, C.R. Taylor and E.R. Weibel (Eds), *Comparative Physiology: Life in Water and on Land*, Liviana Press, Padova, pp. 433–446.

Rahmann, H. (1987b) Brain gangliosides, bio-electrical activity and post-stimulation effects. In: H. Rahmann (Ed.), *Gangliosides and Modulation of Neuronal Functions, NATO ASI Series H7*, Springer, Berlin, Heidelberg, New York, pp. 501–522.

Rahmann, H. (1992) Calcium-ganglioside interactions and modulation of neuronal functions. In: N.N. Osborne (Ed.), *Current Aspects of the Neurosciences, Vol. 4.*, Macmillan Press, London, pp. 87–125.

Rahmann, H., Rösner, H. and Breer, H. (1976) A functional model of sialoglyco-macromolecules in synaptic transmission and memory formation. *J. Theor. Biol.*, 57: 231–237.

Rahmann, H., Schifferer, F. and Beitinger, H. (1992) Calcium-ganglioside interactions and synaptic plasticity: effect of calcium on specific ganglioside/peptide (valinomycin, gramicidin A)-complexes in mixed mono- and bilayers. *Neurochem. Int.*, 20: 323–338.

Rösner, H., Greis, C. and Henke-Fahle, S. (1988) Developmental expression in embryonic rat and chicken brain of a polysialoganglioside-antigen reacting with the monoclonal antibody Q211. *Dev. Brain Res.*, 42: 161–171.

Rösner, H., Al-Aqtum, M., Sonnentag, U., Wurster, A. and Rahmann, H. (1992) Cell surface distribution of endogenous and effects of exogenous gangliosides on neuronal survival, cell shape and growth in vitro. *Neurochem. Int.*, 20: 409–419.

Schifferer, F., Beitinger, H., Rahmann, H. and Möbius, D. (1988) Effect of calcium and temperature on the interaction of gangliosides with valinomycin monolayers: A comparison of glycosphingolipids (ganglioside GT1b, sulphatides) and phosphatidylcholine. *FEBS Lett.*, 233: 963–966.

Schifferer, F., Cordroch, W., Beitinger, H. Möbius, D. and Rahmann, H. (1991) Evidence for the presence of a specific ganglioside GM1/valinomycin complex in mixed monolayers. *J. Biochem.*, 109: 622–626.

Seybold, V. and Rahmann, H. (1985) Changes in developmental profiles of brain gangliosides during ontogeny of a teleost fish (Sarotherodon mossambicus, Cichlidae). *Roux's Arch. Dev. Biol.*, 194: 166–172.

Seybold, V., Rösner, H., Greis, C., Beck, E. and Rahmann, H. (1989) Possible involvement of polysialoganglioside in nerve sprouting and cell contact formation: an ultracytochemical in vitro study. *J. Neurochem.*, 52: 1958–1961.

Slenzka, K., Appel, R. and Rahmann, H. (1990) Influence of exogenous gangliosides (GM1, GD1a, GMix) on a Ca^{2+}-activated Mg^{2+}-dependent ATPase in cellular and subcellular brain fractions of the Djungarian Dwarf Hamster (Phodopus sungorus). *Neurochem. Int.*, 17: 609–614.

Spiegel, S. and Buckley, N.E. (1990) Progress towards understanding to the role of gangliosides in cell growth regulation. *Trends Glycosci. Glycotechn.*, 2: 100–111.

Taki, T. and Handa, S. (1990) Glycolipids as modulators of cell growth and differentiation. *Trends Glycosci. Glycotechn.*, 2: 182–189.

Thomas, P.D. and Brewer, G.J. (1990) Gangliosides and synaptic transmission. *Biochim. Biophys. Acta.*, 1031: 277–289.

Thompson, T.E. and Brown, R.E. (1988) Biophysiological properties of gangliosides. In: R.W. Ledeen, E.L. Hogan, G. Tettamanti, A.J. Yates and R.K. Yu (Eds), *New Trends in Ganglioside Research. FIDIA Res. Series, Vol. 14*, Springer Verlag, Berlin, pp. 65–78.

Varon, B., Pettmann, B. and Manthorpe, M. (1988) Extrinsic regulation of neuronal maintenance and repair. In: R.W. Ledeen, E.L., Hogan, G. Tettamanti, A.J. Yates and R.K. Yu (Eds), *New Trends in Ganglioside Research, FIDIA Res. Series, Vol. 14*, Springer Verlag, Berlin, pp. 607–623.

Veh, R.W. and Sander, M. (1981) Differentiation between gangliosides and sialyllactose sialidases in human tissues. *Perspect. Inher. Met. Dis.*, 4: 71–109.

Vogel, V. and Möbius, D. (1988) Hydrated polar groups in lipid monolayers: effective local dipole moments and dielectric properties. *Thin Solid Films*, 159: 73–81.

Wörner, M., Greiner, G., Rau, H., Rahmann, H. and Probst, W. (1988) Adsorptions- und Grenzflächenverhalten von Gangliosiden an der Phasengrenze Quecksilber/Elektrolyt. *Ber. Bunsenges. Phys. Chem.*, 92: 582–589.

Yates, A.J., Markowitz, D.L., Doreen, L., Stephens, R.E., Pearl, D.K. and Ronald, L. (1988) Growth inhibition of cultured human glioma cells by beta-interferon is not dependent on changes in ganglioside composition. *J. Neuropathol. Exp. Neurol.*, 47: 119–127.

Receptor and Cell Surface Activities of Gangliosides

SECTION III

Receptor and Cell Surface Activities of Gangliosides

L. Svennerholm, A.K. Asbury, R.A. Reisfeld, K. Sandhoff, K. Suzuki, G. Tettamanti and G. Toffano (Eds.)
Progress in Brain Research, Vol. 101

CHAPTER 11

A GM1-Ganglioside-binding protein in rat brain*

Donald A. Siegel[1] and Kunihiko Suzuki[2]

[1]The Howard Hughes Medical Institute, The Rockefeller University, Box 279, 1230 York Avenue, New York, NY 10021, U.S.A and [2]The Brain and Development Research Center, Departments of Neurology and Psychiatry, University of North Carolina School of Medicine, Chapel Hill, NC 27599, U.S.A

Introduction

Gangliosides are a heterogeneous group of glycosphingolipids defined by the presence of one or more sialic acid residues attached to their carbohydrate chains. They are primarily located on the external surface of the plasmalemma (Gahmberg and Hakomori, 1973), and found in greatest abundance in brain (Ledeen and Yu, 1982). This abundance has led to speculations that gangliosides may play an important role in neuronal function. The addition of exogenous gangliosides to primary and transformed neuronal cells in vitro has been shown to stimulate neurite growth (Roisen et al., 1981; Doherty et al., 1985), while in vivo, intraperitoneal injections of gangliosides have been demonstrated to enhance axonal sprouting in the PNS and to retard degenerative processes in the CNS following lesions (Gorio et al., 1983; Sabel et al., 1984). The accumulation of gangliosides within the lysosomes of neurons in ganglioside storage diseases results in elaboration of extensive new and aberrant neurite growth at the axon hillock region of select cortical cells (Purpura and

Suzuki, 1976; Purpura and Baker, 1977). This observation has led to a hypothesis that gangliosides may act as neuritogenic agents (Purpura, 1979).

The influence of gangliosides on cell function is not, however, confined to the nervous system. One of the most common changes in the biochemistry of transformed cells, which are not contact inhibitable and therefore do not stop growing at confluency, is an alteration in their ganglioside pattern (Hakomori et al., 1972). Incorporation of exogenous gangliosides into such cells has been shown to restore the contact inhibition response, with the result that these cells enter a prolonged resting stage and assume a morphology typical of their non-transformed parental cell line (Laine and Hakomori, 1973; Hakomori, 1981).

The observed changes in the behavior of both neuronal and malignant cells after the incorporation of exogenous gangliosides into their plasma membranes has led to a further speculation that gangliosides may play an important role in cell-cell recognition or adhesion phenomena (Hakomori et al., 1972; Mugnai and Culp, 1987) possibly as receptors or receptor modulators. Gangliosides have been shown to function as receptors for bacterial toxins (Holmgren et al., 1973; Cuatrecasas 1973) and viruses (Markwell et al., 1984), and have been reported to bind such endogenous ligands as interferon (Besançon and Ankel, 1974), interleukin 2 (Parker et al., 1982), and several glycoprotein hormones (Ledley et al., 1976). However, with the possible exception of cholera toxin and Sendai virus, the significance of ganglio-

*This work was presented in part at the 11th meeting of the International Society for Neurochemistry, May 31 to June 5, 1987, La Guaira, Venezuela, and was published in a brief abstract form (Siegel et al., 1987). Reprint requests to Kunihiko Suzuki, M.D., Brain and Development Research Center, CB No. 7250, University of North Carolina, Chapel Hill, NC 27599, U.S.A.

side binding to these bioactive substances remains uncertain (Hakomori, 1981, 1984). Thus, existence of an endogenous ligand capable of binding both specifically and avidly to a particular ganglioside would provide an important link between these observations and their underlying molecular mechanism. This study describes characterization and partial purification of an endogenous protein from rat brain that binds with high affinity to monosialoganglioside, GM1.

Establishing the binding assay

Liver membrane preparation

Liver membranes were prepared from female Sprague–Dawley rats by the method of Bennett and Cuatrecasas (1976) with only minor modifications. Livers were sequentially homogenized with a Brinkmann 10/35 Polytron at a setting of 3.5 for 30 sec and then at a setting of 3.0 for 60 sec. After differential centrifugation and washing, the membranes were resuspended at a concentration of 3.8 mg protein/ml in 50 mM Tris–HCl, (pH 7.5). Protein determination was by the method of Lowry et al. (1951) with bovine serum albumin (BSA) as standard. This liver membrane suspension was stored in aliquots at −70°C. For binding assays, a working suspension of the membranes was prepared by diluting the stock suspension to 0.05 mg protein/ml with the Tris buffer containing 0.2 M NaCl and 0.02% NaN$_3$. The working suspension was stored at −20°C. Both the working and stock suspensions retained their binding activity at these temperatures for at least 2 years.

Quantitation of membrane gangliosides

Liver membrane gangliosides were extracted and partially purified by the method of Suzuki (1965), as modified by Ledeen and Yu (1982), and quantitated by scanning densitometry.

Radiolabelling of GM1-ganglioside

GM1-ganglioside was labelled with [^{3}H]KBH$_4$ by the method of Novak et al. (1979) to the final specific activity of 56.2 μCi/μmol.

Preparation of GM1-Spherosil column matrix

The GM1-Spherosil affinity matrix was prepared by the method of Tayot et al. (1981). A ratio of 5 μmol lyso-GM1/g DEAE-Spherosil was used. The maximum theoretical amount of lyso-GM1 that could be bound per column bed volume was 1.8 μmol GM1/ml.

Preparation of CTB-Sepharose 4B matrix

Cholera toxin B subunit (CTB) (Sigma Chemical Co.) was bound to activated Sepharose 4B according to the procedure recommended by the manufacturer and also described by Lowe (1979). The ratio of the toxin to the gel was 2 mg/ml. The binding capacity of the CTB-coupled Sephadex 4B was estimated to be approximately 87 nmol (134 μg) GM1/ml gel.

Preparation of tritiated cholera toxin B subunit

CTB was labelled with N-[^{3}H]-succinimidyl propionate ([^{3}H]NSP) while bound to the GM1-Spherosil affinity column. All procedures were carried out at 4°C unless otherwise indicated. NSP, an acylating agent, specifically reacts with the amino terminus and lysine residues of proteins if they are accessible (Müller, 1980). Since it has been shown that some lysine groups of CTB may play a role in the binding to GM1-ganglioside (Lönnroth and Holmgren, 1975), the toxoid was labelled after it was bound to a GM1-Spherosil affinity column. The rationale for this step was that any residues required for the binding of the toxoid to GM1-ganglioside would be inaccessible to [^{3}H]NSP thus protected from modification which in turn might alter the binding capacity of the toxoid. One milligram of CTB (30 units/μg protein, Schwartz/Mann) was resuspended in 2 ml of ice-cold distilled water and dialyzed against 100 mM sodium phosphate buffer (pH 7.4) containing 0.2 M NaCl and 1 mM sodium-EDTA (buffer A). The dialyzed sample was then brought to a final volume of 5.0 ml with buffer A, and 4.7 ml loaded onto a 0.5 ml GM1-Spherosil affinity column, which contained 0.9 μmol of lyso-GM1 bound to the Spherosil matrix. Approximately 75% of the applied CTB remained bound to the column after extensive washing. The affinity-bound CTB was labelled with [^{3}H]NSP as follows. Approximately 0.25 ml of the GM1-Spherosil matrix which contained about 0.35 mg (6.5 nmol) of bound CTB was removed from the column and incubated with 23.6 nmol [³H]NSP (106 Ci/mmole, Amersham) for 2 h at room temperature. The matrix was again poured into a column and washed extensively with buffer A until the eluent was essentially free of radioactivity. Finally the column was eluted with 7.5 ml of 0.05–0.1 M citrate-phosphate buffer (pH 2.9), and 0.5-ml fractions were collected. All fractions within the eluted peak of radioactivity were combined and dialyzed exhaustively against buffer A. The retentate was concentrated to 2.0 ml using an Amicon ultrafiltration device with a PM10 membrane (molecular size cut-off = 10,000). Although the labelled ligand did not pass through the PM10 membrane, only 56% of the labelled ligand was recovered from the retentate after filtration. Final protein determination was by the method of Lowry et al. (1951) with BSA as standard. The specific activity of the labelled CTB was 45.2 Ci/mmol, assuming a M_r of 56,000 for the B subunit (Lai, 1980). To determine the purity of the labelled ligand, 2.5 pmol (>100,000 cpm) of the [³H]CTB was loaded on a fresh 0.5 ml GM1-Spherosil affinity column. No radioactive material passed through the column even after extensive washing. Finally, SDS-PAGE and fluorography of heat denatured [³H]CTB showed only a single band running close to the dye front indicating that the B subunit had dissociated to monomers, with no contaminating proteins observed. [³H]CTB was stored at 4°C.

SDS-polyacrylamide gel electrophoresis (SDS-PAGE)

SDS-PAGE was performed according to the method of Laemmli (1970) using a 5% stacking gel and a 10% resolving gel.

Binding protein preparation

GM1 binding protein was prepared from 18-day-old Sprague–Dawley rat brains homogenized in 10 vols of ice-cold 50 mM Tris–HCl (pH 7.5) containing 0.2 M NaCl and 0.02% NaN₃ (homogenizing buffer)

and centrifuged at 100,000 g for 60 min. The supernate was carefully transferred to another tube without disturbing the pellet. The binding protein was further fractionated by ammonium sulfate precipitation. Pellets were resuspended in homogenizing buffer when used for direct assays, or in the Tris buffer alone for further purification with DEAE-Sephadex. In either case, the preparation was centrifuged once more at 100,000 g for 60 min to remove any debris.

Standard binding assay

In the standard binding assay, 0.5 μg of liver membrane protein, containing approximately 160 pg GM1, was incubated with 10 ng (0.5 K_D) of [³H]CTB in a final volume of 0.2 ml of the assay buffer (50 mM Tris–HCl (pH 7.5), 0.2 M NaCl, 5 mM Na₂-EDTA, 0.05% BSA and 0.02% NaN₃) for 60 min at 22°C. Nonspecific binding was determined by preincubating liver membranes with 10 μg (approximately 500 K_D) of unlabelled CTB for 30 min followed by a 60-min incubation with the tritiated ligand. GM1 binding protein was measured by its ability to compete against [³H]CTB, and assayed in a manner similar to that of unlabelled CTB. The GM1 binding protein sample was preincubated with the liver membranes for 30 min and then incubated with [³H]CTB for 60 min both at 22°C. In addition, [³H]CTB binding to the binding protein sample itself was controlled for all samples. Here, the binding protein was incubated with [³H]CTB for 60 min in the absence of liver membranes and, as in the standard assay, nonspecific binding was determined by preincubating the binding protein sample with 500 times K_D of unlabelled CTB for 30 min followed by a 60-min incubation with [³H]CTB. Following incubation, 150 μl of the assay mixture were filtered through Millipore Durapore Millititer filters (GVWT) on a Millipore Millititer filtration manifold. Each filter was then washed four times with 0.25 ml of the ice-cold assay buffer. Total filtration time was approximately 20 sec for the 1 ml wash. The filters were punched out and digested with 0.5 ml of Protosol at 50°C for 30 min, neutralized with 20 μl of concentrated acetic acid, mixed with 12 ml of Econofluor and counted.

Results

Characterization of the assay system

When the amount of the liver membrane was varied in the standard assay system with a constant amount of [³H]CTB, a linear relationship was obtained between the amount of ligand bound and the amount of membranes present. As the amount of the membrane increased nearly 100-fold, from 0.13 to 12.5 μg protein/ml, the percentage of toxin bound increased from less than 1% to over 30%. Reasonable linearity of the curve was maintained up to 12.5 μg protein/ml. In contrast, when the amount of liver membranes was held constant and the concentration of the [³H]CTB was varied, the binding was linear up to approximately 175 ng free [³H]CTB/ml, followed by a plateau of the curve at about 375 ng free [³H]CTB/ml, thus indicating a saturable binding process. From the results of these experiments, appropriate amounts of the radioligand and the membrane fragments (GM1) were selected so that, in the standard binding assay, the amount of the liver membrane fraction was well in the linear range of ligand binding, the amount of ligand bound at equilibrium stayed below 10% of total ligand present, and the total ligand concentration was kept below the dissociation constant of the binding system.

Non-specific binding was determined by pre-incubating the membrane fragments with approximately 500 K_D of unlabelled CTB for 30 min prior to the addition of [³H]CTB. At this concentration of unlabelled ligand, virtually no binding of the radiolabelled material was observed. Later experiments showed that preincubation with excess unlabelled toxoid was not necessary for reliable binding assays. When labelled and unlabelled ligand were mixed simultaneously with liver membranes, the results were as predicted from the calculated lower specific activity, supporting the assumption that the radiolabelling did not alter the binding capacity of CTB. However, once [³H]CTB was allowed to reach an equilibrium in binding GM1-ganglioside in the liver membrane fragment, only 50% of bound [³H]CTB could be displaced after 30-min incubation with 1000-fold excess of the unlabelled CTB.

Kinetic studies

Scatchard analysis of the data generated from the [³H]CTB saturation curve showed a non-linear, concave down curve indicative of positive cooperativity in the binding of CTB to its GMl ganglioside receptor (Fig. 1). The free ligand concentration was calculated as the difference between the total ligand present in the assay system and the amount of ligand bound after equilibrium was achieved, (i.e. Ligand $_{free}$ = Ligand $_{total}$ - Ligand $_{bound}$). Positive cooperativity was confirmed by the Hill analysis (Fig. 2) which gave a Hill coefficient of 1.5. The Hill binding constant K_D (Bennett and Yamamura, 1985), measured as the antilog of the X-intercept in the Hill plot was calculated to be 1.8 nM. In a similar experiment using 10 times the amount of liver membrane protein (i.e. 2 μg), and increasing the range of free ligand to over 550 ng [³H]CTB, the Scatchard analysis again revealed a non-linear concave down curve. The Hill coefficient determined from these data was 1.8, and the K_D = 2.2 nM.

GM1 ganglioside binding protein

GM1-ganglioside-binding protein was detected by its ability to compete against CTB for binding to GM1-ganglioside. The relative amount of binding protein present in any given preparation was estimated by

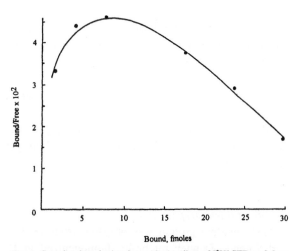

Fig. 1. Scatchard analysis of specific binding of [³H]CTB to 0.2 μg of liver membrane with increasing radioligand concentration. Assay conditions are described in the text.

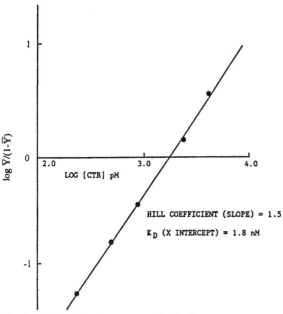

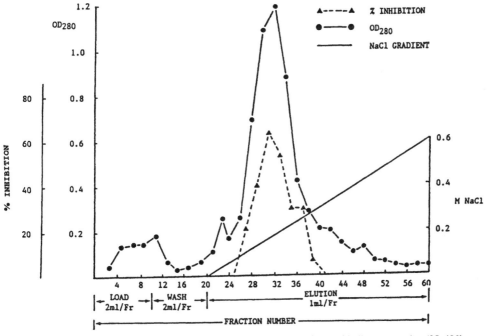

establishing a binding protein concentration curve in which increasing amounts of the binding protein preparation were plotted against the percent of [3H]CTB bound in a standard assay. The IC_{50} from this curve was defined as 1 unit of binding protein activity and thus allowed for direct comparison of different binding protein preparations.

Initial characterization suggested the presence of contaminating GM1-ganglioside in the binding protein preparation. Heating at 55°C for 5 min as well as treatment with trypsin abolished the binding protein's ability to prevent [3H]CTB from binding to liver membrane GM1 ganglioside. However, these treatments increased background binding, suggesting the presence of GM1-ganglioside in the crude binding protein preparation. DEAE Sephadex A-25 anion exchange chromatography of the 25%–40% ammonium sulfate fraction of the crude 100,000 g supernatant was tried to separate the binding protein activi-

Fig. 2. Hill plot of the data presented in Fig. 1.

Fig. 3. DEAE-Sephadex A-25 anion exchange chromatography of 256 units of a binding preparation (25–40% ammonium sulfate fraction of the 100,000 g supernate). Column size: 1.0 × 7.4 cm; bed volume: 5.8 ml; 0.85 g, 3.0 mequiv capacity. Starting conditions: 50 mM Tris–HCl, pH 7.5, 0.02% NaN3. Gradient elution: continuous linear gradient from 0 to 0.6 M NaCl in the starting buffer in 7 bed-volumes at a rate of 6 ml/h. The GM1-binding activity is expressed in percent reduction of [3H]CTB binding in the standard system in absence of the samples.

154

ty from the endogenous GM1 present (Fig. 3). The column, which had a total capacity of about 3.0 mequiv, was loaded with a preparation containing 256 units of the binding protein and eluted with a continuous linear gradient of NaCl from 0 to 0.6 M. The binding protein activity was eluted as a single peak beginning at about 0.1 M NaCl with maximum activity at approximately 0.17 M. Treatment of eluted samples with trypsin completely abolished the binding activity. However, unlike with the treatment of the crude 100,000 g supernatant with trypsin, no increase was observed in the background binding of [^{3}H]CTB to the trypsin treated sample. This suggested that most, if not all, of the contaminating GM1 was removed. Evidence that free GM1-ganglioside can be removed from the solution by anion exchange chromatography was provided by a second DEAE column. In this experiment 2.0 ml of 0.5 mM [^{3}H]GM1-ganglioside (1.6 mg, 0.3 μCi/mM) was loaded onto a 0.074 g DEAE A-50 column (bed volume 0.6 ml) with a maximum theoretical capacity of 0.084 mequiv. The capacity of this column was approximately 1/36 of that of the DEAE column used for the binding protein sample as described above. In addition, the starting conditions of this column included 0.2 M NaCl. However, less than 1% of the total applied radioactivity passed through the column. This experiment demonstrated that, at least in the pure form, GM1-ganglioside is effectively removed from an aqueous solution by DEAE Sephadex chromatography and remains bound to the DEAE matrix at NaCl concentrations as high as 0.2 M. Finally, to determine if the anion exchange chromatography could remove GM1 from binding protein samples as effectively as it does from solutions of pure GM1-ganglioside, a sample of binding protein was combined with [^{3}H]GM1 and chromatographed on a DEAE Sephadex A-25 column similar to that used in Fig. 3. Approximately 0.5 nmol of [^{3}H]GM1 (13.6 Ci/mmol) was mixed with 20 ml of the 25–50% ammonium sulfate fraction of the crude binding protein (100,000 g supernatant) on ice for 2 h. The sample contained 435 units of binding protein, nearly twice as much as the 25–40% ammonium sulfate sample applied in Fig. 3. As shown in Fig. 4, little

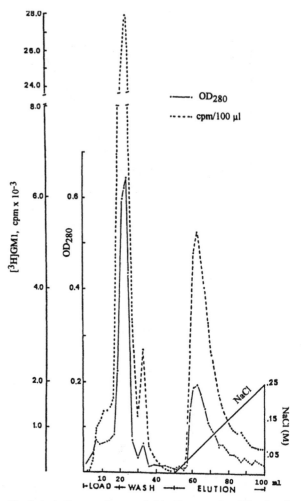

Fig. 4. A similar experiment as in Fig. 3, except that 435 units of the GM1-binding protein preparation in 20 ml was pre-mixed with 0.5 nmol of [^{3}H]GM1(13.5 Ci/mmol). Refer to the text for experimental details. The radioactivity eluted before the introduction of the NaCl gradient represents overloading of the column. A radioactivity peak is eluted coincident with the binding protein. Without the binding protein, no [^{3}H]GM1 bound to DEAE would have been eluted by the NaCl gradient.

radioactivity (<1%) passed through the column on loading until the column was overloaded at fraction 18, after which approximately 40% of the loaded protein but only 28% of loaded [^{3}H]GM1 passed through the column. Following extensive washing (10 bed volumes of the dialysis buffer) the column was eluted with a 0–0.25 M NaCl gradient in 10 bed volumes. Protein and radioactivity coeluted from the column as a single peak beginning at about 50 mM NaCl. Over the entire elution volume, approximately 53% of the remaining protein and 19% of remaining labelled GM1 were released. The elution of significant amounts of GM1 from the column at relatively low salt concentrations combined with the similarity in the elution profiles of both GM1 and protein suggest that the GM1 which eluted from the column may have been bound to the binding protein which consequently prevented the direct interaction of the ganglioside with the DEAE matrix. This conclusion is supported by the two previously described DEAE columns which showed binding protein activity eluting at about 0.1 M NaCl while [³H]GM1 remained bound at 0.2 M NaCl. It should be noted that even after washing the column with 2.5 M NaCl, which effectively removed all remaining protein, 50% of initially bound GM1 remained on the column, further demonstrating the effectiveness of DEAE in binding and removing GM1 from solution even in the presence of the binding protein.

In a further attempt to remove GM1 ganglioside from the binding protein preparation, a second series of experiments were undertaken using a CTB-Sepharose 4B affinity column. The matrix, prepared as described above, had a capacity of about 0.087 μmol (134 μg) GM1/ml column bed. In order to ascertain the specificity of GM1 binding to the CTB column, a deactivated CNBr-Sepharose 4B column was used as a control. Both the control and affinity columns were identical in size and running conditions. Each column was loaded with approximately 24 ng of [³H]GM1 in 2 ml of the assay buffer containing 0.05% BSA. [³H]GM1 passed freely through the control column. In contrast, the appearance of labelled GM1 in the eluate was delayed from the CTB-Sepharose 4B column, and only 2–3% of the total applied sample was recovered in the eluate. These experiments demonstrated the effectiveness with which the CTB affinity matrix removes GM1 ganglioside from solution. However, labelled GM1 was not so easily removed when it is chromatographed through the affinity matrix in the presence of the binding protein. Figure 5 shows the elution pattern of a sample containing approximately 200 units of binding protein and 250 ng of [³H]GM1.

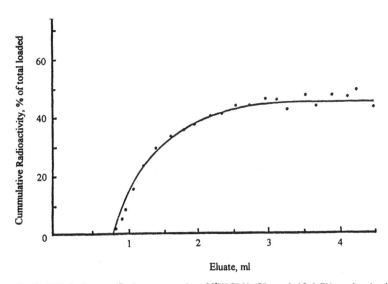

Fig. 5. CTB-Sepharose 4B chromatography of [³H]GM1 (55 nmol, 13.6 Ci/mmol) mixed with approximately 200 units of the GM1-binding protein preparation (30–50% ammonium sulfate fraction of the 100,000 g supernate). Column size: 0.5 × 2.0 cm (bed volume 0.4 ml). The sample was loaded at 5 ml/h, 100-μl fractions collected. Approximately half of the applied radioactive GM1 was eluted from the column after a brief delay. No [³H]GM1 would have been eluted from the column if applied alone without the binding protein.

Approximately 50% of the labelled ganglioside passed through the column even though the amount of labelled GM1 applied was less than 0.5% of the binding capacity of the column. Thus, whereas GM1 alone in solution is nearly completely removed by the CTB affinity column, labelled GM1-ganglioside bound to the binding protein easily passed through the column. Considering the known strong affinity of GM1-ganglioside for cholera toxin, the binding between GM1-ganglioside and the binding proteins is quite strong.

Similar results were obtained from filtration experiments with cellulose ester and fluorocarbon filters which differentially bind the GM1 binding protein. Filtration of the binding protein preparation through 0.22 μm Millipore Millex-GS cellulose ester filters caused a complete loss of binding protein activity in the filtrate, while no binding protein activity is lost when the sample is passed through a 0.22 μm Millex-GV Durapore filter. This differential absorption of the binding protein to different filter materials could be exploited to examine the interactions of GM1 ganglioside with the binding protein in a manner similar to the CTB affinity column. The binding protein and [3H]GM1 were passed through the filters alone and in combination (Table I). Whereas binding protein activity passed freely through the Durapore filter and was completely absorbed by the cellulose ester filter, nearly equal amounts of [3H]GM1 passed through both filters in the absence of binding protein. However, when [3H]GM1 was premixed with the

binding protein, no [3H]GM1 passed through the cellulose ester filter, the filter on which the binding protein was absorbed, but was protected from absorption to the Durapore filter through which the binding protein easily passed. These results complemented and confirmed the results of the CTB affinity column experiments in which GM1 was preferentially absorbed by the CTB affinity matrix. Here it is the binding protein which is preferentially absorbed by the cellulose ester filter material. In both experiments labelled GM1, when pre-mixed with the binding protein, followed the behavior of the binding protein. Both sets of experiments indicated that some GM1 is bound to the binding protein in solution. The CTB affinity column experiment suggested that this interaction is quite strong.

Does contamination of the binding protein preparation by small amounts of endogenous GM1 interfere with the results of the binding assay? This does not appear to be the case. When increasing amounts of GM1 were pre-incubated with the liver membrane in the standard assay system, the amount of labelled CTB which binds to the liver membranes decreased (Fig. 6). However, a 100-fold increase of GM1, from 10^{-10} M to 10^{-8} M, was necessary to reduce binding of [3H]CTB from 87% to 28% of the standard assay. In contrast, only a 3-fold increase in the amount of binding protein was necessary to obtain a similar reduction (from 85% to 35%) (Fig. 7). In a similar experiment in which varying amounts of unlabelled CTB was preincubated with the liver membrane, a

TABLE I

Filtration of [3H]GM1 and the binding protein, alone and combined, through cellulose ester and fluorocarbon membranes

Filter	[H]GM1[a]	Binding Protein[b]	[3H]GM1+Binding Protein[c]
Cellulose ester	32%	0%	8%
Durapore	40%	96%	79%

Results are expressed in percentage of the applied sample that passed through the filter.

[a]0.7 ml of [3H]GM1, 0.17 μCi/ml in the dialysis buffer (approx. 165,000 cpm).

[b]The 20–50% ammonium sulfate fraction of the 100,000 g supernate, 24 units/ml. For the cellulose ester filter, 13 units were filtered, while 24 units were filtered through the fluorocarbon filter.

[c]0.5 ml of (a) pre-mixed with 10.2 units of (b), and then filtered.

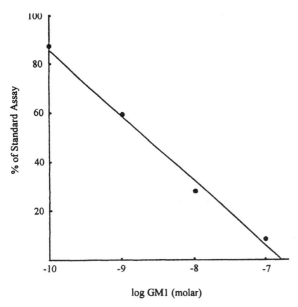

Fig. 6. Effect of preincubation of unlabelled GM1 with liver membranes. Varying amounts of GM1 were pre-mixed with the liver membrane in the standard assay system. Very large increases in the amount of GM1 were necessary to reduce subsequent binding of [³H]CTB.

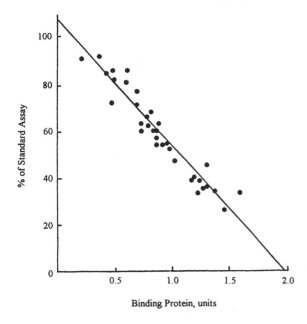

Fig. 7. Equivalent experiment to that shown in Fig. 6, except that the binding protein preparation was preincubated with the liver membrane fraction. Relatively small increments were necessary to effect large reductions in the subsequent binding of [³H]CTB. This is a composite of the data from 10 separate experiments.

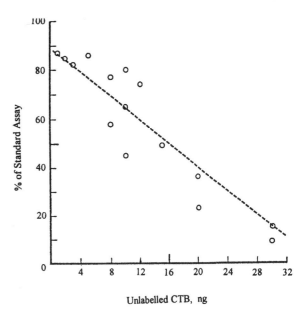

Fig. 8. Equivalent experiment to that shown in Fig. 6, except that unlabelled CTB was preincubated with the liver membrane fraction. Unlabelled CTB is much less effective than the binding protein preparation to reduce the subsequent binding of [³H]CTB. Results in Figs. 7 and 8 suggest that the binding of the binding protein to GM1-ganglioside may be stronger than that between CTB and GM1. Results from 5 separate experiments are combined in this Figure.

greater than 10-fold increase in the amount of unlabelled CTB was necessary to obtain approximately the same 85–35% reduction in binding of labelled CTB (Fig. 8). This amount is nearly 4 times greater than the number of units of binding protein necessary to achieve the same reduction and suggests that the affinity of the endogenous binding protein for GM1 may actually be greater than that of CT.

Tissue distribution study

Seven organs from 60-day-old male Sprague–Dawley rats were examined for their content of the GM1 binding protein. Crude 100,000 g supernatants were prepared from homogenate (v/w = 10) in the homogenizing buffer in a manner similar to the standard binding protein preparation described above. Equal volumes of the 100,000 g supernatants were assayed (Table II). The brain is by far the richest source of the GM1-binding protein. Forty-one µg

TABLE II

Tissue distribution of the GM1-binding protein

Organ	Volume Assayed (µl)	Total protein (µg)	% Reduction from Standard assay	Units/µg protein
Brain	15	41	41	21.7
Heart	75	361	−31	—
Intestine	75	471	67	2.9
Kidney	75	446	−17	—
Liver	75	466	12	0.7
Spleen	75	375	40	2.3
Testis	75	184	−31	—

All samples were the high-speed supernate (100,000 g for 60 min) of 10× homogenates (w/v) of the respective organs from 60-day-old male Sprague–Dawley rats. The results are expressed in % reduction from the cholera toxin binding to the liver membrane preparation in the standard system in the absence of added binding protein preparations. Thus, the greater the number is, the greater the GM1 binding protein activity is. Negative numbers indicate increases in [^{3}H]CTB binding in the presence of the samples. Units of the binding protein were estimated from the standard rat brain binding protein curve.

of the supernate protein reduced the [^{3}H]CTB binding by 41%, and 204 µg virtually abolished all binding. Only the spleen and intestine showed any significant GM1-binding activities. On the basis of units of the binding protein/µg supernate protein, intestine and spleen exhibited approximately 0.14 and 0.11 times, respectively, the level found in the brain. Although the 100,000 g liver supernate did show a very small amount of binding protein activity, it is unlikely to have any effect in the standard assay system which uses 0.5 µg of the liver membrane protein. Even assuming some binding protein activity in the liver membrane preparation, the amount of membrane protein used in the standard assay system is 0.1% of the amount of protein in the high speed supernatant needed to inhibit [^{3}H]CTB binding by 12%.

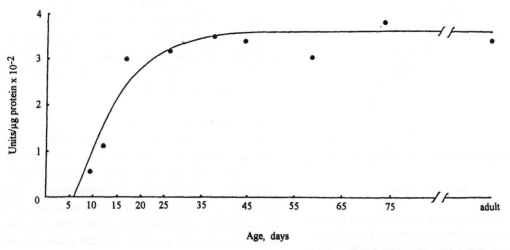

Fig. 9. Developmental profile of rat brain GM1-ganglioside binding protein. Crude preparations (100,000 g supernate of 10 × homogenates) were assayed directly. Unit amounts were calculated from the standard curve obtained from the 18-day-old rat brain preparation.

While the 30% increase in [³H]CTB binding above the standard assay for both heart and testes preparations were reproducible; the reason for this increase is not clear. It cannot, however, be explained by the presence of GM1 ganglioside in the binding protein preparations. There is no detectable GM1 in heart, testes, kidney or intestine and only trace amounts in spleen (Iwamori et al. 1984). Liver and brain contain 0.02 and 1.26 µmol GM1/g dry weight, respectively, approximately 2.5% and 9.4% of the total ganglioside present in each tissue (Iwamori et al., 1984).

Developmental study

A preliminary developmental study was performed to examine the amounts of binding protein activity in rat brain at different ages (Fig. 9). Brains from 2 or 3 littermates of mixed sexes were combined and crude binding protein prepared as described. Units of binding protein activity were determined by comparing the binding protein activity of each sample with the binding protein concentration curve from 18-day-old rats as the standard (Fig. 7). Binding protein activity was not detectable until day 10, rose rapidly and began to reach a plateau by approximately day 28. No binding protein activity could be detected at ages 2 and 6 days even at protein concentrations 2–6 times greater than those necessary for detection of activity at later ages. It should be noted that the developmental changes in the binding protein is roughly coincidental with the developmental changes in the amount of GM1-ganglioside in developing rat brain.

Discussion

Methodological considerations

The present study describes the extraction and characterization of a protein (proteins) from rat brain that binds to GM1 ganglioside with high affinity. The binding assay technique used to detect this protein was carefully developed to ensure proper interpretation of results. Foremost, the cholera toxin, the competitive ligand, must be pure, labelled to a high enough specific radioactivity, and chemically stable. Tritiation of the B subunit of CT while it is bound to a GM1 affinity matrix satisfied these requirements

and represents a technological improvement over the standard iodination procedure of the holotoxin (Cuatrecasas, 1973). While iodination yields high specific radioactivity, labelling occurs almost entirely on the non-binding A subunit (Holmgren and Lönnroth, 1975; Gill, 1976). This subunit has a natural tendency to dissociate from the holotoxin over time (LoSpalluto and Finkelstein, 1972; Finkelstein, 1973), thus resulting in a ligand preparation containing increasing amounts of labelled but non-binding A subunits, and non-labelled but binding B subunits. The consequence is a ligand of constantly changing specific activity. The 50% increase in labelled but non-binding protein observed by Cuatrecasas (1973) over a 3-week period in his sample of ¹²⁵I-labelled CT may have been the result of such dissociation. No deterioration in specific radioactivity was observed for more than 1 year with our labelled [³H]CTB. Furthermore, tritiation of CTB by affinity labelling gives advantages of directly labelling of the binding subunit without modification of the toxoid's binding site, in addition to the reduced radiation hazard and the longer half-life. The parameters of the standard binding assay were established so that liver membrane receptor (GM1) concentration was maintained in the linear range and the ligand concentration kept below the K_D of the binding reaction with the bound ligand not exceeding 10% of the total ligand present. Incubation time, temperature and buffer conditions were also optimized.

Kinetic studies with the affinity labelled toxoid and liver membrane GM1 showed positive cooperativity in the binding of CTB to its receptor GM1 ganglioside. This was demonstrated by a non-linear, concave down Scatchard Plot and Hill analysis with an average Hill coefficient of 1.7. These results are consistent with other studies which showed positive cooperativity in the binding of CT or its B subunit to GM1 or GM1-oligosaccharide in solution (Fishman et al., 1978; Sattler et al., 1978; Schafer and Thakur, 1982). The Hill binding constant K_D determined is also similar to the K_Ds reported by others (Cuatrecasas, 1973; Sattler et al., 1978). The similarities in kinetic properties with those already in the literature further suggest that the labelling of the B subunit with [³H]NSP does not affect its binding characteristics.

Properties of the GM1-binding protein

Through the series of experiments described, we have detected in the rat brain a GM1-ganglioside-binding factor. It is recovered in the 100,000 g supernate as a crude preparation, heat-labile (55°C for 5 min), trypsin-labile, and is anionic at near neutral pH. The apparent molecular size is greater than 100,000. It binds to GM1-ganglioside with high affinity, possibly higher than that of CTB to GM1. CTB binds to the oligosaccharide portion of GM1 with a K_D of approximately 2 nM and $t_{1/2}$ in the presence of 1000-fold excess unlabelled CTB is about 30 min. Thus, it appears to be a high-affinity binding protein. Some important questions remain. They include specificity or lack thereof with respect to different gangliosides, possibility of multiple proteins, and the possibility of its presence also in a membrane-bound form.

Among the rat tissues examined, the brain contained the greatest amount of the binding protein on the units/μg protein basis. Intestine and spleen were the only other organs to contain much lower but still significant activity. It is of interest that ganglioside binding proteins have been reported in each of these tissues. A hormone-like anti-diarrhetic factor that inhibits intestinal hypersecretion induced by cholera toxin has been reported in porcine intestine and pituitary (Lange and Lönnroth, 1984; Lönnroth and Lange, 1984). One possible mechanism proposed for the action of this protein, which is also anionic, is that it binds to GM1-ganglioside and prevents the cholera toxin from binding and, consequently, from initiating the cascade of events which lead to the outpouring of fluids from the intestines. It seems, however, unlikely that this protein is the same as the GM1 binding protein described here. The anti-diarrhetic protein easily passes through a filtration membrane with a molecular exclusion of 50,000 whereas the GM1 binding protein is completely excluded by a membrane filter with an exclusion limit of 100,000 Da. Furthermore, the anti-diarrhetic protein is only observed in rats which have been immunized against cholera toxin. The GM1 binding protein was detected in the intestine of rats which had never been challenged with cholera toxin. Thus, although the anti-diarrhetic protein might inhibit the binding of [³H]CTB in the standard assay system in a manner similar to that of the rat brain binding protein, the two molecules appear to be distinct. Further tests of the intestinal activity found with our assay system will be necessary to determine whether or not it is the same as the binding protein found in the brain.

A ganglioside receptor protein has also been described on the surface of rat macrophages (Boltz-Nitulescu et al., 1984). This protein was assayed for by the ability of macrophages to bind to and form rosettes with sheep erythrocytes which have been coated with various gangliosides. This molecule, however, is also likely to be distinct to the rat brain GM1 binding protein as it showed little activity for GM1 ganglioside. Its greatest affinity was for monosialogangliosides, GM2 and GM3, and polysialogangliosides GD1a, GD1b, and GT1b. Incorporation of GM1 in sheep erythrocytes had no effect on binding. Preliminary experiments carried out with human lymphocytes using our assay system found no GM1 binding protein activity (data not shown).

The properties of the binding protein described in this report suggest that it is distinct from the glycosphingolipid transfer proteins reported from several laboratories (Conzelmann et al., 1982; Wong et al., 1984; Gammon and Ledeen, 1985). The transfer proteins described in the literature generally have much smaller molecular weights, are heat-resistant, and specific for gangliosides other than GM1 (Conzelmann et al., 1982). Tiemeyer et al. (1989, 1990) described a membrane-bound ganglioside binding protein in rat brain and localized it to myelin. Because of the manner we prepared and assayed for our binding protein, it was recovered in the soluble fraction. It is not therefore possible to say whether or not our GM1-binding protein is also present membrane-bound, and, if so, about the relative distribution between the soluble and particulate fractions. On the other hand, their binding protein had relatively low affinity to GM1-ganglioside compared to other gangliosides.

The function of gangliosides have long been a matter of speculation with the greatest interest directed to their possible roles in the nervous system, where they are abundant, and in oncology, where changes in their content on the cell surface occur in oncogenesis (Hakomori, 1975). A very large body of literature now exists describing functions of gangliosides literally in all conceivable manners. They have been shown to bind to a host of natural substances including interferon (Basançon and Ankel, 1974) and interleukin-2 (Parker et al., 1982), although they do not appear to be the specific receptors for any of these molecules (Hakomori, 1981, 1984). Gangliosides stimulate nerve regeneration (Gorio et al., 1983) and their effects on cultured neural cells have been well-established (Roisen et al., 1981). Abnormal meganeurite growth occurs in conjunction with ganglioside accumulation in some lysosomal storage diseases (Purpura and Suzuki, 1976; Purpura and Baker, 1977). Relatively little, however, is known regarding the molecular mechanism underlying such wide and sometimes impressive phenomenology. One obvious possibility is that gangliosides exert their effect through specific receptor/binding proteins. The present study represents an attempt at finding a protein from rat brain which specifically competes against cholera toxin for binding to GM1 ganglioside. This protein appears distinct from the two previously described ganglioside binding proteins found in porcine intestine and pituitary (Lange and Lönnroth, 1984; Lönnroth and Lange, 1984), and on rat macrophages (Boltz-Nitulescu et al., 1984). The membrane-associated ganglioside rat brain binding protein described by Tiemeyer et al. (1989, 1990) may or may not be the same or related to the binding protein described in this report.

References

Bennett, J.P., Jr. and Yamamura, H.I. (1985) Neurotransmitter, hormone, or drug receptor binding methods, In: H.I. Yamamura, S.J. Enna and M.J. Kuhar (Eds), *Neurotransmitter Receptor Binding*, 2nd edn., Raven Press, New York, pp. 61–89.

Bennett, V. and Cuatrecasas, P. (1976) Cholera toxin receptors. In: M. Blecher (Ed.), *Methods in Molecular Biology, Vol. 9, Methods in Receptor Research, Part I*, Marcel Dekker, New York, pp. 73–98.

Besançon, F. and Ankel, H. (1974) Binding of interferon to gangliosides. *Nature*, 252: 478–480.

Boltz-Nitulescu, G., Ortel, B., Riedl, M. and Föster, O. (1984) Ganglioside receptors of rat macrophages. Modulation by enzyme treatment and evidence for its protein nature. *Immunology*, 51: 177–184.

Conzelmann, E., Burg, J., Stephan, G. and Sandhoff, K. (1982) Complexing of glycolipids and their transfer between membranes by the activator protein for degradation of lysosomal ganglioside GM2. *Eur. J. Biochem.*, 123: 455–464.

Cuatrecasas, P. (1973) Gangliosides and membrane receptors for cholera toxin. *Biochemistry*, 12: 3558–3566.

Doherty, P., Dickson, J.G., Flanigan, T.P. and Walsh, F.S. (1985) Ganglioside GM1 does not initiate, but enhances neurite regeneration of nerve growth factor-dependent sensory neurons. *J. Neurochem.*, 44: 1259–1265.

Finkelstein, R.A. (1973) Cholera. *Crit. Rev. Microbiol.*, 2: 553–623.

Fishman, P.H., Moss, J. and Osborne, J.C., Jr. (1978) Interaction of choleragen with the oligosaccharide of ganglioside GM1: Evidence for multiple oligosaccharide binding sites. *Biochemistry*, 17: 711–716.

Gahmberg, C.G. and Hakomori, S.-I. (1973) External labeling of cell surface galactose and galactosamine in glycolipids and glycoproteins of human erythrocytes. *J. Biol. Chem.*, 248: 4311–4317.

Gammon, M. and Ledeen, R.W. (1985) Evidence for the presence of a ganglioside transfer protein in brain. *J. Neurochem.*, 44: 979–982.

Gill, D.M. (1976) The arrangement of subunits in cholera toxin. *Biochemistry*, 15: 1242–1248.

Gorio, A., Martini, P. and Zanoni, R. (1983) Muscle reinervation — III. Motoneuron sprouting capacity, enhanced by exogenous gangliosides. *Neuroscience*, 3: 417–429.

Hakomori, S.-I. (1975) Structures and organization of cell surface glycolipids and glycoproteins, dependency on cell growth and malignant transformation. *Biochim. Biophys. Acta*, 417: 55–89.

Hakomori, S.-I. (1981) Glycosphingolipids in cellular interaction, differentiation, and oncogenesis. *Annu. Rev. Biochem.*, 50: 733–764.

Hakomori, S.-I. (1984) Ganglioside receptors: A brief overview and introductory remarks. *Adv. Exp. Med. Biol.*, 174: 333–339.

Hakomori, S.-I., Kijimoto, S. and Siddiqui, B. (1972) Glycolipids of normal and transformed cells: A difference in structure and dynamic behavior. In: C.F. Fox (Ed.), *Membrane Research*, Academic Press, New York, pp. 253–277.

Holmgren, J., Lönnroth, I. and Svennerholm, L. (1973) Tissue receptor for cholera endotoxin: Postulated structure from studies with GM1 ganglioside and related glycolipids. *Infect. Immun.*, 8: 208–214.

Holmgren, J. and Lönnroth, I. (1975) Oligomeric structure of cholera toxin: Characteristics of the H and L subunits. *J. Gen. Microbiol.*, 86: 49–65.

Iwamori, M., Shimomura, J., Tsuyuhara, S. and Nagai, Y. (1984)

Gangliosides of various rat tissues: Distribution of ganglio-*N*-tetraose-containing gangliosides and tissue-characteristic composition of gangliosides. *J. Biochem.*, 95: 761–770.

Laemmli, U.K. (1970) Cleavage of structural proteins during the assembly of the head of bacteriophage T4. *Nature*, 227: 680–685.

Lai, C-Y. (1980) The chemistry and biology of cholera toxin. *Crit. Rev. Biochem.*, 9: 171–206.

Laine, R.A. and Hakomori, S-I. (1973) Incorporation of exogenous glycosphingolipids in plasma membranes of cultured hamster cells and concurrent change of growth behavior. *Biochem. Biophys. Res. Commun.*, 54: 1039–1045.

Lange, S. and Lönnroth, I. (1984) Passive transfer of protection against cholera toxin in rat intestine. *FEMS Microbiol. Lett.*, 24: 165–168.

Ledeen, R.W. and Yu, R.K. (1982) Gangliosides: structure, isolation and analysis. *Methods Enzymol.*, 83: 139–189.

Ledley, F.D., Mullin, B.R. Lee, G., Aloj, S.M., Fishman, P.H., Hunt, L.T., Dayhoff, M.O. and Kohn, L.D. (1976) Sequence similarity between cholera toxin and glycoprotein hormones: Implications for structure activity relationships and mechanisms of action. *Biochem. Biophys. Res. Commun.*, 69: 852–859.

Lönnroth, I. and Holmgren, J. (1975) Protein reagent modification of cholera toxin: Characterization of effects on antigenic, receptor binding and toxic properties. *J. Gen. Microbiol.*, 91: 263–277.

Lönnroth, I. and Lange, S. (1984) Purification and characterization of a hormone-like factor which inhibits cholera toxin secretion. *FEBS Lett.*, 177: 104–108.

LoSpalluto, J.J. and Finkelstein, R.A. (1972) Chemical and physical properties of cholera exo-enterotoxin (choleragen) and its spontaneously formed toxoid (choleragenoid). *Biochim. Biophys. Acta*, 257: 158–166.

Lowe, C.R. (1979) The chemical technology of affinity chromatography. In: T.S. Work and R.K. Work (Eds), *An Introduction To Affinity Chromatography*, Elsevier/North Holland, pp. 344–400.

Lowry, O.H., Rosebrough, N.J., Farr, A.L. and Randall, R.J. (1951) Protein measurement with the Folin phenol reagent. *J. Biol. Chem.*, 193: 265–275.

Markwell, M.A.K., Fredman, P. and Svennerholm, L. (1984) Receptor ganglioside content of three hosts for Sendai virus MDBK, HeLa, and MDCK cells. *Biochim. Biophys. Acta*, 775: 7–16.

Mugnai, G. and Culp, L.A. (1987) Cooperativity of ganglioside-dependent with protein-dependent substratum adhesion and neurite extenuation of human neuroblastoma cells. *Exp. Cell Res.*, 169: 328–344.

Müller, G.H. (1980) Protein labelling with ^{3}H-NSP (*N*-succinimidyl-[2,3-^{3}H]propionate). *J. Cell Sci.*, 43: 319–328.

Novak, A., Lowden, J.A., Gravel, Y.L. and Wolfe, L., (1979) Preparation of radiolabelled GM2 and GA2 gangliosides. *J. Lipid Res.*, 20: 678–681.

Parker, J., Caldini, G., Kirshnamurti, C., Ahrens, P.B. and Ankel, H. (1982) Binding of interleukin 2 to gangliosides. *FEBS Lett.*, 170: 391–395.

Purpura, D.P. (1979) Pathobiology of cortical neurons in metabolic and unclassified amentias. In R. Katzman (Ed.), *Congenital and Acquired Cognitive Disorders, Vol. 57*, Raven Press, New York, pp. 43–68.

Purpura, D.P. and Baker, H.J. (1977) Neurite induction in mature cortical neurones in feline GM1-ganglioside storage disease. *Nature*, 266: 553–554.

Purpura, D.P. and Suzuki, K. (1976) Distortion of neuronal geometry and formation of aberrant synapses in neuronal storage disease. *Brain Res.*, 116: 1–21.

Roisen, F.J., Bartfeld, H., Nagle, R. and Yorke, G. (1981) Ganglioside stimulation of axonal sprouting in vitro. *Science*, 214: 577–578.

Sabel, B.A., Slavin, M.D. and Stein, D.G. (1984) GM1 ganglioside treatment facilitates behavioral recovery from bilateral brain damage. *Science*, 225: 340–342.

Sattler, J., Schwarzmann, G., Knack, I., Röhm, K-H, and Wiegandt, H. (1978) Studies of ligand binding to cholera toxin. III. Cooperativity of oligosaccharide binding. *Hoppe-Seyler's Z. Physiol. Chem.*, 359: 719–723.

Schafer, D.E. and Thakur, A.K. (1982) Quantitative description of the binding of GM1 oligosaccharide by cholera enterotoxin. *Cell Biophys.*, 4: 25–40.

Siegel, D.A., Budde-Steffen, C. and Suzuki, K., (1987) Characterization and Partial Purification of an Endogenous Rat Brain Protein which Binds to GM1 Ganglioside. *J. Neurochem.*, 48: (suppl.): S040A.

Suzuki, K. (1965) The pattern of mammalian brain gangliosides — III Regional and developmental differences. *J. Neurochem.*, 12: 969–979.

Tayot, J-L, Holmgren, J., Svennerholm, L., Lindblad, M. and Tardy, M. (1981) Receptor-specific large-scale purification of cholera toxin on silica beads derivatized with lysoGM1 ganglioside. *Eur. J. Biochem.*, 113: 249–258.

Tiemeyer, M., Yasuda, Y. and Schnaar, R.L. (1989) Ganglioside-specific binding protein on rat brain membranes. *J. Biol. Chem.*, 264: 1671–1681.

Tiemeyer, M., Swank-Hill, P. and Schnaar, R.L. (1990) A membrane receptor for gangliosides is associated with central nervous system myelin. *J. Biol. Chem.*, 265: 11990–11999.

Wong, M., Brown, R.R., Barenholz, Y. and Thompson, T.E. (1984) Glycolipid transfer protein from bovine brain. *Biochemistry*, 23: 6498–6505.

L. Svennerholm, A.K. Asbury, R.A. Reisfeld, K. Sandhoff, K. Suzuki, G. Tettamanti and G. Toffano (Eds.)
Progress in Brain Research, Vol. 101

CHAPTER 12

Receptors for cholera toxin and *Escherichia coli* heat-labile enterotoxin revisited

Jan Holmgren

Department of Medical Microbiology and Immunology, University of Göteborg, S-413 46 Göteborg, Sweden

Introduction

Cholera, which is the most severe of all diarrhoeal diseases, is caused by *Vibrio cholerae* bacteria of O group 1. These bacteria, irrespective of their biotype (classical or El Tor), can colonize the small intestine of humans and elaborate the protein exotoxin known as cholera toxin. This toxin binds to specific receptors on the mucosal cells and stimulates such excessive secretion of electrolytes and water from the cells that it leads to the severe, often life-threatening diarrhoea, dehydration and metabolic acidosis being characteristic of cholera disease (Carpenter, 1992; Holmgren, 1981).

It is now 20 years since the receptor for cholera toxin was identified as the monosialoganglioside GM1. This finding and the studies carried out in the next several years to characterize, in further detail, this receptor have provided much of the conceptual and methodological basis for later work identifying other specific gangliosides and glycosphingolipids as candidate receptors for numerous other bacterial toxins, bacteria and viruses (Suzuki, K., this volume). More recently, a role has also been proposed for glycolipids as candidate receptors for cell-to-cell contacts between certain mammalian normal and tumour cells (Hakomori, S.-I., this volume; Paulsson, J., ibid) and as receptors for some cell growth factors and signal substances. In addition, the discovery of GM1 as receptor for cholera toxin as well as for the closely related heat-labile enterotoxin (LT) from *Escherichia coli* bacteria (see below) has led to the development of several biotechnological tools used, e.g. for the quantitation and ultrastructural localization of GM1 ganglioside in tissues, sensitive detection of *E. coli* LT and other GM1-binding bacterial enterotoxins and diagnostic laboratory identification of bacteria producing such toxins, and large-scale production of cholera toxin and its B subunit protomer for vaccine manufacturing.

The purpose of this paper is to provide an updated account and evaluation of the properties of GM1 ganglioside as receptor for cholera toxin and *E. coli* LT and to discuss the above mentioned biotechnical applications evolving from this knowledge.

Subunit structure and function of cholera toxin

Cholera toxin and *E. coli* LT are probably the best-defined of all bacterial toxins with regard to their structure and mode of action (Moss and Vaughan, 1980; Holmgren, 1981; Sixma et al., 1991). They are oligomeric proteins, consisting of a binding portion containing five identical B subunits each with a molecular weight of approximately 11,600 (103 amino acid residues), being arranged in a pentameric ring, and a single, toxic-active A subunit attached to this B pentamer ring. The A subunit is a proteolytically 'nicked' polypeptide with a molecular weight of approximately 28,000 comprising two disulphide-linked fragments: the larger fragment, A1, contains the toxic-enzymatic activity whilst the smaller, more hydrophobic fragment, A2, is responsible for the association with the B_5 ring.

Cholera toxin binds to a single class of high affinity receptors being present on the small intestinal brush-border membrane as well as on the plasma membrane of most other mammalian cells (Cuatrecasas, 1973a; Fishman, 1980). It is now well established that these receptors consist of the ganglioside GM1 (Fishman, 1980; Holmgren, 1981). The ceramide moiety of the GM1 molecule is embedded in the lipid matrix of the cell membrane, and the hydrophilic, sialic acid-containing oligosaccharide moiety protrudes from the cell surface in a way allowing interaction with the cholera toxin B subunits.

The rapid, tight binding of cholera toxin to the GM1 receptors (within a few minutes being pentavalent as each of the B subunits gets in contact with a GM1 molecule), leads after a short lag period to translocation of the A subunit through the membrane and release of the A1 fragment into the cytosol. A1 can enzymatically hydrolyze NAD, and in cooperation with still incompletely defined cytosolic factors, A1 at the same time couples the ADP-ribose moiety of the hydrolyzed NAD onto the α subunit of the G_s protein of adenylate cyclase at the inner side of the membrane. This inhibits a normal feed-back regulatory mechanism by which GTP bound to $G_s\alpha$ is rapidly hydrolyzed to inhibitory GDP and P_i and thus 'locks' the adenylate cyclase in an active form. As a result, cyclic AMP accumulates intracellularly and this in its turn is assumed to cause the severe diarrhoeal fluid loss by both inhibiting the absorption of sodium chloride and water by intestinal villus cells and by stimulating chloride, bicarbonate and water secretion from the intestinal crypt cells (Fig. 1).

Identification of GM1 ganglioside as receptor for cholera toxin

The studies resulting in the identification of GM1 as the membrane receptor for cholera toxin took its departure from the observation by van Heyningen et al. (1971) that a crude brain ganglioside mixture could inactivate cholera toxin. This report was published just at the time when we had found that the cholera toxin molecule is composed of the two types of binding and toxic-active subunits now known as B and A, respectively (Lönnroth and Holmgren, 1973). We established a collaboration with Prof. Lars Svennerholm, whose laboratory for many years had characterized and purified all the major and several minor gangliosides. This allowed us to expediously undertake a thorough examination of whether the inactivation of cholera toxin by the crude ganglioside mixture reported by van Heyningen et al. was the result of a specific binding with a defined ganglioside or was due to a more unspecific interaction. Our results (Holmgren et al., 1973), as well as those from the simultaneous less extensive studies by Cuatrecasas (1973b) and King and van Heyningen (1973), demonstrated a highly specific binding between the toxin and ganglioside GM1. Our access

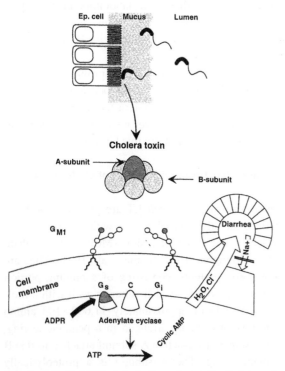

Fig. 1. Role of cholera toxin in the pathogenesis of cholera. For explanations, see text.

to a whole battery of highly purified structurally related other ganglioside and glycolipids also allowed us to identify which sugar residues in the GM1 molecule that were responsible for the receptor specificity (see below).

Subsequent studies in several laboratories confirmed and extended the evidence that the ganglioside GM1 is the natural biological receptor for cholera toxin:

(1) Studies of various cell types, including small intestinal mucosal cells of different species, showed a direct correlation between the cell content of GM1 and the number of toxin molecules that the cells could bind (Holmgren et al., 1975; Hansson et al., 1977; Fishman, 1986). In each cell type examined there were about 5 molecules of GM1 per toxin-binding site, consistent with the data from in vitro fixation studies that each toxin B subunit monomer could bind to one GM1 molecule.

(2) It was shown that exogenous GM1 ganglioside can be incorporated into the membrane of cells and function as biological receptors for cholera toxin. This was first shown by Cuatrecasas (1973b), who observed an increased cholera toxin-binding capacity and glycolytic responsiveness of fat cells which had been soaked in GM1 before the exposure to cholera toxin. Holmgren et al. (1975) demonstrated that when intestinal epithelium from humans and other species was preincubated with tritium-labelled GM1 ganglioside the resulting membrane incorporation of GM1 was associated with a corresponding increase in the cholera toxin-binding capacity of the intestine and, in in vivo tests in rabbits, also led to a proportional increase in susceptibility of the gut to the diarrheogenic action of the toxin. Furthermore, incorporation of GM1 into cholera toxin resistant transformed cell lines lacking GM1 ganglioside was shown by Moss et al. (1976) and Fishman et al. (1980) to restore cell responsiveness to the toxin.

(3) Pretreatment of cell membranes with cholera toxin was found to block specifically the membrane GM1 from reacting with galactose oxidase (Mullin et al., 1976). Chemical modifications of cholera toxin by various reagents have also consistently and proportionally affected the binding to cells and to plastic-adsorbed GM1 ganglioside (Holmgren and Lönnroth, 1976).

(4) Incubation of various tissues with *V. cholerae* sialidase has been found to both increase the number of cholera toxin binding sites in proportion to the additional GM1 produced by the enzyme from more complex gangliotetraose gangliosides in the membrane and to enhance the tissue or cell sensitivity to the toxin action (Holmgren, 1973; Haksar et al., 1974; Révesz et al., 1976; Hansson et al., 1977). However, in the intestinal epithelium of live rabbits *V. cholerae* sialidase failed to significantly affect the content of GM1 or the number of cholera toxin receptors (Holmgren et al., 1975) suggesting that the intestinal epithelium may possess special means of preserving its structural integrity from attack by exogenous enzymes.

Receptor–specific binding epitope in GM1

Already in our first studies in which different, highly purified, glycolipids of known structures were tested for their ability to fix and precipitate and neutralize cholera toxin, it was evident that within the GM1 ganglioside it is the sequence Galβ1-3GalNAcβ1-4(NeuAcα2-3)Gal that contains the binding receptor epitope for cholera toxin (Holmgren et al., 1973). As shown in Table I the removal within this sequence of the terminal Gal (see GM2) or the NeuAc residue (see GA1) or the steric blocking of either of these residues by one or more additional NeuAc residues (see GD1a, GD1b, and GT1) was associated with abolished or dramatically reduced receptor activity. Similar findings were obtained by testing the binding of cholera toxin to the same set of glycolipids attached to polystyrene (Holmgren, 1973) or, later, using the technique developed by Magnani et al. (1980) testing the toxin binding to glycolipids separated by thin-layer chromatography (TLC) (Magnani

TABLE I

Specificity in binding of cholera toxin by ganglioside GMI

Glycolipid	Structure	Rel. conc. for effect		Ref.
		ELISA	Neutral.	
GM1	Gal-GalNAc-Gal-Glu-Cer | NeuAc	1	1	a,b
GM2	GalNAc-Gal-Glu-Cer | NeuAc	>1000	>1000	a,b
GM3	Gal-Glu-Cer | NeuAc	>1000	>1000	a,b
GM1- GlcNAc	Gal-GlcNAc-Gal-Glc-Cer | NeuAc	>1000	>1000	a,b
GD1a	Gal-GalNAc-Gal-Glu-Cer | | NeuAc NeuAc	300	1000	a,b
GD1b	Gal-GalNAc-Gal-Glu-Cer | NeuAc-NeuAc	>1000 (50*)	>1000	a,b d,f
GT1b	Gal-GalNAc-Gal-Glu-Cer | | NeuAc NeuAc-NeuAc	> 1000	>1000	a,b
GA1 (Asialo-GM1)	Gal-GalNAc-Gal-Glu-Cer	1000	1000	a,b
GM1- (ganglio-triaose)	Gal-GalNAc-Gal-GalNAc-Gal-Glu-Cer | NeuAc	1-5	1-5	c
Fuc-GM1	Gal-GalNAc-Gal-Glu-Cer | | Fuc NeuAc	2-5	2-5*	a,e,f

References listed: [a]Holmgren et al., 1973; [b]Holmgren, 1973; [c]Nakamura et al., 1987; [d]Fukuta et al, 1988; [e]Masserini et al., 1992; Holmgren et al., to be published. *TLC binding assay.

et al., 1980; Holmgren et al., 1982; Nakamura et al., 1987; Fukuta et al., 1988). The only difference reported using the different methods has been a variable reactivity of GD1b usually being approximately 20–fold better in the TLC method as compared with, e.g. neutralization tests (although even with the TLC technique the GD1b reactivity has been one to two logs less than that of GM1) (Table I).

The findings and conclusions of receptor specificity reported above were obtained with cholera toxin purified from a single strain: 569B (biotype classical; serotype Inaba). Since then it has been found that there may exist minor sequence variations in cholera toxins from different strains, and especially toxin from strains of El Tor biotype differs appreciably from classical biotype cholera toxin. However, these minor structural variations do not seem to affect the binding specificity for GM1 versus other gangliosides as tested with cholera toxins isolated from various classical and El Tor strains (Holmgren, unpublished data).

The precise interaction between cholera toxin and

GM1 ganglioside remains to be determined at the high level of resolution that could be accomplished by, e.g. X-ray crystallography (cf. Sixma et al., 1992). Even so, however, some further delineation of the specific cholera toxin binding epitope in GM1 now appears possible based on recent studies of the effects of defined modifications in the GM1 ganglioside on the ability to bind cholera toxin. Thus, it is clear that GM1 containing *N*-glycolylneuraminic acid (Fishman et al., 1980), or derivatives of GM1 obtained by mild periodate oxidation of the NeuAc tail (Schengrund and Ringlöv, 1989) have essentially full cholera toxin binding activity. The same is the case for GM1 oxidated in the terminal Gal with galactose oxidase (Karlsson, 1989). In contrast a free carboxyl group on the NeuAc residue appears to be important, if not essential, for binding, since reductive decarboxylation of GM1 has been found to destroy the cholera toxin binding activity either completely (Sattler et al., 1977) or to a large extent (Schengrund and Ringlöv, 1989). Taking also into account the recently published (Sixma et al., 1992) three-dimensional structure of *E. coli* LT complexed with lactose (Galβ1-4Glc) determined by X-ray crystallography at 2.3 Å resolution (and assuming that

corresponding interactions would exist for LT and cholera toxin in their attachment to the terminal galactose in GM1) the specific binding epitope within GM1 might be depicted as suggested in Fig 2. It is then also predicted that there is a corresponding pocket or "canyon" for this receptor epitope in each of the toxin B subunits.

Fuc-GM1 as receptor for cholera toxin?

As mentioned above, ganglioside GM1 has exhibited a very high degree of specificity in its interaction with cholera toxin. Nevertheless, a few other gangliosides containing the same Galβ1-3GalNAcβ1-4(NeuAcα2-3)Gal receptor-active sequence as in GM1 have recently been shown to both bind and neutralize cholera toxin with almost the same activity as GM1 itself (Nakamura et al., 1987; Masserini et al., 1992) (Table I). In particular Fuc-GM1, consisting of the GM1 ganglioside with a single fucosyl residue linked 1-2 to the terminal Gal of GM1 has emerged as a possible alternative receptor for cholera toxin. Masserini et al. (1992) found that the ability of GM1 and of Fuc-GM1 to be incorporated into normally receptor-deficient rat glioma C6 cells and thereby to

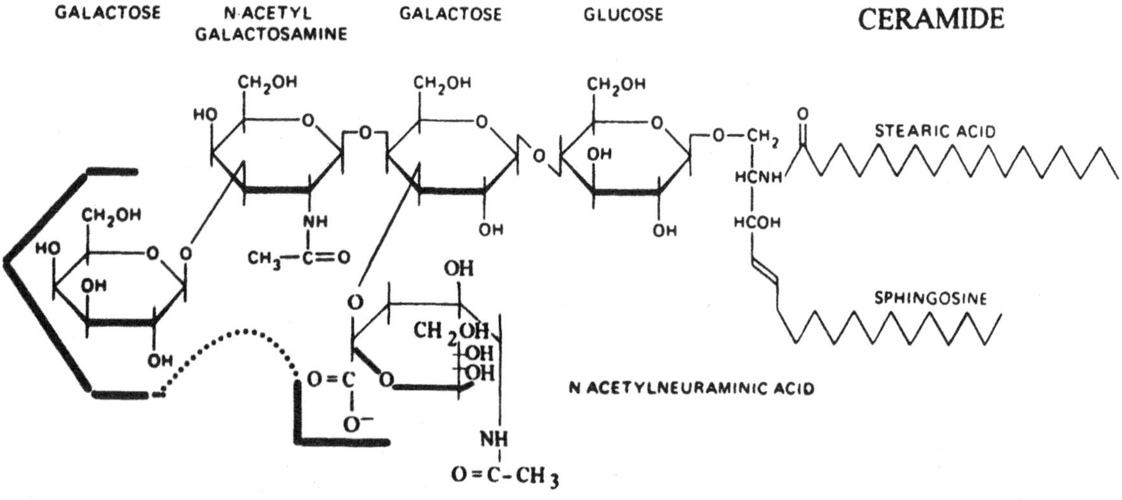

Fig. 2. Proposed cholera toxin binding epitope in ganglioside GM1. The sequence Galβ1-3GalNAcβ1-4(NeuAcα2-3)Gal contains the specific binding epitope. Within this sequence it is postulated that the terminal Gal and the carboxyl group side of NeuAc and possibly also the *N*-acetyl side of GalNAc are directly interacting with the cholera toxin B subunits as indicated in the figure.

sensitize these cells to cholera toxin binding and adenylate cyclase activation was very similar. They concluded that in these cells Fuc-GM1 could function similar to GM1 as receptor for cholera toxin.

An important question is then whether or not the natural target cells for cholera toxin, the enterocytes of the human small intestine, contain Fuc-GM1. Holmgren et al. (1985) showed, using the TLC-immunostaining technique, that human small intestinal epithelial scrapings contained limited amounts of a monosialoganglioside, which (in addition to GM1 and GD1b) could bind both cholera toxin and *E. coli* LT: in subsequent studies this ganglioside was identified as Fuc-GM1 (Fredman et al., unpublished). However, epithelial scrapings notoriously run the risk of being contaminated by intestinal nerve tissue. To better assess the possible role of Fuc-GM1 as receptor in human small intestine enterocytes we have therefore more recently examined the location of this ganglioside in intestinal mucosa using immunohisto-chemical methods (Brezicka et al., to be published). Monoclonal antibodies were raised against Fuc-GM1

(Brezicka, 1991) and their specificity for Fuc-GM1 versus other glycolipids was determined with ELISA and other methods (Fredman et al.; 1986 Brezicka, 1991). Two antibodies, F12 and F15, were found to be specific for Fuc-GM1; these antibodies were used to elaborate an optimally sensitive immunofluorescence staining procedure for the immunohistological detection of Fuc-GM1 in cryostate sections of human small intestine and other tissues. The results are shown in Table II. Fuc-GM1 was found on lymphocyte-like cells in the lamina propria of the small intestine, and in apparently larger amounts as based on the staining intensity in intramural nerve ganglia in both small intestine and colon. In contrast, the enterocytes of both the small intestinal and colonic epithelium did not, however, contain any detectable Fuc-GM1 whatsoever (Table II).

In conclusion, these results suggest that Fuc-GM1 may serve as an additional or alternative receptor to GM1 in cell types containing Fuc-GM1, but, since the human small intestinal epithelium does not contain any detectable Fuc-GM1, ganglioside GM1 remains the only as yet identified cholera toxin receptor molecule in the natural target tissue for the toxin.

TABLE II

Fucosyl-GM1 is missing in small intestinal enterocytes as examined immunohistochemically with Fuc-GM1 specific monoclonal antibodies (from Brezicka, 1991; Brezicka et al., to be published)

Reactivity with glycolipids and intestinal tissues	Monoclonal antibody	
	F12 (or F15)	F4
Glycolipids		
Fuc-GM1 (NeuAc or NeuGc)	+	+
GM1	–	+
GA1 (Asialo-GM1)	–	–
Mixed gangliosides[a]	–	+
Mixed neutral glycolipids[b]	–	–
Small intestine		
Mucosa	–	+
Submucosa ('lymphocytes')	+	+
Smooth muscle	–	+
Intramural ganglia	+	+

[a] (GM3, GM2, GM1, GD3, GD1a, GD1b, GT1b and GQ1b).
[b] (GlcCer, LacCer, GbOse$_3$, GbOse$_4$ and GgOse$_4$)

Receptors for *E. coli* LT

Although there have been reports that cholera toxin may bind to cell surface glycoproteins (Morita et al., 1980) such binding has not been observed upon more careful analysis neither as tested with classical biotype cholera toxin (Beckner et al., 1981; Holmgren et al., 1982; Holmgren et al., 1985; Fishman, 1986; Griffiths et al., 1986) nor with El Tor toxin (Holmgren, J. and Lindblad, M., unpublished). In contrast, it has been convincingly shown that the structurally and functionally similar heat-labile enterotoxin (LT) produced by many strains of *E. coli* causing diarrhoeal disease in humans can bind to either GM1 or glycoprotein receptors in the intestinal epithelium (Holmgren et al., 1982; Holmgren et al., 1985; Griffiths et al., 1986; Zemelman et al., 1989). Furthermore, the binding of LT to either class of receptors was shown to mediate a biologic response resulting in activation of adenylate cyclase and stimulation of intestinal electrolyte and fluid secretion and diarrhoea (Holmgren et al., 1982; Zemelman et al., 1989).

It was first shown by Holmgren (1973) that *E. coli* LT can bind with similar affinity and specificity as cholera toxin to GM1. However, at the same time it was shown that although pretreatment of ligated intestinal segments of rabbits with cholera B subunit (choleragenoid) completely blocked the electrolyte and fluid response to subsequently added cholera toxin, it did not block the secretory response to LT suggesting that either the functional receptors for cholera toxin and LT were different or there was an additional receptor for LT in the rabbit intestine (Holmgren, 1973; Pierce, 1973).

The binding specificity of LT for GM1 ganglioside has been confirmed in many studies (Holmgren et al., 1982; Griffiths et al., 1986; Fukuta et al., 1988). Moss et al. (1979) showed that ganglioside-deficient rat glioma C6 cells when specifically enriched with GM1 became responsive to LT, thus confirming a role of GM1 as biologic receptor for *E. coli* LT.

Direct evidence for the presence of an additional intestinal receptor of glycoprotein nature for *E. coli* LT was provided by Holmgren et al. (1982). These authors demonstrated specific binding of LT to glycoprotein as well as GM1 ganglioside in the small intestine of rabbits, and could also show the significance of the glycoprotein binding sites as functional receptors for LT. Thus, the extraction of lipids from the rabbit small intestine completely removed the

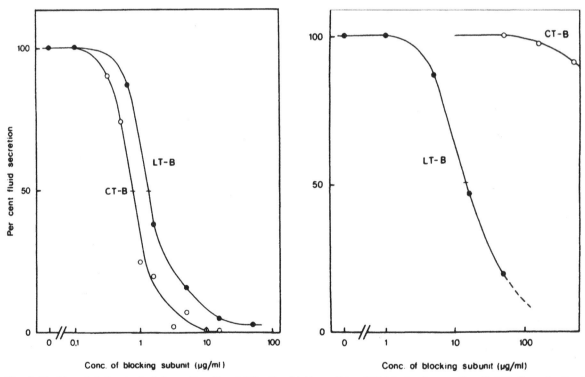

Fig. 3. Blocking of cholera toxin (left panel) or *E. coli* LT action (right panel) in rabbit intestine by use of purified binding subunits of cholera toxin (CT-B) or *E. coli* LT (LT-B). Cholera toxin action, measured as induction of fluid secretion in ligated intestinal segments exposed to toxin after short pre-exposure to buffer only or different concentrations of CT-B or LT-B, was completely blocked by either of CT-B or LT-B (a) in contrast to the *E. coli* LT action which was only marginally inhibited by CT-B (b). The results support the presence of an LT specific receptor in addition to the GM1 ganglioside receptor shared with cholera toxin. Figure reprinted from Holmgren et al. (1982) with permission.

binding sites for cholera toxin; in contrast, the delipidized intestinal tissue retained strong binding activity for *E. coli* LT. The lipid-associated binding activity for both cholera toxin and LT was identified as GM1 ganglioside whereas the residual LT-specific binding activity in the lipid-depleted tissue was shown to reside in the sugar moiety of a glycoprotein (Holmgren et al., 1982). The functional significance of the two types of binding sites for LT was established by blocking experiments using purified B subunit preparations from either cholera toxin or LT. Pretreatment of rabbit intestinal mucosa with B subunit from LT could completely block both cholera toxin and LT, whereas pretreatment with cholera B subunit had only a marginal inhibitory effect on the action of LT although it completely prevented the cholera toxin action (Holmgren et al., 1982) (Fig. 3).

Holmgren et al. (1985) also described the presence of LT-specific glycoprotein receptors in the human small intestine. These receptors appeared to account for approximately 50% of the total binding activity for LT (Holmgren et al., 1985) while in the rabbit intestine approximately 90% of LT-binding appeared to be accounted for by glycoprotein (Holmgren et al., 1982). Recently, Zemelman et al. (1989) have shown that also in the rat intestine the host response to *E. coli* LT takes place via two different microvillus membrane receptors identified as GM1 and a glycoprotein. In their experiments, pretreatment of intestine with the Gal-specific lectin RCA_{60}, could partially block the LT toxin action suggesting that the glycoprotein contains Gal as a critical sugar. Studies of LT binding to microvillus membrane proteins using a Western blot assay also confirmed the presence of several LT-binding galactoglycoproteins having molecular weights ranging from 85,000 to 150,000 (Zemelman et al., 1989). Approximately 70% of the LT binding activity in the rat intestine appeared to be mediated through high affinity (K_{d1}, 0.38 nM) GM1 receptors and 30% through a class of lower affinity (K_{d2}, 3.3 nM) glycoprotein receptors.

The sugar sequence(s) of the LT-specific galactoglycoprotein receptor(s) remain to be determined. It has been suggested by Karlsson (1989) that the binding site for cholera toxin and LT in GM1 may extend beyond the terminal Galβ1-3GalNAc-β1-4 (NeuAcα2-3)Galβ1 sequence to also include the 4Glcβ residues, and that this residue can be replaced by 4GlcNAcβ without affecting the binding of LT whilst the binding of cholera toxin would no longer be possible. It is assumed (Karlsson, 1989) that this carbohydrate sequence should exist in glycoproteins even though no such glycoprotein structure has yet been identified. Personally I prefer to think that the different binding specificity of LT and CT for glycoprotein is mainly due to a higher affinity of LT for galactose. In contrast to cholera toxin, LT can be affinity-purified using agarose columns (Gill et al., 1981), and furthermore, as mentioned above, LT readily forms crystal complexes with lactose (Sixma et al., 1992) while it has not been possible to obtain such crystals between lactose and cholera toxin. It is possible therefore that the suggested tighter binding between LT and the galactose residue(s) of GM1 may allow for some structural ambiguity in the remaining carbohydrate portion of the LT receptor. This could explain the ability of LT, in contrast to cholera toxin to bind also to certain galactoglycoproteins.

In addition to the *E. coli* LT mentioned above (sometimes referred to as LTh-I because of its isolation from strains causing diarrhoea in human and its close immunological relationship to cholera toxin) certain other strains of *E. coli* isolated from water buffalo, cattle and food have been found to produce a serologically different group of LTs (LT-IIa and LT-IIb). Fukuta et al. (1988) recently characterized the binding of the LT-II toxins as compared to LT-I and cholera toxin to different gangliosides. They found that LT-IIa bound well to each of the gangliosides GD1b, GM1, GD1a, GT1b, GQ1b and GD2 and less well but still significantly also to GM2 and GM3 suggesting that it primarily interacts with the Galβ1-3GalNAcβ1-4(NeuAcα2-3)Gal sequence in such a way that in contrast to LT-Ih or cholera toxin its binding is relatively unaffected by the presence of an extra NeuAc attached to the terminal Gal and completely unaffected or even slightly enhanced by the presence of an extra NeuAc to the inner Gal (as in GD1b). LTII-b, on the other hand, bound to GD1a and GT1b but not to GM1 suggesting that it specifically recognizes the sequence (NeuAcα2-3)Galβ1-3GalNAc (Fukuta et al., 1988).

Orientation of the A subunit in relation to the binding site of cholera toxin

Based on a combination of a biochemical and high resolution electron microscopy data it has been proposed that cholera toxin (and then most likely also *E. coli* LT) has a doughnut structure with the A subunit placed on the central axis of a ring of B subunits and with the A2 fragment associating A with the B5 ring (Gill, 1976; Ohtomo et al., 1976; Holmgren and Lönnroth, 1980; Moss and Vaughan, 1980; Gill et al., 1981; Holmgren, 1981; Ribi et al., 1988). Recent X-ray crystallographic examination of *E. coli* LT at a resolution of 2.3 Å has confirmed this structure (Sixma et al., 1991). The B pentamer has been shown to have a very characteristic doughnut shape: the enzymatic A subunit is attached to the B pentamer by means of the A2 fragment which is inserted like a hairpin all the way through a highly charged central pore in the B5 ring (Sixma et al., 1991).

A key question for understanding the toxin action with regard to the mechanism of translocation of the A subunit (or its toxic-active A1 fragment) across the cell membrane into the cell interior is to know the orientation of A1 in relation to the GM1 receptor-binding sites of the B pentamer. Two main proposals have been put forward for the mechanism of membrane translocation by cholera toxin. The first of these implies that the B subunits unfold and extend across the membrane thereby forming a hydrophilic pore through which the A1 fragment can pass; this model assumes that A1 points away from the receptor binding site of the B5 ring. The second proposal states that only the A1 fragment traverses the membrane, either by being driven into the membrane through energy released by the binding of the five B subunits to the GM1 receptors or by reaching the cell interior through a disarray of the membrane lipid bilayer caused by the binding of the B oligomer portion of the toxin with GM1; in this model A1 is assumed to be located on the same side as the GM1 binding sites in the B5 ring. Based on computerized

electron microscopy image processing of cholera toxin crystals formed on GM1-containing lipid layers, Ribi et al. (1988) reached the conclusion that A1 is located on the same side of the B pentamer ring as the GM1-binding sites. However, the recent determination of the three-dimensional structure of *E.coli* LT complexed with lactose as examined by X-ray crystallography at 2.3 Å resolution has seriously challenged this picture (Sixma et al., 1992). Lactose, interacting through its galactose moiety, was found to bind virtually identically to each of the five B subunits. The C1 linker atom of the bound galactose was found in the middle of the convoluted surface of the B pentamer, approximately 8 Å horizontally away from the side of the pentamer and at least 25 Å vertically away from the opposite flat or 'A-subunit binding' surface. This distance is too far for the remaining sugar residues of the GM1 oligosaccharide to span (McDaniel and McIntosh, 1986). As there is no evidence for any large conformational change in the B pentamer upon GM1 binding (Holmgren and Lönnroth, 1976; Fishman, 1986), Sixma et al. (1991, 1992) conclude that the initial binding of LT or cholera toxin to GM1-containing membranes most likely occurs with the A subunit pointing away from the membrane when all five ganglioside binding sites are occupied.

The specificity of the LT-galactose association is based on the fact that every hydrogen-bond donor and acceptor of galactose, except the ether oxygen and the C1 linker to glucose, is engaged in hydrogen bonding to the side chains of residues Glu 51, Gln 61, Asn 90 and Lys 91, as well as to the carbonyl oxygen of Gln 56 and three well defined water molecules in the LTB molecule (Sixma et al., 1992). There are also extensive van der Waals contacts between hydrophobic groups of galactose and Trp 88. His 57 is also in contact with the C6 of galactose, and Lys 91 is also involved which agrees with the key role proposed for lysine residues in GM1 binding. All residues in LT directly interacting with galactose are conserved in cholera toxin and differences occur only in the second electron shell. It is suggested by Sixma et al. (1992) that the difference at position 95 (Ser in LT as compared with Ala in cholera toxin) might explain why LT but not cholera toxin can also bind to certain galactoglycoproteins, although obviously there could be further differences in the binding sites of the other sugar residues of GM1 to the two types of toxin.

This now leaves us with the still unexplained problem how A1 is translocated across the cell membrane in order to exert its intracellular enzymic-toxic section. Does the perturbation of the lipid bilayer underneath the five binding sites of a toxin molecule or the probably even larger disarray of the lipid resulting from several bound toxin molecules forming patches result in micelle formation leading to a rotational tilting of the GM1 cholera toxin complex? Does the translocation of the A subunit involve interaction with a membrane protein? These questions were discussed more than 10 years ago (Holmgren and Lönnroth, 1976; Holmgren, 1981) much along the same lines as they are now being discussed for the entry of different viruses into cells, and they still remain unanswered.

Biotechnological applications of GM1 ganglioside receptors

Based on the highly specific high-affinity binding of cholera toxin and its B subunit pentamer to GM1 ganglioside several new biotechnological tools have been developed and used in a diverse range of biomedical applications.

Ultrastructural localization of cell membrane GM1 ganglioside and sensitive quantitation of GM1.

Hansson et al. (1977) used the high specificity of cholera toxin binding to GM1 ganglioside to develop an immunoelectron microscopic method for sensitive high-resolution localization of cholera toxin and thereby also of GM1 ganglioside. In this method cell or tissue sections are incubated with toxin and then the bound toxin is visualized with the help of toxin-specific peroxidase-conjugated monoclonal antibodies and enzyme substrate and thin cell or tissue sections are examined for electron-opaque precipitates in a transmission electron microscope. By titration of

the limiting concentration of cholera toxin that gives rise to specific precipitates, at least semiquantitative data of the GM1 ganglioside content can be obtained. Furthermore, by combining this technique with prior treatment of cells or tissue sections with *V. cholerae* sialidase before the incubation with cholera toxin it is also possible to determine not only ganglioside GM1 but also higher gangliosides of the same series. Using this technique Hansson et al. (1977) showed that in the central nervous system, GM1 is concentrated in the pre- and postsynaptic membranes of the synaptic terminals; a further increase in reactivity of structures after hydrolysis of the nervous tissue with *V. cholerae* sialidase suggests that higher gangliosides of the same series are also increased in the pre- and postsynaptic junctions.

This work has been followed by many other studies in which cholera toxin or its B subunit protomer has been used to specifically localize and quantitate GM1 ganglioside in different cell types mainly in the central nervous system (Robertson and Grant, 1989).

Diagnostic assay of E. coli LT and other cholera-like enterotoxins.

The ganglioside-enzyme-linked immunosorbent assay (ganglioside-ELISA) system introduced by Holmgren (1973) to compare the binding of *E. coli* LT and cholera toxin to different gangliosides and other glycolipids has with some modifications been widely used as a diagnostic method for the laboratory detection of ganglioside-binding enterotoxins and toxin-producing organisms. The glycolipids in question are first attached to a plastic surface (e.g. a polystyrene or polyvinyl test tube or micro-titer well) and then allowed to react with the test toxins; after washing the bound toxin is detected immunologically using toxin-specific antibody according to the ELISA principle. Svennerholm and Holmgren (1978) described a practical GM1-ELISA method allowing sensitive and reliable identification of LT producing *E. coli* strains ; maximal sensitivity and convenience are obtained by growing the test organisms directly in the GM1 ganglioside-coated micro-titer wells and adding polymixin for the last 2 h of the incubation which releases any cell-associated (periplasmic) toxin and allows it to bind to the GM1 before washing and immunodetection. By combining this method with the use of monoclonal antibodies directed either against an epitope shared by different enterotoxins or against epitopes being specific for cholera toxin, human *E. coli* LT, porcine *E. coli* LT etc. it has been possible to devise versatile diagnostic assays of various enterotoxigenic bacteria (Svennerholm et al., 1986).

Large-scale affinity chromatography techniques for vaccine production.

There has been a growing need for developing industrial systems for the large-scale production and purification of cholera B subunit for use in vaccines.

A new cholera vaccine has been developed, which in extensive clinical studies including a large field trial in Bangladesh involving 90,000 vaccine or placebo recipients has proved to be both safe and protective. The new vaccine contains a mixture of the non-toxic B subunit pentamer of cholera toxin and heat- and formalin-killed *Vibrio cholerae* bacteria (Inaba and Ogawa serotypes and classical and El Tor biotypes), and is administered orally in order to efficiently stimulate the local intestinal mucosal immune system. In the production of this vaccine the B subunit component has been isolated by means of a GM1 receptor-specific affinity chromatography procedure (Tayot et al., 1981). In this method the cholera toxin of 1000-liter fermentor supernatants is first selectively caught in columns of porous silica beads derivatized with covalently coupled GM1 ganglioside, whereafter the A subunit is removed and the B subunits finally isolated by selective elution steps (Tayot et al., 1981). Combining the GM1-affinity purification methods with newly developed recombinant plasmid overexpression systems of cholera B subunit alone (Sanchez and Holmgren, 1989; Lebens, M. et al., to be published) (rather than the holotoxin) it is now possible to produce >500,000 mg (or doses) of cholera B subunit/ 1000-liter culture.

Through its structural and immunological similarity with *E. coli* LT the cholera B subunit has also been shown to induce a high degree of cross-protective immunity against diarrhoeal disease caused by LT producing *E. coli* (Holmgren et al., 1992). Cholera B subunit is therefore an important component (together with formalin-killed colonization fimbriae-expressing *E. coli*) in an oral *E. coli* diarrhoea vaccine which is now in clinical testing. Furthermore, the ability in animal experimental systems to markedly augment induction of IgA mucosal immune responses to various other candidate mucosal vaccine antigens by linking them chemically or genetically to cholera B subunit has stimulated much interest in the possibility to use cholera B subunit as a 'universal' vector system for oral-mucosal vaccines. This interest has been further augmented by demonstrations that such oral vaccination with conjugates between foreign antigens and cholera B subunit can give rise to immune responses not only in the intestinal mucosa but also in distant mucosal tissues such as the salivary glands, the lungs or the genital mucosa as well as in the blood.

Receptor-specific prophylaxis of cholera and *E. coli* LT disease

With the emerging identification of various glycolipids as candidate attachment sites and/or biological receptors for a growing number of bacteria, viruses and bacterial toxins, the possibility to use the corresponding receptor-specific oligosaccharides for the prevention and treatment of disease has attracted much interest. It is therefore appropriate to review the studies in which GM1 ganglioside preparations have been used for the treatment and prevention of cholera or, alternatively, cholera B subunit has been given to specifically block the natural GM1 receptors for similar purposes.

Receptor competition.

Ever since the identity of the cholera toxin receptor first became known, the idea of using GM1 ganglioside for the prevention of cholera has been appealing. In rabbits the development of cholera could be readily prevented by administering purified receptor GM1 ganglioside before or simultaneously with cholera toxin or toxin-producing *V. cholerae* bacteria.

However, to avoid the risk of incorporation of some of the orally given GM1 into intestinal cells (which could increase the susceptibility for cholera) we decided that for human use the ganglioside had better be adsorbed onto medical charcoal. A GM1 ganglioside-charcoal preparation containing 5 μmol of GM1/g of charcoal was tested in cholera patients (Stoll et al., 1980). The study had a dual purpose. The first aim was to establish whether matrix-bound GM1 ganglioside could bind cholera toxin under clinical conditions. Secondly, provided that this was the case, the study was undertaken to investigate whether toxin produced in the gut lumen contributes significantly to clinical illness and if the inactivation of such toxin by binding to GM1 ganglioside-charcoal would alter the clinical course of disease. Forty-six patients with severe cholera were randomly assigned to one of three treatment groups: GM1 ganglioside-charcoal, charcoal alone, or water. Two grams of GM1-charcoal or the placebo treatments were administered every second hour for 2 days, and patients were intra-venously hydrated throughout the study period. The results showed that in patients with severe cholera, the oral administration of the GM1-charcoal preparation completely bound all detectable *V. cholerae* enterotoxin in the gut lumen in contrast to charcoal alone or water (Fig. 4a). Moreover, patients treated with GM1 ganglioside had a greater reduction in purging than patients treated with either charcoal alone or water (Fig. 4b); this difference was statistically significant during the first study period of 8-15 hours and was especially pronounced in patients with very severe initial purging and a short duration of illness prior to admission in which >50% specific reduction in stool volumes was obtained.

As GM1 cannot deactivate toxin that has already bound to the intestine, it would be expected to have only a preventive effect. The most likely explanation for the observed effect of GM1 also in patients with established disease is that, because of the constant and rapid regeneration of the intestinal epithelium, there are always new cells available to bind cholera

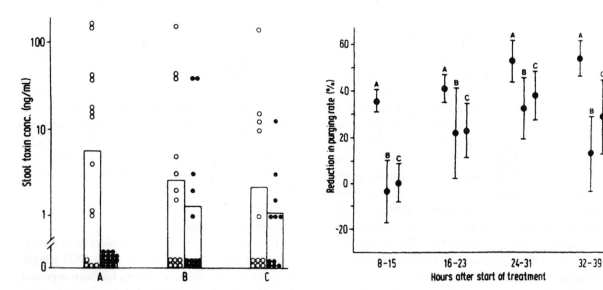

Fig. 4. Clinical trial of GM1 ganglioside-charcoal in cholera patients. Receptor-specific binding of intraluminal toxin (a) is associated with significant reduction in purging rate (b). Treatment groups: A, GM1-charcoal; B, charcoal alone; C, water. Open circles (a) represent toxin levels in stools at the onset of treatment and filled circles toxin levels after 48 h of treatment. Figure reprinted from Stoll et al. (1980) with permission.

toxin (and thus to be protected by an agent that binds the toxin). This is probably the first instance in which a specific receptor has been successfully used to interfere with an infectious disease. These studies stimulated intensified efforts to assess the possible practical usefulness of prophylaxis with GM1 ganglioside among household contacts of cholera cases who are at a particularly higher risk to fall ill with cholera within the next 5-7 days as a result of household cross-infection. For this purpose a further efficient and specific cholera toxin-binder was developed by covalently coupling GM1 ganglioside to fine cellulose powder particles. These particles were administered three times daily to household contacts of cholera, the treatment being initiated on the very day of identification of the household index case. However, the results were disappointing as no significant reduction in incidence or severity of cholera was obtained (Glass, R et al., unpublished data).

Receptor blocking.

It had been shown early that pretreatment of the gut of rabbits with either cholera or *E. coli* LT B subunits could completely protect animals from cholera after challenge with high doses of cholera toxin or live *V. cholerae*. Based on this, clinical studies were performed to test the hypothesis that blocking of toxin receptors would prevent fluid loss also in cholera-infected persons. Oral B subunit, 1.5 or 5 mg/dose, was given to family contacts of cholera patients, who have a more than 100-fold increased risk as compared with the general population (approximately 30%) to be infected with cholera within the first 5–7 days after the first household member became ill. The specific treatment effect was monitored in comparison with placebo treatment through daily surveillance of the treated individuals. The results were not encouraging as a practical intervention even though a statistically significant treatment effect was observed: blocking receptors with B subunit provided up to 40% reduction in illness while previous studies with tetracycline has given >80% reduction (though at the risk of spreading antibiotic resistance!) (Glass et al., 1983).

Based on these results and other considerations it is clear that several important problems need to be overcome before treatment or prevention of infections with carbohydrate receptors is likely to be practically useful. The multivalency of the interactions between target cells and most attacking bacteria and bacterial toxins or viruses makes it difficult to reverse already established binding unless agents of much higher affinity than the natural receptors or binding structures can be designed. Reversal of infection and disease becomes completely impossible if there is intracellular uptake or invasion of the microbe or toxin. A second major problem being fully evident in the cholera studies mentioned above is the difficulty to maintain effective receptor competition or occupancy using competing receptor substances or blocking agents in tissues having a rapid cell renewal rate; e.g. in relation to cholera approximately one-third of the intestinal cells are replaced each day with the new (crypt) cells being especially sensitive to cholera toxin. This problem becomes further enlarged by the ability of most mucosal pathogens to penetrate into or underneath the mucus coat and thus become less accessible for exogenously administered receptor analogue agents. Based on such considerations the best chance for success with carbohydrate receptor treatment of infections should be against non-invasive pathogens interacting with a readily accessible and ideally slowly regenerating tissue: certain types of infections of teeth and other tissues in the oral cavity and perhaps the colonization of the stomach by *Helicobacter pylori* could represent such model infections of some promise for specific receptor interference.

Acknowledgements

I wish to thank my many co-workers cited in the reference list for stimulating collaboration. In particular, it is a pleasure to acknowledge with appreciation the long fruitful and stimulating collaboration I have had with Prof. Lars Svennerholm and his laboratory both on gangliosides as receptors and as tumour-associated antigens. The Swedish Medical Research Council (project 16X–3383) supported most of the work cited from my own laboratory.

176

References

Beckner, S.K., Brady, R.O. and Fishman, P.H. (1981) Reevaluation of the gangliosides in the binding and action of thyrotropin. *Proc. Natl. Acad. USA*, 78: 4848–4852.

Brezicka, F-T. (1991) Fucosyl GM1: a ganglioside antigen associated with small cell lung cancer. *PhD thesis*, University of Göteborg (Medical Faculty), Sweden.

Carpenter, C.C.J. (1992) Treatment of cholera: clinical science at the bedside. *J. Infect. Dis.*, 166: 2–14.

Cuatrecasas, P. (1973a) Interaction of *Vibrio cholerae* enterotoxin with cell membranes. *Biochemistry*, 12: 3547–3558.

Cuatrecasas, P. (1973b) Gangliosides and membrane receptors for cholera toxin. *Biochemistry*, 12: 3558–3566.

Fishman, P.H. (1980) Mechanism of action of cholera toxin: events on the cell surface. In: M. Field, J.S. Fordtran and S.G. Schultz (Eds), *Secretory Diarrhoea, American Physiological Society, Clinical Physiology Series*, Bethesda, Maryland, pp. 85–106.

Fishman, P.H. (1986) Recent advances in identifying the functions of gangliosides. *Chem. Phys. Lipids*, 42: 137–151.

Fishman, P.H., Pacuszka, T., Holm, B. and Moss, J. (1980) Modification of ganglioside GM1. Effect of lipid moiety on choleragen action. *J. Biol. Chem.*, 255: 7657–7664.

Fukuta, S., Magnani, J.L., Twiddy, E.M., Holmes, R.K. and Ginsburg, V. (1988) Comparison of the carbohydrate-binding specificities of cholera toxin and *Escherichia coli* heat-labile enterotoxins LTh-I, LT-IIa, and LT-IIb. *Infect. Immun.*, 56: 1748–1753.

Fredman, P., Brezicka, T., Holmgren, J., Lindholm, L., Nilsson, O. and Svennerholm, L. (1986) Binding specificity of monoclonal antibodies to ganglioside Fuc-GM1. *Biochim. Biophys. Acta*, 875: 316–323.

Gill, D.M. (1976) The arrangement of subunits in cholera toxin. *Biochemistry*, 15: 1242–1248.

Gill, D.M., Clements, J.D., Robertson, D.C. and Finkelstein, R.A. (1981) Subunit number and arrangement in *Escherichia coli* heat-labile enterotoxin. *Infect. Immun.*, 33: 677–682.

Glass, R.I., Holmgren, J., Khan, M.R.B., Hossain, K.M., Huq, M. and Greenough, W.B. (1983) A randomized controlled trial of the toxin-blocking effects of a B subunit in family members of patients with cholera. *J. Infect. Dis.*, 149: 495–500.

Griffiths, S.L., Finkelstein, R.A. and Critchley, R.R. (1986) Characterization of the receptor for cholera toxin and *Escherichia coli* heat-labile toxin in rabbit intestinal brushborders. *Biochem. J.*, 238: 313–322.

Haksar, A., Maudsley, D.V. and Péron, F.G. (1974) Neuraminidase treatment of adrenal cells increases their response to cholera enterotoxin. *Nature*, 251: 514–515.

Hansson, H-A., Holmgren, J. and Svennerholm, L. (1977) Ultrastructural localization of cell membrane GM1 ganglioside by cholera toxin. *Proc. Natl. Acad. Sci. USA*, 74: 3782–3786.

Holmgren, J. (1973) Comparison of the tissue receptors for *Vibrio cholerae* and *Escherichia coli* enterotoxins by means of ganglio-sides and natural cholera toxoid. *Infect. Immun.*, 8: 851–859.

Holmgren, J. (1981) Actions of cholera toxin and the prevention and treatment of cholera. *Nature*, 292: 413–417.

Holmgren, J. and Lönnroth, I. (1976) Cholera toxin and the adenylate cyclase-activating signal. *J. Infect. Dis.*, 133: S64–S74

Holmgren, J. and Lönnroth, I. (1980) Structure and function of enterotoxins and their receptors, In Ö. Ouchterlony and J. Holmgren (Eds.), *Nobel Symposium 43 'Cholera and related Diarrheas'*, Stockholm, Karger, Basel, pp. 88–102.

Holmgren, J., Lönnroth, I. and Svennerholm, L. (1973) Tissue receptor for cholera exotoxin: Postulated structure from studies with GM1 ganglioside and related glycolipids. *Infect. Immun.*, 8: 208–214.

Holmgren, J., Lönnroth, I., Månsson, J-E. and Svennerholm, L. (1975) Interaction of cholera toxin and membrane GM1 ganglioside of small intestine. *Proc. Natl. Acad. Sci. USA*, 72: 2520–2524.

Holmgren, J., Fredman, P., Lindblad, M., Svennerholm, A-M. and Svennerholm, L. (1982) Rabbit intestinal glycoprotein receptor for *Escherichia coli* heat-labile enterotoxin lacking affinity for cholera toxin. *Infect. Immun.*, 38: 424–433.

Holmgren, J., Lindblad, M., Fredman, P., Svennerholm, L. and Myrvold, H. (1985) Comparison of receptors for cholera and *Escherichia coli* enterotoxin in human intestine. *Gastroenterology*, 89 : 27–35.

Holmgren, J., Svennerholm, A-M., Jertborn, J., Clemens, J., Sack, D.A., Salenstedt, R. and Wigzell, H. (1992) An oral B subunit-whole cell vaccine against cholera, *Vaccine*, 10(13) 911–914.

Karlsson, K-A. (1989) Animal glycosphingolipids as membrane attachment sites for bacteria. *Annu. Rev. of Biochem.*, 58: 309–350.

King, C. A. and Van Heyningen, W. E. (1973) Deactivation of cholera toxin by a sialidase-resistant monosialosyl ganglioside. *J. Infect. Dis.*, 127: 639–647.

Lönnroth, I. and Holmgren, J. (1973) Subunit structure of cholera toxin. *J. Gen. Microbiol.*, 76: 417–428.

Magnani, J. L., Smith, D. F. and Ginzburg, V. (1980) Detection of gangliosides that bind cholera toxin: direct binding of ^{125}I-labled toxin to thin–layer chromatograms. *Anal. Biochem.*, 109: 399–402.

Masserini, M., Freire, E., Palestini, P., Calappi, E. and Tettamanti, G. (1992) Fuc-GM1 ganglioside mimics the receptor function of GM1 for cholera toxin. *Biochemistry*, 31: 2422–2426.

McDaniel, R. V. and McIntosh, T. J. (1986) X-ray defraction studies of the cholera toxin receptor, GM1. *Biophys. J.*, 49 : 94–96.

Morita, A., Tsao, D. and Kim, Y.S. (1980) Identification of cholera toxin binding glycoproteins in rat intestinal microvillus membranes. *J. Biol. Chem.*, 255: 2549–2553.

Moss, J. and Vaughan, M. (1980) Mechanism of activation of adenylate cyclase by choleragen and *E. coli* heat-labile enterotoxin, In M. Field, J.S. Fordtran and S.G. Schultz (Eds.), *Secretory Diarrhoea, American Physiological Society, Clinical Physiology Series*, Bethesda, Maryland, pp. 107–126.

Moss, J., Fishman, P.H., Manganiello, V.C., Vaughan, M. and

Brady, R.O. (1976) Functional incorporation of ganglioside into intact cells: induction of choleragen responsiveness. *Proc. Natl. Acad. Sci. USA*, 73: 1034–1037.

Moss, J., Garrison, S., Fishman, P.H. and Richardson, S.H. (1979) Gangliosides sensitize unresponsive fibroblasts to *Escherichia coli* heat-labile enterotoxin. *J. Clin. Invest.*, 64: 381–384.

Mullin, B.R., Aloj, S.M., Fishman, P.H., Lee, G., Kohn, L.D. and Brady, R.O. (1976) Cholera toxin interactions with thyrotropin receptors on thyroid plasma membranes. *Proc. Natn. Acad. Sci. USA* 73: 1679–1683.

Nakamura, K., Suzuki, M., Inagaki, F., Iamakawa, T. and Suzuki, A. (1987) A new ganglioside showing choleragenoid-binding activity in mouse spleen. *J. Biochem.*, 101: 825–835.

Ohtomo, N., Muraoka, T., Tashiro, A., Zinnaka, Y. and Amako, K. (1976) Size and structure of the cholera toxin molecule and its subunits. *J. Infect. Dis.*, 133: S31–S40.

Pierce, N.F. (1973) Differential inhibitory effects of cholera toxoids and ganglioside on the enterotoxins of *Vibrio cholerae* and *Escherichia coli*. *J. Exp. Med.*, 137: 1009–1023.

Révész, T., Greaves, M.F., Capellaro, D. and Murray, R.K. (1976) Differential expression of cell surface binding sites for cholera toxin in acute and chronic leukaemias. *Br. J. Haematol.*, 34: 623–630.

Ribi, H.O., Ludwig, D.S., Mercer, K.L., Schoolnik, G.K. and Kornberg, R.D. (1988) Three-dimensional structure of cholera toxin penetrating a lipid membrane. *Science*, 239: 1272–1276.

Robertson, B. and Grant, G. (1989) Immunocytochemical evidence for the localization of the GM1 ganglioside in carbonic anhydrase-containing and RT97 immunoreactive rat primary sensory neurons. *Neurocytology*. 18: 77–86.

Sanchez, J. and Holmgren, J. (1989) Recombinant system for over-expression of cholera toxin B subunit in *Vibrio cholerae* as a basis for vaccine development. *Proc. Natl. Acad. Sci. USA*, 86: 481–485.

Sattler, J.G., Schwarzmann, J., Staerk, W., Ziegler, W. and Wiegandt, H. (1977) Studies of the ligand binding to cholera toxin, II. The hydrophilic moiety of sialoglycolipids. *Z. Physiol. Chem.*, 358: 159–163.

Schengrund, C-L. and Ringlöv, N.J. (1989) Binding of *Vibrio cholerae* toxin and the heat-labile enterotoxin of *Escherichia coli* to GM1, derivatives of GM1, and non-lipid oligosaccharide polyvalent ligands. *J. Biol. Chem.*, 264: 13233–13247.

Sixma, T.K., Pronk, S.E., Kalk, K.H., Wartna, E.S., Van Zanten, B.A.M., Witholt, B. and Hol, W.G.J. (1991) Crystal structure of a cholera toxin-related heat–labile enterotoxin from *E. coli*. *Nature*, 351: 371–377.

Sixma, T.K., Pronk, S.E., Kalk, K.H., Van Zanten, B.A.M., Berghues, A.M. and Hol, W.G.J. (1992) Lactose binding to heat-labile enterotoxin revealed by X-ray crystallography. *Nature*, 355: 561–564.

Stoll, B., Holmgren, J., Bardhan, P.K., Huq, I., Greenough III, W.B., Fredman, P. and Svennerholm, L. (1980) Binding of intraluminal toxin in cholera: Trial of GM1 ganglioside-charcoal. *Lancet*, II: 888–891.

Svennerholm, A-M. and Holmgren, J. (1978) Identification of *Escherichia coli* heat-labile enterotoxin by means of a ganglioside immunosorbent assay (GM1-ELISA) procedure. *Curr. Microbiol.*, 1: 19–27.

Svennerholm, A-M., Wikström, M., Lindblad, M. and Holmgren, J. (1986) Monoclonal antibodies to *Escherichia coli* heat-labile enterotoxins: neutralizing activity and differentiation of human and porcine LTs and cholera toxin. *Med. Biol.*, 64: 23–30.

Tayot, J-L., Holmgren, J., Svennerholm, L., Lindblad, M. and Tardy, M. (1981) Receptor-specific large scale purification of cholera toxin on silica beads derivatized with lyso-GM1 ganglioside. *Eur. J. Biochem.*, 113: 249–258.

Van Heyningen, W.E., Carpenter, C.C.J., Pierce, N.F. and Greenough III, W.B. (1971) Deactivation of cholera toxin by ganglioside. *J. Infect. Dis.*, 124: 415–418.

Zemelman, B.V., Chu, S-H.W. and Walker, W.A. (1989) Host response to *Escherichia coli* heat-labile enterotoxin by two microvillus membrane receptors in the rat intestine. *Infect. Immun.*, 57: 2947–2952.

L. Svennerholm, A.K. Asbury, R.A. Reisfeld, K. Sandhoff, K. Suzuki, G. Tettamanti and G. Toffano (Eds.)
Progress in Brain Research. Vol. 101

CHAPTER 13

Carbohydrate ligands of leukocyte adhesion molecules and their therapeutic potential

James C. Paulson

Cytel Corporation and Departments of Chemistry and Molecular Biology, Scripps Research Institute, 3525 John Hopkins Court, San Diego 92121, U.S.A.

Introduction

The selectin family of leukocyte adhesion molecules is generally recognized to play key roles in the recruitment of neutrophils and other leukocytes to sites of inflammation and tissue injury, and to mediate trafficking of lymphocytes to peripheral lymph nodes during recirculation between the blood and lymph. These three adhesion receptors are now known by a consensus nomenclature, L-selectin (*L*eukocyte selectin; Me*l*-14; *L*AM-1), E-selectin (*E*ndothelial cell selectin; *E*LAM-1) and P-selectin (*P*latelet selectin; GM*P*-140; *P*ADGEM; CD-62) which is expressed on both platelets and endothelial cells (Bevilacqua et al., 1991).

In a remarkable series of papers in late 1989 the three known selectins were cloned revealing for the first time the structural relationships which grouped them into a new adhesion molecule family (Bevilacqua et al., 1989; Johnston et al., 1989; Lasky et al., 1989; Siegelman et al., 1989). Perhaps the most striking feature of the deduced structures of the selectins was a consensus carbohydrate binding motif (lectin domain) defined previously for other mammalian carbohydrate binding proteins by Drickamer (1988). These findings prompted the prediction that the selectins would recognize carbohydrate ligands, which until then had only been suggested for L-selectin through the work of Rosen and colleagues (see Springer and Lasky, 1991; Lasky, 1992). The

other important prediction from the discovery of this family was that P-selectin would exhibit cell adhesion activity. Until it was cloned it had primarily been studied as a platelet granule membrane protein whose activity as a cell adhesion protein had not been appreciated. However, once the homology to E- and L-selectin was recognized, P-selectin was quickly confirmed to be a leukocyte adhesion molecule by several laboratories (Larsen et al., 1989; Geng et al., 1990; Watanabe et al., 1990).

In the last 3 years, information concerning the roles of the selectins in leukocyte biology has exploded. In particular, carbohydrate ligands for two of the selectins have been identified, and the roles of the selectins in the biochemical mechanisms of leukocyte trafficking are beginning to emerge. Because they participate in a general mechanism for exit of leukocytes from the blood into tissues or lymphoid organs, they are also relevant to disease states in which leukocyte recruitment of these cells in excess leads to injury of normal tissue. For this reason the selectins are considered potential therapeutic targets to prevent leukocyte mediated damage in acute and chronic inflammatory disease. This paper will briefly summarize recent progress in elucidating the roles of carbohydrate ligands in selectin biology with emphasis on E- and P-selectin, and comment on the prospects for carbohydrate based therapeutics in leukocyte mediated disease. Additional information covering L-selectin and a broader scope of subjects relevant to these adhesion receptors can be found in several excellent reviews (Springer and Lasky, 1991; Lasky, 1992; Paulson, 1992; Varki, 1992).

Carbohydrate ligands of the selectins

Within a year of the cloning of the selectin cDNAs several groups identified the carbohydrate ligand of E-selectin as sialyl-Lewis X (SLe^x; Fig. 1, Lowe et al., 1990; Phillips et al., 1990; Walz et al., 1990). Later it was recognized that P-selectin also recognized SLe^x ligands (Polley et al., 1991; Zhou et al., 1991). Recently, L-selectin has also been found to bind SLe^x (Foxall et al., 1992). The key residues for recognition of SLe^x by E- and P-selectin appear to be the sialic acid and fucose residues, since removal of either one dramatically reduces the ability of the carbohydrate structure to support selectin mediated cell adhesion. Several other carbohydrate structures have also been shown to bind the selectins (Fig. 1). These include sialyl-Lewis A (SLe^a) for both E- and P-selectin (Berg et al., 1991; Handa et al., 1991; Tyrrell et al., 1991) and a recently described sulfated derivative of Lewis X for E-selectin and L-selectin (Green et al., 1992; Yuen et al., 1992). Although SLe^x and SLe^a represent different isomers of the same tetrasaccharide structure, analysis of their predicted solution

conformation predicts that their sialic acid and fucose residues are oriented similarly in three dimensional space (Berg et al., 1991; Tyrrell et al., 1991). Thus, E- and P-selectins most likely recognize these two ligands because they are conformationally related.

In vivo the SLe^x ligand appears to be most relevant for leukocyte-endothelial cell interactions involving the E- and P-selectins since the cells they bind, neutrophils, eosinophils, monocytes and a subset of T-cells, contain sialylated Lewis X carbohydrates but not sialyl-Lewis A. Several metastatic carcinomas do express sialyl-Lewis A, however, and it has been suggested that selectins may play a role in the mechanism of metastasis of blood borne tumor cells (Matsushita et al., 1990; Kojima et al., 1992; Majuri et al., 1992).

To put into perspective the nature of selectin-ligand interactions, it is useful to recognize that SLe^x is carried as a terminal structure on the carbohydrate groups of both glycoproteins and glycolipids. Thus, while SLe^x may form the minimal structure for ligand recognition by the E- and P-selectins, other factors may be important in determining the functional receptors on leukocytes. For example, SLe^x bearing glycoproteins and glycolipids are both found on neutrophils and other leukocytes (reviewed in Paulson, 1992). Clearly SLe^x- glycolipids alone carry sufficient information for interaction with the selectins since panels of SLe^x-bearing and related glycolipids have been used as a primary tool for elucidating the specificity of E-selectin and P-selectin (Phillips et al., 1990; Polley et al., 1991; Tyrrell et al., 1991; Handa et al., 1991). Several reports suggest that SLe^x bearing glycoproteins are the primary ligands for selectin mediated adhesion. For example, mild protease treatment of neutrophils, which has no effect on cell surface glycolipids, dramatically reduces selectin mediated neutrophil-endothelial cell adhesion (Jutila et al., 1991). Because certain cell surface glycoproteins are particularly sensitive to proteolytic cleavage, such experiments have supported the hypothesis that specific neutrophil glycoproteins, such as L-selectin, are the predominant carrier of 'functional' E- and P-selectin ligands (Jutila et al., 1991; Picker et al., 1991).

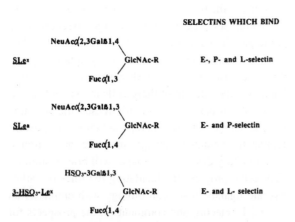

Fig. 1. Selected high affinity ligands of E-, P- and L-selectins. Shown are the structures for sialyl-Lewis X (SLe^x), sialyl-Lewis A (SLe^a) and HSO$_3$-Lewis X (see text for references).

Role of selectins in neutrophil recruitment

The mechanism of recruitment of leukocytes to sites of inflammation has begun to be elucidated on a biochemical level. Particularly good progress has been made for neutrophils, which play a major role in acute inflammation and associated inflammatory disease (see: Butcher, 1991; Lawrence and Springer, 1991; Lorant et al., 1991; von Andrian et al., 1991; Smith, 1992). The diagram in Fig. 2 summarizes an emerging consensus view for the steps involved in recruitment of neutrophils to inflammatory sites. The initial step is the interaction of the neutrophil with the activated endothelium lining the blood vessel wall, resulting in rolling of the cells, which can actually be visualized using intravital microscopy techniques. Once the cells are slowed down they can be activated by inflammatory mediators (e.g. leukotriene B4; platelet activating factor, PAF; fMet-Leu-Phe) resulting in an activation of the cells prior to firm adhesion and extravasation into the surrounding tissue.

All three of the selectins appear to play a critical role in the rolling step in this mechanism, L-selectin on the neutrophil side, and E- and P-selectin on the endothelial cell side (Kishimoto et al., 1990; Lawrence and Springer, 1991; Ley et al., 1991; von Andrian et al., 1991; Smith, 1992). While E-selectin

and P-selectin are both expressed on endothelial cells, they are expressed in response to different activators. E-selectin is expressed in response to inflammatory cytokines (IL1-β, TNF) or bacterial endotoxin while P-selectin is expressed in response to thrombin, oxygen radicals or histamine. As a result, inflammatory conditions may induce one or both of the selectins depending on the activators present. Once expressed the selectins appear to mediate rolling by interacting with their counter ligands containing SLe^x on the neutrophil surface. For L-selectin the primary role may be to present SLe^x (Picker et al., 1991; Smith, 1992), although a role for L-selectin recognition of a ligand has not been ruled out.

Neutrophil activation results in the upregulation of the number and affinity of the leukocyte integrins, LFA-1 and Mac-1. These adhesion molecules are well documented to mediate firm adhesion and extravasation of the cells into the surrounding tissue by interacting with their counter ligands on the endothelial cell, intracellular adhesion molecule-1 and -2 (ICAM-1 and -2) which are members of the immunoglobulin superfamily (Springer, 1990). These adhesion molecules are capable of mounting strong adhesive forces, but do not initiate adhesion under the shear force created by normal rates of blood flow (Lawrence and Springer, 1991; Smith, 1992).

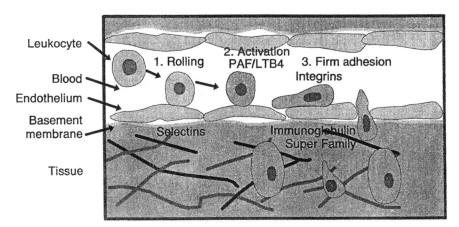

Fig. 2. General mechanism of neutrophil recruitment to sites of inflammation. Three identifiable steps in the recruitment of neutrophils are depicted. At each step the adhesion molecules or inflammatory mediators believed to mediate those steps are indicated. Selectins refer to E-, P- and L-selectins. Integrins refer to the neutrophil integrins LFA-1 and Mac-1, and the immunoglobulin superfamily to ICAM-1 and -2 (adapted from Springer, 1990; see the text for further description).

Evidence for the importance of the rolling and firm adhesion steps in neutrophil recruitment can be seen in known human leukocyte adhesion deficiencies (LAD). Classic LAD patients have been well characterized to be deficient in the leukocyte integrins LFA-1 and Mac-1 (Anderson et al., 1989). As a result the neutrophils are unable to firmly adhere to activated endothelium and leave the blood. Patients typically exhibit high neutrophil counts and have recurrent severe infections which are life threatening. Another LAD (LAD II) has recently been described in two patients resulting from a deficiency in the selectin ligand SLe^x (Etzioni et al., 1992). The neutrophils of these patients are unable to bind to selectins expressed on endothelial cells, and have normal levels of the LFA-1 and Mac-1 integrins which are missing from classic LAD patients. Like the classic LAD patients the LAD II patients exhibit high neutrophil counts and recurrent infections, although the infections typically do not appear to be as severe. Together, the existence of these two leukocyte adhesion deficiencies provide compelling *in vivo* evidence for the importance of leukocyte rolling and firm adhesion as distinct sequential steps required for efficient recruitment to inflammatory sites (Fig. 2).

Selectins as targets for development of anti-inflammatory agents

As a natural defense mechanism, the recruitment of neutrophils to sites of inflammation aids in the destruction of bacteria and other microorganisms. However, in acute inflammatory disease and in reperfusion injury, neutrophils destroy normal tissue as well. The concept that neutrophil mediated tissue injury in acute inflammatory disease could be blocked by preventing exit of neutrophils from the blood was initially demonstrated using blocking monoclonal antibodies to the common β-subunit of the neutrophil integrins (Harlan et al., 1992). More recently, antibodies against E-selectin and P-selectin have also been shown to be efficacious in the inhibition of leukocyte mediated animal models of disease (Gundel et al., 1991; Mulligan et al., 1991; Harlan et al., 1992; Mulligan et al., 1992). Such results are in

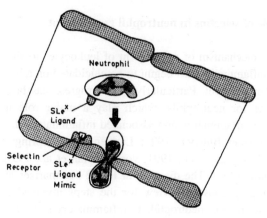

Fig. 3. Selectin ligand based small molecule inhibitors of neutrophil adhesion.

keeping with the predicted role of these adhesion molecules in recruitment of neutrophils to sites of inflammation and tissue injury.

Based on these facts, the selectins represent an ideal target for development of carbohydrate based drugs. Conceptually, a small molecule SLe^x analog or SLe^x-mimic could block neutrophil rolling by occupying the ligand binding site on E-selectin or P-selectin (see Fig. 3). Indeed, SLe^x-based oligosaccharides have been found to inhibit P-selectin mediated rolling of leukocytes in rat mesentery venules *in vivo* using intravital microscopy techniques (Asako, H. and Granger, N., pers. commun.). Providing that such results are predictive of the activity of these compounds as general inhibitors of neutrophil recruitment, SLe^x analogs or mimics may find broad utility as small molecule inhibitors of acute inflammatory disease. The principal challenge for the production of such inhibitors resides in the difficulty in the synthesis of complex carbohydrates. However, recent progress in large scale enzymatic synthesis of carbohydrates using glycosyltransferases as biological catalysts offers promise in solving this problem (Ichikawa et al., 1992; Ito et al., 1992).

Summary and perspective

In the short time that the selectins have been identified as a structurally related family of leukocyte cell adhesion molecules, much has been learned about the carbohydrate ligands which they recognize and the biochemical mechanisms of leukocyte trafficking in which they participate. Despite such rapid advances the overall knowledge of how and when individual selectins are expressed *in vivo* and the extent to which they play important roles in the recruitment of leukocytes other than neutrophils is still fragmentary. Such information is critical to a deeper understanding of the larger roles of selectins in leukocyte trafficking and will be important to guide the development of therapeutics targeting these receptors.

Acknowledgment

I am indebted to my colleagues at Cytel and to Drs. John Harlan, Peter Ward, Neil Granger, John Lowe and Chi-Huey Wong for many insightful discussions relevant to this manuscript.

References

Anderson, D.C., Smith, C.W. and Springer, T.A. (1989) Leukocyte adhesion deficiency and other disorders of leukocyte motility. In C.R. Scriver, A.L. Beaudet, W.S. Sly, and D. Valle (Eds.), *The Metabolic Basis of Inherited Disease, 6th edn.*, McGraw Hill, Inc., USA, pp. 2751–2777.

Berg, E.L., Robinson, M.K., Masson, O., Butcher, E.C. and Mangnani, J.L. (1991) A carbohydrate domain common to both Sialyl Le^a and Sialyl Le^x is recognized by the endothelial cell leukocyte adhesion molecule ELAM-1. *J. Biol. Chem.*, 266: 14869–14872.

Bevilacqua, M.P., Stengelin, S., Gimbrone, M.A. and Seed, B. (1989) Endothelial leukocyte adhesion molecule 1: an inducible receptor for neutrophils related to complement regulatory proteins and lectins. *Science*, 243: 1160–1165.

Bevilacqua, M.P., Butcher, E.C., Furie, B., Gallatin, M., Gimbrone, M., Harlan, J.M., Kishimoto, K., Lasky, L.A., McEver, R. and Paulson, J.C. (1991) Selectins: a family of adhesion receptors. *Cell*, 67: 233.

Butcher, E.C. (1991) Leukocyte-endothelial cell recognition: three (or more) steps to specificity and diversity. *Cell*, 67: 1033–1036.

Drickamer, K. (1988) Two distinct classes of carbohydrate-recognition domains in animal lectins. *J. Biol. Chem.*, 263: 9557–9560.

Etzioni, A., Frydman, M., Pollack, S., Avidor, I., Phillips, M.L., Paulson, J.C. and Gershoni-Baruch, R. (1992) Recurrent severe infections caused by a novel leukocyte adhesion deficiency. *N. Engl. J. Med.*, 327: 1789–1792.

Foxall, C., Watson, S.R., Dowbenko, D., Fennie, C., Lasky, L.A.,

Kiso, M., Hasegawa, A., Asa, D. and Brandley, B.K. (1992) The three members of the selectin receptor family recognize a common carbohydrate epitope, the Sialyl Lewis^x oligosaccharide. *J. Cell Biol.*, 117: 895–902.

Geng, J.G., Bevilacqua, M.P., Moore, K.L., McIntyre, T.M., Prescott, S.M., Kim, J.M., Bliss, G.A., Zimmerman, G.A. and McEver, R.P. (1990) Rapid neutrophil adhesion to activated endothelium mediated by GMP-140. *Nature*, 343: 757–760.

Green, P.J., Tamatani, T., Watanabe, T., Miyasaka, M., Hasegawa, A., Kiso, M., Yuen, C.-T., Stoll, M.S. and Feizi, T. (1992) High affinity binding of the leucocyte adhesion molecule L-selectin to 3'-sulphated-Le^a and -Le^x oligosaccharides and the predominance of sulphate in this interaction demonstrated by binding studies with a series of lipid-linked oligosaccharides. *Biochem. Biophys. Res. Commun.*, 188: 244–251.

Gundel, R.H., Wegner, C.D., Torcellini, C.A., Clarke, C.C., Haynes, N., Rothlein, R., Smith, C.W. and Letts, L.G. (1991) Endothelial leukocyte adhesion molecule-1 mediates antigen-induced acute airway inflammation and late-phase airway obstruction in monkeys. *J. Clin. Invest.*, 88: 1407–1411.

Handa, K., Nudelman, E.D., Stroud, M.R., Shiozawa, T. and Hakomori, S.-I. (1991) Selectin GMP-140 (CD62; PADGEM) binds to Sialosyl-Le^a and Sialosyl-Le^x, and sulfated glycans modulate this binding. *Biochem. Biophys. Res. Commun.*, 181: 1233–1230.

Harlan, J.M., Winn, R.K., Vedder, N.B., Doerschuk, C.M. and Rice, C.L. *In vivo* models of leukocyte adherence to endothelium. (1992) In: J.M. Harlan and D.Y. Liu (Eds.), *Adhesion: Its Role in Inflammatory Disease, Vol. 4*, W.H. Freeman and Company, New York, pp. 117–150.

Ichikawa, Y., Lin, Y.-C., Dumas, D.P., Shen, G.-J., Garcia-Junceda, E., Williams, M.A., Bayer, R., Ketcham, C., Walker, L.E., Paulson, J.C. and Wong, C.-H. (1992) Chemical-enzymatic synthesis and conformational analysis of Sialyl Lewis X and derivatives. *J. Am. Chem. Soc.*, 114: 9283–9298.

Ito, Y., Gaudino, J.J. and Paulson, J.C. (1992) Synthesis of bioactive sialosides. *Pure Appl. Chem.*, 65: 753–762.

Johnston, G.I., Cook, R.G. and McEver, R.P. (1989) Cloning of GMP-140, a granule membrane protein of platelets and endothelium: sequence similarity to proteins involved in cell adhesion and inflammation. *Cell*, 56: 1033–1044.

Jutila, M.A., Kishimoto, T.K. and Finken, M. (1991) Low dose chymotrypsin treatment inhibits neutrophil migration into sites of inflammation *in vivo*: effects on Mac-1 and Mel-14 adhesion protein expression and function. *Cell Immunol.*, 132: 201–214.

Kishimoto, T.K., Warnock, R.A., Jutila, M.A., Butcher, E.C., Anderson, D.C. and Smith, C.W. (1990) Antibodies against human neutrophil LECAM-1 (DREG56/LAM-1/Leu-8 antigen) and endothelial cell ELAM-1 inhibit a common CD18-independent adhesion pathway *in vitro*. *Blood*, 78: 805–811.

Kojima, N., Handa, K., Newman, W. and Hakomori, S.-I. (1992) Inhibition of selectin-dependent tumor cell adhesion to endothelial cells and platelets by blocking O-glycosylation of these cells. *Biochem. Biophys. Res. Commun.*, 182: 1288–1295.

Larsen, E., Celi, A., Gilbert, G.E., Furie, B.C., Erban, J.K., Bonfanti, R., Wagner, D.D. and Furie, B. (1989) PADGEM protein: a receptor that mediates the interaction of activated platelets with neutrophils and monocytes. *Cell*, 59: 305–312.

Lasky, L.A. (1992) The homing receptor (LECAM 1/L Selectin): a carbohydrate-binding mediator of adhesion in the immune system. In: J.M. Harlan and D.Y. Liu (Eds.), *Adhesion: Its Role in Inflammatory Disease, Vol. 4*, W.H. Freeman and Company, New york, pp. 43–63.

Lasky, L.A., Singer, M.S., Yednock, T.A., Dowbenko, D., Fennie, C. et al. (1989) Cloning of a lymphocyte homing receptor reveals a lectin domain. *Cell*, 56: 1045–1055.

Lawrence, M.B. and Springer, T.A. (1991) Leukocyte roll on a selectin at physiologic flow rates: distinction from and prerequisite for adhesion through integrins. *Cell*, 65: 859–873.

Ley, K., Gaehtgens, P., Fennie, C., Singer, M.S., Lasky, L.A. and Rosen, S.D. (1991) Lectin-like cell adhesion molecule 1 mediates leukocyte rolling in mesenteric venules *in vivo*. *Blood*, 77: 2553–2555.

Lorant, D.E., Patel, K.D., McIntyre, T.M., McEver, R.P., Prescott, S.M. and Zimmerman, G.A. (1991) Coexpression of GMP-140 and PAF by endothelium stimulated by histamine or thrombin: a juxtacrine system for adhesion and activation of neutrophils. *J. Cell Biol.*, 115: 223–234.

Lowe, J.B., Stoolman, L.M., Nair, R.P., Larsen, R.D., Berhend, T.L. and Marks, R.M. (1990) ELAM-1-dependent cell adhesion to vascular endothelium determined by a transfected human fucosyltransferase cDNA. *Cell*, 63: 475–484.

Majuri, M.-L., Mattila, P. and Renkonen, R. (1992) Recombinant E-selectin-protein mediates tumor cell adhesion via Sialyl-Le^a and Sialyl-Le^x. *Biochem. Biophys. Res. Commun.*, 182: 1376–1382.

Matsushita, Y., Cleary, K.R., Ota, D.M., Hoff, S.D. and Irimura, T. (1990) Sialyl-dimeric Lewis-X antigen expressed on mucin-like glycoproteins in colorectal cancer metastases. *Lab. Invest.*, 63: 780–791.

Mulligan, M.S., Varani, J., Dame, M.K., Lane, C.L., Smith, C.W., Anderson, D.C. and Ward, P.A. (1991) Role of endothelial-leukocyte adhesion molecule 1 (ELAM-1) in neutrophil-mediated lung injury in rats. *J. Clin. Invest.*, 88: 1396–1406.

Mulligan, M.S., Polley, M.J., Bayer, R.J., Nunn, M.F., Paulson, J.C. and Ward, P.A. (1992) Neutrophil-dependent acute lung injury. *J. Clin. Invest.*, 90: 1600–01607.

Paulson, J.C. (1992) Selectin/carbohydrate-mediated adhesion of leukocytes. In: J.M. Harlan and D.Y. Liu (Eds.), *Adhesion: Its Role in Inflammatory Disease, Vol. 4*, W.H. Freeman and Company, New York, pp. 19–42.

Phillips, M.L., Nudelman, E., Gaeta, F.C.A., Perez, M., Singhal, A.K., Hakomori, S. and Paulson, J.C. (1990) ELAM-1 mediates cell adhesion by recognition of a carbohydrate ligand, Sialyl-Le^x. *Science*, 250: 1130–1132.

Picker, L.J., Warnock, R.A., Burns, A.R., Doerschuk, C.M., Berg,

E.L. and Butcher, E.C. (1991) The neutrophil selectin LECAM-1 presents carbohydrate ligands to the vascular selectins ELAM-1 and GMP-140. *Cell*, 66: 921–933.

Polley, M.J., Phillips, M.L., Wayner, E., Nudelman, E., Singhal, A.K., Hakomori, S. and Paulson, J.C. (1991) CD62 and endothelial cell-leukocyte adhesion molecule 1 (ELAM-1) recognize the same carbohydrate ligand, Sialyl Lewis X. *Proc. Natl. Acad. Sci. U.S.A.*, 88: 6224–6228.

Siegelman, M.H., Van de Rijn, M. and Weissman, I.L. (1989) Mouse lymph node homing receptor cDNA clone encodes a glycoprotein revealing tandem interaction domains. *Science*, 243: 1165–1172.

Smith, C.W. (1992) Transendothelial migration. In: J.M. Harlan and D.Y. Liu (Eds.), *Adhesion: Its Role in Inflammatory Disease, Vol. 4*, W.H. Freeman and Company, New York, pp. 83–115.

Springer, T.A. (1990) The sensation and regulation of interactions with the extracellular environment: the cell biology of lymphocyte adhesion receptors. *Annu. Rev. Cell Biol.*, 6: 359–402.

Springer, T.A. and Lasky, L.A. (1991) Sticky sugars for selectins. *Nature*, 349: 196–197.

Tyrrell, D., James, P., Rao, N., Foxall, C., Abbas, S., Dasgupta, F., Nashed, M., Kasegawa, A., Kiso, M., Asa, D., Kidd, J. and Brandley, B.K. (1991) Structural requirements for the carbohydrate ligand of E-selectin. *Proc. Natl. Acad. Sci. U.S.A.*, 88: 10372–10376.

Varki, A. (1992) Selectins and other mammalian sialic acid-binding lectins. *Current Opin. in Cell Biol.*, 4: 257–266.

von Andrian, U.H., Chambers, J.D., McEvoy, L.M., Bargatze, R.F., Arfors, K.-E. and Butcher, E.C. (1991) Two-step model of leukocyte-endothelial cell interaction in inflammation: distinct roles for LECAM-1 and the leukocyte b$_2$ integrins *in vivo*. *Proc. Natl. Acad. Sci. U.S.A.*, 88: 7538–7542.

Walz, G., Aruffo, A., Kolanus, W., Bevilacqua, M. and Seed, B. (1990) Recognition by ELAM-1 of the Sialyl-Le^x determinant on myeloid and tumor cells. *Science*, 250: 1132–1135.

Watanabe, M., Yagi, M., Omata, M., Hirasawa, N., Mue, S., Tsurufuji S. and Ohuchi, K. (1990) Stimulation of neutrophil adherence to vascular endothelial cells by histamine and thrombine and its inhibition by PAF antagonists and dexamethasone. *Br. J. Pharmacol.*, 102: 239–245.

Yuen, C.-T., Lawson, A.M., Chai, W., Larkin, M., Stoll, M.S., Stuart, A.C., Sullivan, F.X., Ahern, T.J. and Feizi, T. (1992) Novel sulfated ligands for the cell adhesion molecule E-selectin revealed by the neoglycolipid technology among O-linked oligosaccharides on an ovarian cystadenoma glycoprotein. *Biochemistry*, 31: 9126–9131.

Zhou, Q., Moore, K.L., Smith, D.F., Varki, A., McEver, R.P. and Cummings, R.D. (1991) The selectin GMP-140 binds to sialylated, fucosylated lactosaminoglycans on both myeloid and non-myeloid cells. *J. Cell Biol.*, 115: 557–564.

L. Svennerholm, A.K. Asbury, R.A. Reisfeld, K. Sandhoff, K. Suzuki, G. Tettamanti and G. Toffano (Eds.)
Progress in Brain Research, Vol. 101

CHAPTER 14

Receptors for gangliosides and related glycosphingolipids on central and peripheral nervous system cell membranes

Ronald L. Schnaar, James A. Mahoney, Patti Swank-Hill
Michael Tiemeyer and Leila K. Needham

Departments of Pharmacology and Neuroscience, The Johns Hopkins University School of Medicine, Baltimore, MD 21205, USA

Introduction

Advances in glycosphingolipid purification, structural determination and functional analysis have led to a better understanding of the multiple biological roles of this diverse class of molecules. As reviewed elsewhere in this volume, gangliosides and related glycosphingolipids are involved in cell-cell recognition and cellular regulation events in the nervous system and elsewhere (also see Igarashi et al., 1989; Cuello, 1990; Hakomori, 1990; Schnaar, 1991). The molecular mechanisms by which glycosphingolipids control cell behavior are likely to converge on two general motifs (Hakomori, 1990; Schnaar, 1991): Specific binding of cell surface glycosphingolipids to complementary receptors on apposing membranes (*trans* recognition) and modulation of protein activities in the same membrane (*cis* regulation). An example of *trans* recognition is the binding of endothelial cell surface protein E-selectin to complementary fucosylated gangliosides on neutrophils (Brandley et al., 1990). *Cis* regulation is exemplified by the modulation of epidermal growth factor receptor kinase by ganglioside GM3 and related molecules in the same cell membrane (Hanai et al., 1988).

Research in our laboratory focuses on detection and characterization of novel protein receptors for gangliosides and related glycosphingolipids in the nervous system. Our strategy is to identify ganglioside (and related) binding activities using synthetic glycosphingolipid-based radioligands, to characterize the structural and tissue specificity of binding activities, and to purify the responsible protein receptors. Through these efforts, we hope to help elucidate the detailed mechanisms by which gangliosides and related glycosphingolipids modulate cell behaviors in the nervous system. This chapter reviews our progress towards that goal, including identification of a ganglioside receptor in the central nervous system and a receptor for sulfoglucuronyl glycosphingolipids in the peripheral nervous system.

Glycosphingolipid-based Neoglycoconjugates

Neoglycoconjugates are semisynthetic derivatives consisting of a non-glycosylated 'scaffold' (e.g. a carrier protein) covalently derivatized with defined carbohydrate structures. Neoglycoproteins consisting of bovine serum albumin (BSA) derivatized with monosaccharide glycosides have been valuable tools for probing vertebrate cell surface lectins (Stahl et al., 1980; Connolly et al., 1983; Lee and Lee, 1992). Likewise, neo*ganglio*proteins, consisting of BSA covalently derivatized with gangliosides or ganglioside oligosaccharides, have been useful in probing for ganglioside receptors on brain membranes. These conjugates have the advantage of presenting the ganglioside oligosaccharide structures in polyvalent arrays which may be critical to detecting high-affinity receptors (Lee et al., 1984; Lee, 1989). At the same time, the lipid portion is sequestered in covalent association with the carrier protein so that non-specific adsorption and insertion into membranes is minimized. Neoganglioproteins are readily radiolabeled to high specific activity by radioiodination of the carrier.

We have used three synthetic strategies to generate the desired conjugates. The first two approaches retain as much of the intact ganglioside structure as possible by removing the original ganglioside fatty acid amide (by strong alkaline hydrolysis; Neuenhofer et al., 1985) and replacing it, via an

amide bond, with a spacer arm terminated in an activated linker moiety (Fig 1). In each case, the activated ganglioside intermediate is purified to homogeneity prior to covalent linkage to the carrier protein. Our initial synthetic scheme (Tiemeyer et al., 1989) used a sulfosuccinimidyl ester as the activating group for attachment to lysine primary amines on a BSA carrier (Scheme 1, Fig. 1). While this generated the desired ligand, the active ester was susceptible to hydrolysis during purification of the activated ganglioside intermediate. To solve this problem, we now use a maleimidyl activating group, and link to free

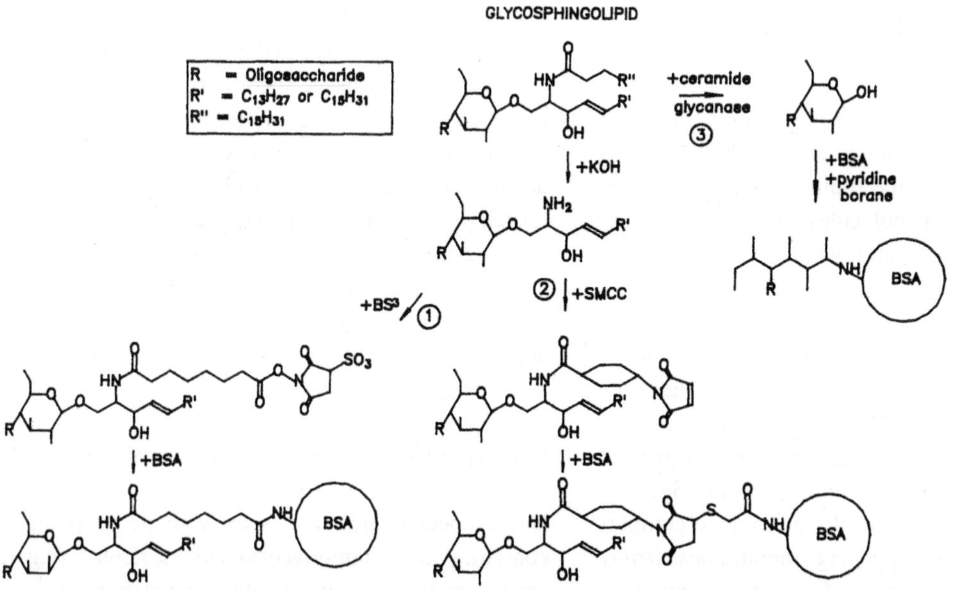

Fig. 1. Preparation of multivalent glycosphingolipid-based derivatives. Three schemes were used to covalently link glycosphingolipids or their oligosaccharides to BSA (a generalized glycosphingolipid is presented at the top of the figure). *Scheme 1* (Tiemeyer et al., 1989): Lyso-GT1b, prepared by alkaline deacylation of the sphingosine amine of purified glycosphingolipid (Neuenhofer et al., 1985), is reacted with excess bis(sulfosuccinimidyl) suberate (BS^3) to generate an active ester intermediate of the lysoglycosphingolipid. The sulfosuccinimidyl-activated intermediate is purified from un-reacted BS^3 by Iatrobeads chromatography before addition to BSA. Reaction of the sulfo-succinimidyl esters of the intermediate with primary amines on BSA results in formation of stable amide bonds linking the derivatized glycosphingolipid to the protein. *Scheme 2*: Lyso-GT1b is reacted with a heterobifunctional crosslinking reagent, succinimidyl 4-(N-maleimidomethyl)cyclohexane-1-carboxylate (SMCC, Pierce Chem. Co., Rockford, IL, USA), and the resulting maleimide-activated intermediate is chromatographically purified. The activated glycosphingolipid is added to BSA containing excess free sulfhydryl groups introduced chemically (at lysine amino groups) with the reagent N-succinimidyl S-acetylthioacetate (Pierce Chemical Co.), followed by hydroxylamine (Duncan et al., 1983). Reaction of the N-maleimidyl group with sulfhydryls on BSA results in formation of stable sulfate ether bonds linking the derivatized glycosphingolipid to the protein. *Scheme 3*: Glycosphingolipid is treated with ceramide glycanase (Zhou et al., 1989) to release the oligosaccharide, which is purified chromatographically and added to BSA in the presence of 10% pyridine borane as mild reducing agent (Cabacungan et al., 1982). Spontaneous reductive amination of lysines on BSA with the reducing terminal glucose of the glycosphingolipid results in a stable secondary amine linkage between the oligosaccharide and the protein.

sulfhydryls introduced into the BSA carrier (Scheme 2, Fig.1). This has led to routine incorporation of ≈ 13 GT1b molecules/BSA.

Both of the above schemes use strong alkaline hydrolysis to generate a lysoganglioside starting material, and are therefore inappropriate for glycosphingolipids retaining important alkali-labile groups, such as O-acetyl groups or sulfates. For these glycoconjugates we use a third method, isolation of the glycosphingolipid oligosaccharide and coupling directly to BSA via reductive amination. This effort has been greatly enhanced by the availability of the enzyme ceramide glycanase (Zhou et al., 1989) which quantitatively releases oligosaccharides from most glycosphingolipids under very mild conditions (Scheme 3, Fig.1). We have used this enzyme to readily generate oligosaccharides from gangliosides

and sulfoglucuronyl glycosphingolipids (see below). The isolated oligosaccharides spontaneously link to BSA under mild conditions in the presence of an appropriate reducing agent (e.g. pyridine-borane complex; Cabacungan et al., 1982). While the resulting conjugate has lost much of the glycosphingolipid structure (the ceramide is removed and the reducing terminal glucose is linearized), it retains the non-reducing saccharides which may be most important for receptor recognition in a *trans* configuration (see above).

The synthesized neoglycoconjugates are purified by gel filtration chromatography and ion exchange HPLC (Tiemeyer et al., 1989), which resolves them from underivatized saccharides or carrier. They are characterized by several techniques including carbohydrate compositional analysis (Table I), and gel

TABLE I

Carbohydrate compositional analysis of various neoganglioproteins

Neoganglio-protein	Sugar (mol/mol)	Sugar/ BSA (mol/mol)	Observed ratio (mol/mol)	Expected ratio (mol/mol)
(GT1b)$_4$BSA[a]	NeuAc	12.8	3.0	3
	GalNAc	4.2	1.0	1
	Gal	8.7	2.0	2
	Glc	4.4	1.0	1
(GT1b)$_{13}$BSA[b]	NeuAc	38.8	3.0	3
	GalNAc	12.2	0.9	1
	Gal	24.5	1.9	2
	Glc	14.2	1.1	1
(oGT1b)$_{11}$BSA[c]	NeuAc	33.0	3.0	3
	GalNAc	13.0	1.2	1
	Gal	19.1	1.7	2
	Glc	13.5	1.2	1

Neoganglioproteins were synthesized using one of three methods (see Fig. 1):

[a] N-sulfosuccinimidyl activation of lyso-GT1b

[b] Maleimidyl activation of lyso-GT1b

[c] Direct coupling of released GT1b oligosaccharide via reductive amination

In each case, the chromatographically purified conjugate was subjected to acid hydrolysis and the released saccharides were quantitated using a Dionex carbohydrate analysis system (strong cation exchange HPLC separation, pulsed amperometric detection; Lee, 1990).

188

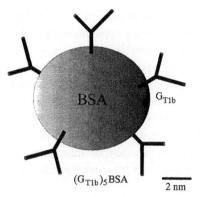

Fig. 2. Neoganglioprotein molecular model. The model represents the relative sizes of covalently bound GT1b molecules (linked as described in Fig.1) and BSA. The approximate size and shape of GT1b were determined using Desktop Molecular Modeler (Oxford University Press, Oxford, UK), while the size of BSA is based on X-ray crystallography studies of the closely related human serum albumin (Carter and He, 1990).

electrophoresis under denaturing conditions. The latter technique reveals that the conjugates migrate less rapidly and more diffusely than the parent BSA. In some cases oligosaccharide-specific reagents (such as labeled cholera B subunit) were used to demonstrate co-localization of glycosphingolipid oligosaccharide and protein on SDS-PAGE. These data demonstrated that the desired covalent conjugate was obtained. Following the nomenclature for neoglycoproteins (Kuhlenschmidt and Lee, 1984), we denote neoganglioproteins as, for instance, $(GT1b)_{13}BSA$, where the glycosphingolipid designation is in parentheses and the average number of molecules attached per carrier is subscripted. Conjugates are readily radioiodinated for use as radioligand probes.

Molecular modeling demonstrates that gangliosides form the major surface structures on the resulting conjugates, even though they constitute a minor portion of the total conjugate mass (Fig. 2). This is because the oligosaccharides (<2 kDa) spread out in space while the BSA carrier protein (>66 kDa) folds into a highly compact structure. Correct models of glycoconjugate-protein interactions must consider this disproportionately large size/mass ratio of complex carbohydrates compared to most proteins (Rademacher et al., 1988).

Ganglioside binding activity in the central nervous system

Our initial efforts to detect ganglioside receptors on brain membranes used $(GT1b)_nBSA$ conjugates. GT1b was chosen because of its relatively high concentration in brain (Yu and Saito, 1989). If structurally specific receptors exist, their concentration may reflect that of the recognized glycosphingolipid. Furthermore, GT1b can be isolated in sufficient quantities to optimize synthetic techniques, and our laboratory had demonstrated that GT1b immobilized in a surface array supported neuronal cell adhesion (Blackburn et al., 1986).

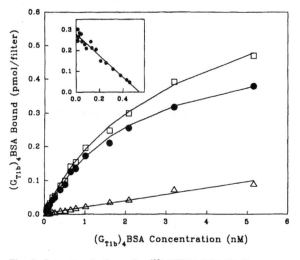

Fig. 3. Saturation isotherm for ^{125}I-$(GT1b)_4BSA$ binding to rat brain membranes (Tiemeyer et al., 1989). Brain 'P2' membranes were pre-treated with 0.3% sodium deoxycholate to remove endogenous glycosphingolipids, and the residual membrane structures (30 μg original membrane protein per reaction) were collected by filtration and incubated with the indicated concentrations of ^{125}I-$(GT1b)_4BSA$ in a total volume of 1 ml of binding buffer (50 mM Hepes pH 7.4, 10 mM calcium chloride, 0.016% Triton X-100, 1 mg/ml BSA) for 90 min at 8°C in the presence or absence of 10 μM GT1b. Unbound radioligand was removed by washing the membranes on filters, and bound radioactivity was determined. Specific binding (filled circles) is defined as total binding (squares) minus non-specific binding (triangles, binding in the presence of 10 μM GT1b). *Inset*: Data from saturation isotherms were transformed by the method of Scatchard (Scatchard, 1949) with the abscissa as bound ligand (pmol/filter) and the ordinate as bound/free ligand. The apparent single binding site has a K_D of 2 nM and a B_{max} of 20 pmol/mg P2 protein.

When ^{125}I-(GT1b)$_4$BSA was incubated with brain membranes (crude 'P2' fraction from rat brain), specific binding was not readily apparent. The possibility that ganglioside receptors might be masked by endogenous gangliosides in the preparation led us to test delipidated brain membrane preparations as a source of binding activity. Two methods for delipidation were successful in revealing ganglioside-specific binding activity: (1) mild detergent extraction; and (2) solvent solubilization followed by chromatographic resolution (see below). Mild detergent treatment of the membranes removed gangliosides and other lipids (as well as many membrane proteins) but left detergent-resistant membrane-like structures which were recovered by ultracentrifugation or filtration. When probed with ^{125}I-(GT1b)$_4$BSA, the residual brain membranes demonstrated high affinity and saturable radioligand binding (Fig. 3). Scatchard transformation of the binding data was linear with a K_D of 2 nM and a B_{max} of 20 pmol/mg 'P2' membrane protein.

Inhibition of ^{125}I-(GT1b)$_4$BSA binding to residual rat brain membranes by various lipids (added as

mixed Triton X-100 micelles) revealed ganglioside specificity (Fig. 4). The most potent ganglioside inhibitors were all of the '1b' series (GD1b, GT1b and GQ1b) with similar IC_{50} values ($\approx$ 100 nM). By comparison, GD1a and GD3 were one sixth as potent inhibitors compared to GD1b, although of equal anionic charge. GM1, which varies from GD1b only by the absence of the NeuAcα2-8 group was the least potent ganglioside inhibitor with an inhibitory potency <4% that of GD1b. Other anionic or neutral sphingolipids including asialo-GM1, globotetraosylceramide, sphingomyelin, sphingosine, and psychosine (galactosylsphingosine) were inefficient at blocking ^{125}I-(GT1b)$_4$BSA binding. Although selected phospholipids also inhibited with IC_{50} values as low as 0.5 μM, kinetic analyses of the inhibition of ^{125}I-(GT1b)$_4$BSA binding by gangliosides and phospholipids demonstrated that ganglioside inhibition was competitive and reversible while phospholipid inhibition was non-competitive and irreversible (Tiemeyer et al., 1989). These data demonstrate that the ^{125}I-(GT1b)$_4$BSA binding protein on brain membranes is selective for gangliosides carrying a defined oligosaccharide structure. This conclusion was supported by inhibition with the oligosaccharide released from GT1b (K_I = 8 μM). In contrast, the oligosaccharide from GM1 inhibited only $\approx$20% at 10 μM (the highest concentration tested). Galactose, glucose, N-acetylgalactosamine, N-acetylglucosamine, N-acetylneuraminic acid, lactose, N-acetyllactosamine, and N-acetylneuraminyllactose were non-inhibitory at 5 mM.

The relatively poor inhibitory potency of GT1b-oligosaccharide (8 μM) compared to GT1b (100 nM added as a mixed detergent micelle) supports a role for the ceramide portion of the ganglioside in receptor binding. The ceramide could act to enhance self-association of GT1b into multivalent arrays, to orient the oligosaccharide appropriately, or as part of the binding determinant. Data to address these alternatives was obtained by comparing the relative inhibitory potencies of GT1b and various GT1b neoganglioproteins (Table II). (GT1b)$_1$BSA and (GT1b)$_2$BSA were an order of magnitude less potent (per saccharide unit) in binding to this receptor than

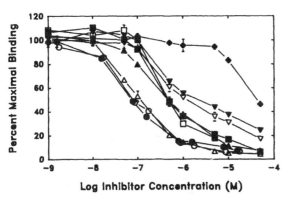

Fig. 4. Inhibition of ^{125}I-(GT1b)$_4$BSA binding to rat brain membranes by underivatized gangliosides (Tiemeyer et al., 1989). Brain 'P2' membranes were pretreated with mild detergent and incubated with 0.5 nM ^{125}I-(GT1b)$_4$BSA as described in Fig.3, except that the binding buffer contained underivatized gangliosides at the indicated concentrations. Binding in the presence of added ganglioside is expressed as a percent of the binding measured in the absence of added inhibitor (maximal binding was comparable to that shown in Fig.3). The gangliosides tested were: ○, GQ1b ; △, GT1b ; ●, GD1b; ▲, GD1a ; □, GD3; ■, GM3; ▽, GM2 ; ▼, GM1 ; ◆, GA1.

TABLE II

Inhibition of ^{125}I-(GT1b)$_4$BSA binding to rat brain membranes by various GT1b conjugates — Effect of valency and molecular form

Ganglioside conjugate	Valency	K_I conjugate (nM)	K_I per saccharide (nM)
GT1b (lipid in Triton micelles)	1–2	100	100–200
(GT1b)$_n$BSA (succinimidyl ester method)	1	81	81
	2	46	92
	4	2.6	10
	8	7.2	58
GT1b-oligosaccharide	1	8000	8000
(oGT1b)BSA (reductive amination)	11	60	660

Inhibitory potencies of GT1b, the oligosaccharide derived from GT1b (oGT1b), and various GT1b related neoganglioproteins were tested by performing inhibition titrations of ^{125}I-(GT1b)$_4$BSA binding to rat brain membranes as described in Fig. 4. The assay was conducted in the presence of 0.016% Triton X-100, resulting in approximately equimolar GT1b to mixed micelle concentration ratio. Due to variability in Triton X-100 critical micelle concentration and micelle size, a reasonable range of valency for GT1b 'lipid in Triton micelles' is given.

(GT1b)$_4$BSA, demonstrating the importance of multivalency (although higher order GT1b conjugates were not better than (GT1b)$_4$BSA). Likewise, GT1b oligosaccharide was an order of magnitude less potent as inhibitor (per saccharide unit) than a multivalent conjugate consisting of 11 GT1b *oligosaccharides*/per BSA, linked via reductive amination (see Fig. 1). However, multivalent oligosaccharide conjugate was far less potent as an inhibitor than multivalent glycosphingolipid. These data suggest that both multivalency and the structural contribution of the ceramide are important to ganglioside receptor recognition. Since gangliosides can cluster in the plane of a cell membrane, these factors may play a role in ganglioside recognition in vivo.

To gather information about the potential functions of this ganglioside binding activity, we determined its histological distribution and ontogeny in rat brain (Tiemeyer et al., 1990). Tissue and regional distribution studies revealed marked specificity for CNS tissue and differences among the different brain regions

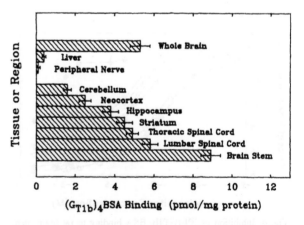

Fig. 5. Tissue and regional distribution of ^{125}I-(GT1b)$_4$BSA binding (Tiemeyer et al., 1990). Membranes were prepared by differential centrifugation of crude homogenates from whole rat brain, peripheral nerve, liver, or the indicated brain regions. All membranes were pre-treated with mild detergent under optimal conditions, collected by filtration, incubated with 0.5 nM ^{125}I-(GT1b)$_4$BSA, and specific binding determined as described in Fig. 3. The means of the specific binding activities per mg protein were calculated using multiple protein concentrations within the linear binding range.

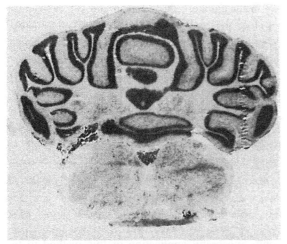

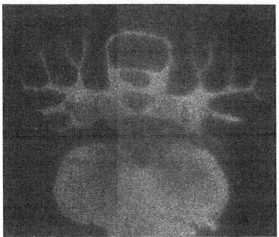

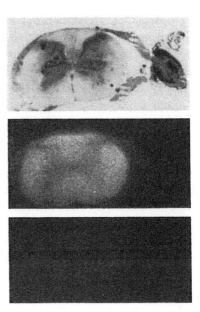

Fig. 7. Histological localization of ^{125}I-(GT1b)$_4$BSA binding to rat spinal cord and peripheral nerve structures (Tiemeyer et al., 1990). Transverse serial sections of frozen rat spinal cord were prepared, overlaid with buffer containing ^{125}I-(GT1b)$_4$BSA, and processed for autoradiographic determination of specific binding as described in Fig. 6. The resulting autoradiographic image (center panel) is presented along with an adjacent section treated identically except for a pre-incubation with 10 μM ganglioside GT1b as inhibitor (bottom panel). For comparison, another adjacent section stained with cresyl violet is presented (top panel). Note that the white matter within the spinal cord was heavily labeled, while dorsal and ventral spinal roots and the dorsal root ganglion (at the right of the section) did not bind radioligand.

Fig. 6. Histological localization of ^{125}I-(GT1b)$_4$BSA binding to coronal rat brain sections (Tiemeyer et al., 1990). Serial coronal sections were prepared from frozen brain, placed on gelatin-coated slides, lightly fixed, treated with 0.3% sodium deoxycholate to remove endogenous lipids, air dried, and overlaid with buffer containing 5 nM ^{125}I-(GT1b)$_4$BSA. After 90 min at 8°C, unbound radioligand was removed by washing, the slides dried, and subjected to autoradiography. The autoradiographic image (lower panel) is presented along with a cresyl violet stained adjacent section (upper panel) for comparison. Note the absence of labeling of cerebellar grey matter, while the white matter tracts within the cerebellum (and in the underlying medulla) are heavily labeled. An adjacent section incubated with radioligand in the presence of 10 μM GT1b as inhibitor was unlabelled (data not shown).

(Fig. 5). Rat brain membranes supported >10-fold higher ^{125}I-(GT1b)$_4$BSA binding compared to liver membranes, and binding of the ligand to peripheral nerve membranes was insignificant. Within the CNS, areas rich in white matter supported more binding than those rich in gray matter. These data suggested that this receptor might be associated with myelinated fibers rather than neuronal soma or synaptosomes. Subcellular fractionation confirmed that ganglioside binding activity is found in myelin, with an 8-fold enrichment of ^{125}I-(GT1b)$_4$BSA binding to myelin (per mg membrane protein) when compared to crude 'P2' membranes. Ontogenic studies further supported the association of this activity with myelin; ^{125}I-(GT1b)$_4$BSA binding appeared first at 15 days postnatal and was half-maximal at ≈ 50 days, paralleling the appearance of myelination. Binding of ^{125}I-(GT1b)$_4$BSA directly to sections of adult rat brain revealed predominant ganglioside-specific binding to white matter tracts throughout the brain. Even in the gray matter-rich cerebellum, which has relatively little neoganglioprotein binding (cf. Fig.5), the white matter tracts are clearly labeled (Fig. 6). Notably, spinal cord white matter displayed substantial binding while attached dorsal root ganglia, as well as dorsal and ventral roots were unstained (Fig. 7). Therefore, the '1b'-directed binding activity we have characterized is specific for CNS myelin, which is elaborated by oligodendrocytes, and is absent from PNS myelin, which is elaborated by Schwann cells. We refer to this activity as the CNS 'myelin ganglioside receptor' (MGR). The observation that myelin is very poor in the '1b' gangliosides while the axons

they ensheathe are rich in GD1b and GT1b have led us to hypothesize that the MGR is involved in 'trans' recognition between oligodendroglia and axons. Further testing of this hypothesis awaits immunological and molecular tools, the development of which are the focus of current research.

A major current goal of our studies is the purification of the MGR. Solubilization and chromatographic resolution of the highly hydrophobic proteins of myelin are technical challenges. Recent studies (Diaz et al., 1991) demonstrate total solubilization of myelin membranes in tetrahydrofuran/water mixtures and bulk resolution of solubilized proteins from extracted lipids by LH-60 size exclusion chromatography. Successful application of this method to the myelin ganglioside receptor is shown in Fig. 8. Binding activity (measured using a receptor-ligand complex precipitation assay) is low in THF/water extracts of myelin membranes (data not shown), presumably due to the presence of inhibitory lipids. However, full activity is revealed in the protein fraction after LH-60 chromatographic resolution from lipids. A comparison of the THF-solubilized activity

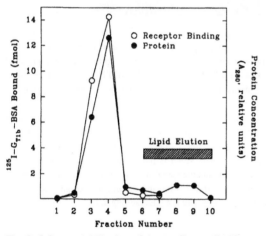

Fig. 8. Solvent-solubilization of the myelin ganglioside receptor: resolution from endogenous lipids via LH-60 column chromatography. Myelin prepared from rat CNS (Norton and Poduslo, 1973) was collected by ultracentrifugation and nearly completely solubilized in 80% tetrahydrofuran, 0.1% trifluoracetic acid (Diaz et al., 1991). The resulting extract was subjected to LH-60 size exclusion chromatography in the same solvent to resolve myelin proteins from lipids. Protein was detected in each fraction by scanning UV spectroscopy, while lipids were detected by thin layer chromatography using a general lipid stain (technique as in Fig. 10). Ganglioside receptor activity was determined by incubating aliquots of each fraction with 50 pM ^{125}I-(GT1b)$_{13}$BSA, essentially as described in Fig.3, except that the reaction was terminated by addition of an equal volume of 15% polyethylene glycol in 50 mM Hepes (pH 7.4) to precipitate the receptor-ligand complex, which was then collected on glass fiber filters. Background binding (in the presence of 10 μM GT1b) was generally <10% of maximal binding, and was subtracted from total binding for each fraction.

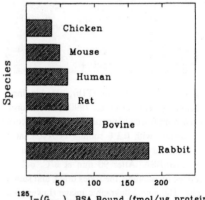

Fig. 9. Ganglioside binding activity in myelin from various species. Myelin was prepared using CNS tissues from the species indicated. The membranes were extracted and the solubilized proteins resolved from lipids by LH-60 column chromatography (see Fig. 8). Protein-containing, lipid-depleted fractions were combined and aliquots removed for determination of protein (using a dye binding assay) and specific radioligand binding (using 60 pM ^{125}I-(GT1b)$_{13}$BSA) as described in Fig. 8. The mean of the specific binding per μg protein was calculated from data using multiple protein concentrations within the linear binding range.

to that in mild detergent-treated membranes demonstrates similar binding kinetics and inhibition by gangliosides (data not shown).

Recovery of total myelin proteins free of lipids aided in the comparison of myelin ganglioside receptor activity from various species. Myelin was prepared from various vertebrate species (including human), solubilized in THF/water, the proteins separated from lipids by LH-60 chromatography, and MGR activity measured. Significant ganglioside-specific binding of ^{125}I-$(GT1b)_{13}BSA$ was detected in all species tested, although at different levels (Fig. 9). Binding activities from all species were similarly inhibited by GT1b (IC_{50} values ≈ 100 nM), and GM1 was a markedly less potent inhibitor in all cases (data not shown).

The '1b'-directed receptor described here is distinct from the major myelin protein, myelin basic protein (MBP), which has been reported to bind to ganglioside GM1 (Yohe et al., 1983). Total brain membranes from normal and 'shiverer' mutant mice had identical levels of ^{125}I-$(GT1b)_{13}BSA$ binding (data not shown). This is notable since 'shiverer' mice carry a large deletion mutation in the MBP gene (Campagnoni and Macklin, 1988). Current efforts are directed at utilizing chromatographic and molecular biological techniques to purify the MGR.

Sulfoglucuronyl glycosphingolipid binding activity in the peripheral nervous system

A striking observation from our studies on GT1b-BSA binding activity is its absence from the peripheral nervous system (PNS), where myelination of

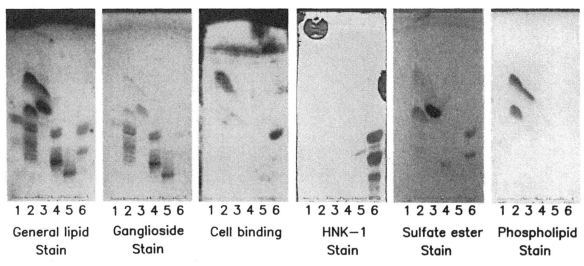

1 2 3 4 5 6	1 2 3 4 5 6	1 2 3 4 5 6	1 2 3 4 5 6	1 2 3 4 5 6	1 2 3 4 5 6
General lipid Stain	Ganglioside Stain	Cell binding	HNK−1 Stain	Sulfate ester Stain	Phospholipid Stain

Fig. 10. Adhesion of primary Schwann cells to TLC-resolved SGNL-lipids (Needham and Schnaar, 1990). Total polar lipids from dog sciatic nerve, isolated using the method of Svennerholm and Fredman (1980), were applied to DEAE-Sepharose in organic solvent and anionic lipid classes eluted stepwise using ammonium acetate in methanol (Magnani et al., 1987). Each of six fractions was subjected to HPTLC in chloroform/methanol/0.25% aqueous KCl (60:35:8) on replicate plates. As indicated, replicate plates were treated as follows: *General Lipid Stain* was via copper sulfate/phosphoric acid char (Yao and Rastetter, 1985). *Ganglioside Stain* was via the resorcinol reagent of Svennerholm (1963) as detailed previously (Dahms and Schnaar, 1983). *Cell binding* was performed in specially designed TLC adhesion chambers as previously described (Swank-Hill et al., 1987). Rat primary Schwann cells were isolated, cultured, and metabolically radiolabeled with inorganic phosphate (Needham et al., 1987). The labeled cells were collected by mild trypsinization, placed in contact with the TLC plate for 30 min at 37°C, then non-adherent cells removed by centrifugation. Adherent cells were fixed in place and detected by autoradiography. *HNK-1 Stain* refers to immunochemical detection using a mouse monoclonal antibody which recognizes SGNL lipids, and was performed as described previously (Magnani et al., 1987). *Sulfate Ester Stain* was via dye binding with Azure A (Iida et al., 1989). *Phospholipids Stain* was via molybdenum blue reagent (Alltech Assoc, Deerfield, IL, USA). Note that Schwann cells bound selectively to the major species of the SGNL lipid, which is a minor component of total PNS lipids (general stain) but are readily detected (along with other minor members of this structural family) by the intensely staining HNK-1 monoclonal antibody.

axons is performed by Schwann cells rather than oligodendroglia. Demyelinating peripheral neuropathies (Quarles et al., 1986) appear in individuals producing monoclonal antibodies reactive with an unusual class of acidic lipids found predominantly in peripheral nerves, the sulfoglucuronyl neolactosylceramides (SGNL-lipids). These glycosphingolipids have neolactotetraosyl or neolactohexosyl cores substituted with a 3-O-sulfated glucuronic acid moiety on the 3-position of the non-reducing terminal galactose (Chou et al., 1986; Ariga et al, 1987).

To test the possibility that SGNL lipids serve as recognition molecules in PNS myelination, we first tested their ability to support intact Schwann cell

recognition using direct cell adhesion to TLC-resolved glycosphingolipids (Needham and Schnaar, 1990; Needham and Schnaar, 1991). Radiolabeled primary Schwann cells readily bound to small amounts of SGNL lipids from peripheral nerve when resolved on TLC plates, while gangliosides were completely non-supportive of adhesion (Fig. 10).

Direct probing and characterization of a potential SGNL receptor on PNS membranes was accomplished (Needham and Schnaar, 1993) by synthesizing a radioiodinated neoglycoconjugate consisting of SGNL oligosaccharides covalently attached, via reductive amination, to BSA (using Scheme 3, Fig. 1). The structure of the resulting ^{125}I-(SGNL)$_7$BSA was confirmed by carbohydrate analysis and SDS-PAGE electrophoresis, with detection using a mouse monoclonal antibody, HNK-1, specific for the sulfoglucuronyl neolacto structure (Ilyas et al., 1990).

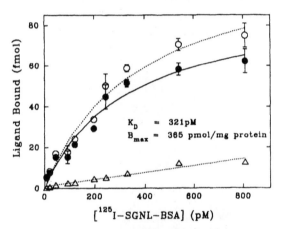

Fig. 11. Saturation isotherm for ^{125}I-SGNL-BSA binding to rat PNS myelin. Rat PNS myelin membranes were isolated from rat sciatic nerve by homogenization and differential centrifugation (Oulton and Mezei, 1976). An aliquot of membrane suspension (containing 0.25 μg membrane protein) was incubated with the indicated concentrations of ^{125}I-SGNL-BSA in 0.25 ml of binding buffer (50 mM imidazole acetate (pH 7.4), 0.032% Triton X-100, 1 mg/ml BSA, 20 mM CaCl$_2$). After 30 min at 4°C membrane-bound radioligand was separated from unbound ligand by filtration (open circles, total bound). Ligand that was non-specifically bound (open triangles), determined in the absence of tissue, was subtracted from the total bound at each ligand concentration to calculate specific bound (filled circles). Values are plotted versus the concentration of free ^{125}I-SGNL-BSA competent to bind PNS myelin membranes (the percent of radioligand which bound to a saturating amount of membrane protein). The solid line was generated by a non-linear fit of the specific binding data using a single site model and yielded the indicated apparent dissociation constant (K$_D$) and total receptor concentration (B$_{max}$) values (Needham and Schnaar, 1993).

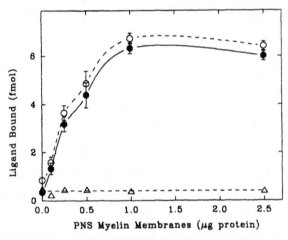

Fig. 12. Binding of ^{125}I-SGNL-BSA to rat PNS myelin: Effect of calcium ions. Rat PNS myelin membranes at the indicated protein amounts were incubated with 0.5 nM ^{125}I-SGNL-BSA for 30 min at 4°C in a total volume of 250 μl binding buffer in the presence or absence of 20 mM CaCl$_2$, and membrane-bound ligand was separated from free ligand by filtration. Calcium-specific binding (closed circles) was calculated by subtracting non-specific binding (open triangles, binding in the absence of CaCl$_2$) from total binding (open circles, binding in the presence of 20 mM CaCl$_2$). Nonspecific binding was essentially identical when defined as binding in the absence of calcium ions, in the presence of calcium plus 10 μM SGNL lipid as inhibitor (see Table III), or in the absence of tissue (see Fig. 11) (Needham and Schnaar, 1993).

TABLE III

Inhibition of ^{125}I-SGNL-BSA binding to PNS Myelin by glyco-sphingolipids

Glycosphingolipid (n)	IC_{50} (range) (μM)
SGNL-lipid, bovine (5)	0.99 (0.55–1.2)
SGNL-lipid, human (7)	0.48 (0.3–0.6)
Desulfated[a] SGNL-lipid, human (2)	1.5 (1–2)
Methyl ester of desulfated SGNL-lipid[a] (1)	>25
Sulfatide (4)	8.2 (4–12)
sialylneolactotetraosylceramide (1)	40
2,3 sialyl Lewis x glycolipid (1)	>10
2,6 sialyl Lewis x glycolipid (1)	>10
GT1b (4)	33 (6–64)
GD1a (2)	58 (15–>100)
GD1b (2)	>100
GM1 (1)	>100
Globotetraosylceramide (1)	>100
Galactosylceramide (1)	>100

Binding studies were performed as described in Fig. 12 using 7.5 μg membrane protein per reaction, except that membranes were preincubated (30 min, 4°C) with various concentrations of the indicated lipid inhibitors. Specific binding at each inhibitor concentration was determined using calcium-free buffer to define non-specific binding. The mean (and range) concentration of each inhibitor resulting in 50% inhibition of specific ^{125}I-SGNL-BSA binding was determined graphically from the indicated number (n) of independent binding experiments performed in triplicate (Needham and Schnaar, 1993).

[a] Prepared as described previously (Ilyas et al., 1990).

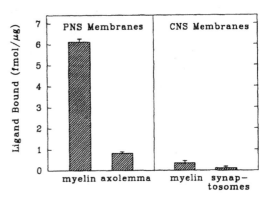

Fig. 13. Tissue and subcellular specificity of ^{125}I-SGNL-BSA binding. *Left panel*: Rat peripheral nerve endoneurium homogenate was fractionated by differential centrifugation and discontinuous sucrose gradient centrifugation to separate PNS myelin from axolemma (Oulton and Mezei, 1976). *Right panel*: Rat brain homogenate was fractionated similarly to isolate CNS myelin and synaptosomes (Gray and Whittaker, 1962). Specific binding of 0.5 nM ^{125}I-SGNL-BSA to aliquots of these fractions containing 1 μg membrane protein was performed as described in Fig. 12.

When peripheral nerve membranes (predominantly PNS myelin) were probed with ^{125}I-SGNL-BSA, saturable binding was detected (Fig. 11). Unlike the CNS myelin ganglioside receptor, SGNL receptor was absolutely dependent on calcium ions for binding (Fig. 12), with maximal binding in the presence of 10 mM calcium. Of various lipids tested as inhibitors of ^{125}I-SGNL-BSA binding to PNS membranes (Table III), SGNL lipid itself was the most potent (IC_{50} < 1 μM), while gangliosides were <3% as effective. Of the glycosphingolipids tested (Table III) only

sulfatide was within an order of magnitude as effective as SGNL lipid as an inhibitor. Structural studies using SGNL lipid derivatives were revealing, in that desulfation (to generate IV^3GlcA-nLc_4Cer) resulted in only a modest (3-fold) decrease in inhibitory potency. In contrast, methyl esterification of the desulfated derivative (at the GlcA 6-position) eliminated inhibitory potency (Table III). These data indicate that the carboxylic acid on the terminal glucuronic acid is a major determinant for SGNL binding to its PNS membrane receptor, while the 3-O-sulfate group contributes but is not critical to binding. It is interesting to note that non-sulfated forms of the SGNL lipid have been found broadly distributed phylogenetically (Dennis et al., 1988), and may be more ancient than gangliosides or SGNL lipids.

When PNS membranes are separated into axolemmal (denser membranes) and myelin fractions, the activity fractionates with the myelin (Fig. 13), suggesting a possible role in 'trans' interactions between Schwann cells and axolemma. Furthermore, limited tissue distribution studies suggest that SGNL receptor serves a specific role in the PNS, in that binding activity is essentially absent from CNS membranes, whether tested with crude 'P2', purified myelin, or synaptosomal membranes.

Conclusions and perspectives

Through use of semi-synthetic glycosphingolipid-based neoglycoconjugates, which combine multivalence, sequestering or removal of the lipid moiety, and facile radioiodination, we have been successful in demonstrating unique binding activities for gangliosides and sulfoglucuronyl glycosphingolipids in the CNS and PNS respectively. Although identification of the functional roles of these glycosphingolipid-specific binding activities await further studies, their differential tissue and membrane distribution suggests that they may be involved in 'trans' interactions between myelinating cells and axons.

Acknowledgements

The work described was supported in part by National Institutes of Health (NIH) grant HD14010, Multiple Sclerosis Society grant RG-2223-A-1, and NIH training grants MH18030 (to MT & LKN) and GM07626 (to JAM).

References

Ariga, T., Kohriyama, T., Freddo, L., Latov, N., Saito, M., Kon, K., Ando, S., Suzuki, M., Hemling, M.E., Rinehart, K.L., Kusunoki, S. and Yu, R.K. (1987) Characterization of sulfated glucuronic acid containing glycolipids reacting with IgM M-proteins in patients with neuropathy. *J. Biol. Chem.*, 262: 848–853.

Blackburn, C.C., Swank-Hill, P., and Schnaar, R.L. (1986) Gangliosides support neural retina cell adhesion. *J. Biol. Chem.*, 261: 2873–2881.

Brandley, B.K., Swiedler, S.J. and Robbins, P.W. (1990) Carbohydrate ligands of the LEC cell adhesion molecules. *Cell*, 63: 861–863.

Cabacungan, J.C., Ahmed, A.I. and Feeney, R.E. (1982) Amine boranes as alternative reducing agents for reductive alkylation of proteins. *Anal. Biochem.*, 124: 272–278.

Campagnoni, A.T. and Macklin, W.B. (1988) Cellular and molecular aspects of myelin protein gene expression. *Mol. Neurobiol.*, 2: 41–89.

Carter, D.C. and He, X.M. (1990) Structure of human serum albumin. *Science*, 249: 302–303.

Chou, D.K.H., Ilyas, A.A., Evans, J.E., Costello, C., Quarles, R.H. and Jungalwala, F.B. (1986) Structure of sulfated glucuronyl glycolipids in the nervous system reacting with HNK-1 antibody and some IgM paraproteins in neuropathy. *J. Biol. Chem.*, 261: 11717–11725.

Connolly, D.T., Townsend, R.R., Kawaguchi, K., Hobish, M.K., Bell, W.R. and Lee, Y.C. (1983) Binding and endocytosis of glycoproteins and neoglycoproteins by isolated rabbit hepatocytes. *Biochem. J.*, 214: 421–431.

Cuello, A.C. (1990) Glycosphingolipids that can regulate nerve growth and repair. *Adv. Pharmacol.*, 21: 1–50.

Dahms, N.M. and Schnaar, R.L. (1983) Ganglioside composition is regulated during differentiation in the neuroblastoma X glioma hybrid cell line NG108–15. *J. Neurosci.*, 3: 806–817.

Dennis, R.D., Antonicek, H., Wiegandt, H. and Schachner, M. (1988) Detection of the L2/HNK-1 carbohydrate epitope on glycoproteins and acidic glycolipids of the insect *Calliphora vicina*. *J. Neurochem.*, 51: 1490–1496.

Diaz, R.S., Regueiro, P., Monreal, J. and Tandler, C.J. (1991) Selective extraction, solubilization, and reversed–phase high-performance liquid chromatography separation of the main proteins from myelin using tetrahydrofuran/water mixtures. *J. Neurosci. Res.*, 29: 114–120.

Duncan, R.J., Weston, P.D. and Wrigglesworth, R. (1983) A new reagent which may be used to introduce sulfhydryl groups into proteins, and its use in the preparation of conjugates for immunoassay. *Anal. Biochem.*, 132: 68–73.

Gray, E.G. and Whittaker, V.P. (1962) The isolation of nerve endings from brain: an electron-microscopic study of cell fragments derived by homogenization and centrifugation. *J. Anat.*, 96: 79–88.

Hakomori, S. (1990) Bifunctional role of glycosphingolipids: Modulators for transmembrane signaling and mediators for cellular interactions. *J. Biol. Chem.*, 265: 18713–18716.

Hanai, N., Nores, G.A., MacLeod, C., Torres-Mendez, C. and Hakomori, S. (1988) Ganglioside-mediated modulation of cell growth. Specific effects of GM3 and Lyso-GM3 in tyrosine phosphorylation of the epidermal growth factor receptor. *J. Biol. Chem.*, 263: 10915–10921.

Igarashi, Y., Nojiri, H., Hanai, N. and Hakomori, S. (1989) Gangliosides that modulate membrane protein function. *Methods Enzymol.*, 179: 521–541.

Iida, N., Toida, T., Kushi, Y., Handa, S., Fredman, P., Svennerholm, L. and Ishizuka, I. (1989) A sulfated glucosylceramide from rat kidney. *J. Biol. Chem.*, 264: 5974–5980.

Ilyas, A.A., Chou, D.K.H., Jungalwala, F.B., Costello, C. and Quarles, R.H. (1990) Variability in the structural requirements for binding of human monoclonal anti-myelin-associated glycoprotein immunoglobulin M antibodies and HNK-1 to sphingoglycolipid antigens. *J. Neurochem.*, 55: 594–601.

Kuhlenschmidt, T.B. and Lee, Y.C. (1984) Specificity of chicken liver carbohydrate binding protein. *Biochemistry*, 23: 3569–3575.

Lee, Y.C. (1989) Binding modes of mammalian hepatic

Gal/GalNAc receptors. *Ciba Found. Symp.*, 145: 80–95.

Lee, Y.C. (1990) High-performance anion-exchange chromatography for carbohydrate analysis. *Anal. Biochem.*, 189: 151–162.

Lee, Y.C. and Lee, R.T. (1992) Synthetic glycoconjugates. In H.J. Allen and E.C. Kisailus (Eds.), *Glycoconjugates: Composition, Structure, and Function*, Marcel Dekker, New York, pp. 121–165

Lee, R.T., Lin, P. and Lee, Y.C. (1984) New synthetic cluster ligands for galactose/*N*-acetylgalactosamine-specific lectin of mammalian liver. *Biochemistry*, 23: 4255–4261.

Magnani, J.L., Spitalnik, S.L. and Ginsburg, V. (1987) Antibodies against cell surface carbohydrates: determination of antigen structure. *Methods Enzymol.*, 138: 195–207.

Needham, L.K. and Schnaar, R.L. (1990) Adhesion of primary Schwann cells to HNK–1 reactive glycolipids. *Ann. N. Y. Acad. Sci.*, 605: 416–419.

Needham, L.K. and Schnaar, R.L. (1991) Adhesion of primary Schwann cells to HNK-1 reactive glycosphingolipids: Cellular specificity. *Ann. N. Y. Acad. Sci.*, 633: 553–555.

Needham, L.K. and Schnaar, R.L. (1993) Carbohydrate recognition in the peripheral nervous system: A calcium-dependent membrane binding site for HNK-1 reactive glycolipids potentially involved in Schwann cell adhesion. *J. Cell Biol.*, 121: 397–408.

Needham, L.K., Tennekoon, G.I. and McKhann, G.M. (1987) Selective growth of rat Schwann cells in neuron-and serum-free primary culture. *J. Neurosci.*, 7: 1–9.

Neuenhofer, S., Schwarzmann, G., Egge, H. and Sandhoff, K. (1985) Synthesis of lysogangliosides. *Biochemistry*, 24: 525–532.

Norton, W.T. and Poduslo, S.E. (1973) Myelination in rat brain: Method of myelin isolation. *J. Neurochem.*, 21: 749–757.

Oulton, M.R. and Mezei, C. (1976) Characterization of myelin of chick sciatic nerve during development. *J. Lipid Res.*, 17: 167–175.

Quarles, R.H., Ilyas, A.A. and Willison, H.J. (1986) Antibodies to glycolipids in demyelinating diseases of the human peripheral nervous system. *Chem. Phys. Lipids*, 42: 235–248.

Rademacher, T.W., Parekh, R.B. and Dwek, R.A. (1988)

Glycobiology. *Annu. Rev. Biochem.*, 57: 785–838.

Scatchard, G. (1949) The attractions of proteins for small molecules and ions. *Ann. N.Y. Acad. Sci.*, 51: 660–672.

Schnaar, R.L. (1991) Glycosphingolipids in cell surface recognition. *Glycobiology*, 1: 477–485.

Stahl, P., Schlesinger, P.H., Sigardson, E., Rodman, J.S. and Lee, Y.C. (1980) Receptor-mediated pinocytosis of mannose glycoconjugates by macrophages: Characterization and evidence for receptor recycling. *Cell*, 19: 207–215.

Svennerholm, L. (1963) Sialic acids and derivatives: Estimation by the ion-exchange method. *Methods Enzymol.*, 6: 459–462.

Svennerholm, L. and Fredman, P. (1980) A procedure for the quantitative isolation of brain gangliosides. *Biochim. Biophys. Acta*, 617: 97–109.

Swank-Hill, P., Needham, L.K. and Schnaar, R.L. (1987) Carbohydrate-specific cell adhesion directly to glycosphingolipids separated on thin-layer chromatography plates. *Anal. Biochem.*, 163: 27–35.

Tiemeyer, M., Yasuda, Y. and Schnaar, R.L. (1989) Ganglioside-specific binding protein on rat brain membranes. *J. Biol. Chem.*, 264: 1671–1681.

Tiemeyer, M., Swank-Hill, P. and Schnaar, R.L. (1990) A membrane receptor for gangliosides is associated with central nervous system myelin. *J. Biol. Chem.*, 265: 11990–11999.

Yao, J.K. and Rastetter, G.M. (1985) Microanalysis of complex tissue lipids by high-performance thin-layer chromatography. *Anal. Biochem.*, 150: 111–116.

Yohe, H.C., Jacobson, R.I. and Yu, R.K. (1983) Ganglioside-basic protein interaction: Protection of gangliosides against neuraminidase action. *J. Neurosci. Res.*, 9: 401–412.

Yu, R.K. and Saito, M. (1989) Structure and localization of gangliosides. In R.U. Margolis and R.K. Margolis (Eds), *Neurobiology of Glycoconjugates*, Plenum Press, New York, pp. 1–42.

Zhou, B., Li, S.-C., Laine, R.A., Huang, R.T.C. and Li, Y.T. (1989) Isolation and characterization of ceramide glycanase from the leech, *Macrobdella decora. J. Biol. Chem.*, 264: 12272–12277.

SECTION IV

Gangliosides and Cancer

L. Svennerholm, A.K. Asbury, R.A. Reisfeld, K. Sandhoff, K. Suzuki, G. Tettamanti and G. Toffano (Eds.)
Progress in Brain Research, Vol. 101
© 1994 Elsevier Science BV. All rights reserved.

CHAPTER 15

Potential of genetically engineered anti-ganglioside GD2 antibodies for cancer immunotheraphy

Ralph A. Reisfeld[1], Barbara M. Mueller[1], Rupert Handgretinger[2],
Alice L. Yu[3] and Stephen D. Gillies[4]

[1]*The Scripps Research Institute, Department of Immunology, La Jolla, CA, U.S.A.* [2]*Universitäts Kinderklinik, University of Tübingen, Tübingen, Germany* [3]*University of California, San Diego, CA, U.S.A.* [4]*Fuji ImmunoPharmaceuticals, Inc., Lexington, MA, U.S.A.*

Introduction

A considerable amount of evidence has accumulated during the last 15 years, which indicates that preferentially expressed antigens on human tumor cells are potentially important targets for cancer therapy. This finding was extended and strengthened by recent, new insights into the complex interactions of host regulatory systems, including the immune system and the control of cell growth and function. Progress was accelerated since these conceptual advances occurred in the same time frame as the generation of monoclonal antibodies (mAbs) (Köhler and Milstein, 1975) and the rapid expansion of novel genetic engineering technologies. These important conceptual and technical advances contributed to the development of genetically engineered mAbs and chemically defined biological response modifiers, including a variety of growth factors and cytokines which were found to act as stimulators or inhibitors of many specialized cellular functions and regulatory mechanisms. This extensive array of chemically defined biological agents catalyzed a number of new approaches to cancer therapy that involved mAbs directed to antigens preferentially expressed on tumor cells. One can group these strategies into the following categories based on their different mechanisms of action:

1. Activation of components or effector cells of the immune system, including triggering of complement- or antibody-dependent cytotoxicity.
2. Tumor site-specific targeting of cytokines and growth factors with genetically engineered fusion proteins involving antibodies or their fragments.
3. Redirection of effector cells to tumor sites by bispecific mAbs, directed both to tumor cell and T cell or monocyte surface markers.
4. Delivery of radionuclides, toxins, or chemotherapeutic drugs by conjugation with mAbs directed to tumor cell antigens.
5. Interference with tumor cell growth or differentiation by antibodies directed against growth factor receptors.

Based on presently available data, it is evident that most of these applications of mAbs for cancer therapy are still very much in their beginning phases of development, and that only very few of them have seen critical evaluation in the clinic. However, some preclinical tests have shown encouraging results, and it is evident that mAbs are here to stay and will contribute to the adjuvant therapy of cancer.

The following chapter deals with only some of these different mechanisms of action by which mAbs may impact on cancer immunotherapy. No attempt is made to provide a comprehensive review and only the first two approaches to immunotherapy listed above are discussed and augmented with data from the authors and several other investigators.

Activation of effector cells of the immune system

For almost a century, attempts have been made to optimize specific passive immunotherapy with xenogeneic and allogeneic antisera made against cancer cells in various animal hosts. However, such studies provided mainly anecdotal results, because the polyclonal antibodies used were generally impure and poorly characterized, often preventing duplication in controlled trials. The more recent development of monoclonal antibodies facilitated a more systematic approach to specific passive immunotherapy, since it provided for relatively large amounts of specific, uniform, and pure antibodies.

Initially, the majority of studies involving monoclonal antibody therapy dealt with hematologic neoplasms involving more than 25 clinical trials in which over 135 patients were treated with mAb. Such studies showed a complete remission rate of 5%, a partial remission rate of 16%, and a minor response rate of 17%, with most of these regressions being of short duration (Janson et al., 1989; Goldenberg et al., 1990). Results from a number of clinical trials were reported involving the treatment of solid tumors with mouse mAb were done with melanoma (Houghton et al., 1985; Cheung et al., 1987; Dippold et al., 1988; Vadhan-Raj et al. 1988) neuroblastoma (Cheung et al., 1987; Handgretinger et al., 1992), and colon carcinoma (Sears et al., 1982; Douillard et al., 1986; Mellstedt et al., 1989). Although the results of most of these studies showed relatively few complete remissions or objective regressions of some duration, it should be emphasized that the majority of patients included in such trials were heavily pretreated and had very advanced disease that no longer responded to conventional therapies. Consequently, many of these patients' immune systems were most likely too heavily compromised to respond to immunotherapy. In addition, the use of mouse antibodies often evoked relatively strong human anti-mouse antibody

(HAMA) responses that severely limited dose and duration of treatments. Other factors that may have limited more rapid progress in cancer therapy with mAbs per se include: (1) the well-known heterogeneity of antigen expression on tumor cells; (2) the presence, in some cases, of free circulating antigen, leading to immune complexes with the injected mAb; (3) modulation or disappearance of target antigens, especially on hematologic tumors; and (4) the relatively short half-life of mouse antibody in the human circulation coupled with a low tumor uptake of injected mAb, often ranging from 0.001% to 0.01% of injected dose per gram of tumor (DeNardo et al., 1986; Dippold et al., 1988). Nonetheless, in spite of all these handicaps, mAbs administered alone have shown some definite antitumor effects, especially in patients with melanoma (Houghton et al, 1985; Dippold et al., 1988; Vadhan-Raj et al., 1988) and neuroblastoma (Cheung et al., 1987).

Based on these clinical findings and with the previously mentioned considerations in mind, we focused our efforts on a monoclonal antibody directed against ganglioside GD2, a carbohydrate antigen, that is preferentially expressed on human melanoma and neuroblastoma cells. This anti-GD2 antibody can kill these tumor cells by activating complement and human effector cells. The following provides an account of some of the successes of these research efforts.

Human/mouse chimeric anti-GD2 antibody ch14.18

One distinct advantage of mAbs directed to carbohydrate antigens, such as disialoganglioside GD2, is their potential use to establish a structure-function relationship for these determinants on the tumor cell surface. This is possible primarily because mAbs can recognize a carbohydrate determinant with known sugar composition and anomeric linkages. Thus, by using mAbs to define oligosaccharide structures on the tumor cell surface, one can pose questions regarding the functional properties of these structures as they relate to the malignant or metastatic phenotype. This is much more difficult to do with mAbs directed to protein or glycoprotein antigens whose epitopes are in most cases structurally ill-defined, since they may depend on yet unresolved conformation and three-dimensional structures, and their complete primary amino acid sequence is frequently not available.

The development of our human/mouse chimeric mAb, ch14.18, was stimulated by two major problems that complicate the critical evaluation of the therapeutic benefits of mouse mAb reactive with human tumor-associated antigens. First, such antibodies are recognized as foreign by the human immune system and thus evoke a HAMA response. Second, depending on isotype, the Fc portion of a murine mAb may not be able to activate human complement and human effector cells as efficiently as a human Fc portion. These limitations, together with data suggesting that complement-dependent cytotoxicity and antibody-dependent cellular cytotoxicity correlate with growth suppression of human tumor xenografts in immunodeficient mice, led to the development of human/mouse chimeric antibodies that usually contain the mouse variable (V) regions joined to human constant (C) regions (Neuberger et al., 1984; Gillies et al, 1989; Mueller et al., 1990a, 1990b).

Our first effort was to construct a chimeric antibody from mouse mAb 14.18 directed against ganglioside GD2, which is preferentially expressed on human melanoma and neuroblastoma cells. GD2 is a good target antigen for chimeric antibody-mediated tumor cell killing, since this chemically defined cell surface molecule is expressed at high levels (up to 10^7 sites/cell) on tumors of neuroectodermal origin (Reisfeld and Cheresh, 1987). Also, lymphocytes that infiltrate primary melanoma and their metastases were reported to express gangliosides (Hersey and Jamal, 1989). In this regard, previous studies indicated that anti-GD2 mAb potentiate lymphocyte responses to various stimuli (Hersey et al., 1987; Welte et al., 1987). Consequently, the potentiation of an immune response to tumor cells may be an additional benefit of the therapeutic application of this antibody.

We initially developed a mouse anti-GD2 mAb 14.18 of IgG3 isotype that mediates lysis of cultured human melanoma and neuroblastoma cells and suppresses the growth of these tumors in athymic mice (Mujoo et al., 1987, 1989). The hybridoma secreting mAb 14.18 was used to construct the chimeric anti-GD2 mAb ch14.18 by combining cDNA sequences encoding the variable portions of murine 14.18 with the constant regions of the human γl H and κL chains. The chimeric IgG construct was expressed at a very high level of 180 mg/l in the spent culture fluid of a murine non-IgG-producing hybridoma cell line (Gillies et al., 1989).

Immunological characterization of mAb CH14

Saturation binding studies with M21 human melanoma cells revealed essentially identical binding for ch14.18 and for mouse 14.G2a, the latter being an isotype switch variant of mouse mAb ch14.18. The average number of binding sites per M21 cell is 1.4×10^7 for ch14.18. The dissociation constants (K_D) were also essentially the same, i.e. 11.9 nmol/l for 14.G2a and 11.2 nmol/l for ch14.18, indicating that these two mAbs bind to GD2 on the surface of M21 melanoma cells with equal affinity. A comparison of the targeting ability of radiolabeled mAb ch14.18 and 14.G2a to M21 melanoma xenografts in athymic mice indicated identical kinetics of blood clearance. Typically, 12% of the injected dose was found 4 hours after injection, a value that decreased to 7%, 4 days after injection. A similar biodistribution of the two iodinated mAbs was found in melanoma tumors and in a variety of other tissues of these athymic mice (Mueller et al., 1990a).

The ability of ch14.18 and 14.G2a to mediate cytolysis with human complement was essentially the same when used at antibody concentrations ranging from 50 μg to 50 ng/ml on M21 cell targets. As little as 1 μg/ml of both mAb achieved 78% lysis of M21 cells. The only clear-cut difference between mAb ch14.18 and 14.G2a was observed in their respective abilities to mediate the cytotoxicity of human effector cells. Thus, specific antibody-dependent lysis by peripheral blood mononuclear cells (PBMC) did occur in a dose-dependent fashion when these two mAbs were tested against M21 melanoma targets, at effector/target cell ratios from 50: 1 to 200: 1, and at concentrations ranging from 50 to 0.001 µg/ml. For each blood donor tested, the higher effector/target cell ratio resulted in a more efficient lysis, although the specific lysis achieved varied from donor to donor. However, the effector cells of each donor at each effector/target cell ratio tested were considerably more potent in mediating antibody-dependent cellular cytotoxicity with ch14.18 than with 14.G2a. Thus, at every given antibody concentration, ch14.18 mediated a higher specific ^{51}Cr release than 14.G2a, and the amount of ch14.18 required to mediate specific lysis was from 50 to 100-fold less than that of 14.G2a. Thus, between 25% and 55%, specific lysis of M21 cells was achieved with ch14.18 at concentrations ranging between 10 and 0.5,µg/ml, whereas it required from 50 to 1 µg of 14.G2a/ml to obtain a maximum specific lysis of 14–33% (Mueller et al., 1990a).

Role of different effector cells and cytokines in the lysis of human neuroblastoma cells coated with ch14.18

MAb ch14.18 was found to bind human neuroblastoma cells equally well as 14.G2a, as determined by indirect immunofluorescence. As with melanoma cells, ch14.18 was more effective in mediating the lysis of neuroblastoma cells with PBMC from normal donors at an effector/target cell ratio of 50:1, because less ch14.18 was required than 14.G2a to achieve the same lytic effect. Similar results as those obtained in antibody-dependent cellular cytotoxicity with PBMC from normal donors were also seen when such effector cells were isolated from the blood of neuroblastoma patients, where again ch14.18 was more effective than 14.G2a (Barker et al., 1991).

A comparison of the effectiveness of ch14.18 vs 14.G2a mediated lysis of neuroblastoma cells by either granulocytes or PBMC indicated that granulocytes were most effective. In a comparison of their respective capabilities to lyse ch14.18-coated neuroblastoma cells, granulocytes also proved superior to PBMC when both were obtained from the same neu-

roblastoma patient. The addition of granulocyte-macrophage colony stimulating factor (GM-CSF) at concentrations as low as 10 ng/ml augmented the ability of granulocytes, but not PBMC obtained from the blood of healthy donors, to mediate lysis of neuroblastoma cells with ch14.18 (Barker et al., 1991). In view of this augmentation of lysis, a combined therapy of neuroblastoma patients with ch14.18 and GM-CSF could possibly optimize the antitumor effect of ch14.18. This contention is supported by results from preliminary studies which demonstrate that granulocytes isolated from the blood of neuroblastoma patients that had been treated with mAb 14.G2a did preferentially migrate to the tumor site following a brief ex vivo activation with GM-CSF. Several neuroblastoma patients also showed some mixed responses and partial tumor regressions following treatment with ch14.18 and GM-CSF.

Specific tumor targeting with genetically engineered monoclonal antibodies

Mutant monoclonal antibodies

The rationale for engineering mutant chimeric mAb directed against human tumor-associated antigens is based on the well-known advantage of antibody fragments over intact antibodies in gaining better tumor access because of their smaller size and more rapid clearance from the circulation. This fact has been well documented, especially in the identification of occult lesions by radioimaging, both in experimental animals (Larson et al., 1983) and in cancer patients (DeNardo et al, 1986). Recombinant DNA techniques facilitate the engineering of antibodies in such a way as to create molecules with novel properties and functions (Neuberger et al., 1984; Morrison, 1985; Gillies et al., 1989), including mutants that simulate antibody fragments (Gillies and Wesolowski, 1990). This approach becomes relevant, because the production of antibody fragments, such as F(ab')$_2$, by digestion with proteolytic enzymes is not applicable to all antibodies. Thus, we were unable to produce active F(ab')$_2$ fragments of either 14.G2a or ch14.18 by proteolytic digestion and also could not engineer a functional $F(ab')_2$ of ch14.18 that lacked the CH2 and CH3 domains (Mueller et al., 1990b). This led to the construction and subsequent analysis of the CH2 deletion mutant of ch14.18 (ch14.18-ΔCH2). This mutant mAb, which bound specifically to purified ganglioside GD2, revealed a molecular mass of 120 kDa by sodium dodecyl sulphate (SDS) gel electrophoresis and by high-pressure exclusion chromatography under non-denaturing conditions. In contrast to intact mAb ch14.18, ch14.18-ΔCH2 does not mediate complement-dependent lysis of GD2+ target cells and mediates only very little cytotoxicity of human mononuclear effector cells. Both ch14.18 and its CH2 deletion mutant stained all M21 melanoma cells with the same intensity by indirect immunofluorescence, indicating no difference in their binding to GD2 (Mueller et al., 1990a).

Studies in athymic mice bearing M21 human melanoma xenografts that compared the blood clearance and biodistribution of ch14.18, ch14.18-ΔCH2, and a $F(ab')_2$ fragment of a non-relevant human IgG indicated some distinct advantages of ch14.18-ΔCH2. Taken together, our data indicate that the CH2-deleted chimeric antibody (ch14.18-ΔCH2) clears from the blood of athymic mice, bearing human melanoma xenografts, with the same kinetics as nonrelevant human IgG $F(ab')_2$ and considerably more rapidly than the intact chimeric antibody ch14.18. Targeting of the human melanoma xenografts by ch14.18-ΔCH2 is specific, and this deletion mutant localizes to the tumor more rapidly and with better localization ratios than the intact ch14.18 antibody. The pharmacokinetic properties of ch14.18-ΔCH2 demonstrate that the catabolic rate of the antibody molecule is controlled by the CH2 domain (Mueller et al., 1990b).

As far as the correlation between these biological functions and immunoglobulin structure is concerned, an obvious difference between the two antibodies is that the chimeric antibody is glycosylated, whereas the mutant lacking Asn-297, the sole site for N-linked glycosylation, is not. Although one could postulate that the rapid elimination of ch14.18 is due to its lack of carbohydrate side chains, this is not the case. Thus, when the carbohydrate attachment site

(Asn-297) is eliminated from ch14.18 by site-directed mutagenesis, the serum half-life in athymic mice bearing M21 melanoma tumors is similar for both the aglycosylated mutant and intact ch14.18, although the clearance rate for the mutant is somewhat more rapid. Biodistribution of the aglycosylated mutant is the same as that of its native analogue. Aglycosylated ch14.18 does not reveal any measurable antibody-dependent cellular cytotoxicity, but exhibits significant complement-dependent cytotoxicity, although it required a ten-fold higher concentration than the intact ch14.18 to achieve equivalent cell lysis (Dorai et al., 1991). Taken together, these results suggest that the carbohydrate side chains in the CH2 domain of human IgGl are necessary to maintain the appropriate structure for the retention of some, but not all, of the effector functions of the CH2 domain.

Genetically engineered antibody fusion proteins with cytokines

In an effort to effectively target cytokines to the tumor site, we constructed and evaluated a genetically engineered fusion protein consisting of ch14.18 and interleukin-2 (IL-2) (Gillies et al., 1992). An underlying rationale for this approach was to achieve an optimal therapy by combining IL-2 activation and tumor antigen presentation together with a tumor-specific antibody such as ch14.18 that mediates both complement-dependent cytotoxicity (CDC) and ADCC activities. By combining such an antibody with IL-2, it was anticipated that relatively large amounts of tumor antigen are presented during IL-2 activation for expansion of cytotoxic T cells, since melanoma cell lines were reported to express as many as 1.5×10^7 sites/cell for ch14.18 (Mueller et al., 1990a). Furthermore, this antibody would also be available to target Fc receptor-bearing cells that have been activated by the targeted IL-2. In this regard, we focused on the ability of the ch14.18-IL-2 fusion protein to stimulate the proliferation and cytolytic activity of a human T cell line against autologous melanoma targets. Such a cell line, 660 TIL, which is CD3+, CD8+ antigen-specific, and MHC class I-restricted (HLA-A2) was originally obtained by outgrowth from a human metastatic melanoma (Reilly et al., 1990). A melanoma line, 660 mel, was derived from the same tumor and served as a source for antigen stimulation, as well as an autologous target for 660 TIL (Reilly and Antognetti, 1991).

In preparing plasmid constructs, the Ig-IL-2 fusion protein expression vector was constructed by fusing a synthetic human IL-2 sequence to the carboxyl end of the human cγl gene. A synthetic DNA linker, extending from the SmaI site near the end of the antibody to the unique PvuII site in IL-2, was used to join the amino terminal residue of mature IL-2 to the exact end of the CH3 exon (CH3-IL-2). The fused gene was inserted into the vector pdHL2-14.18 as was previously described for an antibody/lymphotoxin fusion protein construct (Gillies et al., 1991). Additional constructs were prepared in which the IL-2 sequence was fused to the SacI site in the hinge region of the human Cγ3 gene (Fab-IL-2) or to the end of the CH2 exon at a TaqI site (CH2-IL-2). In both cases, synthetic linkers were used to fuse the antibody and IL-2 sequences directly without introducing any additional amino acid residues (Gillies et al., 1992).

An analysis of the antigen binding and IL-2 activities of these proteins revealed no binding activity for the Fab-IL-2 protein containing the 14.18 V regions, while the CH2-IL-2 fusion protein was strongly positive. The 14.18 Fab, produced by genetic engineering, had greatly reduced antigen binding, suggesting that bivalency is required for full activity. Since we were puzzled by these findings, we constructed a second Fab-IL-2 fusion protein using the V regions of the anti-TAG 72 antibody, B72.3 to test whether a molecule that retained both IL-2 activity and antigen binding could indeed be constructed. In fact, the B72 3 Fab-IL-2 protein revealed normal antigen binding activity and like all of the other fusion proteins, had IL-2 specific activities ranging from 5 to 6.5 $\times 10^6$ units/mg when normalized for IL-2 content. Based on our findings with different domains of ch14.18, we next constructed a whole antibody/IL-2 fusion protein by fusing the coding sequence of IL-2 to the end of the H chain CH3 exon (CH3-IL-2). In this way we hoped to produce fully assembled IL-2 fusion proteins, possibly with more favorable pharmacokinetic properties in vivo. The CH3-IL-2 protein, constructed with the 14.18 anti-GD2 V regions could indeed be expressed as a fully assembled antibody fusion protein which showed full IL-2 activity, as well as enhanced antigen binding activity.

The availability of a matched set of TIL and its autologous tumor cell line, expressing the GD2 antigen, allowed us to exploit this system as a model for testing the biological properties of antibody-targeted IL-2. When testing the IL-2 activity of the CH3-IL-2 (14.18) fusion protein in a standard T cell proliferation assay, we found the activity of the fusion protein to be somewhat less than that of a recombinant IL-2 made in bacteria; however, it was identical in activity to a recombinant IL-2 preparation produced in yeast $(2.5 \times 10^6$ units/mg) when either murine or human T cells were used as targets in the assay. Thus, it appears that fusion of IL-2 at the carboxyl terminus of an antibody or antibody fragment does not significantly reduce its activity. In addition, the effector functions of the CH3-IL-2 protein, i.e. the ability to mediate complement and Fc receptor-dependent lysis, were found to be maintained, although somewhat decreased when compared to the chimeric 14.18 antibody.

The CH3-IL-2 fusion protein produced enhanced cytotoxic activity of tumor infiltrating lymphocytes (TIL) against their autologous human melanoma targets. Specifically, the lytic activity of the CD8+ human TIL line 660 against autologous 660 melanoma cells was increased considerably more when these tumor cells were coated with the CH3-IL-2 fusion protein, rather than with the ch14.18 antibody. The effect of the CH3-IL-2 fusion protein was even more pronounced with TILs that were deprived of IL-2 for 4 days (Gillies et al., 1992). When similar experiments comparing the fusion protein and exogenously added IL-2 was done 1 week later when the autologous killing activity had declined. In this case, the addition of IL-2 (100 units/ml) to the assay had little effect. However, the stimulatory effect of CH3-IL-2 in these experiments was quite striking when IL-2 depleted effector cells were used, especially at low effector to target ratios, i.e. 10:1. However, in all cases tested, the amount of stimulation obtained by coating the tumor cells with the CH3-IL-2 fusion protein exceeded that obtained by adding equivalent levels of IL-2 (Gillies et al., 1992).

In another effort to target cytokines to tumor cells and thereby mediate effective cell killing, TNF-β, also known as lymphotoxin (LT), was genetically conjugated to the constant region of human IgGl gene at the end of either the second (CH2-TNF-β) or the third (CH3-TNF-β) constant region domain. The altered heavy-chain constant regions were combined in a plasmid vector with the variable regions of a mouse antibody (14.18) against ganglioside GD2 and the human kappa constant region. When the resulting immunoconjugate constructs were expressed in transfected hybridoma cells and tested for both their antibody and their TNF-β activities, the two constructs were assembled to various degrees, depending on whether the third heavy-chain constant region domain was present. Although both forms retained their ability to bind antigen and mediate antibody-dependent cellular cytotoxicity, only CH3-TNF-β was able to mediate the lysis of melanoma target cells in the presence of human complement. TNF-β activity increased significantly as a function of heavy-chain assembly and was equivalent to the activity of unconjugated TNF-β. We thus showed that the chimeric anti-GD2 antibody, ch14.18, can be genetically fused to TNFβ without loss of the antibody's antigen-binding activity or effector functions or loss of the receptor-binding and biological activity of TNF-β (Gillies et al., 1991). The ability of such a conjugate to target TNF-β to tumors in vivo remains to be determined. The fact that the conjugate does not directly kill the GD2-bearing tumor cells used in the present study will make it easier to interpret studies comparing the anti-tumor activities of the conjugate and chimeric antibodies.

Phase I clinical trials with murine and chimeric human/mouse anti-GD2 monoclonal antibodies

It is quite evident at this time that the overall long-term survival of neuroblastoma patients has not improved significantly during the last 20 years, despite the introduction of highly aggressive therapeutic modalities into the treatment of stage IV patients, i.e. high-dose chemotherapy with subsequent allogeneic or autologous bone marrow transplantation or application of high-dose radiotherapy with m-[^{131}I]iodobenzylguanidine (mIBG) (Treuner et al., 1987).

Therefore, new therapeutic strategies are very much needed to achieve a more effective treatment of neuroblastoma. Among antibodies that are most suitable for the treatment of neuroblastoma are those directed against ganglioside GD2, which is expressed at high density on these tumor cells and is present, to a much lesser extent, on normal tissues (Schulz et al., 1984). Consequently, the GD2 antigen served as a tumor target in our initial phase I study initiated at the University of Tübingen, Germany, where neuroblastoma patients were initially injected with murine anti-GD2 monoclonal antibody, 14.G2a of IgG2a isotype. The major aim of this phase I clinical study was to determine the safety of this antibody, assess its clinical side effects, and evaluate possible anti-tumor effects.

Nine patients with stage IV neuroblastoma were injected with mAb 14.G2a at least 4 weeks after they were treated with chemotherapy, high-dose [^{131}I]mIBG, allogeneic bone marrow transplantation, single or double autologous bone marrow transplantation and/or several cycles of interleukin-2 therapy. During the antibody treatment, all patients received steroids (1–2 mg/kg) for alleviation of allergic side-effects. Varying amounts of mAb 14.G2a were administered for time periods that varied from 5 to 10 days, owing to clinical side-effects. The time of treatment was reduced only in those patients where side-effects such as pain, hypertension, or allergy were no longer treatable; however, side-effects that resulted in reduction of treatment time were not life-threatening. The cumulative dose of the antibody injected varied from 100 mg/m^2 to 400 mg/m^2. The serum concentration of mAb 14.G2a determined in nine patients twice daily before and after every infusion ranged from 11 to 61 µg/ml, respectively.

Pharmacokinetic studies of mAb 14.G2a in serum determined for three of the patients indicated that the elimination of 14.G2a in the serum best fits a two-compartment model with a rapid elimination ($t1/2\alpha$) and a slower elimination ($t1/2\beta$). Accordingly, $t1/2\alpha$,

$t1/2\beta$ and the decrease in the antibody's serum concentration D ($=t1/2\alpha + t1/2\beta$) ranged from 0.66 h to 1.98 h for $t1/2\alpha$, 30.13 h to 53.33 h for $t1/2\beta$ and from 32.11 h to 54.25 h for the decrease in serum concentration D. An anti-mouse IgG response (HAMA) to mAb 14.G2a was detected in all nine patients, varying widely and ranging from 1.6 to 22 μg of human anti-mouse IgG/ml of serum. As far as activation of complement by mAb 14.G2a was concerned, all patients revealed a continuous decrease in the serum concentration of complement component C4 and an initial decrease of C3c. After a constant initial decrease of C3c, there was either a further decrease or an increase of this complement component during the course of therapy. There was also an initial rapid increase in C3a (Handgretinger et al., 1992).

As far as tumor responses were concerned, only six patients could be evaluated, since three of nine patients treated were in clinical remission at the time of antibody treatment. Two patients had a complete remission (CR) of a single localized tumor, as determined by a positive [^{125}I]mIBG scintigraphy before and a negative [^{125}I]mIBG scintigraphy after antibody treatment. One of these patients has remained in complete remission for more than 104 weeks, while the other is in complete remission for more than 72 weeks. Another two patients showed a partial response (PR), as indicated by more than 50% reduction of the [^{125}I]mIBG uptake. After 6 weeks, tumor progression was seen in one of these patients who died 1 month later. The other patient presented initially with extensive disseminated disease, as determined by uptake of [^{125}I]mIBG. After antibody treatment, a partial remission, i.e. more than 50% reduction of mIBG uptake in the skull metastasis, was seen with the same detection system and equal amounts of [^{125}I]mIBG as used for tumor detection prior to treatment with mAb 14.G2a. The [^{125}I]mIBG uptake in both humeri was negative and the presacral tumor showed no change. Several weeks after antibody treatment, the presacral tumor could be resected in toto and its histological examination revealed a ganglioneuroma. The patient has now had stable disease for more than 60 weeks. He still shows a faint [^{125}I]mIBG uptake in the previous tumor sites in the skull, but is otherwise in excellent clinical condition. Two patients with extensive tumor burden did not respond to therapy and showed progressive disease (Handgretinger et al., 1992).

As far as the side effects produced by the treatment with murine mAb 14.G2a are concerned, the rapid increase of C3a, a known anaphylotoxin, may offer an explanation for the allergic reactions occurring within 1 h after the start of the first antibody infusion. Human anti-mouse IgG development, observed in all patients, could, in part, be another reason for the allergic reactions that occurred later during therapy or even thereafter. Thus, the anti-mouse IgG level started to rise within 4–10 days after the start of therapy. In three patients, serum-sickness-like symptoms were seen even 1 week after the termination of therapy. However, there was no clear correlation between the extent of the anti-mouse IgG level and the severity of the allergic reactions.

Side effects produced by treatment with mAb 14.G2a also included severe pain and hypertension. These pains were preferentially located in the bowels and lower extremities, requiring the use of morphine for relief in all patients. Although the reason for this pain is not entirely clear, immunohistochemical studies of bowel tissue with mAb 14.G2a revealed a staining of fine nerve fibers, suggesting the expression of GD2 on nerve fibers. Pain is likely to result from the binding of mAb 14.G2a to those nerve fibers with a subsequent inflammation. However, this pain did not present a serious clinical problem, since it could be adequately controlled by the infusion of morphine during the entire period of antibody treatment. Hypertension, which was observed in some patients, could be caused by the binding of circulating immune complexes to the glomerula, resulting in an immune complex nephritis; however, this seems unlikely, since we did not observe a measurable increase in the amount of such complexes in the neuroblastoma patients during therapy with mAb 14.G2a. Since the anaphylatoxin C3a is known to cause smooth muscle contraction, a possible explanation for the hypertension occurring in four of the patients may be the binding of this complement component to smooth muscles of capillaries, resulting in a subsequent contraction that is reflected by an elevation in blood pressure. Finally, non-specific binding of mAb 14.G2a at low density on normal tissues may be another reason for some of the observed side effects.

Although it is not clear which specific mechanism(s) actually contributes to the therapeutic effect of mAb 14.G2a in patients, one mechanism that may lead to tumor destruction are CDC-like reactions. In this regard, our previous in vitro studies revealed that mAb 14.G2a is able to mediate an effective CDC that results in the lysis of neuroblastoma cells (Barker et al., 1991). Evidence that CDC may play a role in the clinical use of mAb 14.G2a comes from our clinical finding that all patients studied showed a continuous decrease of C4, an initial decrease of C3c and an initial increase of C3a, suggesting an activation of the complement cascade (Handgretinger et al., 1992). Overall, results of our study indicate that each patient's stage of disease at the time of the initial antibody injection is the most important criterion for successful treatment. In this regard, it is well known that residual or relapsed neuroblastoma is very resistant to every form of therapy, including immunotherapy. Although treatment with mAb 14.G2a resulted in a partial remission, even in patients with disseminated relapse and high tumor burden, we conclude from our clinical data that this antibody is likely to be most effective in patients with a lower tumor burden. Therefore, it is clear that the optimal time for its application is during minimal residual disease.

Consequently, we used mAb 14.G2a to treat three patients while in clinical remission, who originally presented with disseminated disease and were considered at high risk for relapse. Two of these patients relapsed 52 weeks and 60 weeks, respectively, after antibody treatment, while one of them still remains in continuous, complete remission. Since the high relapse rate of stage IV neuroblastoma is well known, we conclude that the application of mAb 14.G2a is justified in those patients who are in apparent remission. In this regard, it is known that up to 70% of stage IV neuroblastoma patients relapse within 2 years. In order to maintain a remission, it should,

therefore, be an aim to treat the children for at least 2 years after their first remission is achieved with anti-GD2 antibody. Realistically, however, this aim can only be attained by the repeated application of chimeric human/mouse antibodies, which are far less immunogenic than the corresponding murine mAb (Handgretinger et al., 1992).

Indeed, the first clinical experiences with mAb ch14.18, which represent the chimeric counterpart of mAb 14.G2a, have indicated that such repeated applications are quite feasible. In this regard, the following results were obtained in a phase I clinical trial at the University of California, San Diego, which involved nine patients with evaluable stage IV neuroblastoma who were treated with ch14.18. Overall, there were one CR, one PR, three mixed responses, two stable disease and two patients with progressive disease. When the clinical responses of these nine patients were grouped according to site of neuroblastoma, an interesting pattern became apparent. Thus, three of seven patients with bone marrow disease achieved CR and two patients improved with a reduction in the percentage of neuroblastoma cells in the marrow from 80% to 18% and 89% to 47%, respectively. One of eight patients with bone disease had a CR, three patients presented with stable disease, while four patients showed progression of disease. One of seven patients with a bulky tumor mass had a partial remission, while three stabilized and three progressed.

A comparison with an earlier clinical phase I trial that was conducted with murine mAb 14.G2a at the University of California, San Diego, revealed a somewhat different picture. Thus, two of eight neuroblastoma patients with bone marrow disease who were treated with murine anti-GD2 mAb 14.G2a achieved CR and one had PR. Furthermore, one of 12 patients with bone disease presented with stable disease, while the 11 remaining patients showed disease progression. Also, three of 11 patients with bulky tumors had stable disease. In addition, one of three patients with lymph node metastasis had a complete remission.

Although the clinical side effects of murine mAb 14.G2a and chimeric mAb ch14.18 were somewhat similar, treatment with ch14.18 was associated with more tolerable and reversible toxicities. A clear-cut difference was observed in the occurrence of HAMA, since 11 neuroblastoma patients who were treated with murine mAb 14.G2a presented with relatively high HAMA titers, i.e. up to 22 µg antibody/ml serum. This required a waiting period of 2–3 months before a second dose regimen of antibody could be given. In contrast, the human/mouse chimeric mAb ch14.18 did not result in the typical HAMA response, but produced an antibody titer approximately three logs less than that observed with the murine mAb. The antibody response against the chimeric mAb was only directed against the variable and not the framework region of the anti-GD2 antibody, i.e. resembling an anti-idiotype response.

Taken together, the data from these phase I clinical trials indicate that both murine and chimeric anti-GD2 mAbs produce some definite therapeutic effects in neuroblastoma, especially in those patients with bone marrow metastasis. The chimeric mAb seems to be advantageous, since it does not produce HAMA and several different dose regimens can be given without any serious complications.

Perspectives

This article summarizes our experience with chimeric mouse/human monoclonal antiganglioside GD2 antibodies, including its mutant deleted of the CH2 domain and several genetically engineered fusion proteins of this antibody with cytokines. These molecular constructs could be produced by combining hybridoma methodology with recombinant DNA technology. Although it seems that most of the HAMA problems occurring with murine mAbs have been solved by the development of chimeric human/murine mAbs, a number of challenges have to be met before monoclonal antibodies against tumor-associated antigens can be most effectively targeted to tumor sites and thereby make their optimal contribution to cancer therapy. Among these challenges are the many problems associated with tumor cell heterogeneity and the inability to adequately penetrate the tumor vasculature. Thus, human/mouse chimeric mAbs will have to be genetically engineered in such a way that they will exhibit greatly improved targeting capabilities and perform better in penetrating tumor vasculatures than presently available constructs. Mutant ch14.18 with deleted CH2 domains represent one such effort that may increase the potential of these reagents for radioimaging and possibly even for tumor targeting of radionuclides. Another possibility is the use of genetically engineered antibody fusion proteins with cytokines. Although this approach still has to be optimized to achieve effective in vivo results, it seems likely that the specific targeting of cytokines at physiologically relevant concentrations may aid in improving currently available cancer immunotherapies.

Finally, however, one has to be aware that evolution over millions of years has perfected the three-dimensional structure of the antibody molecule to provide it with optimal functional properties. It is evident that at present, the engineering of antibody structures by recombinant DNA technologies has not yet improved upon this design, but has provided, at best, compromise solutions that may prove useful in certain applications for the treatment of cancer.

Acknowledgements

The authors thank Ms. Lynne Kottel for the preparation of this manuscript. This work was supported by the National Cancer Institute Grants CA42508 and CA51946.

References

Barker, E., Mueller, B.M., Handgretinger, R., Herter, M., Yu, A.L. and Reisfeld, R.A. (1991) Effect of a chimeric antiganglioside GD2 antibody on cell-mediated lysis of human neuroblastoma cells. *Cancer Res.*, 51: 144–149.

Cheung, N.-KA., Lazarus, H., Miraldi, F.D., Abramowski, C., Kallik, S., Saarinen, S., Spitzer, D., Strandjord, S., Coccia, B. and Berger, N. (1987) Ganglioside GD2 specific monoclonal antibody 3F8: A phase I study in patients with neuroblastoma and malignant melanoma. *J. Clin. Oncol.*, 5: 1430–1440.

DeNardo, S.J., DeNardo, G.L. and O'Grady, L.F. (1986) Radioimmunotherapy of patients with B-cell lymphoma using I-131 LYM-l mAb. *J. Nucl. Med.*, 27: 903–904.

Dippold, W.G., Bernhard, H., Dienes, H. and Meyer zum Büschenfelde, K.-H. (1988) Treatment of patients with malignant melanoma by monoclonal ganglioside antibodies. *Eur. J. Cancer Clin. Oncol.*, 24: (suppl) S65–67.

Dorai, H., Mueller, B.M., Reisfeld, R.A. and Gillies, S.D. (1991) Aglycosylated chimeric mouse/human antibody retains some effector function. *Hybridoma*, 10: 211–217.

Douillard, J.Y., Lehur, P.A., Vignoud, J., Blottiere, H., Maurel, C., Thedrez, P., Kremer, M. and LeMevel, B. (1986) Monoclonal antibodies specific immunotherapy of gastrointestinal tumor. *Hybridoma*, 5:(suppl.1): S139–149.

Gillies, S.D. and Wesolowski, J.S. (1990) Antigen binding and biological activities of engineered mutant chimeric antibodies with human tumor specificities. *Hum. Antibodies Hybridomas* 1: 47–54.

Gillies, S.D., Lo, K.M. and Wesolowski, J. (1989) High level expression of chimeric antibodies using adopted cDNA variable region cassettes. *J. Immunol. Methods*, 125: 191–202.

Gillies, S.D., Young, D., Lo, K.-M., Foley, S.F. and Reisfeld, R.A. (1991) Expression of genetically engineered immunoconjugates of lymphotoxin and chimeric anti-ganglioside GD2 antibody. *Hybridoma*, 10: 347–356.

Gillies, S.D., Reilly, E.G., Lo, K.-M. and Reisfeld, R.A. (1992) Antibody-targeted interleukin 2 stimulates the T-cell killing of autologous tumor cells. *Proc. Natl. Acad. Sci. USA*, 89: 1428–1432.

Goldenberg, D.M., Sharkey, R.M., Goldenberg, H., Hall, T.C., Murthy, S., Izon, D.O., Gascon, P. and Swayne, L.C. (1990) Monoclonal antibody therapy of cancer. *N.J. Med.*, 87: 913–918.

Handgretinger, R., Baader, P., Dopfer, R., Klingebiel, T., Reuland, P., Reisfeld, R.A., Treuner, J. and Niethammer, D. (1992) A phase I study of neuroblastoma with the anti-ganglioside GD2 antibody 14.G2a. *Cancer Immunol. Immunother.*, in press.

Hersey, P., Townsend, P., Burns, C. and Cheresh, D.A. (1987) Enhancement of cytotoxic and proliferative responses of lymphocytes from melanoma patients by incubation with monoclonal antibodies, organs, and ganglioside GD3. *Cancer Immunol. Immunother.*, 24: 144–150.

Hersey, P. and Jamal, O. (1989) Expression of gangliosides GD3 and GD2 on lymphocytes in tissue sections of melanoma. *Pathology*, 21: 51–58.

Houghton, A.M., Mintzer, D., Cordon-Cardo, C., Welt, S., Fliegel, B., Vadhan, S., Carswell, E., Melamed, M., Dettgen, F. and Old, L.J. (1985) Mouse monoclonal IgG3 antibody detecting GD3 ganglioside-A phase I trial in patients with malignant melanoma. *Proc. Natl. Acad. Sci. USA*, 82: 1242–1246.

Janson, C.H., Tehrani, M., Wigzell, H. and Mellstedt, H. (1989) Rational use of biological response modifiers in hematological malignancies: A review of treatment with interferon, cytotoxic cells and antibodies. *Leukemia Res.*, 13: 1039–1046.

Köhler, G. and Milstein, C. (1975) Continuous cultures of fused cells secreting antibody of predefined specificity. *Nature*, 256: 495–497.

Larson, S.M., Brown, J.P., Wright, P.W., Carasquillo, J.A., Hellström, I. and Hellström, K.E. (1983) Imaging of melanoma with I-131 labeled monoclonal antibodies. *J. Nucl. Med.*, 24: 123–129.

Mellstedt, H., Frodin, J.-E. and Masucci, G. (1989) Clinical status of monoclonal antibodies in the treatment of colorectal carcinoma. *Oncology*, 3: 25–32.

Morrison, S.L. (1985) Transfectomas provide novel chimeric antibodies. *Science*, 229: 1202–1207.

Mueller, B.M., Romerdahl, C.A. and Reisfeld, R.A. (1990a) Enhancement of antibody-dependent cytotoxicity with a chimeric anti-GD2 antibody. *J. Immunol.*, 144: 1382–1386.

Mueller, B.M., Reisfeld, R.A. and Gillies, S.D. (1990b) Serum half-life and tumor localization of a chimeric antibody deleted of the CH2 domain and directed against the disialoganglioside GD2. *Proc. Natl. Acad. Sci. USA*, 87: 5702–5705.

Mujoo, K., Cheresh, D.A., Yang, H.M. and Reisfeld, R.A. (1987) Disialoganglioside GD2 on human neuroblastoma cells: Target antigen for monoclonal antibody-mediated cytolysis and suppression of tumor growth. *Cancer Res.*, 47: 1098–1104.

Mujoo, K., Kipps, T.J., Yang, H.M., Cheresh, D.A., Wargalla, W., Sander, D. and Reisfeld, R.A. (1989) Functional properties and effect on growth suppression of human neuroblastoma tumors by isotype switch variants of monoclonal anti-ganglioside GD2 antibody 14.18. *Cancer Res.*, 49: 2857–2861.

Neuberger, M.S., Williams, G.T. and Fox, R.O. (1984) Recombinant antibodies possessing novel effector functions. *Nature*, 312: 604–608.

Reilly, E.B. and Antognetti, G. (1991) Increased tumor-specific CTL activity in human tumor-infiltrating lymphocytes stimulated with autologous tumor lines. *Cell. Immunol.*, 135: 526–533.

Reilly, E.B., Antognetti, G., Wesolowski, J.S. and Sakorafas, P. (1990) The use of microcapsules for high density growth of human tumor infiltrating lymphocytes and other immune reactive T cells. *J. Immunol. Methods*, 126: 273–279.

Reisfeld, R.A. and Cheresh, D.A. (1987) Human tumor antigens. In: F.J. Dixon (Ed.), *Advances in Immunology,* Academic Press, New York, pp. 323–377.

Schulz, G., Cheresh, D.A., Varki, N.M., Yu, A., Staffileno, L.K. and Reisfeld, R.A. (1984) Detection of ganglioside GD2 in tumor tissues and sera of neuroblastoma patients. *Cancer Res.*, 44: 5914–5920.

Sears, H.G., Atkinson, B., Herlyn, D., Hayry, B., Ernst, C., Steplewski, Z. and Koprowski, H. (1982) The use of monoclonal antibody in a phase I clinical trial of human gastrointestinal tumor. *Lancet*, 1: 762–765.

Treuner, J., Klingbiel, T., Bruchelt, G., Feine, U. and Niethammer D. (1987) Treatment of neuroblastoma with metaiodobenzylguanidine: Results and side effects. *Med. Pediatr. Oncol.*, 15: 199–202.

Vadhan-Raj, S., Cordon-Cardo, C., Carswell, E., Duteau, C., John, M., Dantis, L., Templeton, M.A., Minzer, D., Oettgen, H.F.,

Old, L.J. and Houghton, A.N. (1988) Phase I trial of mouse monoclonal antibody against GD3 ganglioside in patients with melanoma: Industion of inflammatory responses at tumor sites. *J. Clin. Oncol.*, 6: 1636–1648.

Welte, K., Miller, G., Chapman, P.B., Yuasa, H., Natoli, E., Kunick, J.E., Cordon-Cardo., C., Buhrer, C., Old. L.J. and Houghton, A.N. (1987) Stimulation of T lymphocyte proliferation by monoclonal antibodies against GD3 ganglioside. *J. Immunol.*, 139: 1763–1768.

L. Svennerholm, A.K. Asbury, R.A. Reisfeld, K. Sandhoff, K. Suzuki, G. Tettamanti and G. Toffano (Eds.)
Progress in Brain Research, Vol. 101

CHAPTER 16

Detection of glioma-associated gangliosides GM2, GD2, GD3, 3'-isoLM1 and 3',6'-isoLD1 in central nervous system tumors in vitro and in vivo using epitope-defined monoclonal antibodies

C. J. Wikstrand,[1] P. Fredman,[2] L. Svennerholm[2] and D. D. Bigner[1,3]

[1]*Department of Pathology and the* [3]*Preuss Laboratory for Brain Tumor Research, Duke University Medical Center, Box 3156, Durham, NC 27710 USA, and* [2]*Department of Psychiatry and Neurochemistry, University of Göteborg, Göteborg, Sweden*

Introduction

The appearance of novel or quantitative shifts in ganglioside phenotype following neoplastic transformation has been reported in tumors of diverse origin. Quantitatively increased 'normal' ganglioside epitopes on tumor cells are effective targets for both localizing (GD2, GM3) and therapeutic (GD2, GD3) applications, especially with tumors of neuroectodermal or neural origin (Houghton et al., 1985; Herberman et al., 1985; Irie and Morton, 1986; Heiner et al., 1987; Dohi et al., 1988; Yu et al., 1991; Reisfeld, 1993). Among tumors of neuroectodermal origin, gliomas have been reported to demonstrate a marked shift from the expression of the more complex oligosialylated gangliosides such as GD1a, GD1b and GT1b found in normal brain, to the simpler, less polar gangliosides. Epitope-defined monoclonal antibodies (MAbs) recognizing GM2, GD2 and GD3 have been described by this group and applied to the phenotypic analysis of neuroectodermal tumors and cell lines (He et al., 1989; Vrionis et al., 1989; Longee et al., 1991).

As the infiltrative nature of growth of gliomas almost ensures the 'contamination' of human tumor samples with relatively ganglioside-rich normal central nervous system (CNS) elements, human glioma xenografts, which contain no normal brain elements, grown in athymic rodents, have been used to study glioma-associated gangliosides. With such an approach, we described the lactotetraose series gangliosides IV3NeuAc-LcOse4Cer (3'-isoLM1) and IV3NeuAc, III6NeuAc-LcOse4Cer (3',6'-isoLD1) (Svennerholm, 1963; IUPAC-IUB, 1977) as the major mono- and oligo-sialogangliosides, respectively, of the human glioma-derived xenograft D-54 MG (Månsson et al., 1986). We have recently reported the distribution of these gangliosides in human gliomas in vitro and in vivo using MAbs defining 3'-isoLM1 and 3',6'-isoLD1, i.e., SL-50 and DMAb-22 (Wikstrand et al., 1991, 1992). The apparent absence of these gangliosides in the normal adult human CNS, as opposed to their presence in developing and neonatal brain (periods of astroglial proliferation) (Svennerholm et al., 1989), strongly recommends them as targets for therapeutic approaches. In this review, we will summarize the phenotypic distribution of GM2, GD2, GD3, 3'-isoLM1 and 3',6'-isoLD1 in tumors of the CNS. The relevance of these moieties to the development of targeted immunotherapeutic approaches to CNS tumors is demonstrated by the therapeutic applicability of GD3 and GD2; the involvement of GM2, GD2 and GD3 in the target structures of human natural killer and polymorphonu-

214

clear leukocytes (Thurin et al., 1987; Ando et al., 1987; Barker et al., 1991; Kushner and Cheung, 1991); and the description of operationally specific glioma-associated gangliosides and defining MAbs (Månsson et al., 1986; Fredman et al., 1988; Wikstrand et al., 1991).

Development and specificity of epitope-defined MAbs to GM2, GD2, GD3, 3′-isoLM1 and 3′,6′-isoLD1

Complete descriptions of the isolation and characterization of each of the MAbs in these analyses are found in the initial reference cited for each MAb. In general, following the immunization regimen described, hybridomas of interest were initially identified by screening against the purified target ganglioside, when available, or a ganglioside extract mixture of the target tumor of interest. Cross-reactivity with reference standard gangliosides was determined in the laboratory at Duke University. Purified MAbs of interest were then definitively analyzed with fast atom bombardment-mass spectrometry defined ganglioside and neutral glycolipid standards at Göteborg University. Once epitopic specificity was determined, the appropriate MAbs were used for analysis of tumor cell lines and xenograft ganglioside extractions and for immunohistochemical analysis of cytological preparations and tumor biopsies. As we stressed in a report of the isolation of the ganglioside GalNAc-3′-isoLM1 (Fredman et al., 1989), even the use of epitope-defined MAbs can be misleading if not combined with methods of molecular characterization of the target, since minimum defining epitopes may be found on a variety of glycoconjugates, some of which may comigrate in chromatographic analyses. Immunohistochemical analyses, therefore, identify the presence of the defining epitope targeted and not the molecular species upon which that epitope resides.

DMAb-1, DMAb-2, DMAb-3, DMAb-4 and DMAb-5 were isolated following immunization of spleen cell donors with the GM2-expressing human glioma line D-54 MG (Vrionis et al., 1989; Fig. 1). Of the five MAbs, DMAb-3 was found optimal for

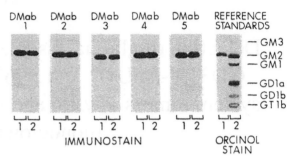

ANTI-GM2 Mabs

Fig. 1. HPTLC analysis of anti-GM2 MAbs. Lane 1: purified GM2. Lane 2: reference standard cocktail, as shown by orcinol stain, containing GM3, GM2, GM1, GD1a, GD1b and GT1b. DMAb-1, DMAb-2, DMAb-3, DMAb-4 and DMAb-5 are unreactive with all but GM2.

the detection of the GalNAcβ1-4(NeuAcα2-3)Gal-terminal epitope common to GM2 and GalNAc-GD1a in a variety of assays (high-performance thin-layer chromatography [HPTLC], solid phase radioimmunoassay and immunohistochemistry; Fig. 2). DMAb-7 and DMAb-8 were isolated following immunization of spleen cell donors with the GD3-rich melanoma cell line SK-MEL 28 (He et al., 1989). While several hybridomas isolated from these fusions were found to react with GD3, 3′,8′-LD1 and to various degrees with GD2, GT1b and GD1b, only DMAb-7 and DMAb-8 confined reactivity to the terminal tetrasaccharide NeuAcα2-8NeuAcα2-3Galβ1-4-(Glc or GlcNAc) found on GD3 and 3′,8′-LD1 (Figs. 2 and 3). GT1a, which differs in the penultimate sugar in the neutral backbone (N-acetylgalactosamine vs. N-acetylglucosamine), binds these MAbs only weakly. DMAb-20 was isolated following immunization of spleen cell donors by a mixed immunization regimen of purified GD2 coupled to *Salmonella minnesota* and GD2-expressing LAN-1 cells (Longee et al., 1991). DMAb-20 was found to bind to GD2 (Fig. 4). Its failure to bind GD3, GD1b, GT1b or GQ1b established the need for an unbound, terminal N-acetylgalactosamine, and failure to bind GM2 indicated the requirement for the disialylation of galactose. Thus the minimum binding epitope for DMAb-20 was determined to be GalNAcβ1-4(NeuAcα2-8NeuAcα2-3)Gal-.

SPECIFICITY ANALYSIS WITH DEFINED GANGLIOSIDES

Ganglioside		* Reactivity With:			
		DMAb-3 (Anti-GM2)	DMAb-20 (Anti-GD2)	DMAb-7 (Anti-GD3)	DMAb-8 (Anti-GD3)
GM2 (NeuAc)		+ + + +	-	-	-
GM2 (NeuGc)		+	NT	NT	NT
GalNAc-GD1a		+ +	NT	NT	NT
GD2		-	+ + + +	-	-
GD3		-	-	+ + + +	+ + + +
3',8'-LD1		NT	NT	+ +	+ +
GT1a		NT	NT	±	±
GM3		-	-	-	-
GM1		-	-	-	-
GD1a		-	-	-	-
GD1b		-	-	-	-
GT1b		-	-	-	-
GQ1b		-	-	NT	NT

* As assessed by quantitative ELISA and HPTLC immunostain

Fig. 2. Summary of the specificity analysis of DMAb-3, DMAb-7, DMAb-8 and DMAb-20 for gangliosides defined by fast atom bombardment-mass spectrometry as assessed by quantitative enzyme-linked immunosorbent assay and HPTLC immunostain. For symbol identification, see Fig. 6.

MAb SL-50 was generated following immunization with purified 3'-isoLM1 coupled to *S. minnesota* (Davidsson et al., 1989; Fredman et al., 1990; Fig. 5). DMAb-14 was isolated following a mixed immunization regimen incorporating multiple doses of trypsinized D-54 MG xenograft cells and oligosialoganglioside fractions thereof coupled to *S. minnesota* (Wikstrand et al., 1992; Fig. 6). DMAb-21 and DMAb-22 were isolated following immunization with trypsinized PA-1 human teratoma cells grown in xenograft form in athymic mice. PA-1 cells in both cultured and xenograft form express 3'-isoLM1 and 3',6'-isoLD1 (Fukuda et al., 1986; Wikstrand et al., 1991). Both SL-50 and DMAb-14 were found to bind to both *N*-glycolylneuraminic and

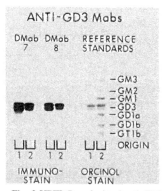

Fig. 3 HPTLC analysis of anti-GD3 MAbs. Lane 1: purified GD3. Lane 2: reference standard cocktail, as shown by orcinol stain, containing GM3, GM2, GM1, GD3, GD1a, GD1b and GT1b. DMAb-7 and DMAb-8 are unreactive with all but GD3; reactivity with 3',8'-LD1 (Fig. 2) not illustrated here.

ANTI - GD2 Mabs

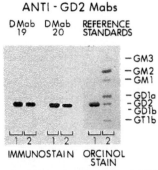

Fig. 4. HPTLC analysis of anti-GD2 MAbs. Lane 1: purified GD2. Lane 2: reference standard cocktail, as shown by orcinol stain, containing GM3, GM2, GM1, GD1a, GD2, GD1b and GT1b. DMAb-20 was found to be specific for GD2; by quantitative enzyme-linked immunosorbent assay (Fig. 2), DMAb-19 was marginally reactive with GQ1b, GT1b, GD1b and GD3 (Longee et al., 1991).

N-acetylneuraminic acid containing 3'-isoLM1 (Figs. 5 and 7). SL-50 requires an unsubstituted GlcNAc residue, while DMAb-14 will accept the α2–6-linked sialic acid to GlcNAc of 3',6'-isoLD1. DMAb-21 and DMAb-22 react only with 3',6'-isoLD1 (Figs. 6 and 7). None of the MAbs react with 3'-LM1, which has a β1–4 rather than a β1–3 linkage between Gal and GlcNAc (Wikstrand et al., 1991). The reactivity spectrum of the MAbs in these studies of CNS tumors are summarized in Table I.

ANTI-3' ISO-LM1 /3',6' ISO-LD1 Mabs

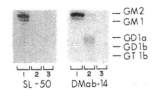

1:3' ISO-LM1   2:3',6' ISO-LD1   3:Reference Standard

Fig. 5. HPTLC analysis of SL-50 and DMAb-14. Lane 1: D-54 MG xenograft monosialoganglioside fraction. Lane 2: D-54 MG xenograft oligosialoganglioside fraction. Lane 3: reference standard cocktail, as shown by orcinol stain, containing GM2, GM1, GD1a, GD1b and GT1b.

ANTI-3',6' ISO-LD1 Mabs

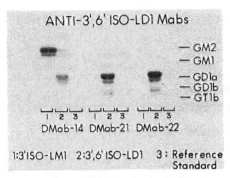

1:3'ISO-LM1   2:3',6' ISO-LD1   3: Reference Standard

Fig. 6. HPTLC analysis of DMAb-14, DMAb-21, and DMAb-22. Lane 1: D-54 MG xenograft monosialoganglioside fraction. Lane 2: D-54 MG xenograft oligosialoganglioside fraction. Lane 3: reference standard cocktail, as shown by orcinol stain, containing GM2, GM1, GD1a, GD1b and GT1b.

GANGLIOSIDE/GLYCOLIPID		REACTIVITY WITH:			
		DMAb-14	SL-50	DMAb-21	DMAb-22
LACTOTETRAOSE SERIES					
isoLA1		−	−	−	−
3'-isoLM1		+	+	−	−
		+	+	−	−
GalNAc-3'-isoLM1		−	−	−	−
Fuc-3'-isoLM1		−	−	−	−
3',6'-isoLD1		+	−	+	+
		+	−	+	+
NEOLACTOTETRAOSE SERIES					
3'-LM1		−	−	−	−
3',8'-LD1		−	−	−	−
GANGLIO SERIES					
GM2		−	−	−	−
GD2		−	−	−	−
GD1a		−	−	−	−
GD1b		−	−	−	−
GT1b		−	−	−	−
LACTOSYL SERIES					
GD3		−	−	−	−

□ GALACTOSE ◇ N-ACETYLGLUCOSAMINE ■ GLUCOSE ↓ N-ACETYLNEURAMINIC ACID ⇣ N-GLYCOLYLNEURAMINIC ACID ○ N-ACETYLGALACTOSAMINE ↑ FUCOSE − β1-3 LINKAGE ~ β1-4 LINKAGE

Fig. 7. Summary of the specificity analysis of SL-50 and DMAb-21 and DMAb-22 with gangliosides defined by fast atom bombardment-mass spectrometry as assessed by quantitative enzyme-linked immunosorbent assay and HPTLC immunostain.

TABLE I

Panel of epitope-defined antiganglioside MAbs (by permission, Wikstrand, et al., 1992)

MAb	Minimum binding epitope[a]	Defined gangliosides expressing epitope[b]	Reference
DMAb-3	○–□– (with ▼ below)	GM2, GalNAc-GD1a	Vrionis et al., 1989
DMAb-20	○–□– (with ▼ below)	GD2	Longee et al., 1991
DMAb-7	□– (with ▼ below)	GD3, GT1a, 3′,8′-LD1	He et al., 1989
SL-50	□–◇– (with ▼ below □)	3′-isoLM1	Davidson et al., 1989 Fredman et al., 1990
DMAb-22	□–◇– (with ▼ above and ▼ below)	3′,6′-isoLD1	Wikstrand et al., 1992

[a]As defined by comparative analysis of binding to FAB-MS analyzed, purified ganglioside standards, and described thoroughly in reference given. O: *N*-acetylgalactosamine; □: galactose; ◇: *N*-acetylglucosamine; ▼: NeuAc.
[b]Reactivity spectrum of antibody designated.

Reactivity of antiganglioside MAb panel with cultured human cell lines of neuroectodermal and neural origin

Table II summarizes the results obtained in the original descriptive studies for the component MAbs (Vrionis et al., 1989; He et al., 1989; Longee et al., 1991; Wikstrand et al., 1991). Each MAb-cell population pair was analyzed by titration of antibody versus live cells in either radioimmunoassay or indirect membrane immunofluorescence as assessed by direct visual inspection or fluorescence-activated cell sorting scan and by HPTLC analysis of separate mono- and oligo-sialoganglioside fractions prepared from cultured cell pellets as previously described (Svennerholm and Fredman, 1980; Fredman et al., 1980; Vrionis et al., 1989). Within the limits of HPTLC analysis using epitope-defined MAbs, the activity detected by DMAb-3 in these preparations migrated as GM2, and that by DMAb-7 migrated as GD3. As summarized in Table II, of the 15 glioma-derived cell lines examined, 14 (93%) expressed GM2, and 11 (73%) expressed GD2 or GD3. Only 2 (13%) expressed the oncofetally expressed 3′-isoLM1, and none expressed 3′,6′-isoLD1. GD3, GM2 and GD2 were co-expressed by 10 (67%) of the glioma lines, a pattern repeated by 3/4 of the medulloblastoma cell lines and 2/2 of the neuroblastoma cell lines studied. These results are consistent with previous individual reports of the presence of GD3, GM2 and GD3 in cultured glioma, neuroblastoma and melanoma lines (Dippold et al., 1984; Schultz et al., 1984; Natoli et al., 1986; Cheresh et al., 1986). The observations concerning the distribution of 3′-isoLM1 and the ganglioside phenotype of the medulloblastoma cell lines examined are unique to this series of studies. In summary, in glioma and other neuroectodermal and neural tumor-derived cell lines, GM2 is the most frequently observed monosialoganglioside (93 and 83%, respectively), while GD3 and GD2 demonstrate equivalent frequencies (73–83%, and 73–100%, respectively). 3′-isoLM1 was found

TABLE II

Ganglioside phenotype of 21 human cultured cell lines of neuroectodermal or neural origin

Cell line origin	Defined MAb (epitope detected)				No. of cell lines (%)	
	DMAb-7 (GD3)	DMAb-3 (GM2)	DMAb-20 (GD2)	SL-50 (3'-isoLM1)		
Gliomas	+	+	+	+	1/15	(7)
	+	+	+	−	9/15	(60)
	−	+	+	−	1/15	(7)
	−	+	−	−	3/15	(20)
	+	−	−	+	1/15	(7)
Medulloblastomas	+	+	+	−	3/4	(75)
(Daoy)	−	+	+	−	1/4	(25)
Neuroblastomas	+	+	+	−	2/2	
Total number (%)	16/21 (76)	20/21 (95)	17/21 (81)	2/21 (10)		

Data presented summarize radioimmunoassay, immunofluorescence and HPTLC results obtained with MAb defined for detected epitope (in parentheses); complete data available in Vrionis et al. (1989), He et al. (1989), Longee et al. (1991), and Wikstrand et al. (1991, 1992). DMAb-22 did not react with any of the 21 cell lines in Table II.

only in 2/15 cultured cell lines derived from gliomas; it has also been detected in embryonal carcinoma cell lines (Fukuda et al., 1986; Wikstrand et al., 1991) and a pancreatic adenocarcinoma cell line (Wikstrand et al., 1991). 3',6'-isoLD1 was found to be expressed in 3/3 teratoma cell lines and 1 pancreatic carcinoma cell line (Wikstrand et al., 1992); it has previously been reported to be expressed by a human colonic adenocarcinoma cell line (Fukushi et al., 1986) and by teratomas (Fukuda et al., 1986). With the exception of 3'-isoLM1, the expression of GM2, GD2 and GD3 is relatively consistent in tumor cell lines of neuroectodermal and neural origin.

Reactivity of the antiganglioside MAb panel with human glial tumor specimens

Table III summarizes the immunohistochemical analysis of frozen sections of 24 specimens of pathologically verified human gliomas; the tissue panel includes 21 glioblastomas, 2 gliosarcomas, and 1 giant cell glioblastoma. No chromatographic analysis

of these tumor tissues was performed. However, in a separate study (Fredman et al., 1988), 14/14 glioma (Grades III and IV) but only 1/4 astrocytoma (Grade II) samples contained 3'-isoLM1 as detected by HPTLC-enzyme-linked immunosorbent assay with the 3'-isoLM1 cross-reactive MAb C-50. In the series presented here analyzing 24 gliomas, 24 were reactive with the anti-GD3 MAb DMAb-7, and 20 (83%) expressed both GD3 and GM2. The frequency of GD2 expression in this series was essentially identical to that observed in vitro (83 vs. 81%). Most significant was the increase in the frequency of 3'-isoLM1 expression; 13 samples (54%) were dramatically reactive with the anti-3'-isoLM1 defining MAb SL-50 (Wikstrand et al., 1991). 3',6'-isoLD1 was detected by DMAb-22 in 15 (63%) of the cases examined; it was notably positive in six cases in which 3'-isoLM1 was not detectable by immunohistochemistry with MAb SL-50. The most probable explanation for the discrepancy between the frequency of 3'-isoLM1 detection in the study by Fredman et al. (1988) and the results here (100% vs. 54% of

TABLE III

Ganglioside phenotype of 24 biopsied human tumors of glial origin

	Defined MAb (epitope detected)					No. of tumors (%) (n = 24)
	DMAb-7 (GD3)	DMAb-3 (GM2)	DMAb-20 (GD2)	SL-50 (3'-isoLM1)	DMAb-22 (3',6'-isoLD1)	
Glioma	+	+	+	+	+	9 (37.5)[a]
	+	+	+	+	−	3 (12.5)
	+	+	+	−	+	3 (12.5)
	+	+	+	−	−	3 (12.5)
	+	−	−	−	+	2 (8)
	+	+	−	−	−	1 (4)
	+	+	−	−	+	1 (4)
	+	−	+	+	−	1 (4)
	+	−	+	−	−	1 (4)
Total number (%)	24/24 (100)	20/24 (83)	20/24 (83)	13/24 (54)	15/24 (63)	

Immunohistochemical analysis with MAb defined for detected epitope (in parentheses); complete data available in Longee et al. (1991) and Wikstrand et al. (1991, 1992).

[a] The 9 gliomas with this pattern include 2 gliosarcomas and 1 giant cell glioblastoma.

malignant gliomas) is the relatively higher sensitivity of the HPTLC-enzyme-linked immunosorbent assay employed as opposed to the immunohistochemical analysis summarized here.

Recently we reported the application of a subset of this MAb panel to the evaluation of cytologic specimens of poorly differentiated tumors (Vick et al., 1992). Cytologic specimens, consisting of air-dried cytospins from fine-needle aspirate or cerebrospinal fluid samples, were analyzed for the expression of GM2, GD2 and GD3, as defined by MAbs DMAb-3, DMAb-20 and DMAb-7. Only the results obtained with samples from patients with a diagnosed glial tumor are presented in Table IV. In addition, the results of HPTLC analysis of mono- and oligo-sialo-ganglioside fractions prepared from human glioma and medulloblastoma xenografts grown subcutaneously in athymic rodents are also presented in Table IV. In general, the immunohistochemical evaluation of cytologic specimens revealed a distribution of positive staining similar to the observations with frozen sections, although the number of positive cases was lower for the cytologic preparations. This discrepancy could well be a reflection of the small number of cells available for evaluation per fine-needle aspirate or cerebrospinal fluid sample. Although the number of samples investigated is small, the data obtained by evaluation of tumor xenografts grown in athymic rodents by HPTLC of ganglioside extracts is generally considered more accurate, reflecting its sensitivity and relative migration information. In this regard, the xenograft extraction data are consistent with the cytological sample data for GM2 and GD2, but not at all so for GD3. Values obtained for 3'-isoLM1 expression in xenografts fall between those in the frozen section series (54%) and the biopsy extract series (100%) reported earlier (Fredman et al., 1988). The incidence of 3',6'-isoLD1 in xenografts (40%) was less than that observed in frozen sections; this discrepancy may be a reflection of the smaller sample number (10) or may suggest an inhibitory effect of growth in athymic rodents.

The relevance of such xenograft data to the accurate portrayal of the ganglioside phenotype of human

TABLE IV

Summary of human cytologic and tissue sample reactivity in tumors of glial origin

MAb (primary specificity)	Cytologic specimens[a]		Frozen sections[a] from primary CNS tumors (%)	Xenografts[b]	
	Glial tumors (%)	Medulloblastomas (%)		Gliomas (%)	Medulloblastomas (%)
DMAb-3 (GM2)	9/14 (64)	3/6 (50)	26/31 (84)	5/8 (62.5)	3/4 (75)
DMAb-20 (GD2)	6/14 (43)	1/6 (17)	24/30 (80)	3/7 (43)	3/3 (100)
DMAb-7 (GD3)	14/14 (100)	4/6 (67)	31/31 (100)	4/7 (57)	2/3 (67)
SL-50 (3'-isoLM1)	ND[c]	ND	15/31 (48)	6/8 (75)	1/5 (20)
DMAb-22 (3,6'-isoLD1)	ND	ND	15/24 (63)	4/10 (40)	0/4 (0)

[a]A panel of cytologic specimens consisting of air-dried cytospins from fine-needle aspirates or cerebrospinal fluid samples of glial tumors.
[b]Human tumor xenografts grown in athymic mice or rats analyzed by HPTLC.
[c]ND, Not done.

gliomas is unresolved. Several studies, however, have addressed this question. In previous studies we have shown that cells grown in culture had a less complex pattern than the cells grown as xenografts. By HPTLC analysis of cell lines and xenografts derived from them, GM2 and GD2 are predominant ganglio-sides of cultured cells (Vrionis et al., 1989; Fredman et al., 1990). In xenograft form, there is a general reduction in GM2, an increased complexity of ganglioside pattern (Fredman et al., 1990), and/or the appearance of gangliosides such as 3'-isoLM1 and 3',6'-isoLD1 not found in the cultured cells

TABLE V

Patterns of ganglioside expression in tumors of glial origin in vitro and in vivo

Defined epitope	% Population expressing epitope	
	Cultured cell lines ($n = 15$)	Biopsies ($n = 24$)
GD3	67	100
GM2	93	83
GD2	73	83
3'-isoLM1	13	54
3',6'-isoLD1	0	63

(Månsson et al., 1986). These observations are in complete accord with the report by Tsuchida et al. (1987). These authors demonstrated, by HPTLC analysis of melanoma cell lines grown in culture or xenograft form, that not only were GM2 and GD2 consistently more abundant in cultured cells than in xenograft form, but that the ganglioside phenotypes of the xenografts more closely resembled those of the biopsy samples from which the cultures and xenografts were derived. If the HPTLC analysis of xenografts is indeed more indicative of the original human tumor, this series of studies then would support observations previously made (Fredman et al., 1990) concerning the appearance of lacto and neolacto series gangliosides and complex branched structures in gliomas. Such a comparison is provided in Table V. Although the proportion of cell lines and glioma tissue samples expressing GM2, GD2 and GD3 are approximately the same, the incidence of 3'-isoLM1 and 3',6'-isoLD1 is significantly higher in vivo.

Summary

In this study, MAbs to the 'conventional' gangliosides expressed by human gliomas were generated and used to detect ganglioside species previously unisolated or defined in normal adult CNS tissue. Despite the marked phenotypic and genotypic heterogeneity shown by glioma cell lines (Bigner et al., 1981), the ganglioside phenotype of these cell lines is remarkably consistent qualitatively, if not quantitatively, in the ganglioside species expressed (Table V). The majority of cell lines and tumor samples express GM2, GD2 and GD3; this does not provide a diagnostic advantage (Vick et al., 1992). Nevertheless, as the relative amounts of these gangliosides in tumor as compared with normal adult CNS tissue is considerable, such reagents might be considered in compartmental immunotherapeutic approaches. Since GD2 and GD3 have been determined to mediate tumoricidal activity with human effector cells via specific antiganglioside epitope MAbs (Thurin et al., 1987; Kushner and Cheung, 1991; Barker et al., 1991; Reisfeld, 1993), cell-mediated approaches, as well as targeted immunoglobulin therapies, are also possible.

The prospect of a more targeted approach with little or no effect on normal CNS tissue is now possible via the 'oncofetal' epitopes characteristic of 3'-isoLM1 and 3',6'-isoLD1. Several factors recommend the use of these moieties for compartmental immunotherapy: the inability to detect them within the adult CNS; the relatively high frequency of expression of 3'-isoLM1 and 3',6'-isoLD1, especially in human tumor samples (50–100%, depending upon the series and assay); and the existence of specific MAbs reactive with these epitopes. Current technology is being applied to these MAbs to transfer the specific recognition capacity of existing murine MAbs into various human framework structures of any desired immunoglobulin class, and thereby, biologic function. The variety of effector functions, the stability in affinity, labeling capacity, and the exquisite sensitivity of these MAbs for these glioma-distinctive epitopes is an exciting and promising approach for immunotherapy of human CNS tumors.

Acknowledgements

The authors wish to express their appreciation to Laura Shaughnessy and Beatrice Brewington for their technical expertise and to Ann S. Tamariz, E.L.S., for editorial assistance on this manuscript. This work was supported by NIH grants CA 11898, NS 20023, and CA 56115; by the Swedish Medical Research Council (Project No. 03X-09909-01; and by the National Swedish Board for Technical Development (Project No. 84-4667P).

References

Ando, I., Hoon, D.S.B., Suzuki, Y., Saxton, R.E., Golub, S.H. and Irie, R.F. (1987) Ganglioside GM2 on the K562 cell line is recognized as a target structure by human natural killer cells. *Int. J. Cancer*, 40: 12–17.

Barker, E., Mueller, B.M., Handgretinger, R., Herter, M., Yu, A.L. and Reisfeld, R.A. (1991) Effect of a chimeric anti-ganglioside GD2 antibody on cell-mediated lysis of human neuroblastoma cells. *Cancer Res.*, 51: 144–149.

222

Bigner, D.D., Bigner, S.H., Pontén, J., Westermark, B., Mahaley, M.S., Ruoslahti, E., Herschman, H., Eng, L.F. and Wikstrand, C.J. (1981) Heterogeneity of genotypic and phenotypic characteristics of fifteen permanent cell lines derived from human gliomas. *J. Neuropathol. Exp. Neurol.*, 40: 201–229.

Cheresh, D.A., Rosenberg, J., Mujoo, K., Hirschowitz, L. and Reisfeld, R.A. (1986) Biosynthesis and expression of the disialoganglioside GD2, a relevant target antigen on small cell lung carcinoma for monoclonal antibody-mediated cytolysis. *Cancer Res.*, 46: 5112–5118.

Davidsson, P., Fredman, P. and Svennerholm., L. (1989) Gangliosides and sulphatide in human cerebrospinal fluid: quantitation with immunoaffinity techniques. *J. Chromatogr.*, 496: 279–289.

Dippold, W.G., Knuth, A. and Buschenfelde, K.H.M. (1984) Inhibition of human melanoma cell growth *in vitro* by monoclonal anti-GD3-ganglioside antibody. *Cancer Res.*, 44: 806–810.

Dohi, T., Nores, G.A. and Hakomori, S. (1988) An IgG3 monoclonal antibody established after immunization with GM3 lactone: immunochemical specificity and inhibition of melanoma cell growth *in vitro* and *in vivo*. *Cancer Res.*, 48: 5680–5685.

Fredman, P., Nilsson, O., Tayot, J.L. and Svennerholm, L. (1980) Separation of gangliosides on a new type of anion exchange resin. *Biochim. Biophys. Acta*, 618: 42–52.

Fredman, P., von Holst, H., Collins, V.P., Granholm, L. and Svennerholm, L. (1988) Sialyllactotetraosylceramide, a ganglioside marker for human malignant gliomas. *J. Neurochem.*, 50: 912–919.

Fredman, P., Månsson, J.E., Wikstrand, C.J., Vrionis, F.D., Rynmark, B.M., Bigner, D.D. and Svennerholm, L. (1989) A new ganglioside of the lactotetraose series, GalNAc-3'-isoLM1, detected in human meconium. *J. Biol. Chem.*, 264: 12122–12125.

Fredman, P., Månsson, J.E., Bigner, S.H., Wikstrand, C.J., Bigner, D.D. and Svennerholm, L. (1990) Gangliosides in the human glioma cell line U-118 MG grown in culture or as xenografts in nude rats. *Biochim. Biophys. Acta*, 1045: 239–244.

Fukuda, M.N., Bothner, B., Lloyd, K.O., Rettig, W.J., Tiller, P.R. and Dell, A. (1986) Structures of glycosphingolipids isolated from human embryonal carcinoma cells. *J. Biol. Chem.*, 261: 5145–5153.

Fukushi, Y., Nudelman, E., Levery, S. T., Higuchi, T., and Hakomori, S.-I. (1986) A novel disialoganglioside (IV3NeuAcIII6NeuAcLc4) of human adenocarcinoma and the monoclonal antibody (FH9) defining this disialosyl structure. *Biochemistry*, 25: 2859–2866.

He, X., Wikstrand, C.J., Fredman, P., Månsson, J.E., Svennerholm, L. and Bigner, D.D. (1989) GD3 expression by cultured human tumor cells of neuroectodermal origin. *Acta Neuropathol.*, 79: 317–325.

Heiner, J.P., Miraldi, F., Kalliek, S., Makely, J., Neely, J., Smith-Mensah, W.H. and Cheung, N.-K.V. (1987) Localization of GD2-specific monoclonal antibody 3F8 in human osteosarcoma.

Cancer Res., 47: 5377–5381.

Herberman, R.B., Morgan, A.C., Reisfeld, R.A., Cheresh, D.A. and Ortaldo, J.R. (1985) Antibody-dependent cellular cytotoxicity (ADCC) against human melanoma by human effector cells in cooperation with mouse monoclonal antibodies. In: R.A. Reisfeld and S. Sell (Eds.), *Monoclonal Antibodies and Cancer Therapy*, Alan R. Liss, New York, pp. 193–203.

Houghton, A.N., Mintzer, D., Cordon-Cardo, C., Welt, S.W., Fliegel, B., Vadhan, S., Carswell, E., Melamed, M.R., Oettgen, H.F. and Old, L.J. (1985) Mouse monoclonal IgG3 antibody detecting GD3 ganglioside: a phase I trial in patients with malignant melanoma. *Proc. Natl. Acad. Sci. USA*, 82: 1242–1246.

Irie, R.F. and Morton, D.L. (1986) Regression of cutaneous metastatic melanoma by intralesional injection with human monoclonal antibody to ganglioside GD2. *Proc. Natl. Acad. Sci. USA*, 83: 8694–8698.

IUPAC-IUB Commission on Biochemical Nomenclature (1977) The nomenclature of lipids. *Eur. J. Biochem.*, 79: 11–21.

Kushner, B.H. and Cheung, N.K.V. (1991) Clinically effective monoclonal antibody 3F8 mediates nonoxidative lysis of human neuroectodermal tumor cells by polymorphonuclear leukocytes. *Cancer Res.*, 51: 4865–4870.

Longee, D.C., Wikstrand, C.J., Månsson, J.E., He, X., Fuller, G.N., Bigner, S.H., Fredman, P. , Svennerholm, L., and Bigner, D.D. (1991) Disialoganglioside GD2 in human neuroectodermal tumor cell lines and gliomas. *Acta Neuropathol.*, 82: 45–54.

Månsson, J.E., Fredman, P., Bigner, D.D., Molin, K., Rosengren, B., Friedman, H.S. and Svennerholm, L. (1986) Characterization of new gangliosides of the lactotetraose series in murine xenografts of a human glioma cell line. *FEBS Lett.*, 201: 109–113.

Natoli, E.J., Livingston, P.O., Pukel, C.S., Lloyd, K.W., Wiegandt, H., Szalay, J., Oettgen, H.F. and Old, L.J. (1986) A murine monoclonal antibody detecting *N*-acetyl- and *N*-glycolyl-GM2: characterization of cell surface reactivity. *Cancer Res.*, 46: 4116–4120.

Reisfeld, R.A. (1993) Potential of genetically engineered antiganglioside GD2 antibodies for immunotherapy of melanoma and neuroblastoma. In: L. Svennerholm (Ed.), *Progress in Brain Research*, Elsevier Science Publishers, Amsterdam, this issue.

Schultz, G., Cheresh, D.Z., Varki, N.M., Yu, A., Staffileno, L.K. and Reisfeld, R.A. (1984) Detection of ganglioside GD2 in tumor tissue and sera of neuroblastoma patients. *Cancer Res.*,44: 5914–5920.

Svennerholm, L. (1963) Chromatographic separation of human brain gangliosides. *J. Neurochem.*, 10: 613–623.

Svennerholm, L. and Fredman, P. (1980) A procedure for the quantitative isolation of brain gangliosides. *Biochim. Biophys. Acta*, 617: 97–109.

Svennerholm, L., Bostrom, K., Fredman, P., Månsson, J.E., Rosengren, B. and Rynmark, B.M. (1989) Human brain gangliosides: developmental changes from early fetal stage to advanced age. *Biochim. Biophys. Acta*, 1005: 109–117.

Thurin, J., Thurin, M., Kimoto, Y., Herlyn, M., Lubeck, M.D.,

Elder, D.E., Smereczynska, M., Karlsson, K.A., Clark, W.M., Steplewski, Z. and Kprowski, H. (1987) Monoclonal antibody-defined correlations in melanoma between levels of GD2 and GD3 antigens and antibody-mediated cytotoxicity. *Cancer Res.*, 47: 1229–1233.

Tsuchida, T., Ravindranath, M.H., Saxton, R.E. and Irie, R.F. (1987) Gangliosides of human melanoma: altered expression *in vivo* and *in vitro. Cancer Res.*, 47: 1278–1281.

Vick, W.W., Tello, J.W., Wikstrand, C.J., He, X., Longee, D.C., Fredman, P., Svennerholm, L., Bigner, D.D., Johnston, W.W. and Bigner, S.H. (1992) The application of a panel of anti-ganglioside monoclonal antibodies to cytologic specimens. *Acta Cytol.*, in press.

Vrionis, F.D., Wikstrand, C.J., Fredman, P., Mansson, J.E., Svennerholm, L. and Bigner, D.D. (1989) Five new epitope-defined monoclonal antibodies reactive with GM2 and human glioma and medulloblastoma cell lines. *Cancer Res.*, 49:

6645–6651.

Wikstrand, C.J., He, X., Fuller, G.N., Bigner, S.H., Fredman, P., Svennerholm, L. and Bigner, D.D. (1991) Occurrence of lacto series gangliosides 3'-isoLM1 and 3',6'-isoLD1 in human gliomas *in vitro* and *in vivo. J. Neuropathol. Exp. Neurol.*, 50: 756–769.

Wikstrand, C.J., Fredman, P., Svennerholm, L., Humphrey, P.A., Bigner, S.H. and Bigner, D.D. (1992) Monoclonal antibodies to malignant human glliomas. *Mol. Chem. Neuropathol.*, in press.

Wikstrand, C.J., Longee, D.C. Fuller, G.N., McLendon, R.E., Friedman, H.S., Fredman, P., Svennerholm, L., and Bigner, D.D. (1993) Lactotetraose series ganglioside 3',6'-isoLD1 in tumors of central nervous and other systems *in vitro* and *in vivo. Cancer Res.*, 53: 120–126.

Yu, A.L., Reisfeld, R.A. and Gillies, S.D. (1991) Immune response to monoclonal anti-GD2 antibody therapy. *Proc. Am. Assoc. Cancer Res.*, 32: 263 (abstr.).

L. Svennerholm, A.K. Asbury, R.A. Reisfeld, K. Sandhoff, K. Suzuki, G. Tettamanti and G. Toffano (Eds.)
Progress in Brain Research, Vol. 101

CHAPTER 17

Gangliosides associated with primary brain tumors and their expression in cell lines established from these tumors

Pam Fredman

Department of Psychiatry and Neurochemistry, University of Göteborg, Sweden

Introduction

Tumor-associated gangliosides have attracted a great deal of interest in tumor research. Most studies have focused on their potential as targets for monoclonal antibodies in diagnosis and therapy of cancer, and a large number of tumor-associated ganglioside structures and monoclonal antibodies binding to epitopes of such antigens have been described (reviewed recently by Hakomori 1989; Alhadeff 1989; Fredman 1993). The altered glycosylation resulting in tumor-associated gangliosides also involves other glycoconjugates (Hakomori, 1989). Monoclonal antibodies recognizing carbohydrate epitopes of gangliosides might thus also react with other glycoconjugates. The role of gangliosides has not been completely elucidated. However, there are many studies that lend support for their involvement in transmembrane signaling, and in cell to cell and cell to matrices interactions (Hakomori, 1981, 1990; Svennerholm 1984; Fenderson et al., 1990). The uncontrolled growth and the metastatic and invasive properties of tumor cells may thus be a consequence of an abberant expression of gangliosides.

Only a few of many studies on tumor-associated gangliosides and their potential role in tumor cells have been performed on primary brain tumors. This may possibly be due to some extent to the limited availability of tumor tissue. Autopsies are rarely performed on brain tumor patients, and brain tumor biopsies obtained at open surgery do not generally provide enough tissue for biochemical analyses of tumor-associated gangliosides. This is particularly the case when isolation of gangliosides and characterization of unknown structures has to be accomplished. Another contributing factor that hampers biochemical studies, particularly malignant gliomas, is that these tumors grow infiltratively into adjacent normal brain tissues. The proportion of tumor cells in biopsy specimens varies therefore and renders the interpretation of biochemical analyses difficult.

This chapter will provide a brief overview of gangliosides that have been found in primary brain tumor tissues and describe their distribution and expression in cell lines derived from these primary brain tumors.

Gangliosides associated with human primary brain tumor tissues

All the tumor-associated gangliosides that have been reported so far (for review see Hakomori 1989; Alhadeff, 1989; Fredman, 1993) belong to the lacto or ganglio series (Fig. 1). The lacto series consists of two series, the type 1 and the type 2 chain, also named lacto and neolacto series, respectively, depending on the linkage of the terminal galactose to glucosamine. A β1-3 linkage specifies the type 1 and a β1-4 linkage the type 2 series, respectively. The lacto series is characterized by its content of *N*-

acetylglucosamine in contrast to the ganglio series, which contains *N*-acetylgalactosamine. The structures shown in Fig. 1 are the core structures to which one or more sugar residues can be added. The tumor-associated gangliosides GM3 and GD3 (Fig. 1, Table I) do not contain the *N*-acetylgalactosamine residue which is required in the ganglio series gangliosides. However, since they are precursors in the synthesis of the ganglio series this chapter includes them in this series.

No ganglioside structure reported to date was found to be exclusively expressed by tumor cells. Gangliosides assigned to be associated with primary brain tumors in Table I are: (1) not detectable by immunological methods in the surrounding tissue; and (2) not found in significantly increased proportion and/or amount in tumor cells as compared to surrounding brain tissue. It should be noted that the occurrence of such gangliosides has been investigated in only a limited number of normal and neoplastic tissues.

Immunological methods using monoclonal antibodies have been one of the most important technical improvements in the search for tumor-associated gangliosides and other glycoconjugates. Monoclonal antibodies to carbohydrates can be characterized with regard to their binding epitope and highly specific antibodies can be obtained. Several monoclonal antibodies directed against defined carbohydrate epitopes of tumor-associated gangliosides have been described (for review see Svennerholm, 1988; Hakomori 1989; Fredman, 1993). An up to ten-fold higher amount of gangliosides in normal brain tissue than in tumors (Fredman et al., 1986, 1988) often makes the detection of tumor-associated gangliosides impossible. This problem might be circumvented by the use of immunological methods. Monoclonal antibodies can also be used for immunohistochemical analyses of the tissue distribution of ganglioside antigens and they are invaluable tools to study the biological role of tumor-associated gangliosides. However, the major aim in raising antibodies to tumor-associated gangliosides has been to apply them for diagnosis and therapy of tumors and their use for brain tumors is discussed by Wikstrand et al. (1993).

The first report of brain tumor-associated gangliosides came from Seifert and Uhlenbruck (1965) who showed that human meningiomas contained large

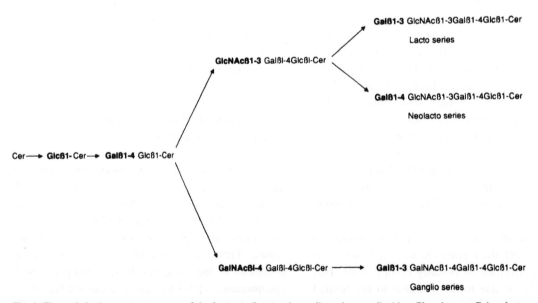

Fig. 1. The carbohydrate core structures of the lacto, neolacto and ganglio series gangliosides. Glc, glucose; Gal, galactose; GlcNAc, glucosamine; GalNAc, galactosamine; Cer, ceramide.

TABLE I

Gangliosides associated with primary brain tumors

Gangliosides	Designation	Main tumor reactivity	References
Ganglioseries			
GalNAcβ1-4Galβ1-R	GM2	Glioma	Yates et al. 1979; Traylor and Hogan, 1980; Irie et al., 1982;
3		Melanoma	Tai et al., 1983; Fredman et al., 1986; Tsuchida et al., 1987a
2		Germ cell tumors	Miyake et al., 1990
NeuAcα			
GalNAcβ1-4Galβ1-R	GD2	Glioma	Fredman et al., 1986
3		Neuroblastoma	Schulz et al., 1984; Cheung et al., 1985; Wu et al., 1986
2		Melanoma	Cahan et al., 1982
NeuAcα2-8NeuAcα		Small cell lung carcinoma	Cheresh et al., 1986a
Galβ1-R	GM3	Glioma	Yates et al., 1979; Traylor and Hogan, 1980;
3			Fredman et al., 1986, 1988; Jenneman et al., 1990
2		Meningioma	Seifert and Ohlenbruck, 1965; Aruna et al., 1973;
NeuAcα			Sunder-Plassman and Bernheimer, 1974; Davidsson et al., 1989; Eto and Shinoda 1992
		Melanoma	Tuschida et al., 1987a; Furukawa et al., 1989; Yamamato et al., 1990
		Medulloblastoma	Gottfries et al., 1990
Galβ1-R	GD3	Glioma	Seifert, 1966; Eto and Shinoda, 1982;
3			Berra et al., 1985; Fredman et al., 1986, 1988; Jennemann et al.,1990
2		Medulloblastoma	Gottfries et al., 1990
NeuAcα2-8NeuAcα		Meningioma	Davidsson et al., 1989; Seifert 1966; Sunder-Plassman and Bernheimer 1974; Yates et al., 1979; Berra et al., 1990
		Leukemia	Siddiqui et al., 1984
		Melanoma	Pukel et al., 1982
Lactoseries (Type 1 chain)			
Galβ1-3GlcNAcβ1-3Galβ1-R	3′-isoLM1	Glioma	Fredman et al., 1988; Wikstrand et al., 1991, this volume
3		Medulloblastoma	Gottfries et al., 1990
2		Teratocarcinoma cells	Fukuda et al., 1986
NeuAcα		Small cell lung carcinoma	Nilsson et al., 1985a
NeuAcα			
2			
6		Glioma	Wikstrand et al., 1992, this volume
Galβ1-3GlcNAcβ1-3Galβ1-R	3′6′-isoLD1	Liver metastasis of colon cancer	Fukushi et al., 1986
3			
2		Embryonal carcinoma cells	Fukuda et al., 1986
NeuAcα			
		Epithelial cancer	Wikstrand et al., 1992

amounts of ganglioside GM3. With the use of two-dimensional thin-layer chromatography, a second ganglioside was structurally characterized as GD3 (Seifert 1966). Since then other studies on the ganglioside composition of human brain tumors have been reported (see below). The results in many of these studies were based solely on thin-layer chromatographic (TLC) migration of the gangliosides as compared to known ganglioside standards. Partial or complete structural characterization of the tumor-associated gangliosides was only performed in a few of these studies (Berra et al., 1983, 1985; Fredman et al., 1986, 1988). However, identification of gangliosides by TLC migration can be misleading as co-migration of different ganglioside often occurs (Fig. 2).

Meningioma

A common finding in all studies on gangliosides in human meningiomas (Seifert and Uhlenbruck, 1965; Aruna et al., 1973; Sunder-Plassmann and Bernheimer, 1974; Eto and Shinoda, 1982; Berra et al., 1983; Davidsson et al., 1989) was the large proportion of ganglioside GM3, irrespective of the histological classification. The only exception documented in the literature is the report by Kostic and Buchheit (1970) who found an increased proportion of a ganglioside named G_{gal}; however, no characterization of this ganglioside was performed. The large proportion of ganglioside GD3, first reported by Seifert (1966), has been documented in several studies (Sunder-Plassmann and Bernheimer 1974; Yates et al., 1979; Davidsson et al., 1989; Berra et al., 1990).

Gangliosides of the gangliotetraose series, GM1, GD1a, GD1b, and GT1b, were detected as minor constituents in most studies. However, one study (Sunder-Plassmann and Bernheimer, 1974) reported GD1a to be a major ganglioside and in another (Berra et al., 1983) GM1 was a major component in some of the meningiomas. The gangliotetraose series gangliosides are major ganglioside constituents in normal brain tissue (Svennerholm, 1963; Svennerholm et al., 1989), and a slight contamination of the tumor specimen with brain tissue might thus be responsible for the minor amount detected. However, except in rare malignant cases, meningiomas do not generally grow infiltratively into the brain and it would be possible to get tumor specimens uncontaminated with normal brain tissue. Gangliotetraose series gangliosides were found in histopathologically well characterized meningioma tissue (Davidsson et al., 1989) but the proportions of these gangliosides were different from those of brain tissue. This finding led to the conclusion that the gangliotetraose series gangliosides were most likely derived from the tumor tissue.

We (Davidsson et al., 1989) also investigated the occurrence of gangliosides of the lacto series, i.e. gangliosides that are commonly found to be associated with tumors outside the brain (Hakomori, 1989; Alhadeff. 1989; Fredman, 1993). Our data indicated that 8 of 20 meningioma specimens contained the lacto series ganglioside with the type 2 chain ganglioside 3'-LM1. The corresponding isomer with the type 1 chain, 3'-isoLM1, a ganglioside found in all malignant gliomas (Fredman et al., 1988) and in medulloblastomas showing astrocytic differentiation (Gottfries et al., 1990), could not be detected in any specimen. Ganglioside 3'-isoLM1 was suggested to be a marker for proliferating normal and neoplastic astrocytes (Fredman et al., 1988; Gottfries et al., 1990; Fredman et al, 1993) and its absence in meningiomas provides further support for the contention that meningiomas are of mesenchymal origin.

In agreement with results of previous studies (Sunder-Plassmann and Bernheimer, 1974; Berra et al., 1983) we (Davidssson et al., 1989) could not find any correlation between ganglioside composition and histological classifications of meningiomas. However, a subdivision of meningiomas based on ganglioside composition was proposed (Davidsson et al., 1989), namely a 'GM3-rich' and a 'GD3-rich' group. Berra et al. (1983) suggested a tentative classification of the meningiomas into a 'GM3-rich' and a 'GM1-rich' group, but neither we (Davidsson et al., 1989) nor Sunder-Plassmann and Bernheimer (1974) found any 'GM1-rich' meningiomas.

In a recent study (Fredman et al., 1990b) we showed that the expression of GD3 was associated with monosomy of chromosome 22, which occurs in approximately half of the meningiomas. Logistic regression revealed a 66% probability in predicting monosomy of chromosome 22 by the proportion of GD3. Similar results were reported by Berra et al. (1991). Ganglioside GD3 was also found in increased concentrations in cerebrospinal fluid (CSF) from patients with meningiomas (Davidsson et al., 1990), most likely due to shedding from tumor cells. A shedding of gangliosides to plasma from tumor cells has previously been described in association with neuroblastoma (Schulz et al., 1984; Ladisch et al., 1987) and melanoma (Portoukalian et al., 1978). Recently Nakamura et al. (1991) described an increased level of GD3 in sera from patients with advanced stage glioma.

Medulloblastoma

We have recently analysed the ganglioside content of six human medulloblastomas (Gottfries et al., 1990). This is, to the best of our knowledge, except for the single specimen analysed by TLC in the study by Yates et al. (1979), the only study performed on medulloblastomas. We found both quantitative and qualitative variations between the specimens. Tumors relatively rich in ganglioside contained a larger proportion of the gangliotetraose series gangliosides, GM1, GD1a, GD1b, GT1b, and also GQ1b, than those with a low ganglioside content, in which the gangliosides GM3 and GD3 dominated and constituted up to 40% of the total ganglioside sialic acid. The gangliotetraose series gangliosides may have been derived from contaminating normal brain tissue in which they represent major components. However, histological examinations of the specimens did not reveal any signs of brain tissue, a finding which suggests that these gangliosides were really tumor derived. Gangliotetraose series gangliosides thus seemed to be synthesized both in meningiomas of mesenchymal origin and in medulloblastomas of neuroectodermal origin.

Ganglioside 3'-LM1, of the lacto series type 2 chain, was found in all specimens, but like the gangliotetraose series gangliosides this ganglioside is a normal constituent of the adult brain tissue (Li et al., 1973; Svennerholm et al., 1989). However, gan-glioside 3'-isoLM1, of the lacto series type 1 chain has not been detected in normal adult brain tissue (Fredman et al., 1988; Svennerholm et al., 1989). This ganglioside was found to be associated in all cases with astrocytic differentiation, in agreement with the previously raised hypothesis that ganglioside 3'-isoLM1 is related to proliferating astrocytes (Fredman et al., 1988; Gottfries et al., 1990; Svennerholm et al., 1987).

Glioma

The diffuse infiltrative growth of gliomas into adjacent normal brain tissue makes it difficult to interpret their biochemical analysis. In this regard, the proportion of tumor cells can vary from 10% or less to almost 100%. Histopathological analyses will only comprise a very small area of the specimen and thus only provide a rough estimation of the number of tumor cells in the remaining tumor specimen used for biochemical investigation. The concentration of gangliosides in normal white and grey matter ranges from 1 to 3 μmol ganglioside sialic acid/g wet weight of tissue, respectively, while glioma specimens with a large proportion of tumor cells have a ganglioside content of around 0.3 μmol ganglioside sialic acid/g wet weight (Fredman et al., 1986). The proportion of normal brain tissue in the tumor specimens thus strongly influences any results.

Gangliosides of the gangliotetraose series, the dominating gangliosides in normal brain, were found in meningioma and medulloblastoma specimens and, as discussed above, are likely to be tumor derived. In the case of gliomas there is no possibility to verify the origin of gangliotetraose series gangliosides because contamination with normal brain tissue can never be excluded. In our studies on human glioma biopsies, obtained at open surgery, the most striking difference observed by TLC-analyses between normal brain and glioma tissue was the increased proportions of the monosialylated gangliosides GM3 and GM2 and their disialylated derivatives GD2 and GD3 (Fredman et al., 1986, 1988) (Table I). The proportion of these gangliosides were previously found to be elevated when compared to normal brain tissue by Yates et al. (1979) and Traylor and Hogan (1980)

230

and recently also by Jennemann et al. (1990). This increase was most pronounced for ganglioside GD3, a finding first reported by Seifert (1966) and later also by Eto and Shinoda (1982) and Berra et al. (1985). Among the glioma-associated gangliosides mentioned above we only detected a quantitative increase in ganglioside GD3 (Fredman et al., 1988; Fredman et al., 1993) and there was no correlation between the content of GD3 and the degree of malignancy (Fredman et al., 1988); however, such a correlation was reported by others (Berra et al., 1985; Traylor and Hogan, 1980; Eto and Shinoda, 1982; Nakamura et al., 1987).

The glioma-associated ganglioside 3′-isoLM1 was first isolated from the human glioma cell line D-54 MG grown as xenografts in nude mice (Månsson et al., 1986). The characterization of ganglioside 3′-isoLM1 then made it possible to determine its occurrence in human glioma tissue with a monoclonal antibody, C-50, that cross-reacts with this ganglioside (Månsson et al., 1985). A pool of ganglioside extracts from glioma biopsies served to verify the existence of ganglioside 3′-isoLM1 (Fredman et al., 1988).

Analyses of human glioma biopsies obtained at open surgery showed that 14/14 specimens contained 3′-isoLM1 (Fredman et al., 1988). To date, more than 50 specimens of malignant gliomas have been investigated by HPTLC-ELISA with a new specific monoclonal antibody, SL-50 (Wikstrand et al., 1993; Fredman et al., to be published) and all of these have been postive for this antigen (to be published). Investigations of autopsy brains from cases with malignant gliomas with the SL-50 antibody showed that 3′-isoLM1 was widely distributed in the brain (Fredman et al., 1993). It was mainly found in tissue specimens adjacent to macroscopic tumors and in the highest concentration in the region of the opposite hemisphere corresponding to the tumor. The corpus callosum, histopathologically shown to contain tumor cells, also contained 3′-isoLM1. These results have lead us to suggest that 3′-isoLM1 could be related to a certain clone or clones of glioma cells that are invading tumor cells. Another possible explanation is that 3′-isoLM1 is produced by normal cells as a reaction to the presence of the tumor cells.

Analyses of the human glioma D-54 MG cell line grown as xenografts in nude mice revealed the occurrence of ganglioside 3′6′-isoLD1, the disialylated

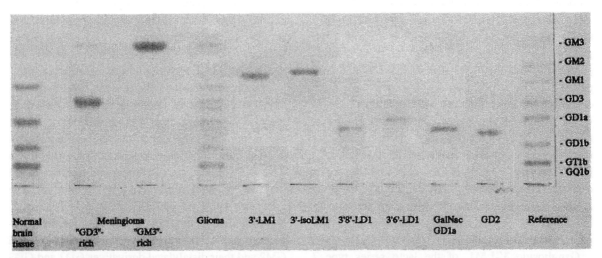

Fig. 2. Thin-layer chromatographic separation of gangliosides in human normal brain, meningiomas and gliomas and their migration relative to characterized ganglioside standards. The separation was performed on HPTLC-plates (Merck AG) and the developing solvent was chloroform/methanol/0.25% aqueous KCl (50:40:10, by vol.). The gangliosides were visualized with resorcinol. All standards were characterized by fast atom bombardment-mass spectrometry. The structures for all gangliosides standards, except for GalNAc-GD1a, GalNAcβ1-(NeuAcα2-3)4Galβ1-3GalNAcβ1-(NeuAcα2-3)4Galβ1-4GlcβCer, are shown in Fig. 3 and in Table I.

form of 3'-isoLM1 (Månsson et al., 1986). Recently, two monoclonal antibodies directed against 3'6'-isoLD1 were described (Wikstrand et al., 1993). Immunohistochemical localization of the epitope for these antibodies, NeuAc(or NeuGc)α2-3Galβ1-3(NeuAc or NeuGc)α2-6GlcNAc in frozen sections of primary central nervous system neoplasms (Wikstrand et al., 1992, 1993) revealed that 20/30 of glial tumors were positive. TLC-immunostaining of 3'6'-isoLD1 ganglioside antigen with one of these antibodies (DMAb 22) gave a positive reaction in 20/38 malignant glioma biopsies (Fredman et al., to be published). This ganglioside antigen was not detected in adult normal brain tissue but through the first trimester (Svennerholm et al., 1989).

Expression of primary brain tumor-associated gangliosides in normal brain and other tumors

As discussed above, no ganglioside structure described so far turned out to be tumor specific. In fact, these molecules are considered tumor-associated when a significant increase is found in tumor cells as compared to other tissues, i.e. in the case of brain tumors versus their expression in normal human brain tissue.

GM3 was found to be a common ganglioside found in medulloblastomas and gliomas as well as in meningiomas. This ganglioside is also a major component of melanomas (Ravindrath and Irie, 1988) and commonly occurs in a large variety of cells and tissues, including normal brain tissue (Vanier et al., 1973) and human red blood cells (Ando and Yamakawa, 1973).

The most apparent increase of a ganglio series ganglioside in glioma biopsies and autopsy brains (Fredman et al., 1986, 1988, 1993) was that of GD3. Besides its association with gliomas, meningioma and medulloblastoma, discussed above, GD3 was also described as a tumor-associated ganglioside in melanoma (Pukel et al., 1982; Ravindrath and Irie, 1988) and in leukemia (Siddiqui et al., 1984). In a recent study (Fredman et al., 1993) a high proportion of GD3 was detected in metastases of breast, lung

and low differentiated cancers, as well as adenocarcinoma obtained by excision biopsy. However, in none of these cases did the tissue concentration of GD3 reach that observed in glioma tissues.

Ganglioside GM2, found in increased proportion in medulloblastomas and gliomas, is a minor ganglioside in normal adult brain tissue (Svennerholm, 1963; Vanier et al., 1973) but is found in higher concentration in newborns (Svennerholm et al., 1989). It is a major ganglioside component of human melanomas (Irie et al., 1982; Tai et al., 1983; Tsuchida et al., 1987a), lung cancers of various histological types (Miyake et al., 1988), and germ cell tumors (Miyake et al., 1990). Ganglioside GM2, being present in endodermal tumors, adenocarcinoma and squamous cell carcinoma of the lung (Miyake et al., 1988), is thus not restricted to neuroectodermal tumors, as previously suggested (Natoli et al., 1986).

Ganglioside GD2 has been described as a tumor-associated ganglioside in melanoma (Cahan et al., 1982), neuroblastoma (Schultz et al., 1984; Wu et al., 1986), and small cell lung carcinoma (Cheresh et al., 1986a). It has been suggested, therefore, that GD2 is a marker for neuroectodermal tumors. Hypotheses were also advanced that GD2 is related to differentiation as the proportion of this ganglioside is high in neuroblastoma but much lower or not detectable in the more differentiated forms of this tumor, i.e. ganglioneuroblastoma and ganglioneurinoma (Wu et al., 1986).

Ganglioside 3'-LM1, of the lacto series type 2 chain, was found in all three types of brain tumors discussed. However, this ganglioside is a normal constituent of the brain (Li et al., 1973; Svennerholm et al., 1989), and constitutes the major ganglioside in peripheral nerve and normal red blood cells (Li et al., 1973). Thus blood contamination, which is common in human tumor biopsies, might add to an apparent increase in 3'-LM1 concentration in tumors.

The monosialylated lacto series gangliosides of the type 1 chain, 3'-isoLM1, are not found in normal human adult brain (Svennerholm et al., 1989) or in red blood cells. However, 3'-isoLM1 is present in gliomas and in medulloblastomas with astrocytic differentiation, and was first isolated and characterized from small cell lung carcinoma (Nilsson et al., 1985). It has also been established in teratocarcinoma cells (Fukuda et al., 1986, Wikstrand et al., 1991). Ganglioside 3'-isoLM1 was also detected in human fetal brains (Svennerholm et al., 1989), in brains of infants (Molin et al., 1987) and in brains from children who succumbed from infantile neuronal ceroid lipofuscinosis (Svennerholm et al., 1987), a disease associated with intense reactive astrocytosis and astrogliosis. These findings have lead us to suggest that 3'-isoLM1 is a marker for proliferating astrocytes, both normal and neoplastic.

The disialylated form of the lacto series type 1 chain gangliosides, 3'6'-isoLD1, has been reported as a tumor-associated ganglioside in colonic adenocarcinoma (Fukushi et al., 1986) and found in embryonal carcinoma cells (Fukushi et al., 1986). This ganglioside was detected only in fetal human brain during the first trimester (Svennerholm et al., 1989).

Gangliosides in cell lines established from human gliomas and medulloblastomas

Only limited quantities of tumor material are provided by biopsy and autopsy material, and their usefulness for biochemical studies is hampered by the large contamination with normal brain tissue. One way to circumvent these difficulties is to use tumor cell lines established from these tumors. However, since tumor specimens contain several clones of tumor cells with different phenotypic characteristics (Bigner, 1981; Bigner et al., 1981a; Collins, 1983), it is likely that during the establishment of a cell line one or a few of these clones that are most adaptable to in vitro culture may be selected (Bigner et al., 1981a,b; Wikstrand et al., 1985). Malignant cells can also alter their original characteristics because of environmental factors introduced by the culture system (Hoshino et al., 1978; Bigner et al., 1981a). Provided the cell lines retain those phenotypic characteristics that are important for the studies to be performed, they constitute an invaluble source for isolation of tumor-associated antigens and for studies concerning their biological functions. Many established cell lines can also be grown as solid tumors, i.e. xenografts in nude mice or rats. Such tumors provide important models for tumor-imaging studies with labelled antibodies.

The phenotypic heterogeneity of cell lines was also found to be reflected in their ganglioside expression. Thus in a preliminary study (Fredman, 1988) where the ganglioside patterns of eleven human glioma cell lines were investigated a large variation was observed. The simplest pattern was found in the D-54 MG cell line in which ganglioside GM2 dominated and the most complex one was detected in the U-118 MG cell line (Fredman et al., 1990a). This variation is not restricted to glioma cell lines since our study of two medulloblastoma cell lines showed large discrepancies, i.e. ganglioside patterns (Gottfries et al., 1991). Tsuchida et al. (1987b) reported that TLC analyses indicated variations in ganglioside patterns between human melanoma cell lines.

A common finding among the glioma cell lines grown in monolayers was a high expression of the ganglio series ganglioside GM2 (Fredman, 1988; Vrionis et al., 1989; Fredman et al., 1990a), as its proportion is generally higher than in glioma tissues. A large proportion of ganglioside GM2 was also demonstrated in cell lines established from human medulloblastomas (Vrionis et al., 1989; Gottfries et al., 1991), melanomas (Tsuchida et al., 1987a,b) and from urothelium (Ugorski et al., 1989). In the early studies on Simian Virus 40 (SV-40)-transformed rodent cell lines, a lack of more complex gangliosides was described, including GM2 of the ganglio series (Brady and Fishman, 1974; Hakomori, 1973, 1975). This finding led to the general hypothesis that there is a block in the synthesis of more complex gangliosides; however, recent studies (Hoffman et al., 1991) have shown, that in agreement with the earlier studies, SV-40 transformation of fibroblastic as well as neural cells from rodents results in increased expression of GM3. In contrast SV-40 transformation of human cells of both neural and fibroblastic origin led to an increase of GM2. These results indicate that a high expression of GM2 may be common to human tumor cells.

In the studies described in the previous paragraph the tumor cells were grown in monolayer culture. We also studied ganglioside expression in glioma and medulloblastoma cell lines grown as solid tumors in nude mice or rats. The first glioma xenograft to be investigated was established from the human glioma cell line D-54 MG which was serially transplanted into nude mice (Månsson et al., 1986). The major ganglioside was 3'-isoLM1, which comprised approximately 50% of the total ganglioside sialic acid, and the second major ganglioside was 3'6'-isoLD1, comprising approximately 20%. The lacto series gangliosides were thus most prominent in the ganglioside fraction. However, D-54 MG cells grown in monolayer culture did not express any detectable amounts of these gangliosides (Fredman, 1988). Instead ganglioside GM2 of the ganglio series comprised around 90% of the ganglioside sialic acid. Similar to the D-54 MG cells, another human glioma cell line, U-118 MG, grown in xenograft form showed a loss of the ganglio series ganglioside GM2 together with a proportial increse in total ganglioside sialic acid from 40 to 66%. This was also reflected by a quantitative increase of the lacto series ganglioside 3'-isoLM1 from 74 to 132 nmol ganglioside sialic acid/g wet weight (Fredman et al., 1990a).

A common phenomenon of a recent study on glioma cell lines was a relatively infrequent expression of the lacto series gangliosides 3'-isoLM1 and 3'6'-isoLD1 in glioma cells grown in vitro as compared to cells grown in vivo (Wikstrand et al., 1991; Wikstrand et al., 1993). Among the cell lines grown in vitro, 2/16 expressed 3'-isoLM1 as compared to 6/8 grown as xenografts. A proportional increase of the lacto series gangliosides was also observed in human medulloblastoma cells when these were transferred from monolayer culture to nude rats (Gottfries et al., 1991). TLC analyses of melanoma cells indicated a more abundant expression of ganglio series gangliosides GM2 and GD2 in monolayer cultured cells than in solid tumor cells (Tsuchida et al., 1987b). Lacto series gangliosides were not investigated, since they were thus far never described for human melanoma cell lines.

A switch from a high expression of the ganglio series gangliosides in vitro to a high expression of those of the lacto series in vivo thus appears to be a feature common to glioma and medulloblastoma

cells. Tsuchida et al. (1987b) reported that the ganglioside composition of melanoma cells grown as solid tumors in nude mice reflects the ganglioside pattern of the original tumors better than cultured monolayer cells. The expression of 3'-isoLM1, the suggested marker for both neoplastic and normal proliferating astrocytes, that is also found in malignant glioma (Fredman et al., 1988; Fredman et al., 1993), support the contention that there is a resemblance in solid tumors between xenografts and original tumor. The retainment of other phenotypic characteristics of the original tumors in tumor cells grown as xenografts has also been reported (Schold et al., 1982).

These results imply that environmental factors influence the expression of gangliosides. In this regard previous studies indicated that altered conditions of the cell culture system may cause changes in the ganglioside composition of cells (Liepkalns et al., 1981, 1983; Markwell et al., 1984). Environmental factors that have to be considered are soluble substances and/or the influence of membrane components associated with neighbouring cells. A recent study (Bjerkvig et al., 1991) demonstrated that D-54 MG spheroids grown in a three-dimensional in vitro system, still retained GM2 as the major ganglioside. Three-dimensional growth as such seems thus unlikely to be responsible for the switch from ganglio to lacto series gangliosides.

Only the N-acetyl form of sialic acid was detected in normal human tissues, in human glioma specimens or in D-54 MG cells grown in monolayer (Fredman, 1988). However, 90% of sialic acid attributable to gangliosides of D-54 MG cells grown as xenografts in nude mice was expressed as N-glycolylneuraminic acid. The conversion from the N-acetyl to the N-glycolyl form of sialic acid does not have to be solely an intracellular process but can occur at the cell membrane. In fact, the appearance of N-glycolylneuraminic acid in D-54 MG gangliosides support the involvement of factors introduced by the host system. Mouse gangliosides predominantly contain N-glycolylneuraminic acid. The hypothesis that environmental factors are involved in the phenotypic expression was advanced by Bigner et al. (1990) who found high levels of expression and structural abnormalities of the EGF receptor proteins in two human glioma cell lines grown as xenografts but not when those cells were grown in monolayer culture.

Another possible explanation for the switch in ganglioside expression from the ganglio series in monolayer culture to the lacto series in solid tumors is a clonal selection. Established cell lines often consist of several cell populations and in vivo growth conditions may favor one or more of these, the establishment of certain cell clones. Treatment of spheroids derived from the D-54MG human glioma cell line with anti-GM2 antibodies resulted in the isolation of a population of cells expressing 50% of the GM2 elaborated by the parent cells (Bjerkvig et al., 1991). The influence of genetic factors on ganglioside expression was indicated by early studies on transformed cells (Brady and Fishman, 1974; Hakomori, 1973, 1975) and also more recent transfection studies. In c-*myc* transfected fibroblastic 3Y1 cells, where the proposed localization of the oncogene is the nucleus, an increase of GD3 was observed, while v-*fes*, v-*ras*, v-*crs* and v-*fps* transfection, oncogene products which localize to the cell membrane, led to

an increase of 3'-LM1 (Sanai and Nagai, 1989). Studies showing an association between the GD3 level and loss of chromosome 22 in meningioma (Fredman et al., 1990b; Berra et al., 1991) gives further support for genetic influences.

Regardless as to whether genetic and/or environmental factors are responsible for the altered ganglioside expression that may be due to changes in the activity of the glycosyltransferases involved in their biosynthesis (Fig. 3), no glycosyltranferase has thus far proven to be an oncogene product. However, genetic alterations need not directly influence the production of glycosyltransferase. Alternatively genetic changes could result in the production or supression of proteins that in one way or another inhibit or potentiate glycosylation. Measurement of the activity of specific glycosyltransferases should then reflect levels of gangliosides synthesized. Holmes and Hakomori (1983) found that the accumulation α-L-fucosylated gangliosides in precancerous liver of rats that were fed carcinogens was accompanied by an increased activity of α-L-fucosyltransferase. Holmes et al. (1987) also demonstrated activity of the normally unexpressed β-1-3*N*-acetylglucos-

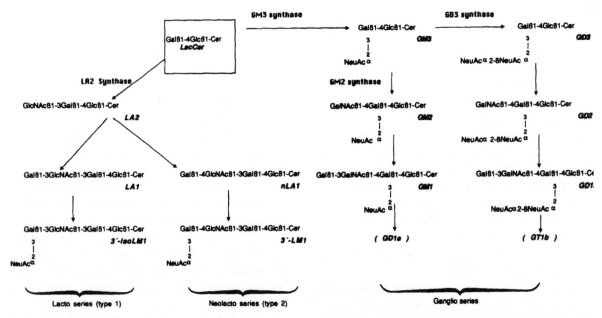

Fig. 3. Biosynthetic pathways for lacto, neolacto and ganglio series gangliosides. GM3-synthase (EC 2.4.99.9), GM2-synthase (EC 2.4.1.92); Glc, glucose; Gal, galactose; GlcNAc, glucosamine; GalNAc, galactosamine, NeuAc, *N*-acetylneuraminic acid; Cer, ceramide.

aminyltransferase (LA2-synthase) in colonic adeno-carcinoma, as the possible explanation for the accumulation of lacto series glycolipids of both type 1 and type 2. We were not able to establish any correlation between ganglioside expression and glycosyltransferase activity in our studies on medulloblastoma cell lines (Gottfries et al., 1991).

Lactosylceramide is the precursor for both the lacto and the ganglio series (Figs 1 and 3). The relatively high proportion of GM2 and low amounts of lacto series gangliosides in medulloblastoma cells grown in vitro was expected to be reflected by relatively higher values of the GM3 and/or GM2/LA2 synthase activity (Fig. 3), especially when compared to that in these cells grown as xenografts, where the proportion of lacto series was higher than that of GM2. In contradiction, this same study indicated a good correlation between GD3-synthase activity and expression of GM2 in cell lines propagated either in vitro or in vivo. In this regard in vitro assays may suffer from artifacts like detergents and may lack factors that could potentiate or inhibit the activity of these enzymes in vivo. Consequently, in vivo assays of glycosyltransferase activities most likely offer a more reliable system; however, in this case the administration of labelled sugars to tumor-bearing mice is difficult to optimize as far as dose and time schedules are concerned.

Potential role of primary brain tumor gangliosides

The functional role of glycosphingolipids in tumor progression is discussed in more detail by Hakomori (this volume). This chapter only describes a potential role for tumor-associated gangliosides expressed by primary brain tumors. Support for the involvement of the gangliosides GD3 and GD2 in cell to substratum interactions comes from studies performed with melanoma and neuroblastoma cell lines. Thus, Cheresh et al. (1984) used monoclonal antibodies to GD2 and GD3 to define the topographical distribution of these respective gangliosides on the surface of human melanoma cells and defined their localization in adhesion plaques at the interface between cells and their substratum. Ganglioside GD2 was also demon-strated to redistribute to the microprocesses of melanoma cells during the cell attachment process (Cheresh and Klier, 1986). Cheresh et al. (1986b) also reported that anti-GD2 and anti-GD3 monoclonal antibodies inhibited attachment of melanoma cells to extracellular matrix proteins. A similar inhibitory effect was observed with anti-GD2 antibodies on neuroblastoma cells containing high levels of GD2 but no detectable GD3. Two other antibodies reacting with neuroblastoma cell surface had no effect. These results indicate that the potential role of GD3 and GD2 gangliosides in cell/matrix interactions is not restricted to melanomas and may also be relevant for primary brain tumor cells.

Studies of melanoma cells (Dippold et al., 1984; Cheresh et al., 1985; Hellström et al., 1985) have also shown that antibodies against GD3 can produce cell lysis. No growth inhibitory effect was noticed in melanoma cells with low or non-detectable amounts of GD3 (Dippold et al., 1984; Welt et al., 1987). It has also been reported that anti-GD3 (Cheresh et al., 1985; Hellström et al., 1985) and anti-GD2 (Katano et al., 1984) monoclonal antibodies can inhibit the growth of human melanoma xenografts in nude mice. Antibodies to gangliosides have also been used successfully to reduce tumor growth in clinical trials on melanoma and neuroblastoma, results that are discussed in this volume by Reisfeld and Morton.

To the best of our knowledge, there is but one report (Bjerkvig et al., 1991) on the role of tumor-associated gangliosides in human glioma cells or in cell lines derived from primary brain tumor. This particular study investigated the effect of anti-GM2 antibodies on the human glioma cell line D-54 MG (Bjerkvig et al., 1991), which was reported to contain a high proportion of ganglioside GM2 (Fredman, 1988; Vrionis et al., 1989). In this case the cells were grown as spheroids in a three-dimensional in vitro culture system (Sutherland, 1988; Lund-Johansen et al., 1989). After 48 h exposure to anti-GM2 antibody the spheroids developed necrosis, which started at the center. Anti-GM2 antibodies absorbed with GM2 as well as monoclonal antibodies to GD3, both of IgM iso-type, and IgM isolated from a mouse myeloma cell line had no effect. A small portion of the cells in spheroids that mainly localized to its periphery, survived the anti-GM2 treatment. These cells were found to constitute a subpopulation that had a low GM2 content as verified by flow cytometry and biochemical ganglioside determination. New spheroids obtained from this subpopulation of D-54 MG cells were not affected by the anti-GM2 antibody. These results indicate that anti-GM2 antibodies may cause necrosis of glioma cells in vitro provided a threshold number of ganglioside antigen is expressed in the cells. The cytotoxic effect of anti-GD3 antibodies on melanoma cells also required a certain density of antigen (Dippold et al., 1984; Welt et al., 1987). The amounts of tumor-associated ganglioside antigens expressed on tumor cells are thus of utmost importance for the optimal effect of monoclonal antibodies on tumor growth.

An in vitro study has not yet been performed to delineate the possible function(s) of the lacto series gangliosides. Results obtained from autopsy material of malignant glioma patients revealed a different distribution of 3′-isoLM1 than of GD3, and it was suggested that this could reflect different functions of these gangliosides (Fredman et al., 1993). Because 3′-isoLM1 was located mainly adjacent to macroscopic tumor tissue, in particular in the region of the opposite hemisphere corresponding to the tumor, one could hypothesize that expression of this ganglioside may be related to tumor cell migration. This hypothesis is supported by relatively high levels of 3′-isoLM1 in the corpus callosum, shown by histopathology to contain tumor cells. However, the role of lacto series gangliosides remains yet to be elucidated.

Summary and conclusion

Human primary brain tumors differ in their ganglioside composition when compared to adjacent tissues. One ganglioside found in all malignant glioma specimens, but not detected in normal adult brain, is 3′-isoLM1, a ganglioside of the lacto series. This ganglioside was also identified in medulloblastomas with astrocytic differentitation and in brain tissues containing benign proliferating astrocytes.

The appeearence of 3′-isoLM1 was seen over large regions of brain from glioma but was found mainly in areas either adjacent to the macroscopic tumor or areas correspondning to the tumor in the opposite hemisphere. A high concentration of 3′-isoLM1 was also seen in the corpus callosum, the anatomical structure along which glioma cells may migrate to the opposite brain hemisphere.

Ganglioside expressed by cell lines established from primary malignant brain tumors varied widely among cell lines and within a given cell line propagated under different conditions. In in vitro-cultured glioma and medulloblastoma cell lines, gangliosides of the ganglio series dominated and the expression of the lacto series gangliosides, including 3′-isoLM1 was low if at all detectable. However, in vivo growth of solid subcutaneous tumors in nude mice or rats led to a significantly increased expression of the often dominant gangliosides of the lacto series and revealed a decreased expression of ganglio series gangliosides.

In conclusion, these findings indicate that environmental factors could strongly influence the expression of gangliosides that may lead to a switch from the ganglio to the lacto series. These results also suggest that ganglioside 3′-isoLM1 is associated with proliferating astrocytes, of both neoplastic and non-neoplastic origin and that this ganglioside may be involved in cell-cell recognition and attachment during development and tumor cell migration.

Acknowledgements

I would like to express my gratitude to Professor Lars Svennerholm, Department of Psychiatry and Neurochemistry, University of Göteborg, Sweden, for introducing me to the field of gangliosides, and for his scientific guidance and never-failing stimulation through all the years we have been collaborating. I would also like to thank all collaborators of the brain tumor project. This includes Professor Darell Bigner and his group, of which I also like to mention Dr Carol Wikstrand, at Department of Pathology and Duke Comprehensive Cancer Center, Preuss Laboratory for Brain Tumor Research, Duke University Medical Center, Durham, NC, USA;

Professor Didrik Laerum and Dr Rolf Bjerkvig, Department of Pathology, the Gade Institute, University of Bergen, Norway; Professor Peter Collins, Department of Pathology, University of Göteborg, Sweden; Dr Hans von Holst, Department of Neurosurgery, Karolinska Hospital, Sweden; and Dr Jan-Eric Månsson, Department of Psychiatry and Neurochemistry, University of Göteborg, Sweden. I also wish to express my gratitude to Barbro Lundmark for invaluble expert secretarial work. These studies were supported by grants from the Swedish Medical Research Council (Project No. 03X-627, B86-03-07462-01); NIH grants TR37 CA11898, NS20023 and CA32672; the Swedish Cancer Society (Project No. 2260-B88-01X, B89-02X, 2986-B91-01XAC); University of Göteborg, the Ingabritt and Arne Lundberg Foundation; the Cancer Society in Stockholm (Project No. 85-47, 86-63).

References

Alhadeff, J.A. (1989) Malignant cell glycoproteins and glycolipids. *CRC Crit. Rev. Oncol. Hematol.*, 9: 37–107.

Ando, S. and Yamakawa, T. (1973) Separation of polar glycolipids from human red blood cells with special reference to blood group A activity. *J. Biochem.*, 73: 287–396.

Aruna, R.M., Balasubramanian, K.A., Mathai, K.V. and Basu, D. (1973) Isolation and characterization of glycolipids and glycosaminoglycans from meningiomas. *Indian J. Med. Res.*, 61: 1688–1693.

Berra, B., Riboni, L., De Gasperi, R., Gaini, S.M. and Ragnotti, G. (1983) Modifications of ganglioside patterns in human meningiomas. *J. Neurochem.*, 40: 777–782.

Berra, B., Gaini, S.M. and Riboni, L. (1985) Correlation between ganglioside distribution and histological grading of human astrocytomas. *Int. J. Cancer*, 36: 363–366.

Berra, B., Rapelli, S., Brivio, M., Gornati, R. and Omodeo Sale, M.F. (1990) Modification of glycoconjugate content and distribution in human meningiomas. *Clin. Chem. Enzym. Comm.*, 2: 383–390.

Berra, B., Papi, L., Bigozzi, U., Serino, D., Morichi, R., Mennonna, P., Rapelli, S., Cogliati, T. and Montali, E. (1991) Correlation between cytogenetic data and ganglioside pattern in human meningiomas. *Int. J. Cancer Res.*, 47: 329–333.

Bigner, D.D. (1981) Biology of gliomas: potential clinical implications of cellular heterogeneity. *Neurosurgery*, 9: 320–326.

Bigner, D.D., Bigner, S.H., Pontén, J., Westermark, B., Mahaley, M.S., Rouslahi, E., Herschman, H., Eng, L.F. and Wikstrand, C.J. (1981a) Heterogeneity of genotypic and phenotypic charac-

teristics of fifteen permanent cell lines derived from human gliomas. *J. Neuropathol. Exp. Neurol.*, 40: 201–229.

Bigner, S.H., Bullard, D.E., Pegram, C.E., Wikstrand, C.J. and Bigner, D.D. (1981b) Relationship of *in vitro* morphologic and growth characteristics of established human glioma-derived cell lines to their tumorogenecity in athymic mice. *J. Neuropathol. Exp. Neurol.*, 40: 390–409.

Bigner, S.H., Humphrey, P.A., Wong, A.J., Vogelstein, B., Mark, J., Friedman, H.S. and Bigner, D.D. (1990) Characterization of the epidermal growth factor receptor in human glioma cell lines and xenografts. *Cancer Res.*, 50: 8017–8022.

Bjerkvig, R., Engebraaten, O., Laerum, O.D., Fredman, P., Svennerholm, L. and Bigner, D.D. (1991) Anti-GM2 monoclonal antibodies induce necrosis in GM2-rich cultures of a human glioma cell line. *Cancer Res.*, 51: 4643–4648.

Brady, R.O. and Fishman, P. (1974) Biosynthesis of glycolipids in virus-transformed cells. *Biochim. Biophys. Acta*, 355: 121–148.

Cahan, L.D., Irie, R.F., Singh, R., Cassidenti, A. and Paulson, J.C. (1982) Identification of a human neuroectodermal tumor antigen (OFA-I-2) as ganglioside GD2. *Proc. Natl. Acad. Sci. USA*, 79: 7629–7633.

Cheresh, D.A. and Klier, F.G.(1986) Disialoganglioside GD2 distributes preferentially into substrate-associated microprocesses on human melanoma cells during their attachment to fibronectin. *J. Cell Biol.*, 102: 1887–1897.

Cheresh, D.A., Harper, J.R., Schultz, G. and Resifeld, R.A. (1984) Localization of the gangliosides GD2 and GD3 in adhesion plaques and on the surface of human glioma cells *Proc. Natl. Acad Sci. USA*, 81: 5767–5771.

Cheresh, D.A., Honsik, C.J., Staffileno, L.K., Jung, G. and Reisfeld, R.A. (1985) Disialoganglioside GD3 on human melanoma cells serves as a relevant target antigen for monoclonal antibody-dependent cytotoxicity and immuno-therapies. *Proc. Natl. Acad. Sci. USA*, 82: 5155–5159.

Cheresh, D.A., Rosenberg, J., Mujoo, K., Hirschowitz, L. and Resifeld, R. (1986a) Biosynthesis and expression of disialoganglioside GD2, a relevant target antigen on small cell lung carcinoma for monoclonal antibody-mediated cytolysis. *Cancer Res.*, 46: 5112–5118.

Cheresh, D.A., Pierschbacher, M.D., Herzig, M.A. and Mujoo, K.(1986b) Disialogangliosides GD2 and GD3 are involved in the attachement of human melanoma and neuroblastoma cells to extracellular matrix proteins. *J. Cell Biol.*, 102: 688–696.

Collins, V.P. (1983) Cultured human glial and glioma cells. *Exp. Rev. Exp. Pathol.*, 24: 135–202.

Davidsson, P., Fredman, P., Collins, V.P., von Holst, H., Månsson, J.-E. and Svennerholm, L. (1989) Ganglioside composition in human meningiomas. *J. Neurochem.*, 53: 705–709.

Davidsson, P., Fredman, P., von Holst, H., Wikstrand, C.J., He, X., Bigner, D.D. and Svennerholm, L.(1990) Circulating glycoconjugates in CSF of meningioma patients. *Acta Neurol. Scand.*, 82: 203–208.

Dippold, W.G., Knuth, A and zum Büschenfelde, K.H.M. (1984) Inhibition of human melanoma cell growth *in vitro* by monoclonal anti-GD3 ganglioside antibody. *Cancer Res.*, 44: 806-810.

Eto, Y. and Shinoda, S. (1982) Gangliosides and neutral glycosphingolipids in human brain tumors: specificity and significance. *Adv. Exp. Med. Biol.*, 152: 279-290.

Fenderson, A.B. Eddy, E.M and Hakomori, S.-I. (1990) Glycoconjugate expression during embryogenesis and its biological significance. *BioEssays*, 12: 173-179.

Fredman, P. (1988) Gangliosides in human malignant gliomas. In: R.W. Ledeen, E.L. Hogan, G. Tettamanti, A.J. Yates and R.K. Yu (Eds), *New Trends in Ganglioside Research. Fidia Research Series, Vol. 14*, Liviana Press, Springer Verlag, pp. 151-161.

Fredman, P. (1993). Glycosphingolipid tumor antigen. *Adv. Lipid Res.* (in press).

Fredman, P., von Holst, H., Collins, V.P., Ammar, A., Dellheden,B., Wahren, B., Granholm, L. and Svennerholm, L. (1986) Potential ganglioside antigens associated with human gliomas. *Neurol. Res.*, 8: 123-126.

Fredman, P., von Holst, H., Collins, V.P., Granholm, L. and Svennerholm, L. (1988) Sialyllactotetraosylceramide, a ganglioside marker for human malignant gliomas. *J. Neurochem.*, 50: 912-919.

Fredman, P., Månsson, J.-E., Bigner, S.H., Wikstrand, C.J., Bigner, D.D. and Svennerholm, L. (1990a) Gangliosides in the human glioma cell line U-118MG grown in culture or as xenografts in nude rats. *Biochim. Biophys. Acta*, 1045: 239-244.

Fredman, P., Dumanski, J., Davidsson, P., Svennerholm, L. and Collins, V.P. (1990b) Expression of ganglioside GD3 in human meningiomas is associated with monosomy of chromosome 22. *J. Neurochem.*, 55: 1838-1840.

Fredman, P., von Holst, H., Collins, V.P., Dellheden, B. and Svennerholm, L. (1993) Expression of gangliosides GD3 and 3'-isoLM1 in autopsy brains from patients with malignant tumors *J. Neurochem.*, (in press).

Fukuda, M., Bothner, B., Lloyd, K.O., Rettig, W.J., Tiller, R.P. and Dell, A. (1986) Structures of glycosphingolipids isolated from human embryonal carcinoma cells. *J. Biol. Chem.* 261, 5145-5153.

Fukushi, Y., Nudelman, E., Levery, S.B., Higuchi, T. and Hakomori, S.-I. (1986). A novel disialoganglioside (IV^{3}NeuAcIII^{6}NeuAcLc$_4$) of human adenocarcinoma and the monoclonal antibody (FH9) defining this disialosyl structure. *Biochemistry*, 25, 2859-2866.

Furukawa, K., Yamaguchi, H., Oettgen, H.F., Old, L.J. and Lloyd, K.O. (1989) Two human monoclonal antibodies reacting with the major gangliosides of human melanomas and comparison with corresponding mouse monoclonal antibodies. *Cancer Res.*, 49: 191-196.

Gottfries, J., Fredman, P., Månsson, J.-E., Collins, V.P., von Holst, H., Armstrong, D.D., Percy, A.K., Wikstrand, C.J., Bigner, D.D. and Svennerholm, L. (1990) Determination of gangliosides in six human primary medulloblastomas. *J. Neurochem.*, 55: 1322-1326.

Gottfries, J., Percy, A.K., Månsson, J.-E., Fredman, P., Wikstrand,

C.J., Friedman, H.S., Bigner, D.D. and Svennerholm, L. (1991) Glycolipids and glycosyltransferases in permanent cell lines established from human medulloblastomas. *Biochim. Biophys. Acta*, 1081: 253-261.

Hakomori, S.-I. (1973) Glycolipids of tumor cell membrane. *Adv. Cancer Res.*, 18: 265-315.

Hakomori, S.-I. (1975) Structures and organization of cell surface glycolipids dependency on cell growth and malignant transformation. *Biochim. Biophys. Acta*, 417: 55-89.

Hakomori, S.-I. (1981) Glycosphingolipids in cellular interaction, differentiation, and oncogenesis. *Annu. Rev. Biochem.*, 50: 733-764.

Hakomori, S.-I. (1989) Abberant glycosylation in tumors and tumor-associated carbohydrate antigens. *Adv. Cancer Res.*, 52: 257-331.

Hakomori, S.-I. (1990) Bifunctional role of glycosphingolipids. Modulators for transmembrane signaling and mediators for cellular interactions. *J. Biol. Chem.*, 265: 18713-18716.

Hellström, I., Brankovan, V. and Hellström, K.E. (1985) Strong antitumor activities of IgG 3 antibodies to a human melanoma-associated ganglioside. *Proc. Natl. Acad. Sci. USA*, 82: 1499-1502.

Hoffman, L.M., Brooks, S.E., Stein, M.R. and Schneck, L. (1991) SV-40 transformation: effect on GM2 ganglioside in cultured cell lines. *Biochim. Biophys. Acta*, 1084: 94-100.

Holmes, E.H. and Hakomori, S.-I. (1983) Enzymatic basis for changes in fucoganglioside during chemical carcinogenesis. *J. Biol. Chem.*, 258: 3706-3713.

Holmes, E.H., Hakomori, S.-I. and Ostrander, G.K. (1987) Synthesis of type 1 and 2 lacto series glycolipid antigens in human colonic adenocarcinoma and derived cell lines is due to activation of a normally unexpressed 1-3-*N*-acetyl-glucosaminyltransferase. *J. Biol. Chem.*, 262: 15649-15658.

Hoshino, T. Namura, K., Wilson, K.B., Knebel, K.P. and Gray, J.W. (1978) The distribution of nuclear DNA from human brain tumor cells. *J. Neurosurg.*, 49: 13-21.

Irie, R.F., Sze, L.L. and Saxton, R.E. (1982) Human antibody to OFA-I, a tumor antigen, produced *in vitro* by Epstein-Barr virus-transformed human B-lymphoid cell lines. *Proc. Natl. Acad. Sci. U.S.A.*, 79: 5666-5670.

Jenneman, R., Rodden, A., Bauer, B.L., Mennel, H.-D. and Wiegandt, H. (1990) Glycosphingolipids of human gliomas. *Cancer Res.*, 50: 7444-7449.

Katano, M., Jien, M. and Irie, R.F. (1984) Human monoclonal antibody to ganglioside GD2-inhibited human melanoma xenograft. *Eur. J. Clin. Oncol.*, 20: 1053-1059.

Kostic, D. and Buchheit, F. (1970) Gangliosides in human brain tumors. *Life Sci.*, 9: 589-598.

Ladisch, S., Wu, Z.-L., Feig, S., Ulsh, L., Schwartz, E., Floutsis, G., Wiley, F., Lenansky, C. and Seeger, R. (1987) Shedding of GD2 ganglioside by human neuroblastoma. *Int. J. Cancer*, 39: 73-76.

Li, Y.-T., Månsson, J.-E., Vanier, M.-T. and Svennerholm, L. (1973) Structure of the major glucosamine-containing ganglioside of human tissues. *J. Biol. Chem.*, 248: 2634–2636.

Liepkalns, V.A., Icard, C., Yates, A.J, Thompson, D.K. and Hart, R.W. (1981) Effects of cell density on lipids of human glioma and fetal neural cells. *J. Neurochem.*, 36: 1959–1965.

Liepkalns, V.A., Icard-Liepkalns, C., Yates, A.J., Mattison, S. and Stephens, R.E. (1983) Effects of human brain cell culture conditions on ^{14}C-glucosamine radioactivity incorporation into gangliosides. *J. Lipid Res.*, 24: 533–540.

Lund-Johansen, M., Bjerkvig, R. and Endersen, K.-J.(1989) Multicellular tumors pheroids in serum-free culture. *Anticancer Res.*, 9: 413–420.

Markwell, M., Fredman, P. and Svennerholm, L. (1984) Receptor ganglioside content of three hosts for Sendai virus MDBK, HeLa and MDCK cells. *Biochim. Biophys. Acta,* 775: 7–16.

Månsson, J.-E., Fredman, P., Nilsson, O., Lindholm, L., Holmgren, J. and Svennerholm, L. (1985) Chemical structure of carcinoma ganglioside antigens defined by monoclonal antibody C-50 and some allied gangliosides of human pancreatic adenocarcinoma. *Biochim. Biophys. Acta*, 834: 110–117.

Månsson, J.-E., Fredman, P., Bigner, D.D., Molin, K., Rosengren, B., Friedman, H.S. and Svennerholm, L. (1986) Characterization of new gangliosides of the lactotetraose series in murine xenografts of a human glioma cell line. *FEBS Lett.*, 201: 109–113.

Miyake, M., Ito, M., Hitomi, S., Ikeda, S., Taki, T., Kurata, M., Hino, A., Miyake, N. and Kannagi, R. (1988) Generation of two monoclonal antibodies that can discriminate *N*-glycolyl and *N*-acetyl neuraminic acid residues of GM2 gangliosides. *Cancer Res.*, 48: 6154–6160.

Miyake, M., Hashimoto, K., Ito, M., Ogawa, O., Arai, E., Hitomi, S. and Kannagi, R. (1990) The abnormal occurence and the differentiation-dependent distribution of *N*-acetyl and *N*-glycolyl species of the ganglioside GM2 in human germ cell tumors. *Cancer*, 65: 499–505.

Molin, K., Månsson, J.-E., Fredman, P. and Svennerholm, L. (1987) Sialyllactotetraosylceramide, 3′-isoLM1, a ganglioside of the lactotetraose series isolated from human infant brain. *J. Neurochem.*, 49: 216-219.

Nakamura, O., Ishihara, E., Iwamori, M, Nagai, Y., Matsutani, M., Nomura, K. and Takakura, K. (1987) Lipid composition of human malignant brain tumors. *Brain Nerve*, 39: 221–226.

Nakamura, O., Iwamori, M., Matsutani, M. and Takakura, K. (1991) GD3 shedding by human gliomas. *Acta Neurochir.*, 109: 34–36.

Natoli, E.J., Livingstone, P.O.,Pukel, C., Lloyd, K.O., Wiegandt, H., Szalay, J., Oettgen H.F. and Lloyd, L.J. (1986) A murine monoclonal antibody detecting *N*-acetyl- and *N*-glycolyl-GM2: characterization of cell surface reactivity. *Cancer Res.*, 46: 4116–4120.

Nilsson, O., Månsson, J.-E., Lindholm, L., Holmgren, J. and Svennerholm, L. (1985) Sialosyllactotetraosylceramide, a novel ganglioside antigen detected in human carcinomas by a monoclonal antibody. *FEBS Lett.*, 182: 398–402.

Portoukalian, J., Zvingelstein, G., Abdul-Maluk, N. and Doré, J.F.

(1978) Alteration of gangliosides in plasma and red cells of humans bearing melanoma tumors. *Biochim. Biophys. Acta*, 185: 916–920.

Pukel, C.S., Lloyd, K.O., Travassos, L.R., Diuppold, W.G., Oettgen, H.F. and Old, L.J. (1982) GD3, a prominent ganglioside of human melanomas: Detection and characterization by mouse monoclonal antibody. *J. Exp. Med.*, 155: 1133–1147.

Ravindrath, M.H. and Irie, R.F. (1988) Gangliosides as antigens of human melanoma. In: L. Nathansson (Ed.), 'Malignant Melanoma: Biology, Diagnoses and Therapy', Kluwer Academic Publishers, Boston, pp. 17–43.

Sanai, Y. and Nagai, Y. (1989) Cellular expression of glycolipids after oncogene transfection; Oncogene type specific changes of gangliosides in rat 3Y1 cells. In: H.F. Oettgen (Ed.), *Gangliosides and Cancer*, VCH Verlagsgesellschaft, Weinheim, Germany, pp. 69–78.

Schold, S.C., Bullard, D.E., Bigner, S.H., Jones, T.R. and Bigner, D.D. (1982) Growth, morphology, and serial transplantation of anaplastic human gliomas in athymic mice. *J. Neurooncol.*, 1: 5–14.

Schulz, G., Cheresh, D.A., Varki, N.M., Yu, A., Staffileno, L.K. and Reisfeld, R.A. (1984) Detection of ganglioside GD2 in tumor tissues and sera of neuroblastoma patients. *Cancer Res.*, 44: 5914–5920.

Seifert, H. (1966) Über ein weiters hirntumorcharacteristisches gangliosid. *Klin. Wochenschr.*, 44: 469–470.

Seifert, H. and Uhlenbruck, G. (1965) Über Ganglioside von Hirntumoren. *NaturWissenschaften*, 52: 190.

Siddiqu, B., Buehler, J., DeGrehorio, M.W. and Macher, B.A. (1984) Differential expression of ganglioside GD3 by human leukocytes and leukemia cells. *Cancer Res.*, 44: 5262–5265.

Sunder-Plassmann, M. and Bernheimer, H. (1974) Ganglioside in Meningiomen and Hirnhäuten. *Acta Neuropathol.*, 27: 289–297.

Sutherland, R.M. (1988) Cell and environment interactions in tumor microregions: the multicell spheroid model. *Science*, 240: 177–184.

Svennerholm, L., (1963) Chromatographic separation of human brain gangliosides. *J. Neurochem.*, 10: 613–623.

Svennerholm, L. (1984) Biological significance of gangliosides. In: H. Dreyfus, R. Massarelli, L. Freysz and G. Rebel (Eds), *Cellular and Pathological Aspects of Glycoconjugate Metabolism, Inserm, Vol.126,* Inserm, Paris, pp. 21–47.

Svennerholm, L. (1988) Immunological and tumoral aspects of gangliosides. In: R.W. Ledeen, E.L. Hogan, G. Tettamanti, A.J. Yates and R.K. Yu (Eds), *New trends in Ganglioside Research: Neurochemical and Neuroregenerative Aspects, Fidia Research Series, Vol 14,* Liviana Press, pp. 135–150.

Svennerholm, L., Fredman, P., Jungbjer, B., Månsson, J.-E., Rynmark, B.-M., Boström, K., Hagberg, B., Norén, L. and Santavouri, P.(1987) Large alteration in ganglioside and neutral glycosphingolipid pattern in brains from cases with infantile neuronal ceroid lipofuscinosis/polyunsaturated fatty acid lipidosis. *J. Neurochem.*, 49: 1772–1783.

Svennerholm, L., Boström, K., Fredman, P., Månsson, J.-E.,

Rosengren, B. and Rynmark, B.-M. (1989) Human brain ganglioside: developmental changes from early fetal stage to advanced age. *Biochim. Biophys. Acta,* 1005: 109–117.

Tai, T., Paulsson, J., Cahan, L.D. and Irie, R.F. (1983) Ganglioside GM2 as a human antigen (OFA-I-1). *Proc. Natl. Acad Sci. U.S.A.,* 80: 5392–5396.

Traylor, D.T. and Hogan, E.L. (1980) Gangliosides of human cerebral astrocytomas. *J. Neurochem.,* 34: 126–131.

Tsuchida, T., Saxton, R.E., Morton, D.L. and Irie, R.F. (1987a) Gangliosides of human melanoma *J. Natl. Cancer Inst.,* 78: 45–54

Tsuchida, T., Ravindrath, M.H., Saxton, R.E. and Irie, R.F. (1987b) Gangliosides of human melanoma: Altered expression in vivo and in vitro. *Cancer Res.,* 47: 1278–1281.

Ugorski, M., Dus, D., Skouv, J., Påhlsson, P. and Radzikowski, C. (1989) Gangliosides in human urothelial cell lines of different transformed phenotypes: The effect of v-*ras* oncogene transfection. *Anticancer Res.,* 9: 1583–1586.

Vanier, M.T., Holm, M., Månsson, J.-E. and Svennerholm, L. (1973) The distribution of lipids in the human nervous system. V. Gangliosides and allied neutral glycolipids of infant brain. *J. Neurochem.,* 21: 1375–1384.

Vrionis, F., Wikstrand, C.J., Fredman, P., Månsson, J.-E., Svennerholm, L. and Bigner, D.D. (1989) Five new eiptope-defined monoclonal antibodies reactive with GM2 and human glioma and medulloblastoma cell lines. *Cancer Res.,* 49: 6645–6651.

Welt, S.W., Carswell, E.A., Vogel, C.-V., Oettgen, H.F. and Old, L.J. (1987) Immune and non immune effector functions of IgG3 mouse monoclonal antibody R24 detecting the disialoganglioside GD3 on the surface of melanoma cells. *Clin. Immunol. Immunopathol.,* 45: 214–229.

Wikstrand, C.J., Grahmann, F.C., McComb, R.D. and Bigner, D.D. (1985) Antigenic heterogeneity of human anaplastic gliomas and glioma-derived cell lines defined by monoclonal antibodies. *J. Neuropathol. Exp. Neurol.,* 44: 229–241.

Wikstrand, C.J., He, X., Fuller, G.N.,Bigner, S.H., Fredman, P., Svennerholm, L. and Bigner, D.D. (1991) Occurrence of lacto series gangliosides 3′-isoLM1 and 3′6′-isoLD1 in human gliomas *in vitro* and *in vivo* . *J. Neuropathol. Exp. Neurol.,* 50: 756–769.

Wikstrand, C.J., Fredman, P., Svennerholm, L. and Bigner, D.D. (1992) Detection of glioma-associated gangliosides GM2, GD2, GD3, 3′-isoLM1 and 3′6′-isoLD1 in central nervous system tumors *in vitro* and *in vivo* using epitope-defined monoclonal antibodies. In: L. Svennerholm, A.K. Ashbury, R.A. Reisfield, K. Sandhoff, K. Suzuki, G. Tettamanti and G. Toffano (Eds), *Progress in Brain Research, Vol. 101, Biological Function of Gangliosides,* Elsevier Science Publishers, Amsterdam, This volume.

Wikstrand, C.J., Longee, D.C., Fuller, G.N., McLendon, R.E., Friedman, H.S., Fredman, P., Svennerholm, L. and Bigner, D.D. (1993) Lactotetraose series ganglioside 3′6′-isoLD1 in tumors of central nervous and other systems *in vitro* and *in vivo. Cancer Res.,* 33: 1–7.

Wu, Z.-I., Schwartz, E., Seeger, R. and Ladisch, S. (1986) Expression of GD2 ganglioside by untreated primary human neuroblastomas. *Cancer Res.,* 46: 440–443.

Yamamoto, S., Yamamoto, T., Saxton, R.E., Hoon, D.S.B., Irie, R.F. (1990) Anti-idiotype monoclonal antibody carrying the internal image of ganglioside GM3.

Yates. A.J., Thompson, D.K., Boesel, C.P., Albrightson, C. and Hart, R.W. (1979) Lipid composition of human neural tumors. *J. Lipid Res.,* 20: 428–436.

L. Svennerholm, A.K. Asbury, R.A. Reisfeld, K. Sandhoff, K. Suzuki, G. Tettamanti and G. Toffano (Eds.)
Progress in Brain Research, Vol. 101

CHAPTER 18

Role of gangliosides in tumor progression

Sen-itiroh Hakomori

The Biomembrane Institute, 201 Elliott Ave W, Seattle, WA 98119 and Departments of Pathobiology and Microbiology, University of Washington, Seattle, WA 98195, U.S.A.

Introduction

Glycosphingolipids (GSLs), especially gangliosides and their breakdown products, play two important roles in functioning of normal cells: (i) they define specificity of cell-cell or cell-substratum interaction; and (ii) they control transmembrane signaling via modulation of receptor kinases, protein kinase C, or other kinases (Hakomori, 1990). Certain GSLs highly expressed in tumor cells or tissues have been defined by specific monoclonal antibodies (MAbs) and thereby identified as tumor-associated carbohydrate antigens (TACAs) resulting from aberrant GSL synthesis in tumors (Hakomori, 1985, 1989). Aberrant expression of GSLs in primary tumors has been correlated with metastatic potential and invasiveness of tumors (Hakomori, 1991b; Miyake et al., 1992). Our recent studies, therefore, have focused on the functional role of gangliosides in tumor progression.

Tumor-associated carbohydrate antigens as adhesion molecules

Expression of such TACAs as H/Le^y/Le^b in primary lung carcinoma (particularly squamous cell carcinoma) (Miyake et al., 1992), sialosyl-Tn in colorectal carcinoma (Itzkowitz et al., 1990), sialosyl-Le^x in colonic carcinoma (Irimura et al., unpublished), and GM3 in murine B16 melanoma (Kojima et al., 1992b) is closely correlated with invasiveness, metastatic potential, and degree of tumor progression. Typical examples illustrating the relationship

between expression of these antigens in primary tumor and patient survival are shown in Fig. 1. Metastatic properties of tumor cells depend closely on: (i) adhesiveness of tumor cells to specific areas of microvascular endothelial cells (ECs); and (ii) ability of tumor cells to activate platelets, leading to platelet-tumor cell adhesion and microembolism of tumor cell aggregates. Therefore, we investigated the possible role of TACAs as adhesion molecules. Some (if not all) of the TACAs shown in Fig. 1 have been identified as adhesion molecules, recognized by complementary carbohydrate (CHO) structures on target cells via CHO–CHO interaction (Hakomori, 1991a) or interaction between CHO and lectin (including selectin) (Phillips et al., 1990; Polley et al., 1991; Handa et al., 1991b; Berg et al., 1991). That is, tumor cells showing high surface expression of these TACAs have a higher probability of interacting with the complementary CHO or lectin on the target cell. An example is described in detail in the following section.

Adhesion of melanoma cells to endothelial cells through GM3–LacCer interaction: correlation with metastatic potential

GM3 ganglioside, which is highly expressed at the surface of mouse B16 melanoma cells, is recognized as a melanoma-associated antigen defined by MAb M2590 (Hirabayashi et al., 1985; Nores et al., 1987). While GM3 is ubiquitous in normal cells and tissues, it occurs at a much higher density on B16 cells and

242

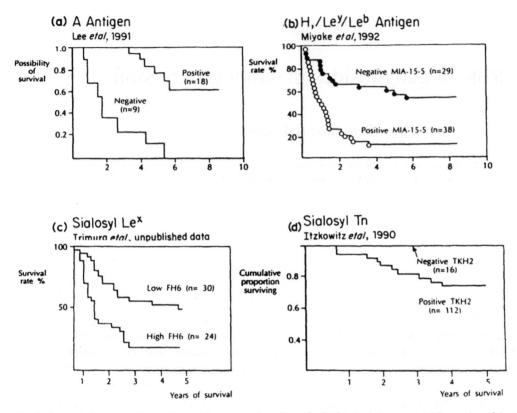

Fig. 1. Survival of cancer patients with or without expression of specific TACAs in their tumors. (a) Expression of A antigen in human lung cancer (Lee et al., 1991). (b) Expression of H/Le^y/Le^b (precursor of A antigen) in human lung cancer (Miyake et al., 1992). (c) Expression of SLe^x antigen in human colonic cancer (Irimura, T. et al., unpublished). (d) Expression of sialosyl-Tn antigen in human colonic cancer (Itzkowitz et al., 1990).

human melanoma cells. MAb M2590 has a 'density threshold' for recognition of GM3. GM3 expression is high in highly-metastatic B16 variants BL6 and F10, low in less-metastatic variant B16/F1, and minimal in non-metastatic variant B16/WA4 (Fig. 2 insert). Several lines of evidence show that Bl6-EC adhesion is mediated by GM3–lactosylceramide (LacCer) interaction: (i) The order of adhesion of B16 variants to LacCer-coated plates (Fig. 2A) or LacCer/fibronectin (FN) co-coated plates (Fig. 2B) is the same as the order of metastatic potential, whereas the variants were indistinguishable in terms of adhesion to FN-coated plates (Fig. 2C). (ii) LacCer, a simple GSL highly expressed on mouse and human microvascular ECs, can be surface-labeled by galactose oxidase/NaB^{3}H$_4$ treatment (Gillard et al., 1990; Kojima et al., 1992b) (data not shown). Relative

degree of adhesion of the B16 variants to non-activated ECs is in the same order as their metastatic potential, i.e. BL6> F10> F1> WA4 (Fig. 3A). The differences among B16 variants in adhesion to activated ECs are much smaller than differences in adhesion to non-activated ECs (Fig. 3B). (iii) Adhesion of B16/BL6 or B16/F10 cells to ECs is inhibited by liposomes containing LacCer, GM3, or methyl-β-lactoside, but not methyl-β-N-acetyllactosaminide (Fig. 4). Adhesion of GM3-liposomes to LacCer-coated plates, or vice-versa, has been clearly observed, and its degree is correlated with density of both GSL species (Kojima and Hakomori, 1991). (iv) Involvement of other adhesion molecules in adhesion of non-activated ECs is minimal. For example, adhesion of B16 cells to plates coated with FN, integrin, or Ig family receptors did not vary significantly

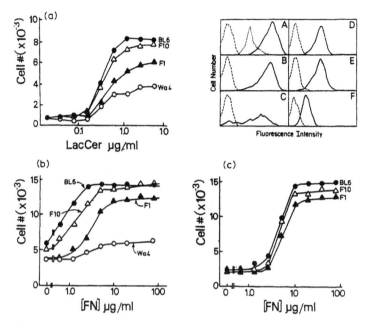

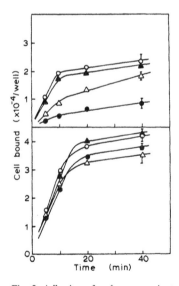

Fig. 2. Degree of GM3 expression in B16 melanoma variants with different metastatic potentials, and their adhesiveness to LacCer-coated, FN-coated, and FN/LacCer co-coated plates. Insert (upper right): GM3 expression as revealed by cytofluorometry with anti-GM3 MAb DH2. A, B16/BL6 cells. B, B16/F10 cells. C, B16/F1 cells. D, A431 cells. E, 3T3 cells. F, B16/WA4 cells. Panel A: Adhesion of BL6 ●, F10 △, F1 ▲, and WA4 ○ cells to LacCer-coated plates. Panel B: Adhesion to FN/LacCer co-coated plates in the co-presence of 1 μg/well of LacCer. Panel C: Adhesion to FN-coated plates.

Fig. 3. Adhesion of melanoma variants to non-activated and activated ECs. ○, BL6 cells. ▲, F10. △ F1. ●, WA4. Adhesion to non-activated ECs (upper panel) was strongly inhibited when ECs were pre-treated with anti-LacCer MAb T5A7, or when B16 cells were treated with MAb DH2 or sialidase (data not shown). Adhesion to activated ECs (lower panel) was unaffected by treatment with these reagents (data not shown).

among the four melanoma variants (Fig. 2C), and required at least 20–30 min incubation. In contrast, adhesion to LacCer-coated plates required <10 min (Fig. 5). This time-dependent difference between melanoma cell adhesion to GSL-coated vs. FN-coated plates provides an explanation for observed differences in a dynamic flow adhesion system (see below).

Unique characteristics of melanoma adhesion based on GM3–LacCer interaction, as compared with lectin- or integrin-based adhesion, in a dynamic flow system

Adhesion of B16 cells in a dynamic flow system is distinct in many respects from adhesion in a static system; importantly, in the dynamic system adhesion to LacCer- or Gg3-coated surfaces predominates over adhesion to integrin- or lectin-coated surfaces. We constructed a dynamic flow assembly similar to that desribed by Lawrence et al. (1990). In this assembly, a tumor cell suspension is pumped through a parallel-plate laminar flow chamber, and flows over a glass

244

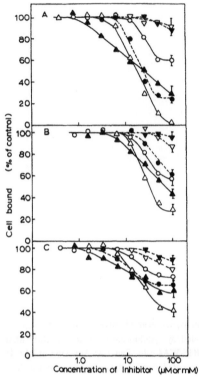

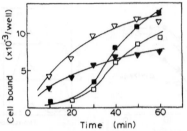

Fig. 5. Time-dependent difference of BL6 cell ad
coated and adhesive protein-coated plates. ▽, Gg3
LacCer (1 µg/well). □, FN (5 µg/well). ■, LN (5
that adhesion to LacCer or Gg3 occurs much m
adhesion to FN or LN.

Fig. 4. Inhibitory effect of oligosaccharides and GSL-liposomes on adhesion of BL6 cells to LacCer-coated plates (A), human ECs (B), and mouse ECs (C). Significant inhibition was observed for LacCer-liposome △, Gg3-liposome ▲, and methyl-β-lactoside ●. GM3-liposome ○ was moderately inhibitory. Methyl-β-*N*-acetyl-lactosaminide ▼ and lactose ▽ had no effect. Concentrations of liposomes are expressed in µM on the abscissa, while concentrations of oligosaccharides are in mM.

tion (Fig. 6B), in striking contrast to a s
in which the reverse was true (data
Adhesion of BL6 cells to LacCer-coate

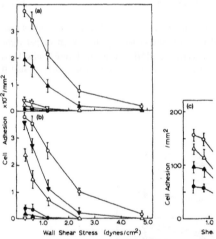

plate coated with the suspected adhesion molecule. Adhesion of tumor cells to the plate is recorded on videotape and quantified. These conditions are designed to mimic the microvascular environment in which metastatic deposition of tumor cells occurs. In contrast to static (non-flow) model adhesion systems, degree of adhesion in the dynamic system was greatly dependent on the time required for adhesion. That is, integrin-dependent adhesion (which requires a much longer incubation period than does GSL–GSL interaction in a static system) was minimal in the dynamic system (Fig. 6A). Even Con A- or *Erythrina* lectin-dependent adhesion were found to be less prominent than adhesion based on GSL–GSL interac-

Fig. 6. Adhesion of BL6 cells to glass plates coat GSLs, lectins, and adhesive proteins in a dynam Panel A: BL6 cells suspended in phosphate-buffe 10[5] cells/ml) were introduced into the dynamic flo ious wall shear stress values (abscissa). GSL-lipc GSL/ml) were coated on a circular area (0.5 cm di plates. GSL absorbed on this area was calculated Number of adherent cells after 3 min flow was c ▲, LacCer. ●, GM3. ▽, paragloboside. △, control ing GSL). Panel B: procedure as in Panel A but instead of GSL-liposomes. ▼, Con A lectin, *Erythrina* lectin, 200 µg/ml. ●, FN, 100 µg/ml. ▽ ▲, Con A, 20 µg/ml. ○, Gg3-liposome (for compa A). Panel C: inhibition of BL6 cell adhesion to plates. ○, untreated BL6 cells (control). △, cells M ethyl-β-lactoside. ▲, sialidase. ●, MAb DH2.

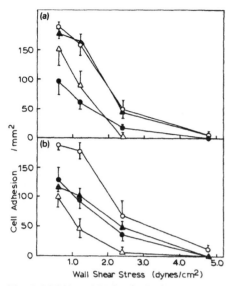

Fig. 7. Inhibition of BL6 cell adhesion to mouse ECs by various reagents in a dynamic flow system. Panel A: B16 melanoma variants (10^5 cells/ml) were passed over a glass plate on which mouse ECs had been grown. ○, BL6. ▲, F10. ●, F1. △, WA4. Panel B: Adhesion of BL6 cells treated with various reagents. △, MAb DH2. ▲, lactose, 100 mM. ●, ethyl-β-lactoside, 50 mM. ○, control.

inhibited by ethyl-β-lactoside, sialidase, and anti-GM3 MAb (Fig. 6C). Adhesion of BL6, F10, F1, and WA4 cells to ECs in the dynamic system, and its inhibitability by various reagents, is shown in Fig. 7A,B. These findings suggest that B16 melanoma cell adhesion to ECs in vivo is based on GM3–LacCer interaction, and initiates metastatic deposition. This process may in turn trigger a series of 'cascade reactions' leading to activation of ECs, expression of selectins and Ig family receptors, and enhanced tumor cell adhesion and migration (Kojima et al., 1992b).

Assuming the above hypothesis is true, melanoma cell metastasis could be inhibited by oligosaccharides representing GM3 or LacCer. Indeed, methyl- or ethyl-β-lactoside co-injected or separately injected with tumor cells did inhibit melanoma cell metastasis to lung (Oguchi et al., 1990). More recently, we observed that BL6 cell adhesion to LacCer-coated plates or ECs was more strongly inhibited by 6-deoxy-6-fluoro-Galβ1 → 4Glcβ1-O-Me than by methyl-β-lactoside (Cai et al., 1992) (Fig. 8). Molecular modeling experiments indicate that the 6-

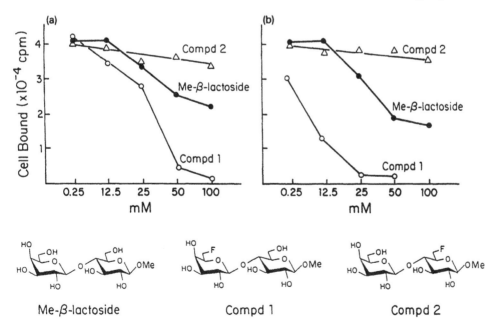

Fig. 8 Inhibitory effect of fluorinated compounds 1 and 2, as compared to methyl-β-lactoside, on BL6 cell adhesion. Panel A, adhesion to mouse ECs. Panel B, adhesion to LacCer-coated plates. Note that only compound 1 (6-deoxy-6-fluoro-Galβ1 → 4Glc) strongly inhibited BL6 cell adhesion.

hydroxymethyl group of the galactopyranosyl residue of LacCer is involved in hydrogen bond formation during GM3-LacCer interaction.

Possible functional significance of GSL and CHO antigens in human cancer

A mechanism similar to that described above for mouse B16 melanoma metastasis may well operate in human cancers, since high GM3 expression has been reported for human melanoma and other cancers, and human ECs express LacCer. Similarly, expression of H/Le^y/Le^b antigen, defined by MAb MIA-15-5, showed a correlation with human lung carcinoma malignancy (Miyake et al., 1992; see Fig. 1B). This phenomenon could reflect interaction of H/Le^y/Le^b-expressing tumor cells with H-expressing ECs, based on H–H, H–Le^y, and/or H–Le^b interaction, as illustrated in Fig. 9. Expression of H antigen on microvascular ECs of various human organs has been well established on histochemical (Holthofer et al., 1982) and immunochemical (Handa, K, Tashiro, K and Hakomori, S, unpublished) bases. It is also possible that recognition of sialosyl-Le^x (SLe^x) or sialosyl-Le^a (SLe^a) by selectins, and activation of platelets by tumor cells inducing P-selectin expression, could facilitate tumor cell aggregation and deposition of these aggregates on ECs. The evidence for these hypotheses was recently reviewed (Hakomori, 1991c).

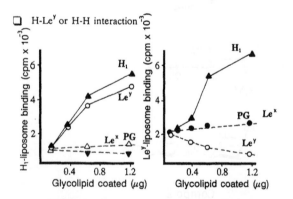

Fig. 9. Binding of H$_1$-liposomes (left panel) and Le^y-liposomes (right panel) containing [^{14}C]-cholesterol to plates coated with various GSLs. Strongest binding was observed for H–H and H–Le^y interaction.

Recently, we identified the P-selectin epitope expressed upon platelet activation as SLe^a, in addition to SLe^x. Binding of P-selectin to these two epitopes is strongly inhibited by a low concentration of sulfated glycan (e.g., 1 µg/ml dextran sulfate or fucoidin). Binding of SLe^x or SLe^a to the lectin domain of P-selectin is conformationally regulated by non-specific sulfated glycan (Handa et al., 1991b). A similar trend is clearly observed for L-selectin, but less so for E-selectin. The proposed modulatory effect of sulfated glycan on selectin binding to SLe^x and SLe^a is illustrated in Fig. 10.

The epitopes recognized by selectins are carried mainly by O-linked CHO chains. We observed that adhesion of HL60, U937, or Colo205 tumor cells to ECs or platelets was abolished or greatly reduced if O-glycosylation extension was inhibited by benzyl-α-GalNAc (Kojima et al., 1992a). Expression of P-selectin is strongly down-regulated by N,N-dimethyl- or N,N,N-trimethyl-sphingosine, or calphostin-C (all of which inhibit protein kinase C), thus inhibiting selectin-mediated adhesion of tumor cells to platelets (Handa et al., 1991a).

Although the mechanism by which tumor cell metastasis is mediated by selectins is not well understood, there are obviously a number of factors involved besides expression of selectin on ECs and platelets, and mode of presentation of SLe^x or SLe^a at the tumor cell surface (density, O-linked, N-linked, lipid-linked, molecular carrier species). Close access and contact between tumor cells and ECs/platelets is a prerequisite for activation of and selectin expression on ECs/platelets. Our findings clearly indicate that certain tumor cells showing high expression of GM3 or H/Le^y/Le^b adhere to non-activated ECs through GSL–GSL interaction, which may trigger activation of ECs. Tumor cells showing high expression of Le^x may autoaggregate through Le^x-Le^x interaction (Eggens et al, 1989), leading to microembolism and activation of ECs/platelets. This suggests the following possible sequence of events: (1) tumor cells adhere to non-activated ECs, or autoaggregate through CHO–CHO interaction; (2) subsequently, ECs or platelets are activated by tumor cells, which lead to surface expression of E- or P-selectin; (3)

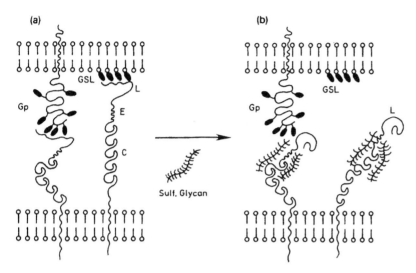

Fig. 10. Proposed effect of sulfated glycans on P-selectin binding to SLe^a or SLe^x presented on glycoprotein or GSL. Gp, glycoprotein. Solid oval, SLe^a or SLe^x. Sulfated glycans may interact with the EGF domain (E) or complement-regulatory sequence repeat (C), thereby inducing conformational changes of the lectin domain (L) of P-selectin. Sulfated glycans also have a clear inhibitory effect on L-selectin, but have less effect on E-selectin binding to SLe^a or SLe^x (Shiozawa, T., Handa, K., Nudelman, E.D. and Hakomori, S., unpublised).

SLe^x or SLe^a epitope is appropriately presented at the tumor cell surface via clustered, O-linked membrane proteins; and (4) additional stronger adhesion and enhancement of cell motility may take place through mechanisms involving integrin and Ig family receptors. GSLs are organized in clusters at the cell surface

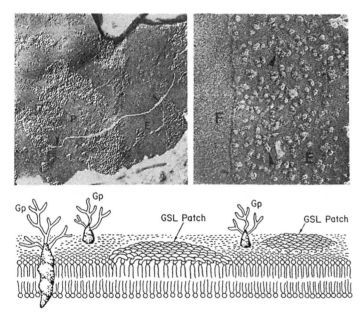

Fig. 11. Proposed organization of GSLs at the cell surface. Upper left: clustering of globoside on the human erythrocyte surface (Tillack et al., 1983). Upper right: clustering of Forssman GSLs on Forssman-liposome. E, external surface. F, fractured internal surface. Lower panel: conceptual scheme of GSL patches expressed on plasma membrane. Gp, glycoprotein.

248

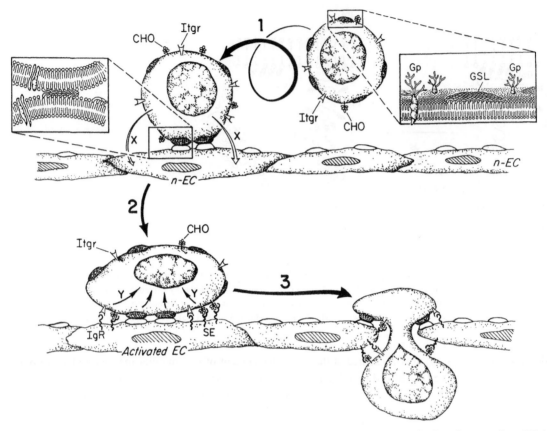

Fig. 12. Proposed scheme for stepwise progression of adhesion between tumor cells and ECs. Adhesion of tumor cells to ECs in a dynamic flow system is initially mediated via GSL patches on both plasma membranes (GM3–LacCer, H–Le^y, or H–H) (step 1), followed by activation of ECs (step 2), leading to expression of selectin (SE) and binding of selectin to SLe^a or SLe^x on the tumor cell surface. This sends signal 'Y' to activate cell motility (step 3), leading to reinforcement of adhesion by integrin receptor (Itgr) or immunoglobulin receptor (IgR) mechanisms.

(Tillack et al., 1983; Rock et al., 1990, 1991) (Fig. 11). Initial interaction between GSLs expressed on tumor cells and those on ECs may take place via these GSL clusters (Fig. 12). Thus, if initial interaction between GSL clusters were blocked, the whole adhesion process would be blocked. The process could also be blocked by inhibiting transmembrane signaling involved in selectin expression. During induction of P-selectin expression by tumor cells, the presence of sulfated proteoglycans may modulate P-selectin binding activity.

The metastatic process is complex, involving many different factors and steps. We may therefore be able to block metastasis at various steps by application of several different types of reagents: (i) oligosaccharides or their derivatives; (ii) anti-oligosaccharide MAbs; (iii) inhibitors of O-glycosylation or O-glycosylation extension; and (iv) blockers of transmembrane signaling leading to selectin expression. In fact, we have successfully blocked metastasis using some of these reagents (Oguchi et al., 1990; Okoshi et al., 1991). Further studies along these lines are in progress.

Acknowledgment

I thank Dr. Stephen Anderson for scientific editing and preparation of the manuscript.

References

Berg, E.L., Robinson, M.K, Mansson, O., Butcher, E.C. and Magnani, J.L. (1991) A carbohydrate domain common to both sialyl Le^a and sialyl Le^x is recognized by the endothelial cell leukocyte adhesion molecule ELAM-1. *J. Biol. Chem.*, 266: 14869–14872.

Cai, S., Hakomori, S. and Toyokuni, T. (1992) Application of protease-catalyzed regioselective esterification in synthesis of 6'-deoxy-6'-fluoro- and 6-deoxy-6-fluorolactosides. *J. Org. Chem.*, 57: 3431–3437.

Eggens, I., Fenderson, B.A., Toyokuni, T., Dean, B., Stroud, M.R. and Hakomori, S. (1989) Specific interaction between Le^x and Le^x determinants: A possible basis for cell recognition in preimplantation embryos and in embryonal carcinoma cells. *J. Biol. Chem.*, 264: 9476–9484.

Gillard, B.K, Jones, M.A., Turner, A.A., Lewis, D.E. and Marcus, D.M. (1990) Interferon-gamma alters expression of endothelial cell-surface glycosphingolipids. *Arch. Biochem. Biophys.*, 279: 122–129.

Hakomori, S. (1985) Aberrant glycosylation in cancer cell membranes as focused on glycolipids: Overview and perspectives. *Cancer Res.*, 45: 2405–2414.

Hakomori, S. (1989) Aberrant glycosylation in tumors and tumor-associated carbohydrate antigens. *Adv. Cancer Res.*, 52: 257–331.

Hakomori S, (1990) Bifunctional role of glycosphingolipids: Modulators for transmembrane signaling and mediators for cellular interactians. *J. Biol. Chem.*, 265: 18713–18716.

Hakomori, S. (1991a) Carbohydrate-carbohydrate interaction as an initial step in cell recognition. *Pure Appl. Chem.*, 63: 473–482.

Hakomori, S. (1991b) Possible functions of tumor-associated carbohydrate antigens. *Curr. Opin. Immunol.*, 3: 646–653.

Hakomori, S. (1991c) Possible new directions in cancer therapy based on aberrant expression of glycosphingolipids: Anti-adhesion and ortho-signaling therapy. *Cancer Cells*, 3: 461–470.

Handa, K, Igarashi, Y., Nisar, M. and Hakomori, S. (1991a) Down-regulation of GMP-140 expression on platelets by *N,N*-dimethyl and *N,N,N*-trimethyl derivatives of sphingosine. *Biochemistry*, 30: 11682–11686.

Handa, K, Nudelman, E.D., Stroud, M.R., Shiozawa, T. and Hakomori, S. (1991b) Selectin GMP-140 (CD62; PADGEM) binds to sialosyl-Le^a and sialosyl-Le^x, and sulfated glycans modulate this binding. *Biochem. Biophys. Res. Commun.*, 181: 1223–1230.

Hirabayashi, Y., Hanaoka, A., Matsumoto, M., Matsubara, T., Tagawa, M., Wakabayashi, S. and Taniguchi, M. (1985) Syngeneic monoclonal antibody against melanoma antigen with interspecies cross-reactivity recognizes GM3, a prominent ganglioside of B16 melanoma. *J. Biol. Chem.*, 260: 13328–13333.

Holthofer, H., Virtanen, I., Kariniemi, A.-L., Hormia, M., Linder, E. and Miettinen A. (1982) *Ulex europaeus* I lectin as a marker for vascular endothelium in human tissues. *Lab. Invest.*, 47: 60–67.

Itzkowitz, S.H., Bloom, E.J., Kokal, W.A., Modin, G., Hakomori, S. and Kim, Y.S. (1990) Sialosyl-Tn: A novel mucin antigen associated with prognosis in colorectal cancer patients. *Cancer*, 66: 1960–1966.

Kojima, N., Handa, K, Newman, W. and Hakomori, S. (1992a) Inhibition of selectin-dependent tumor cell adhesion to endothelial cells and platelets by blocking O-glycosylation of these cells. *Biochem. Biophys. Res. Commun.*, 182: 1288–1295.

Kojima, N., Shiota, M., Sadahira, Y. and Hakomori, S. (1992b) Cell adhesion in a dynamic flow system as compared to static system: Glycosphingolipid-glycosphingolipid interaction in the dynamic system predominates over lectin- or integrin-based mechanisms in adhesion of B16 melanoma cells to non-activated endothelial cells. *J. Biol. Chem.*, 267: in press.

Kojima N. and Hakomori S. (1991) Cell adhesion, spreading, and motility of GM3-expressing cells based on glycolipid-glycolipid interaction. *J. Biol. Chem.*, 266: 17552–17558.

Lawrence, M.B., Smith, C.W., Eskin, S.G. and McIntire, L.V. (1990) Effect of venous shear stress on CD18-mediated neutrophil adhesion to cultured endothelium. *Blood*, 75: 227–237.

Lee, J.S., Ro, J.Y., Sahin, A.A., Hong, W.K., Brown B.W., Mountain, C.F. and Hittelman, W.N. (1991) Expression of blood-group antigen A: A favorable prognostic factor in non-small-cell lung cancer. *N. Engl. J. Med.*, 324: 1084–1090.

Miyake, M., Taki, T., Hitomi, S. and Hakomori, S. (1992) Correlation of expression of H/Le^y/Le^b antigens with survival in patients with carcinoma of the lung. *N. Engl. J. Med.*, 327: 14–18.

Nores, G.A., Dohi, T., Taniguchi, M. and Hakomori, S. (1987) Density-dependent recognition of cell surface GM$_3$ by a certain anti-melanoma antibody, and GM$_3$ lactone as a possible immunogen: Requirements for tumor-associated antigen and immunogen. *J. Immunol.*, 139: 3171–3176.

Oguchi, H., Toyokuni, T., Dean, B., Ito, H., Otsuji, E., Jones, V.L., Sadozai, K.K. and Hakomori, S. (1990) Effect of lactose derivatives on metastatic potential of B16 melanoma cells. *Cancer Commun.*, 2: 311–316.

Okoshi, H., Hakomori, S., Nisar, M., Zhou, Q., Kimura, S., Tashiro, K. and Igarashi, Y. (1991) Cell membrane signaling as target in cancer therapy II: Inhibitory effect of *N,N,N*-trimethyl-sphingosine on metastatic potential of murine B16 melanoma cell line through blocking of tumor cell-dependent platelet aggregation. *Cancer Res.*, 51: 6019–6024.

Phillips, M.L., Nudelman, E.D., Gaeta, F.C.A., Perez, M., Singhal, A.K, Hakomori, S. and Paulson, J.C. (1990) ELAM-1 mediates cell adhesion by recognition of a carbohydrate ligand, sialyl-Le^x. *Science*, 250: 1130–1132.

Polley, M.J., Phillips, M.L., Wayner, E.A., Nudelman, E.D., Singhal, A.K, Hakomori, S. and Paulson, J.C. (1991) CD62 and endothelial cell-leukocyte adhesion molecule 1 (ELAM-1) recognize the same carbohydrate ligand, sialyl-Lewis x. *Proc. Natl. Acad. Sci. USA*, 88: 6224–6228.

Rock, P., Allietta, M., Young, W.W. Jr., Thompson, T.E. and Tillack, T.W. (1990) Organization of glycosphingolipids in phosphatidylcholine bilayers: Use of antibody molecules and Fab fragments as morphologic markers. *Biochemistry,* 29: 8484–8490.

Rock P., Allietta, M., Young, W.W. Jr., Thompson, T.E. and Tillack, T.W. (1991) Ganglioside G_{M1} and asialo-G_{M1} at low concentration are preferentially incorporated into the gel phase in two-component, two-phase phosphatidylcholine bilayers. *Biochemistry,* 30: 19–25.

Tillack, T.W., Allietta, M., Moran, R.E. and Young, W.W. Jr. (1983) Localization of globoside and Forssman glycolipids on erythrocyte membranes. *Biochim. Biophys. Acta,* 733: 15–24.

L. Svennerholm, A.K. Asbury, R.A. Reisfeld, K. Sandhoff, K. Suzuki, G. Tettamanti and G. Toffano (Eds.)
Progress in Brain Research, Vol. 101
© 1994 Elsevier Science BV. All rights reserved.

CHAPTER 19

Tumor gangliosides as targets for active specific immunotherapy of melanoma in man

Donald L. Morton, Mepur H. Ravindranath, and Reiko F. Irie

John Wayne Cancer Institute, 2200 Santa Monica Blvd., Santa Monica, California 90404, U.S.A.

Introduction

Our interest in active immunotherapy for melanoma began with our finding that the serum of melanoma patients contains antibodies which react with membrane antigens on melanoma cells (Morton et al., 1968). Shortly thereafter, we reported that intratumoral injection of *Bacillus* Calmette-Guerin (BCG) into cutaneous melanoma metastases enhanced systemic immunity, as shown by an elevation in the titers of antimelanoma antibodies, and by clinical regression of uninjected skin metastases in 17% of patients (Morton et al., 1970, 1974). Biopsy of these uninjected melanoma lesions revealed intense lymphocytic infiltration.

Our initial attempts to reproduce these observations by active immunotherapy used an intradermally injected whole-cell tumor vaccine (TCV) composed of irradiated allogeneic melanoma cells mixed with BCG (Morton et al., 1978; 1987). This vaccine had limited success, but about one-third of the patients developed high titers of IgM antibodies reactive with melanoma membrane antigens. Importantly, these patients enjoyed prolonged survival (Jones et al., 1981).

The identification of gangliosides as possible cell membrane antigens of melanoma was first suggested by our finding that the melanoma-associated antigens recognized by the IgM antibodies were also present in ganglioside-rich neural tissues of fetal brain (Irie et al., 1976, 1979b). These antigens were named oncofetal antigen-immunogenic (OFA-I), to indicate their immunogenicity in man (Sidell et al., 1979a, 1979b, 1980). Later, we found that the oncofetal antigen complex consisted of two distinct reactive epitopes: OFA-I-1 and OFA-I-2 (Irie et al., 1982; Rees et al., 1981).

We were able to identify these antigens as gangliosides GM2 (Tai et al., 1983) and GD2 (Cahan et al., 1982) by showing their reaction with specific human monoclonal antibodies produced by Epstein-Barr virus transformation of peripheral blood lymphocytes from melanoma patients (Irie et al., 1981, 1982). Shortly thereafter, we found that the IgM antibodies which became elevated in the sera of patients immunized with TCV were directed predominantly against GM2 and GD2 (Tai et al., 1985). Direct evidence for the importance of GM2 and GD2 as target antigens for immunotherapy was provided by our observation that human monoclonal antibodies against these two gangliosides induce regression of human cutaneous melanoma metastases only if the melanoma possesses the corresponding ganglioside antigens (Irie et al., 1986, 1989a,b).

These findings motivated us to focus on gangliosides and protein antigens newly discovered in our laboratory as potential targets for active specific immunotherapy of melanoma (Morton et al., 1987, 1992). We reformulated our vaccine, using polyvalent allogeneic melanoma cell lines specifically selected for their high concentrations of GM2 and GD2. The reformulated melanoma cell vaccine

(MCV) also contains O-acetyl GD3 (O-AcGD3), which we later discovered was an important immunogenic ganglioside (Ravindranath et al., 1989).

In 1985, we began a phase II clinical trial to evaluate a polyvalent, ganglioside-rich MCV in patients with advanced metastatic melanoma. Active specific immunotherapy with MCV produced dramatic regressions of metastatic melanoma in some patients and an overall threefold prolongation in median survival (Morton et al., 1992).

This paper reviews some of the important aspects of gangliosides as vaccines for active specific immunotherapy of melanoma, and presents new data on the antibody response to gangliosides induced in patients receiving MCV.

Gangliosides of melanocytes and melanoma cells

Melanoma is the result of neoplastic transformation of melanin-producing cells called melanocytes, which are found in skin, uveal tract, meninges, ectodermal mucosa and internal viscera. Gangliosides, the glycosphingolipids that contain sialic acids, are important components of the outer lipid bilayer of cancer cell membranes. Figure 1 shows the carbohydrate moiety of the major gangliosides of melanoma.

GM3 is the major ganglioside of normal melanocytes. The diploid melanocytes obtained from foreskins of newborns and grown for two passages in vitro contain 86–90% of GM3 (Carubia et al., 1984). Other gangliosides include GD3 (2–6%), GM2 (1%), GD1a (2%) and GQ1b (1%). GM3 is the common progenitor of GD3 and GM2, which are the result of 2,8-sialyl transferase activity and GalNAc transferase activity, respectively (Fig. 1).

Melanoma cells in the radial growth phase have a GM3 and GD3 profile similar to that of normal melanocytes; however, the vertical growth phase is marked by a shift towards GD3, demonstrated on chromatograms of gangliosides purified from cultures of melanoma cells derived from tissue biopsied during these growth phases (Herlyn et al., 1985). GD3 is in fact the major ganglioside of metastatic cutaneous melanoma; its derivatives include GD2 and O-AcGD3.

Gangliosides of Human Melanoma

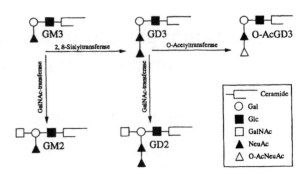

Fig. 1. The gangliosides of human melanoma. The carbohydrate epitopes of various melanoma-associated major gangliosides are illustrated. The enzymes involved in GM2, GD2, GD3, and O-AcGD3 are indicated.

Most metastatic melanomas are marked by elevated levels of GD3, GM2 and GD2, and by markedly decreased levels of GM3. However, melanomas developing in 'immunologically privileged' sites such as the eye may have a preponderance of GM3 (Kanda et al., 1992). Furthermore, the expression of ganglioside antigens on human cancer cells varies among different patients and different cell lines, and with different stages of disease (Tsuchida et al., 1987c, 1989). This heterogeneity is partly the result of changes in milieu interna, tumor stage, and possibly the host's genetic composition. It could also reflect a complex interaction between metastatic tumor spread and host immune surveillance.

Changes in ganglioside composition of melanoma cells have also been proposed to explain various changes associated with neoplastic transformation, particularly cell motility. Melanocytes of the embryonic neural crest are actively motile. Their motility ceases after differentiation but reappears after neoplastic transformation. This behavioral alteration may reflect changes in the cell-surface ganglioside profile. In support of this is the documented role of disialogangliosides in the spreading and infiltration of tumor cells (Cheresh and Klier, 1986; Cheresh et al., 1986). Also, monoclonal antibodies directed against the carbohydrate

TABLE I

Amounts of five major gangliosides found in human melanoma, expressed as mean percent of lipid-bound sialic acid (range in parentheses)[a]

Gangliosides	Biopsied uveal (N = 14)	Biopsied cutaneous (N = 52)	Cultured cutaneous (N = 28)
GM3	89.2	43.2 (14–90)	38.6 (11–84)
GD3	9.5	47.7 (6–73)	34.6 (4–63)
GM2	0.5	3.2 (1–14)	13.6 (2–47)
GD2	0.0	2.0 (0–10)	7.8 (0–32)
Alkali-labile GD3	0.0	3.7 (0–13)	1.6 (0–6)
Others	0.8	3.9	5.1
Total lipid-bound (μg/g)	456 (289–1378)	100 (33–302)	104 (34–183)

[a]Values determined as described by Ledeen and Yu (1982).

moieties of GD3 and GD2 inhibit attachment of melanoma cells to various basement membrane components.

Importance of the GM3:GD3 ratio in clinical course

The heterogeneous expression of GM3 and GD3 and the functional variations associated with malignant transformation suggest that the phenotypic expression of these gangliosides may reflect the natural history of melanoma. We examined this possibility by analyzing the ganglioside profile of metastatic tumors excised from 52 patients with cutaneous melanoma (Tsuchida et al., 1987c; Ravindranath et al., 1991) (Table I). The ganglioside profiles of melanoma biopsies from this series generally showed a preponderance of GM3 and GD3 (Figs. 2 through 6). Although GD3 levels varied markedly in various tumor specimens, the overall level was considerably higher than that in various normal tissues (Table II). Interestingly, high GD3 levels were followed by

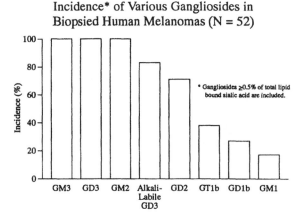

Fig. 2. Incidence of various gangliosides in biopsied human melanomas, based on analysis of 52 tumor specimens. Note that GM3, GD3, and GM2 are found in all the biopsies.

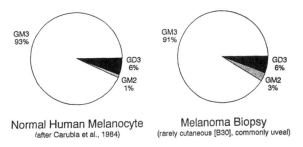

Fig. 3. The relative proportions of gangliosides in normal melanocytes, and uveal melanoma. Note that the ganglioside profile of uveal melanoma is similar to normal melanocytes. Such a profile is rare in cutaneous melanoma. In our study, we observed one such biopsy in malignant cutaneous melanoma (B30).

Group IA: high GM3 (with low GM2), low GD3

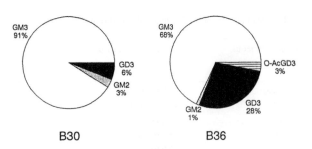

B30　　　　　　　　B36

Fig. 4. The ganglioside proportions in group IA biopsies show high GM3 and low GD3. B30 and B36 are biopsies showing the profile characteristic of group IA.

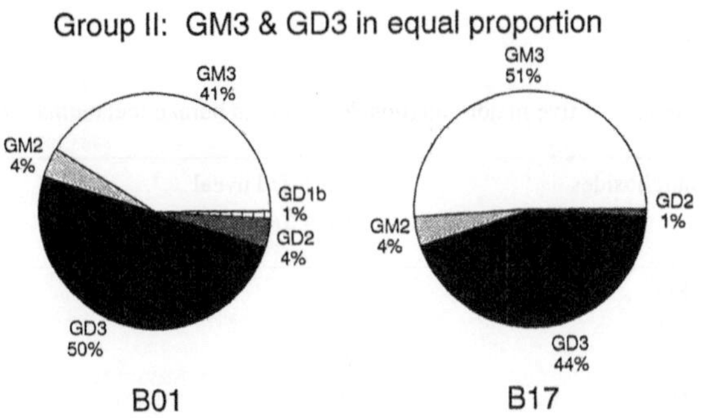

Group II: GM3 & GD3 in equal proportion

B01　　　　　　　　B17

Fig. 6. The ganglioside profile of group II shows equal proportions of GM3 and GD2.

increased expression of GD2 and O-AcGD3 (Ravindranath et al., 1988) (Figs 7 and 8).

We also analyzed ganglioside profiles of 14 cases of uveal melanoma, including 13 primary lesions and one metastatic tumor (Kanda et al., 1992). Figure 3 shows that the ganglioside pattern observed in 11 of these uveal melanoma specimens is strikingly similar to that of normal melanocytes (Carubia et al., 1984). In three other cases, GD3 was increased to 19.5–46.0%. Histologic examination of these three biopsy specimens yielded results suggesting that an increase in GD3 may be related to tumor infiltration by lymphocytes and macrophages.

On the assumption that changes in cell ganglioside

profile might reflect the chronology of neoplastic transformation, we developed four arbitrary categories for malignant cutaneous melanoma (Ravindranath et al., 1991). These categories, corresponding to Figs 4–7, are based on GM3:GD3 ratio and GM2 level. The GM3:GD3 ratio in group IA and group IB ranged from 15:1 to 1.5:1; in group II, it ranged from 1.4:1 to 1:1.4; and in group III, the ratio ranged from 1:1.5 to 1:5. Group IB biopsies had unusually high GM2 levels (Fig. 5), the significance of which is unclear. The tumor of these patients may express more GM3:GalNAc transferase activity and may reflect a genetically different population of melanoma patients. We have found that increased

Group IB: high GM3 (with high GM2), low GD3

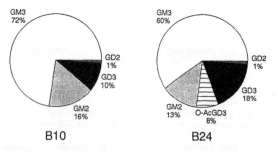

B10　　　　　　　　B24

Fig. 5. The ganglioside profile of group IB biopsies is distinct from group IA in expressing a high level of GM2. The significance of this is far from clear; it may reflect genetic differences between the two groups. B10 and B24 are examples of this group.

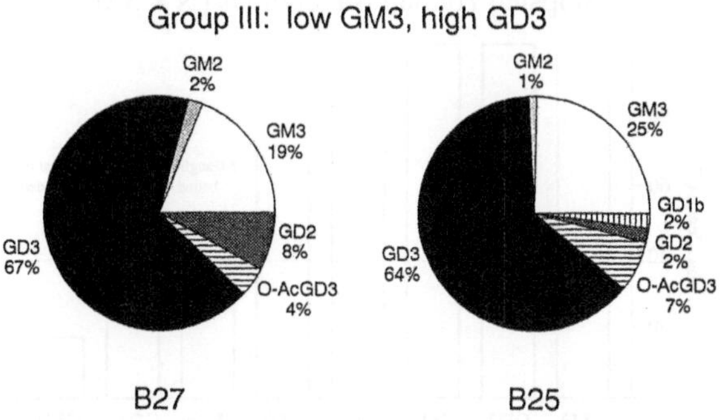

Group III: low GM3, high GD3

B27　　　　　　　　B25

Fig. 7. The ganglioside profile of group III is unique in expressing a high level of GD3. Our study shows that this pattern may reflect poor prognosis.

TABLE II

Distribution of major melanoma-associated gangliosides in normal human tissues, expressed as percent of total lipid-bound sialic acid or relative distribution (+ or -)

Tissue	GM3	GM2	GD3	GD2	Reference
Brain					
Gray	<1	2	5	0	Traylor and Hogan, 1980
White	0	1	4	0	Traylor and Hogan, 1980
Neuron	+	0	3+	+	Dreyfus et al., 1984
Spinal Cord	+	+	+	+	Ando, 1983; Sastry, 1985
Spleen	>95	0	0	0	Svennerholm, 1963
Retina	10	0	50	0	Holm et al., 1972
Serum	48	6	13	0	Senn et al., 1989
Liver	>90	<1	0	0	Seyfried et al., 1978
Erythrocytes	–	+	+	0	Kundu et al., 1983
Aorta	82	0	5	0	Prokazova et al., 1987
Thyroid	40	0	25	0	Bouchon et al., 1985
Lung	>50	<10	0	0	Narasimhan and Murray, 1979
Melanocytes	92	2	6	0	Carubia et al., 1984
Kidney	74	0	19	0	Rauvala, 1976

GM2 expression in melanoma cell lines is associated with the capacity for growth upon transplantation into nude mice (Tsuchida et al., 1987b).

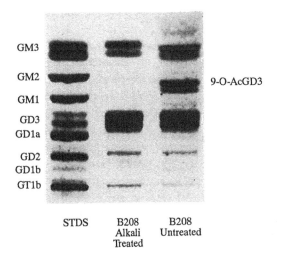

Fig. 8. The ganglioside profile of a unique biopsy showing a high amount of alkali labile O-AcGD3. We have identified the ganglioside by chromatographic mobility, alkali lability and affinity of an O-acetyl sialic acid-specific lectin (Ravindranath et al., 1988.

Figure 9 shows the survival differences among these groups after lymphadenectomy treatment of stage II disease (sample size = 28 patients with known date of diagnosis). Since group I represents tumors with ganglioside profiles closest to those of normal melanocytes, it is perhaps not suprising that survival of group I patients is significantly better than that of other groups. The difference among groups is significant at $P = 0.05$ from a log-rank test with two degrees of freedom. Mean survival time (in months) of patients in group I (both IA and IB) is 85 ± 24 (S.E.); group II, 29 ± 7.9; group III, 34.4 ± 6.6. The differences in survival time between group I and the other two groups are statistically significant*. Interestingly, the lipid-bound sialic acid level is lower in group I.

This study suggests that the ganglioside profile, particularly the GM3:GD3 ratio, of tumor biopsy

---

*The unadjusted P values are 0.0169 and 0.0139, respectively. The difference in survival time among groups, as calculated using Mantel-Cox formula, is significant, with a P value of 0.0446.

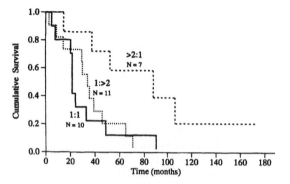

Fig. 9. Probability of survival from stage II melanoma, based on tumor tissue GM3:GD3 % LBSA ratio. The Kaplan–Meier method was employed to calculate the survival curves of patients with different ganglioside ratios. The difference between the curves is significant with a *P* value of 0.0446 and 2 d.f. from a log-rank determined by a generalized Mantel–Cox test.

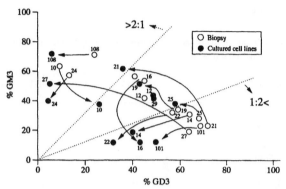

Fig. 10. This figure illustrates the change observed in the melanoma cells from biopsy to culture. Note that biopsies 12 and 29 show no change in the GM3 and GD3 percentage, indicating a stable nature. On the other hand, most of the cells show plasticity in their profile, in that the pattern observed in the culture is entirely different from that observed in their original biopsy. The differences could be due to selective growth of specific clones under culture conditions or to expression of enzymes affecting the ganglioside profile. The reversal of ganglioside pattern in culture cell lines transplanted in nude mice suggests that enzymes such as GalNAc transferase may be activated under culture conditions.

specimens has prognostic relevance. In support of this hypothesis is the low concentration of GD3 and the absence of GD2 in uveal melanoma (Table I), which has a correspondingly high frequency of survival and low incidence of metastasis (Kanda et al., 1992). Screening the GM3:GD3 ratio is feasible since melanoma expresses a simple profile of gangliosides, unlike other forms of cancer. Moreover, classifying patients based on ganglioside profile of tumor biopsies may be useful in selecting appropriate ganglioside-targeted therapies. For example, group III expressed elevated levels of gangliosides that are highly immunogenic in man (GM2, GD2, O-AcGD3); patients in this group might therefore benefit from therapies targeted against these ganglioside antigens. However, more studies are needed, with larger sample sizes, before recommending routine analysis of ganglioside levels from excised tumors.

In vivo versus in vitro ganglioside profiles

The relative proportions of various gangliosides in melanoma biopsies could reflect one of two possibilities. Either the ganglioside ratio is the same for all cells in a tumor specimen, or it varies for each clone within a specimen, in which case the relative ganglioside proportion would be an overall mean for all cell

populations in that specimen. Karyotypic profiles of metastatic melanoma biopsy tissues support the second possibility, revealing a heterogeneous and genetically unstable cell population (Parmiter and Nowell, 1988).

When we established melanoma cell lines* from the biopsied specimens of metastatic melanoma patients, we discovered that the ganglioside profile of melanoma cells grown in vitro is unique and distinct from that of biopsies (Tsuchida et al., 1987a,c). Figures 10 and 11 illustrate the alterations observed in GM3:GD3 ratio in vitro, offering further evidence of the heterogeneity and genetic instability of the melanoma cell population. Interestingly, some of the biopsies (12 and 29) showed remarkable stability, whereas others (16, 21 and 27) showed remarkable

---

*The cells were maintained in RPMI-1640 supplemented with 5–10% fetal calf serum and antibiotic antimycotic mixture containing penicillin, Fungizone, and streptomycin.

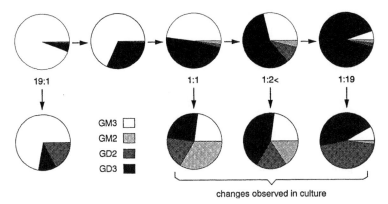

19:1   1:1   1:2<   1:19

GM3 □
GM2 ▨
GD2 ▨
GD3 ■

changes observed in culture

Fig. 11. The changes and evolution of gangliosides during neoplastic transformation, radial and vertical migration, metastasis of human melanocytes. The different pie charts are based on the ganglioside profiles observed in various biopsies. Note that placing the cells under the culture condition results in an alteration not frequently seen in vivo.

plasticity. The most uniform feature of melanoma cells grown in vitro is increased expression of GalNAc-containing gangliosides such as GM2 and GD2. This could be due to favorable growth conditions for tumor cells expressing these gangliosides, or to enhanced expression of GalNAc-transferase(s), which would augment the expression of GM2 and GD2 (Tsuchida et al., 1987a,b). In either case, the finding is serendipitous, since melanoma cells grown in tissue culture are the preferred source of ganglioside antigens for a vaccine. Interestingly, the increased expression of GM2 and GD2 is reversed when cell lines are transplanted into nude mice for in vivo growth (Tsuchida et al., 1987a).

Ganglioside-targeted passive immunotherapy by monoclonal antibodies: implications for active immunotherapy

Passive immunotherapy involves administration of monoclonal antiganglioside antibodies capable of mediating antibody-dependent or complement-dependent cytotoxicity. Two kinds of passive immunotherapeutic trials using antiganglioside antibodies will be briefly discussed: (1) injection of human monoclonal antiganglioside antibodies directly into the tumor (intralesional administration); and (2) systemic

administration of murine or chimerized antiganglioside antibodies.

Intralesional administration

Human monoclonal antibodies against gangliosides were first produced in our laboratory (Irie et al., 1981, 1982). In brief, lymphocytes were isolated from the peripheral blood of melanoma patients who had autoantibodies against melanoma-associated gangliosides. These lymphocytes were immortalized in vitro using Epstein-Barr virus, and cloned. Two cell lines secreting monoclonal antibodies reacting to GM2 and GD2, respectively, were established.

Eight patients with cutaneous melanoma cells were given intralesional injections daily or weekly (Irie and Morton, 1986). (The intralesional approach was necessary because at this time human monoclonal antibodies are too expensive to produce in quantities appropriate for systemic administration.) Regression was seen in all tumors except two that had low amounts of GD2. Histopathological analysis showed tumor degeneration, fibrosis, free melanin, and some degree of lymphocyte or macrophage infiltration. One patient with melanoma satellitosis had a complete regression with no sign of recurrence 20 months after the initial treatment. With the exception of mild erythema, no side effects were observed in any patients. The paucity of side effects can be attributed to the administration of small doses and the fine specificity of the antibodies. Another factor responsible for the negligible side effects could be the low density of GD2 antigen in normal tissues.

A second human monoclonal antibody (L55) against GM2 was also tested intralesionally in two patients with malignant melanoma (Irie et al., 1989b). Patients received both L55 and L72. In vitro tests for antigen expression revealed that the first patient reacted strongly to L55 (70% of positive cells) and L72 (56% of positive cells). The biopsied tissues of the second patient, on the other hand, did not react with either of the antibodies (6% for L55 and 0% for L72). Punch biopsies done after day 35 of the treatment showed total regression in the first patient. Histological sections displayed tumor degeneration, fibrosis, free melanin and some degree of lymphocyte or macrophage infiltration. Thus, no significant response was seen to either antibody, unless the patient's melanoma possessed the target antigen to which the human monoclonal antibody was directed.

Systemic administration

Despite direct documentation of the antitumor activity of human monoclonal antiganglioside antibodies following intralesional administration, systemic administration of murine monoclonal antibodies has been generally unsuccessful. Long-term, repeated administration of murine monoclonal antibodies is not possible, since patients develop antimurine immunoglobulins, which cause hypersensitivity against the foreign proteins. Therefore, regression is dramatic but temporary, and recurrence is common. One intriguing clinical study (Vadhan-Raj et al., 1988) linked clinical response to a change in the ganglioside profile of biopsy specimens obtained before, during, and after administration of R24, a murine monoclonal anti-GD3 antibody. Before treatment, 13 of 21 patients expressed GD3 on almost 100% of their melanoma cells; seven patients had 70–90% GD3-positive melanoma cells; only one patient had less than 50% GD3-positive cells. R24 treatment was associated with a localized inflamma-

tory response at tumor sites. After treatment, biopsies in 15 patients demonstrated that the majority of tumor cells still expressed GD3. However, the authors postulate that cell populations low in GD3 survived R24 treatment, subsequently giving rise to tumor cells with high GD3 levels. In support of this possibility was biopsy evidence from one of the patients who had 100% GD3 expression before treatment. Three weeks after ending R24 treatment, an inflammed subcutaneous lesion in this patient showed necrosis, inflammatory cell infiltrates, and small nests of melanoma cells that had no detectable GD3; analysis of a melanoma cell line established from this lesion confirmed the absence of GD3. However, during subsequent long-term culture of these cells, a GD3-positive population reappeared. These findings suggest that clones expressing little or no GD3 can escape destruction by R24 and subsequently give rise to a GD3-positive population.

Thus another cause for the failure of passive immunotherapy appears to be directly related to the ganglioside-based heterogeneity of the tumor cell population. The heterogeneity of ganglioside expression in melanoma makes it highly unlikely that a single monoclonal antibody will have a significant, lasting therapeutic effect. A better approach would be to use multiple monoclonal antibodies directed against different ganglioside antigens, or to stimulate the patient's immune system to manufacture its own specific antibodies against the ganglioside antigens of tumor cells.

Immunosuppressive effects of gangliosides shed from tumor cells: implications for active immunotherapy

Several lines of evidence indicate that gangliosides modulate immune functions. Among the documented immunoregulatory properties of gangliosides are: (1) inhibition of normal human lymphoproliferative responses to mitogens and antigens (Miller and Esselman, 1975; Lengle et al., 1979; Ryan and Shinitzky, 1979; Whisler and Yates, 1980; Gonwa et al., 1984; Ladisch et al., 1984, 1991; Prokazova et al., 1988); (2) inhibition of interleukin-2 dependent cell proliferation (Merritt et al., 1984); (3) inhibition of cytotoxic effector functions (Prokazova et al., 1988); and (4) binding to helper cells and modulated expression of CD4 (Offner et al., 1987; Kawaguchi et al., 1989; Morrison et al., 1989). Gangliosides have also been shown to modulate the humoral immune response (Agarwal and Neter, 1971; Esselman and Miller, 1977).

The immunomodulatory effect of the melanoma gangliosides was first described in our laboratory, and is currently the subject of several investigations. Hoon et al. (1988) in our laboratory observed that GM3 and GD3 inhibited the proliferation of human lymphocytes induced by phytohemagglutinin (PHA) and interleukin-2. Similarly, Portoukalian (1989) observed that gangliosides, particularly GM3, significantly inhibited the concanavalin A-induced proliferation of peripheral blood lymphocytes. The lymphocytes of melanoma patients seemed to be more sensitive to the influence of exogenous gangliosides than did those of healthy controls. The same was true for mixed lymphocyte reactions.

Portoukalian et al. (1978) has further demonstrated that both production and activity of interleukin-1 (IL-1) are markedly inhibited by melanoma gangliosides. He suggests that ganglioside-induced inhibition may be related to the production of prostaglandins by macrophages, which reportedly inhibit IL-1. In support of this is his observation that the incubation of monocytes with GM3 purified from melanoma cells strongly enhanced the release of arachidonic acid, the direct precursor of prostaglandins.

Ando and co-workers (1987a,b) in our laboratory reported several pieces of evidence highlighting the role of the melanoma-associated ganglioside GM2 in natural killer (NK) recognition of target cells. First, only target cells with a high GM2 content showed a high susceptibility to lysis by NK cells; this correlation did not exist for other gangliosides. Second, GM2 and GM3 inhibited the cytotoxicity of NK cells for several cell lines (K562, SKW-4, U937, and Molt 4), particularly in the presence of short-chain fatty acids purified from K562. Finally, GM2 purified

from melanoma cell line M14 also prevented the binding of NK cells. We also found GM3 and GM2 to be effective inhibitory agents of lymphokine-activated killer (LAK) cell cytotoxicity (Hoon et al., 1989).

Prokazova and co-workers (1987, 1988) showed that GD3 strongly suppressed PHA-induced lymphocyte activation of peripheral blood lymphocytes, and that GM3 actively inhibited lectin-induced lymphocyte proliferation by stimulating T-cell suppressor activity. They also tested the effect of exogenous gangliosides on NK cell cytotoxicity: when incubated with NK cells prior to the addition of target tumor cells, GD3 and GM3 had a remarkably high, dose-dependent inhibitory effect on NK cytotoxicity. This effect was maximal in cancer patients with high serum concentrations of gangliosides, a finding that supports an immunosuppressive role for shed gangliosides.

Takahashi et al. (1988) provided a new line of evidence for preferential immune suppression by soluble gangliosides. They demonstrated that GM3 may form an immunosuppressive complex with proteins. The GM3-protein complex purified from the spent culture medium of B16 melanoma (derived from C57BL/6) effectively augmented melanoma cell growth, inhibited antimelanoma cytotoxic lymphocyte (CTL) activity, and induced specific T-suppressor cells that blocked CTL generation in the induction phase. Interestingly, the antimelanoma T-suppressor cells recognized NeuAc-GM2 but not NeuGl-GM3. Based on these findings, they proposed that tumor cells can escape immune surveillance by stimulating the repertoire of T-suppressor cells for GM3. These studies point out that tumor gangliosides existing free or in micellar form or in a complex with proteins can be detrimental to the normal functioning of different compartments of the immune system. Therefore, one of the goals of ganglioside-targeted active immunotherapy should be to reduce the concentration of shed gangliosides in the tumor vicinity, in the lymph nodes, and in the circulation.

In neuroblastoma, another tumor of neuroectodermal origin, the level of shed ganglioside(s) may vary with the stage of tumor progression. Ladisch et al.

(1987) observed that the mean serum GD2 concentration was 603 ± 136 pmol/ml in patients with widespread disease (stages III and IV), compared to 198 ± 54 pmol/ml in the circulation of patients with localized disease (stages I and II); this suggests that mean ganglioside values could be proportional to the clinical stage of tumor progression. Although there are no comparable studies on serum ganglioside concentrations during the clinical course of melanoma, Portoukalian et al. (1978) reported that plasma GD3 concentration was significantly lower after surgical reduction of tumor burden in melanoma patients. Also, we recently documented a correlation between GM3:GD3 biopsy ratio and survival (Ravindranath et al., 1991). These findings warrant a careful study of gangliosides derived from the sera of patients in various clinical stages of melanoma.

The dose-dependent immunosuppression by gangliosides, the relationship between circulating ganglioside levels and tumor burden, and the possible correlations between circulating ganglioside levels and clinical stage of disease all underline the importance of clearing shed gangliosides from the circulation. Thus, antiganglioside antibodies induced by active immunotherapy may have the dual function of clearing immunosuppressive gangliosides from the circulation and the vicinity of the tumor.

Ganglioside-targeted active specific immunotherapy

To circumvent the problem of antigenic modulation, an ideal vaccine for melanoma should express most of the melanoma-associated antigens (MAA). Also, to facilitate antigen presentation, it should be HLA-compatible with the recipient (in the case of protein-associated tumor antigens). For this reason, autologous tumor cells might be preferred. However, the difficulty of obtaining sufficient quantities of autologous tumor cells from biopsies, and the failure to routinely develop melanoma cell lines from a number of tumor biopsies have limited the scope of an autologous vaccine. For these reasons, the best way to maximize the host's immune response to autolo-

gous melanoma is to formulate allogeneic cell vaccines that share HLA antigens with the host and optimally express most of the melanoma-associated tumor antigens, which include proteins as well as gangliosides.

Clinical trials with purified ganglioside (Livingston et al., 1987) showed that the ganglioside alone does not possess intrinsic adjuvanticity. Purified gangliosides such as GD3 fail to elicit an antibody response in humans, even after combination with BCG or *Salmonella minnesota* R595. However, five of six patients immunized with a whole-cell vaccine rich in GM2, and ten of 12 patients pretreated with cyclophosphamide (cy) and immunized with GM2 combined with BCG, produced high titers of anti-GM2 IgM antibodies. Anti-GM2 antibodies produced by the GM2-BCG combination were highly cytotoxic for GM2-positive tumor cells in the presence of complement. At 1 year, the recurrence rate of 17 patients producing high titers of anti-GM2 IgM antibodies in response to vaccination was significantly less than the recurrence rate in patients who were antibody-negative after immunization. Moreover, high titers of anti-GM2 antibody were associated with improved survival (Livingston et al., 1989).

In our laboratory, we screened the ganglioside profiles of a variety of melanoma cell lines, in order to prepare a suitable vaccine. Our initial vaccine, TCV, contained three melanoma cell lines: M7, M12, and M14 (Table III, first column). We measured the serum titers of antiganglioside IgM and IgG in 25 stage II (lymph-node positive) melanoma patients who received TCV intradermally or intralymphatically (Tai et al., 1985). Immunization with TCV enhanced anti-GM2 IgM titers in three patients and induced anti-GM2 IgM in seven others; two patients developed anti-GM2 IgG antibodies. Also, two patients developed anti-GD2 IgM antibodies.

Interestingly, neither our group (Tai et al., 1985) nor Livingston's group (1989) documented antibodies against the major melanoma gangliosides, GM3 and GD3, which suggests that these gangliosides may not be immunogenic, or that the antibodies if formed may occur as immune complexes. Indeed, gangliosides have been identified in serum-immune com-

TABLE III

Distribution of gangliosides in melanoma cell vaccines (minor gangliosides such as GT1b, GD1a, and GM1 are excluded)[a]

| | Ganglioside concentration ($\mu g/g$) | | |
	Vaccines: M7/M12/M14	M7/M20/M14	M10/M24/M101
GM3	57	44	34
GM2	15	14	33
GD3	58	47	38
GD2	19	18	17
O-AcGD3	03	01	04
Lipid-bound			
sialic acid ($\mu g/g$)	153	124	131
GM3:GM2:GD3:GD2 ratio	4:1:4:1	3:1:3:1	2:2:2:1

[a]Values determined as described by Ledeen and Yu (1982).

plexes from tumor-bearing patients (Hakansson et al., 1985).

We later demonstrated that melanoma patients who received TCV developed autoantibodies against a derivative of GD3, O-AcGD3 (a minor ganglioside expressed on human melanoma cells), and that these antibodies cross-reacted with GD3 (Ravindranath et al., 1988, 1989). None of the sera responded only to GD3, although the vaccine contained seven- to twelvefold higher levels of GD3 than O-AcGD3. Furthermore, anti-GD3 activity was completely abolished by absorption with animal erythrocytes expressing O-acetylated disialoganglioside. Endoglycoceramidase treatment of GD3 showed the epitope to be the sialyl-oligosaccharide moiety. Periodate oxidation abolished antibody reactivity to GD3 but not to O-AcGD3, revealing that the glycerol side chain of the sialic acids in both was an important determinant of this epitope.

These results suggest that O-AcGD3 may be the immunogen responsible for anti-GD3 antibodies. In fact, of the five gangliosides commonly expressed by melanoma, GM2 and 9-OAcGD3 appear to be the most immunogenic in man, followed by GD2. GD3 and GM3 may not be very immunogenic in humans per se. Alternatively, these two gangliosides may be

very immunogenic, producing antibodies which form immune complexes due to the abundance of these gangliosides in serum; this would explain why the antibodies are normally not detectable by in vitro assays.

Phase II trial of active specific immunotherapy with a new polyvalent melanoma cell vaccine

In 1985 we began a phase II trial to evaluate a new polyvalent melanoma cell vaccine (MCV) in patients with advanced melanoma metastatic to regional skin and subcutaneous sites (AJCC stage IIIA) or distant sites (AJCC stage IV). Our results, outlined in the study described below, show that patients receiving MCV survive significantly longer than those receiving other regimens of immunotherapy or chemotherapy (Morton et al., 1992). Moreover, the level of humoral antibody and/or cell-mediated immunity to melanoma membrane antigens or the ganglioside GM2, as induced by MCV, appears to be directly related to survival time.

Patients and methods
Control group. To establish a reliable control group, we reviewed the clinical records of all melanoma

patients treated by John Wayne Cancer Institute (JWCI) staff during the 20-year period between April 1, 1971, and November 1991. We found that 1275 patients had melanoma recurring to a distant site (AJCC stage IV), specifically the brain, liver, lung, bone, GI, skin, soft tissue or distant nodes.

Because the JWCI staff has been uniquely stable, patients in this retrospective 20-year review were managed by uniform work-up and treatment criteria. Patients with solitary metastases in the skin and sub-cutaneous tissues were treated with intralesional BCG or human monoclonal antibody to ganglioside antigens. If BCG did not induce complete regression, patients were treated by excision or hyperthermic perfusion (Morton et al., 1974, 1991b; Storm et al., 1979). Those with metastases to visceral sites were usually managed with chemotherapy, but some underwent surgical resection followed by immunotherapy either with BCG by the tine tech-nique (Morton et al., 1974; Morton, 1986a, 1986b) or with a prior TCV (Morton et al., 1978). Patients whose disease progressed while on immunotherapy were treated with systemic chemotherapy.

MCV immunotherapy patients. Eligible patients were those who had regional (AJCC stage IIIA) or remote (AJCC stage IV) soft-tissue or visceral metastases, either with objectively measurable disease, or with no evidence of disease following excisional biopsy or resection of metastatic lesions. No candidate for MCV treatment had received immuno-, chemo-, or radiotherapy within 30 days. Patients with brain metastases were not considered unless their metas-tases had been resected, or brain radiation had been completed, and they had stopped taking immunosup-pressive steroid medications for treatment of brain edema for at least 30 days.

During the study period, 136 patients received MCV. Of these, 61 had stage IIIA melanoma, and 75 had stage IV melanoma.

MCV treatment protocol. Patients were stratified by stage and disease status, and randomly administered either MCV alone or MCV plus one of the biologic response modifiers (BRMs) that downregulate sup-pressor cell activity (Morton et al., 1989a,b; Hoon et al., 1990). These BRMs included cimetidine (CIM; 1200 mg/day) (Smith/Kline, PA); indomethacin (IND; 150 mg/day) (Lederle, NJ); or low-dose cyclophosphamide (CYP; 75, 150 or 300 mg/m²) (Johnson/Mead, NJ). Since the clinical results of active immunotherapy with vaccine alone were simi-lar to those for patients receiving BRMs, the data have been pooled for this analysis.

MCV consisted of three human melanoma cell lines (M10, M24, and M101; Table III, column 3), which were selected from a series of melanoma cell lines after careful examination for the high expres-sion of ganglioside antigens and other MAA immunogenic in melanoma patients (Table IV). Cells were grown and prepared for administration, as pre-viously described (Hoon et al., 1990). MCV was pro-duced in large batches and analyzed for MAA expression to determine variance between lots. An outside laboratory screened the MCV for viral (HIV, hepatitis), bacterial, and fungal infectious organisms. Equal amounts of each line were pooled to a total of 24×10^6 cells in serum-free medium containing 10% dimethylsulfoxide and frozen in liquid nitrogen. Before administration, the cells were irradiated to 100 Gy.

Immediately prior to treatment, MCV was thawed and washed three times in phosphate-buffered saline. The vaccine was then injected intradermally in axil-lary and inguinal regions every 2 weeks for 6 weeks, and then every month for 1 year. For the first two treatments, MCV was mixed with BCG (Glaxo, England) (24×10^6 organisms/vial). (Since 1989 we have used tice strain BCG (8×10^6 organisms) due to non-availability of Glaxo BCG.) After 1 year, the immunization interval was increased to every 3 months for 1 year, and then every 6 months. Follow-up clinical and laboratory evaluations were repeated monthly, with chest X-rays every 3 months.

Laboratory evaluation. To evaluate the humoral anti-body and cell-mediated immune response to MCV therapy, patients were evaluated prior to and at monthly intervals following immunization. The fol-lowing assays were performed:

TABLE IV

Immunogenic melanoma-associated antigens of MCV (melanoma cell lines M10, M24, and M101)

Antigen	Antibody response Ig Class[a]	Cytotoxicity[b]
Ganglioside		
GM2	IgM	Yes
GD2	IgM	Yes
O-Acetylated GD3	IgM	Yes
Glycoprotein		
Fetal Ag (69.5 kd)	IgG, IgM	Yes
Urine Ag (90 kd)	IgG, IgM	Yes
Lipoprotein		
M-TAA (180 kd)	IgG, IgM	Yes

[a]Predominant Ig class in sera.

[b]Antibody-dependent complement cytolysis.

Humoral immune response. The antibody response to melanoma cell-surface antigens following MCV immunization was evaluated by indirect membrane immunofluorescence (IMIF) assay as previously described (Morton et al., 1968). Sera were tested against the M14 melanoma cell line after preabsorption with matched lymphoblastoid cells autologous to the test melanoma line, in order to remove antibodies to HLA antigens (Irie et al., 1979a). Autologous melanoma cell lines were established from patient biopsy specimens obtained prior to therapy and were substituted for M14 in IMIF when available. The antibody response to specific ganglioside antigens was monitored by a modification of the enzyme-linked immunosorbent assay (ELISA) (Tai et al., 1984).

Delayed cutaneous hypersensitivity (DCH). Intradermal skin tests with MCV were performed before and during therapy. One-tenth of the pooled MCV (2.4×10^6 cells) was administered at a remote site on the forearm. After 48 h, the average diameter of the induration was recorded as the DCH response. The Student *t*-test was used to compare the absolute values of DCH from week zero to weeks four and 16.

General immunocompetence was evaluated by sensitization and challenge to DNCB and response to common skin test antigens such as mumps and *Candida*. The responses to purified protein derivative (PPD) antigen to which the patient became sensitized as a result of immunization with BCG in the vaccine served as additional controls.

Statistical methods. Survival time was defined as the time a patient remained alive after the documented date of regional skin and soft-tissue metastases (stage IIIA) or site-specific distant metastases (stage IV). The estimated survival rates were obtained by the non-parametric Kaplan-Meier method and have previously been described in detail (Morton et al., 1992).

Results

Humoral and cell-mediated immune assays showed that a specific antimelanoma response was often induced within 2 weeks, peaking in 4–8 weeks and then gradually declining in most patients to a level significantly above the preimmunization baseline.

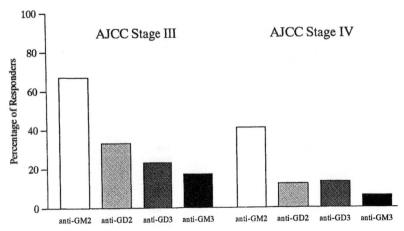

Fig. 12. The antiganglioside IgM response to MCV in AJCC stage III and IV patients. Responder postimmune sera show twofold or more increase in ELISA absorbency (compared to preimmune sera) for antibodies to that particular ganglioside, within 12 weeks after immunization.

Antiganglioside antibody detected by ELISA. Figure 12 shows the antiganglioside IgM response to MCV in 12 patients with stage III melanoma and 17 patients with stage IV melanoma. Using a criterion of 0.200 or higher in a 1:300 serum dilution, few patients had pre-existing antiganglioside antibodies prior to immunization. If MCV responders are defined as those with a twofold or more rise in

ELISA absorbency within 12 weeks of immunization (compared to preimmune serum levels), then the most immunogenic ganglioside was GM2, with 69% of stage III and 41% of stage IV patients responding. As expected, GD2 was second in immunogenicity, followed closely by GD3. Antibodies to 9-OAcGD3 were not tested in this study; however, some of the anti-GD3 reactivity may have been due to cross-reac-

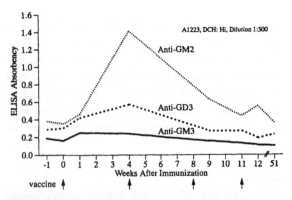

Fig. 13. The antiganglioside antibody response to melanoma cell vaccine in a melanoma patient (AJCC stage III) who has survived more than 5 years. Antibody response to GM2 is significantly higher than other antiganglioside antibodies. Our observations indicate that anti-GM2 antibody may have prognostic relevance.

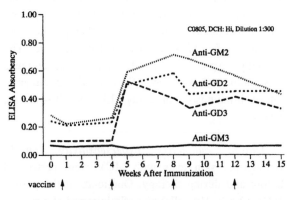

Fig. 14. The antiganglioside antibody response to melanoma cell vaccine in a melanoma patient (AJCC stage IV) who survived more than 36 months. Note that antibodies to GM2, GD2, and GD3 are much more pronounced than anti-GM3 antibody.

tive antibodies between 9-OAcGD3 and GD3. Surprisingly, we found that 17% of the stage III patients developed antibodies to GM3. Both the incidence and titer of antibodies were higher with MCV than its predecessor TCV (Tai et al., 1985).

The antiganglioside titers of two representative patients are illustrated in Figs 13 and 14. IgM titers against most ganglioside antigens peaked within 4 weeks of initial immunization, although occasionally the peak was delayed eight to 12 weeks.

The association between anti-GM2 IgM antibody and survival was significant ($P < 0.04$). Thirteen (87%) of 15 high responders survived longer than 18 months, compared to five (42%) of 12 patients having a low response to MCV. (A high antibody response was defined as a twofold increase in anti-GM2 titer within 12 weeks following immunotherapy.) Although this must be confirmed by study of additional patients, these results with MCV appear to confirm our previous observations (Jones et al., 1981) and those of Livingston et al. (1989).

Humoral immune response to whole melanoma cells. Anti-MAA IgM correlated best with survival, as previously discussed (Jones et al., 1981). There was no significant correlation between anti-MAA IgG and survival. As shown in Table V, 62% of patients immunized with MCV developed IgM directed

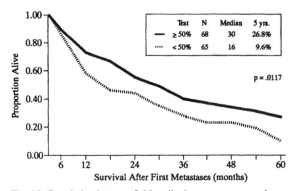

Fig. 15. Correlation between IgM antibody response to membrane melanoma-associated antigens and survival following active immunotherapy with melanoma vaccine. Reprinted with permission from Morton et al. (1992), page 468.

against the MAA of M14 melanoma cells. This was significantly better than the 35% response to our previous TCV (Morton et al., 1978). In 26 patients receiving MCV, autologous melanoma cells were available for use as targets in the in vitro assay. Prior to active immunotherapy with MCV, four of these 26 patients had pre-existing antibody to autologous melanoma cells, at titers below 1:10. Following immunotherapy, 17 of the 26 patients exhibited antibodies to autologous MAA, at a mean titer of 1:29. This was similar to the response observed against

TABLE V

Comparison of the IgM antibody response to membrane-associated antigens on autologous and allogeneic melanoma cells following active immunotherapy with allogeneic MCV

Serum tested for IgM antibody[a]	Target cell lines	No. positive[b]/ No. tested	Mean titer
Preimmunization	M14 allogeneic	0/26	<1:10
Postimmunization	M14 allogeneic	16/26 (62%)	1:38
Preimmunization	Autologous	4/26 (15%)	<1:10
Postimmunization	Autologous	17/26 (65%)	1:29

[a]Sera were preabsorbed with L-14 lymphoblastoid cells, which are autologous to the M14 melanoma, to remove antibodies to HLA antigens prior to immunofluorescence assays.
[b]MIF index above 0.20.

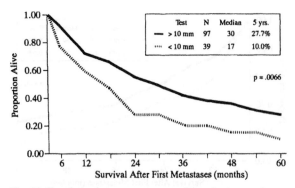

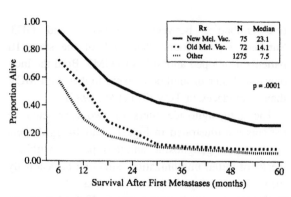

Fig. 16. Correlation between response to delayed cutaneous hypersensitivity tests and survival in advanced-stage melanoma patients receiving melanoma vaccine. Reprinted with permission from Morton et al. (1992), page 468.

Fig. 17. Survival of stage IV melanoma patients treated with old and new vaccines versus other therapy.

M14-associated MAA. Although the M14 melanoma cell line is not a component of MCV, most patients developed increases in anti-M14 antibody which matched the increases in antibody to autologous melanoma cells. These data clearly indicate sharing of common MAA between allogeneic and autologous melanoma cells, thus confirming our previous observations for the humoral response to allogeneic TCV (Irie et al., 1979a).

High levels of antimelanoma antibodies (membrane immunofluorescence indices above 50%) were associated with significant ($P < 0.01$) improvement in survival, as illustrated in Fig. 15 and previously reported with TCV (Jones et al., 1981). For these patients, the rate of 5-year survival increased almost threefold (9.6–26.8%), while median survival time increased twofold (16–30 months).

Delayed cutaneous hypersensitivity reactions to melanoma vaccine. Most patients were judged generally immunocompetent by their response to PPD, DNCB, and/or common skin test antigens.

When all patients were stratified by their maximum skin test reactions to MCV, there was a highly significant ($P = 0.0066$) correlation between survival following treatment and a DCH reaction >10 mm during the first 12 weeks after initial MCV therapy. The median survival was 30 months for those >10.0

mm and only 17 months for those <10.0 mm (Fig. 16). Five-year survival increased from 10% to 27.7%.

Clinical results of patients with evaluable disease. Forty stage IV patients with evaluable metastatic disease were observed for 12 weeks after beginning MCV immunotherapy. Nine (22%) experienced regressions, three of which were complete (Morton et al., 1992). Although most of the 40 patients required chemotherapy and/or other follow-up treatment for progressive disease, we sometimes observed stabilization of metastatic growth. In some cases, objective assessment by tumor doubling time was possible (Morton et al., 1973); this showed clear reduction in the growth rate of metastatic disease.

Overall survival. Stage IV patients receiving active immunotherapy with MCV survived significantly longer than those treated by other methods, including TCV (Fig. 17). The median survival time of stage IV patients increased threefold, from 7.5 to 23.1 months (or from 14.1 to 23.1 months in those patients treated by surgery/chemotherapy or TCV); the rate of 5-year survival increased fourfold, from 6% for patients treated by other modalities and TCV, to 26% ($P = 0.0001$). Improvement occurred regardless of metastatic site.

TABLE VI

Normal human tissues in which GD3 represents more than 15% of total gangliosides

Tissue	Percent GD3	Reference
Milk[a]	53-74	Takamizawa et al., 1986
Retina	50	Holm et al., 1972
Dura mater	50	Berra et al., 1983
	39	Sunder-Plassman and Bernheimer, 1974
Thymus		
Fetal	22	Etienne-Decerf et al., 1986
2 months	31	Etienne-Decerf et al., 1986
Dermis	24	Tsuchida et al., 1984
Subcutaneous	26	Tsuchida et al., 1984
Thyroid	25	Bouchon et al., 1985
Aorta intima	23	Prokazova et al., 1987
Colonic mucosa	20	Hakomori et al., 1983
Kidney	19	Rauvala, 1976

[a]Days 2-40 of lactation.

Discussion

One of the perplexing questions raised by our current findings relates to the observation that some melanoma patients form antibodies to GD3 and GM3 after MCV immunization. Since normal tissues and human sera both contain substantial amounts of these ganglioside antigens (Tables II and VI), we assumed that they would not be immunogenic. If this were not the case, then the antibodies subsequently produced would be neutralized by binding to circulating or tissue-bound GD3 or GM3, resulting in a negative ELISA.

However, our prior results and present investigations indicate that this assumption may not be valid. In patients immunized with MCV, the O-AcGD3 antigen of MCV induces an antibody that cross-reacts with GD3 (Ravindranath et al., 1989). Moreover, we have produced a human monoclonal antibody to GM3 from the peripheral lymphocytes of immunized patients (Yamamoto et al., 1990). Also, our recent preliminary findings indicate that 17% of AJCC stage III patients are capable of responding to MCV with antibodies to GM3.

The apparent paradox may be explained in several ways. GM3 and GD3 may have a cryptic form of expression on normal cell membranes and sera, so that they are not accessible to the antibody. Also, there may be subtle differences in ganglioside epitopes between normal tissues and melanoma cells, making melanoma gangliosides more immunogenic to the human immune system. This might be a result of heterogeneity in the ceramide portion, particularly the fatty acid or carbohydrate moiety (Nudelman et al., 1982); alternatively, the response to GM3 may be induced by a GM3-like antigen, as suggested by the interesting work of Takahashi et al. (1988) with the B16 melanoma of mice. Finally, Nores et al. (1987) have pointed out the importance of increased ganglioside GM3 density on the cell surface as a factor determining the specificity of a murine monoclonal antibody for mouse melanoma B16, as compared to normal mouse cells. This is obviously an important area for further study, since the presence of substantial amounts of both GM3 and GD3 on human melanoma cells makes these gangliosides important potential targets of active immunotherapy — if their immunogenicity can be increased.

Clinical implications

At the John Wayne Cancer Institute, our primary research goal during the past 25 years has been to develop more effective methods for the active specific immunotherapy of melanoma.

Our clinical phase II study clearly indicates that MCV significantly prolongs the survival of patients with AJCC stage IV metastatic melanoma when compared to other therapies, including our previous TCV (Morton et al., 1978). Differences in patient mix and/or time intervals of treatment cannot account for survival differences between control and MCV groups. In fact, when all patients were analyzed by univariate and multivariate analysis, only two factors significantly influenced prognosis: initial site of metastasis and MCV immunotherapy.

If melanoma metastasizes to distant sites, average life-expectancy drops to less than a year (Balch et al., 1982; Berdeaux et al., 1985), and long-term survival is extremely rare. In a series from the Mayo Clinic reviewed by Ahman et al. (1989), there were only ten (2%) 5-year survivors among 502 patients with advanced melanoma treated by chemotherapy.

High response rates initially reported for various chemotherapy regimens quite often cannot be reproduced or translated into prolonged survival. Usually these reports are based on small groups of patients, the majority of whom usually have skin and subcutaneous metastases — sites more responsive to therapy than are visceral sites. By contrast, our control and treatment groups were drawn from a huge melanoma database that has been remarkably constant for 20 years. Extensive statistical review failed to detect any bias favoring MCV treatment.

By all objective criteria, our results consistently indicate that MCV immunization significantly enhances survival of stage IV melanoma patients, by a factor of threefold or fourfold. However, even if a subsequent stratified and randomized phase III trial further validates MCV immunization as superior to the best currently available alternative therapy, it is clear that MCV is not yet the optimal melanoma vaccine. We are now attempting to increase its immunogenicity through cytokine modulation, biologic

response modifiers, and other mechanisms (Roth et al., 1978; Hoover et al., 1985; Wallack et al., 1986; Bystryn J-C et al., 1988; Mitchell et al., 1988; Naito et al., 1988; Rosenberg et al., 1989; Arroyo et al., 1990; Hoon et al., 1991a–c; Ravindranath and Morton, 1991; Shibata et al., 1992). Chemical modification of non-immunogenic gangliosides or the admixture of extrinsic adjuvants with a whole-cell vaccine could be invaluable for augmenting antiganglioside antibody response in melanoma patients (Ravindranath et al., unpublished observations). Also, the composition of gangliosides in the vaccine might be altered by further changing the cell lines.

It is possible that MCV may prove useful for active immunotherapy of non-melanoma cancers, since many of MCV's ganglioside and protein antigens are present in other types of human neoplasms, particularly those of neural origin. Although 9-O-AcGD3 may be restricted to melanoma, it induces cross-reacting antibodies to GD3, which is more widely distributed in other types of human neoplasms.

The low toxicity of MCV might justify its use as the first treatment for recurrent melanoma, before considering more toxic regimens such as IL-2, lymphokine-activated killer cells or tumor-infiltrating lymphocytes (Rosenberg et al., 1988; Shibata et al., 1992), and chemotherapy (McClay and Mastrangelo, 1988), since the overall long-term survival associated with these regimens has not been superior to that following active immunotherapy with MCV. The low toxicity of MCV also makes it reasonable to consider using this vaccine as an adjuvant in stage II and III patients who are clinically free of disease following surgery.

Rationale for a ganglioside-targeted whole-cell MCV

Active immunotherapy against cancer became a realistic strategy with the demonstration that DCH reactions induced by the intralesional injection of BCG (an attenuated strain of *Mycobacterium*) (Morton et al., 1991a) resulted in the regression and eradication of injected cutaneous melanoma metastases and occasionally uninjected metastases. These reports rekindled interest in the concept of a vaccine for cancer and revived efforts to find the crucial formulas for effective vaccine therapy.

Cancer vaccines differ from vaccines against infectious diseases in that they are administered as therapy after the advent of disease, rather than prophylactically before the disease develops. However, the theory behind vaccines for cancer and infectious diseases is similar. Both seek to stimulate the patient's immune system through introduction of attenuated whole organisms or cells, specific subcellular antigens, and non-pathologic strains of living organisms or tumor cells (Morton, 1986a).

In animal tumor-host systems, whole-cell preparations are invariably more effective than extracts of tumor cells, soluble membrane antigens, or highly purified tumor antigens. For this reason, we developed our polyvalent vaccine from viable melanoma tumor cells grown in tissue culture but rendered incapable of prolonged growth by radiation. We did not pursue a single purified ganglioside vaccine because our data suggest that only 65%–80% of autologous melanomas will have high concentrations of any single immunogenic ganglioside antigen, whereas virtually all melanomas will have one or more of the three major ganglioside antigens of our MCV. Also, we do yet not know which of the melanoma ganglioside antigens are most important for host rejection and protective immunity; MCV may even have additional, unidentified ganglioside MAA important in antitumor immunity.

Use of a standardized allogeneic polyvalent melanoma cell vaccine is supported by extensive, if not conclusive, evidence that patients so treated become sensitized to their own melanoma as well as the allogeneic melanoma: our polyvalent MCV produces both humoral and cell-mediated in vivo immunity to MAA after two treatments; sensitization is maintained after five treatments. In vivo evidence for cross-reactivity is found in the strong correlation between the level of immune response and survival, and in the complete and partial regressions observed in patients with evaluable disease. In vitro evidence is manifest by the concomitant increase in reactivity to allogeneic and autologous melanomas using humoral antibody assays. Finally, our in vitro studies show that allogeneic melanoma cells can render autologous melanoma cells susceptible to cytotoxic T cells, as long as the target and stimulator cell lines share MHC Class I antigens (Hayashi et al., 1992a,b).

Potential problems and possible solutions associated with gangliosides as immunogens for active specific immunotherapy of melanoma

Successful use of gangliosides as target antigens for active specific immunotherapy assumes that several potential problems have been resolved (Table VII). First is the heterogeneous and often unstable expression of melanoma antigens within the same patient (Berd et al., 1991). If an immune attack is directed against only one antigen, natural selection or antigenic modulation may allow certain tumor cells to escape detection (Vadan-Raj et al., 1988). We circumvented this problem by selecting a polyvalent vaccine that induces a simultaneous immune response to multiple ganglioside-associated antigens, thereby blocking the escape of non-antigenic clones. Had we used a vaccine which was directed at only one MAA, it might have missed an antigen-negative tumor-cell clone.

A second potential problem is the weak immunogenicity of isolated and purified gangliosides in the host. This can be circumvented by developing a vaccine containing adequate amounts of highly immunogenic gangliosides, particularly GM2, GD2, and O-AcGD3 (which cross-reacts with GD3). Alternatively, Livingston's group (1989) joined highly immunogenic carriers and adjuvants to the ganglioside antigen.

T-cell independent immunity, which is typically induced by gangliosides, is a problem because it is not long-lasting. However, it should be possible to induce T-cell dependent immunity by using an anti-idiotypic antibody (anti-id) that mimics the ganglioside antigen which might be used as the immunogen. We have already developed a murine monoclonal anti-id which mimics GM3 (Yamamoto et al., 1990). This has been converted into a human chimeric anti-id (Hastings et al., 1992). This anti-id was recently

TABLE VII

Problems associated with ganglioside antigens as targets of active immunotherapy

---

1. Heterogeneous expression on tumor cells
        Solution: Generate broad-spectrum coverage using a polyvalent whole-cell vaccine

2. Short-term (T-cell independent) immunity
        Solution: Use beta-anti-ids to evoke long-lasting (T-cell dependent) immunity

3. Low or non-existent immunogenicity of purified gangliosides
        Solutions:
                  a. Use a vaccine containing high levels of gangliosides
                  b. Chemically modify gangliosides, e.g. GM3 to lactone GM3
                  c. Use adjuvants or carriers

4. Immunosuppression by shed gangliosides
        Solutions:
                  a. Decrease tumor burden (and therefore ganglioside levels) by cytoreductive surgery
                  b. Administer prostaglandin E inhibitors such as indomethacin

5. Mass production prohibitively expensive
        Solution: Use anti-id monoclonal antibodies that mimic ganglioside epitopes

---

found to stimulate in vitro proliferation of lymphocytes from immunized melanoma patients (Hoon et al., unpublished observations). Furthermore, we have found that intralesional injection of human monoclonal antibody induces anti-ids that activate the idiotypic network, resulting in production of anti-GD2 IgG, which we had not previously observed in melanoma patients (Irie et al., 1989a). Also, recent findings suggest that IgG anti-ids may be formed following induction of antiganglioside IgM in response to MCV active immunotherapy (Irie et al., unpublished observations).

Shed gangliosides may suppress immune function locally (at the tumor site) as well as systemically. Suppressive effects of the shed antigen can be reduced by cytoreductive surgery that decreases the tumor burden and thereby the level of circulating gangliosides. Activation of T-suppressor cells by circulating gangliosides or antigen-antibody complexes may be counteracted by neutralizing prostaglandin E with inhibitors such as indomethacin. Even so, immunosuppression secondary to shed gangliosides remains problematic.

From a production standpoint, the use of ganglioside immunogens is currently not very practical. Tumor-associated gangliosides exist at very low concentrations in nature, and ganglioside epitopes cannot be produced by gene technology or synthesized de novo. Producing large quantitites of a ganglioside-based vaccine therefore becomes quite expensive. A solution is to use anti-id monoclonal antibodies containing the internal image of tumor antigens. Large quantities of these anti-ids can be manufactured at a relatively low cost.

A final potential problem of active immunotherapy with ganglioside immunogens relates to the presence of many of these antigens in normal host tissues (Tables II and VI), and the possible induction of autoimmune disease by activation of antiganglioside humoral and cell-mediated immune responses. Such side effects have in fact been observed by the induction of severe pain in children receiving anti-GD2 murine monoclonal antibody treatment for neuroblastoma. Fortunately, autoimmune disease has thus far not been an observed complication of active immunotherapy with MCV, for reasons that are not really understood.

In conclusion, neoplastic transformation of normal melanocytes is accompanied by a progressive increase in the most immunogenic gangliosides: GM2, GD2, and O-AcGD3. This serendipitous finding has allowed us to successfully attack melanoma cells using a polyvalent whole-cell vaccine directed against these ganglioside antigens. The promising results of our phase II trial encourage us to continue our investigations; melanoma gangliosides are very attractive immunogens for active specific immunotherapy of metastatic disease.

Acknowledgements

Supported by grants (CA 12582, CA 30647, and CA 29605) from the National Cancer Institute, DHHS, the Ben and Joyce Eisenberg Fund, and the Steele Foundation. These investigations were conducted with permission of Human Subjects Protection Committees, the John Wayne Cancer Institute, and the Jonsson Comprehensive Cancer Center.

References

Agarwal, M.K. and Neter, E. (1971) Effect of selected lipids and surfactants on immunogenicity of several bacterial antigens. *J. Immunol.*, 107: 1448–1456.

Ahman, D.L., Creagan, E.T., Hahn, R.G., Edmonson, J.H., Bisel, H.F. and Schaid D.J. (1989) Complete responses and long-term survivals after systemic chemotherapy for patients with advanced malignant melanoma. *Cancer*, 63: 224–227.

Ando, I., Hoon, D.S.B., Pattengale, P.K., Golub, S.H. and Irie, R.F. (1987a) Ganglioside GM2 as a target structure recognized by human natural killer cells. *J. Clin. Lab. Anal.*, 1: 209–213.

Ando, I., Hoon D.S.B., Suzuki, Y., Saxton, R.E., Golub, S.H. and Irie, R.F. (1987b) Ganglioside GM2 on the K562 cell line is recognized as a target structure by human natural killer cells. *Int. J. Cancer*, 40: 12–17.

Ando, S. (1983) Gangliosides in the nervous system. *Neurochem. Int.*, 5: 507–537.

Arroyo, P.J., Bash, J.A. and Wallack, M.K. (1990) Active specific immunotherapy with vaccine colon oncolysate enhances the immunomodulatory and antitumor effects of interleukin-2 and interferon-alpha in a murine hepatic metastasis model. *Cancer*

Immunol. Immunother., 31: 305–311.

Balch, C.M., Soong, S-J., Murad, T.M., Casidenti, A. and Paulson J.C. (1982) A multifactorial analysis of melanoma: IV. Prognostic factors in 200 melanoma patients with distant metastases (stage III). *J. Clin. Oncol.*, 1: 126–134.

Berd, D., Murphy, G., Maguire, H. and Mastrangelo, M. (1991) Immunization with haptenized, autologous tumor cells induces inflammation of human melanoma metastases. *Cancer Res.*, 51: 2731–2734.

Berdeaux, D.H., Moon, T.E. and Meyskens Jr., F.L. (1985) Clinical-biologic patterns of metastatic melanoma and their effect on treatment. *Cancer Treat. Rep.*, 69: 397–401.

Berra, B., Riboni, L., De Gasperi, R., Gaini, S.M. and Ragnotti, G. (1983) Modifications of ganglioside patterns in human meningiomas. *J. Neurochem.*, 40: 777–782.

Bouchon, B., Portoukalian, J. and Bornet, H. (1985) Major gangliosides in normal and pathological human thyroids. *Biochem. Int.*, 10: 531–538.

Bystryn, J-C., Oratz, R., Harris, M.N., Roses, D.F., Golomb, F.M. and Speyer J.L. (1988) Immunogenicity of a polyvalent melanoma antigen vaccine in humans. *Cancer*, 61: 1065–1070.

Cahan, L.D., Irie, R.F., Singh R., Cassidenti, R. and Paulson, J.C. (1982) Identification of a human neuroectodermal tumor antigen (OFA-1-2) as ganglioside GD2. *Proc. Natl. Acad. Sci. USA*, 79: 7629–7633.

Carubia, J.M., Yu, R.K., Macala, L.J., Kirkwood, J.M. and Varga, J.M. (1984) Gangliosides of normal and neoplastic human melanocytes. *Biochem. Biophys. Res. Commun.*, 120: 500–504.

Cheresh, D.A. and Klier, F.G. (1986) Disialoganglioside GD3 distributes preferentially into substrate associated microprocesses on human melanoma cells during their attachment to fibronectin. *J. Cell Biol.*, 102: 1887–1897.

Cheresh, D.A., Pierschbacher, M.D., Herzig, M.A. and Mujoo, K. (1986) Disialogangliosides GD2 and GD3 are involved in the attachment of human melanoma and neuroblastoma cells to extracellular matrix proteins. *J. Cell Biol.*, 102: 688–696.

Dreyfus, H., Ferret, B., Harth, S., Gorio, A., Durand, M., Freysz, L. and Massarelli, R. (1984) Metabolism and function of gangliosides in developing neurons. *J. Neurosci. Res.*, 12: 311–322.

Esselman, W.J. and Miller, H.C. (1977) Modulation of B cell responses by glycolipid released from antigen-stimulated T cells. *J. Immunol.*, 19: 1994–2000.

Etienne-Decerf, J., Gosselin-Rey, C., Gosselin, L. and Winand, R. (1986) Effects of thyrotropin and thyroid stimulating immunoglobulins on ganglioside labelling of human thymus cultured pathological cells. *J. Endocrinol. Invest.*, 9: 57–63.

Gonwa, T.A., Westrick, M.A. and Macher, B.A. (1984) Inhibition of mitogen- and antigen-induced lymphocyte activation by human leukemia cell gangliosides. *Cancer Res.*, 44: 3467–3470.

Hakansson, L., Fredman, P. and Svennerholm, L. (1985) Gangliosides in serum immune complexes from tumor-bearing patients. *J. Biochem.*, 98: 843–849.

Hastings, A., Morrison, S.L., Kanda, S., Saxton, R.D. and Irie, R.F. (1992) Production and characterization of a murine/human chimeric anti-idiotype antibody that mimics ganglioside. *Cancer Res.*, 52: 1681–1686.

Hayashi, Y., Hoon, D.S.B., Park, M.S., Terasaki, P.I., Foshag, L.J. and Morton, D.L. (1992a) Induction of CD4$^+$ cytotoxic T cells by sensitization with allogeneic melanomas bearing shared or cross-reactive HLA-A. *Cell. Immunol.*, 139: 411–425.

Hayashi, Y., Hoon, D.S.B., Park, M.S., Terasaki, P.I. and Morton, D.L. (1992b) Cytotoxic T cell lines recognize autologous and allogeneic melanomas with shared or cross-reactive HLA-A. *Cancer Immunol. Immunother.*, 34: 419–423.

Herlyn, M., Thurin, J., Balaban, G., Bennicelli, J.L., Herlyn, D., Elder, D.E., Bondi, E., Guerry, D., Nowell, P., Clark, W.H. and Koprowski, H. (1985) Characteristics of cultured human melanocytes isolated from different stages of tumor progression. *Cancer Res.*, 45: 5670–5676.

Holm, M., Mansson, J-E., Vanier, M-T. and Svennerholm, L. (1972) Gangliosides of human, bovine and rabbit retina. *Biochim. Biophys. Acta.*, 280: 356–364.

Hoon, D.S.B., Irie, R.F. and Cochran, A.J. (1988) Gangliosides from human melanoma immunomodulate response of T-cells to IL-2. *Cell. Immunol.*, 111: 1–10.

Hoon, D.S.B., Ando, I., Sviland, G., Tsuchida, T., Okun, E., Morton, D.L. and Irie, R.F. (1989) Ganglioside GM2 expression of human melanoma cells correlates with sensitivity to lymphokine activated killer cells. *Int. J. Cancer*, 43: 857–862.

Hoon, D.S.B., Foshag, L.J., Nizze, A.S., Bohman, R. and Morton D.L. (1990) Suppressor cell activity in a randomized trial of patients receiving active specific immunotherapy with melanoma cell vaccine and low dosages of cyclophosphamide. *Cancer Res.*, 50: 5358–5364.

Hoon, D.S.B., Banez, M., Okun, E., Morton, D.L. and Irie, R.F. (1991a) Modulation of human melanoma cells by interleukin-4 and in combination with gamma-interferon or alpha-tumor necrosis factor. *Cancer Res.*, 51: 2002–2008.

Hoon, D.S.B., Hayashi, Y., Banez, M. and Morton, D.L. (1991b) Specific cytotoxic T cells (CTL) to human melanoma. *Proc. Am. Assoc. Cancer Res.*, 32: 236.

Hoon, D.S.B., Okun, E., Banez, M., Irie, R.F. and Morton, D.L. (1991c) Interleukin-4 alone and with gamma-interferon or alpha-tumor necrosis factor inhibits cell growth and modulates cell surface antigens on human renal cell carcinomas. *Cancer Res.*, 51: 5687–5693.

Hoover Jr., H.C., Surdyke, M.G., Dengel, R.B., Peters, L.C. and Hanna Jr., M.C. (1985) Prospectively randomized trial of adjuvant active specific immunotherapy for human colorectal cancer. *Cancer*, 55: 1236–1243.

Irie, R.F., Irie, K. and Morton, D.L. (1976) A membrane antigen common to human cancer and fetal brain tissues. *Cancer Res.*, 36: 3510–3517.

Irie, K., Irie, R.F. and Morton, D.L. (1979a) Humoral immune response to melanoma-associated membrane antigen and fetal brain antigen demonstrable by indirect membrane immunofluorescence. *Cancer Immunol. Immunother.*, 6: 33–39.

Irie, R.F., Giuliano, A.E. and Morton, D.L. (1979b) Oncofetal anti-gen (OFA): a tumor-associated fetal antigen immunogenic in man. *J. Natl. Cancer Inst.*, 63: 367–373.

Irie, R.F., Jones, P.C., Morton, D.L. and Sidell, N. (1981) In vitro production of human antibody to a tumor-associated fetal antigen. *Br. J. Cancer*, 44: 262–266.

Irie, R.F., Sze, L.L. and Saxton, R.E. (1982) Human antibody to OFA-1, a tumor antigen, produced in vitro by EBV-transformed human B-lymphoblastoid cell lines. *Proc. Natl. Acad. Sci. USA*, 79: 5666–5670.

Irie, R.F. and Morton, D.L. (1986) Regression of cutaneous metastatic melanoma by intralesional injection with human monoclonal antibody to ganglioside GD2. *Proc. Natl. Acad. Sci. USA*, 83: 8694–8698.

Irie R.F., Chandler P.J. and Morton D.L. (1989a) Melanoma gangliosides and human monoclonal antibody. In R.S. Metzgar and M.S. Mitchell (Eds.), *Human Tumor Antigens and Specific Tumor Therapy*, Alan R. Liss Inc., New York, pp. 115–126.

Irie, R.F., Matsuki, T. and Morton, D.L. (1989b) Human monoclonal antibody to ganglioside GM2 for melanoma treatment. *Lancet*, 1: 786–787.

Jones P.C., Sze L.L., Liu P.Y., Morton D.L. and Irie R.F. (1981) Prolonged survival for melanoma patients with elevated IgM antibody to oncofetal antigen. *J. Natl. Cancer Inst.*, 66: 249–254.

Kanda S., Cochran A.J., Lee W., Morton, D.L. and Irie, R.F. Variations in the ganglioside profile of uveal melanoma correlate with cytologic heterogeneity and malignant potential. *Int. J. Cancer*, 52: 682–687.

Kawaguchi, T., Nakakuma, H., Kagimoto, T., Shirono, K., Horikawa, K., Hidaka, M., Iwamori, M., Nagai, Y. and Takatsuki, K. (1989) Characteristic mode of action of gangliosides in selective modulation of CD4 on human T lymphocytes. *Biochem. Biophys. Res. Commun.*, 158: 1050–1059.

Kundu, S.K., Samuelsson, B.E., Pascher, I. and Marcus, D.M. (1983) New gangliosides from human erythrocytes. *J. Biol. Chem.*, 258: 13857–13866.

Ladisch, S., Ulsh, L., Gillard, B. and Wong, C. (1984) Modulation of the immune response by gangliosides: inhibition of adherent monocyte accessory function in vitro. *J. Clin. Invest.*, 74: 2074–2081.

Ladisch, S., Wu, Z.L., Feig, S., Ulsh, L., Schwartz, E., Floutsis, G., Wiley, E., Lenarsky, C. and Seeger, R. (1987) Shedding of GD2 gangliosides by human neuroblastoma. *Int. J. Cancer*, 39: 73–76.

Ladisch, S., Becker, H. and Ulsh, L. (1991) Carbohydrate structure and immunosuppression by human gangliosides. *Proc. Am. Assoc. Cancer Res.*, 32: 226.

Ledeen, R.W. and Yu, R.K. (1982) Gangliosides: structure, isolation, and analysis. *Methods Enzymol.* 83: 139–191.

Lengle, E.E., Krishnaraj, R. and Kemp, R.G. (1979) Inhibition of the lectin-induced mitogenic response of thymocytes by glycolipids. *Cancer Res.*, 39: 817–822.

Livingston, P.O., Natoli, E.J., Calves, M.J., Stockert, E., Oettgen, H.F. and Old, L.J. (1987) Vaccines containing purified GM3 ganglioside elicit GM2 antibodies in melanoma patients. *Proc. Natl. Acad. Sci. USA.*, 84: 2911–2915.

Livingston, P.O., Ritter, G., Oettgen, H.F. and Old, L.J. (1989) Immunization of melanoma patients with purified gangliosides. In H.F. Oettgen (Ed.), *Gangliosides and Cancer*, VCH Publishers, New York, pp. 293–300.

McClay, E.F. and Mastrangelo, M.J. (1988) Systemic chemotherapy for metastatic melanoma. *Semin. Oncol.*, 15: 569–577.

Merritt, W.D., Bailey, M. and Pluznik, D.H. (1984) Inhibition of interleukin-2 dependent cytotoxic T-lymphocyte growth by gangliosides. *Cell. Immunol.*, 89: 1–10.

Miller, H.C. and Esselman, W.J. (1975) Modulation of immune response by antigen reactive lymphocytes after cultivation with gangliosides. *J. Immunol.*, 115: 839–843.

Mitchell, M.S., Kan-Mitchell, J., Kempf, R.A., Harel, W., Shau, H. and Lind, S. (1988) Active specific immunotherapy for melanoma: phase I trial of allogeneic lysates and a novel adjuvant. *Cancer Res.*, 48: 5883–5893.

Morrison, W.J., Offner, H. and Vandenbark, A.A. (1989) Specific ganglioside binding to receptor sites on T lymphocytes that couple to ganglioside-induced decrease of CD4 expression. *Life Sci.*, 45: 1219–1226.

Morton, D.L. (1986a) Active immunotherapy against cancer: present status. *Semin. Oncol.*, 13: 180–185.

Morton, D.L. (1986b) Adjuvant immunotherapy of malignant melanoma: status of clinical trials at UCLA. *Int. J. Immunother.*, 2: 31–36.

Morton, D.L., Malmgren, R.A., Holmes, E.C. and Ketcham, A.S. (1968) Demonstration of antibodies against human malignant melanoma by immunofluorescence. *Surgery*, 64: 233–240.

Morton, D.L., Eilber, F.R., Malmgren, R.A. and Wood, W.C. (1970) Immunological factors which influence response to immunotherapy in malignant melanoma. *Surgery*, 68: 158–164.

Morton, D.L., Joseph, W.L., Ketcham, A.S., Geelhoed, G.W. and Adkins, P.C. (1973) Surgical resection and adjunctive immunotherapy for selected patients with multiple pulmonary metastases. *Ann. Surg.*, 178: 360–366.

Morton, D.L., Eilber, F.R., Holmes, E.C., Hunt, J.S., Ketcham, A.S., Silverstein, M.J. and Sparks, F.C. (1974) BCG immunotherapy of malignant melanoma: summary of a seven-year experience. *Ann. Surg.*, 180: 635–643.

Morton, D.L., Eilber, F.R., Holmes, E.C. and Ramming, K.P. (1978) Preliminary results of a randomized trial of adjuvant immunotherapy in patients with malignant melanoma who have lymph node metastases. *Aust. N.Z. J. Surg.*, 48: 49–52.

Morton, D.L., Nizze, J.A., Gupta, R.K., Famatiga, E., Hoon, D.S.B. and Irie, R.F. (1987) Active specific immunotherapy of malignant melanoma. In: J.P. Kim, B.S. Kim and J-G. Park (Eds.), *Current Status of Cancer Control and Immunobiology*, Seoul, pp. 152–161.

Morton, D.L., Foshag, L.J., Nizze, A.J., Famatiga, E., Hoon, D.S.B. and Irie, R.F. (1989a) Active specific immunotherapy in malignant melanoma. *Semin. Surg. Oncol.*, 5: 420–435.

Morton, D.L., Hoon, D.S.B., Gupta, R.K., Nizze, A.J., Furutani, S.,

Foshag, L.J., Famatiga, E. and Irie, R.F. (1989b) Treatment of malignant melanoma by active specific immunotherapy in combination with biological response modifiers. In: M. Torisu and T. Yoshida (Eds), *New Horizons of Tumor Immunotherapy*, Elsevier, Amsterdam, pp. 665–683.

Morton, D.L., Hoon, D.S.B., Foshag, L.J., Nizze, J.A., Gupta, R.K., Famatiga, E. and Irie, R.F. (1991a) Active immunotherapy of metastatic melanoma with melanoma vaccine immunomodulation. *Proc. Am. Assoc. Cancer Res.*, 32: 492–494.

Morton, D.L., Hunt, K.K., Bauer, R.L. and Lee, J.D. (1991b) Immunotherapy by active immunization of the host using nonspecific agents. In: V.T. DeVita Jr., S. Hellman and S.A. Rosenberg (Eds.), *Biologic Therapy of Cancer*, J.B. Lippincott Co., Philadelphia, pp. 627–642.

Morton, D.L., Foshag, L.J., Hoon, D.S.B., Nizze, J.A., Famatiga, E., Wanek, L.A., Chang, C., Davtyan, D.G., Gupta, R.K. and Elashoff, R. (1992) Prolongation of survival in metastatic melanoma following active specific immunotherapy with a new polyvalent melanoma vaccine. *Ann. Surg.*, 216: 463–482.

Naito, K., Pellis, N.R. and Kahan, B.D. (1988) Effect of continuous administration of interleukin-2 on active specific chemoimmunotherapy with extracted tumor-specific transplantation antigen and cyclophosphamide. *Cancer Res.*, 48: 101–108.

Narasimhan, R. and Murray, R.K. (1979) Neutral glycosphingolipids and gangliosides of human lung and lung tumors. *Biochem. J.*, 179: 199–211.

Nores, G.A., Dohi, T., Taniguchi, M. and Hakomori, S. (1987) Density-dependent recognition of cell surface GM3 by a certain anti-melanoma antibody, and GM3 lactone as a possible immunogen: requirements for tumor-associated antigen and immunogen. *J. Immunol.*, 139: 3171–3176.

Nudelman, E., Hakomori, S., Kannagi, R., Levery, S., Yeh, M-Y., Hellstrom, K.E. and Hellstrom, I. (1982) Characterization of a human melanoma-associated ganglioside antigen defined by a monoclonal antibody. *J. Biol. Chem.*, 257: 12752–12756.

Offner, H., Thieme, T. and Vandenbark, A.A. (1987) Gangliosides induce selective modulation of CD4 from helper T lymphocytes. *J. Immunol.*, 139: 3295–3305.

Parmiter, A.H. and Nowell, P.C. (1988) The cytogenetics of human malignant melanoma and premalignant lesions. In: L. Nathanson (Ed.), *Malignant Melanoma: Biology, Diagnosis and Therapy*, Kluwer Academic Publishers, Boston, pp. 47–62.

Portoukalian, J. (1989) Immunoregulatory activity of gangliosides shed by melanoma tumors. In: H.F. Oettgen (Ed.), *Gangliosides and Cancer*, VCH Publishers, New York, pp. 207–216.

Portoukalian, J., Zwinglestein, G., Abdul-Malek, N. and Dore, J.F. (1978) Alteration of gangliosides in plasma and red cells of human bearing melanoma tumors. *Biochem. Biophys. Res. Commun.*, 85:916–920.

Prokazova, N.V., Orekhov, A.N., Mukin, D.N., Mikhailenko, I.A., Kogter, L.S., Sadovskaya, V.L., Golovanova, N.K. and Bergelson, L.D. (1987) The gangliosides of adult human aorta: intima, media, and plaque. *Eur. J. Biochem.*, 167: 349–352.

Prokazova, N.V., Dyatlovitskaya, E.V. and Bergelson, L.D. (1988)

Sialylated lactosylceramides. Possible inducers of non-specific immunosuppression and atherosclerotic lesions. *Eur. J. Biochem.*, 171: 1–6.

Rauvala, H. (1976) Gangliosides of human kidney. *J. Biol Chem.*, 251: 7517–7520.

Ravindranath, M.H. and Morton, D.L. (1991) Role of gangliosides in active immunotherapy with melanoma vaccine. *Int. Rev. Immunol.*, 7: 303–329.

Ravindranath, M.H., Morton, D.L. and Irie, R.F. (1989) An epitope common to gangliosides O-acetyl-GD3 and GD3 recognized by antibodies in melanoma patients after active specific immunotherapy. *Cancer Res.*, 49: 3891–3897.

Ravindranath, M.H., Paulson, J.C. and Irie, R.F. (1988) Human melanoma antigen *O*-acetylated ganglioside GD3 is recognized by *Cancer antennarius* lectin. *J. Biol. Chem.*, 263: 2079–2086.

Ravindranath, M.H., Tsuchida, T., Morton, D.L. and Irie, R.F. (1991) Ganglioside GM3:GD3 ratio as an index for the management of melanoma. *Cancer*, 67: 3029–3035.

Rees, W.V., Irie, R.F. and Morton, D.L. (1981) Oncofetal antigen-1 (OFA-1): distribution in human tumors. *J. Natl. Cancer Inst.*, 67: 557–562

Rosenberg, S.A., Packard, B.S., Aebersold, P.M., Solomon, D., Topalian, S.L., Toy, S.T., Simon, P., Lotze, M.T., Yang, J.C. and Seipp, C.A. (1988) Use of tumor-infiltrating lymphocytes and interleukin-2 in the immunotherapy of patients with metastatic melanoma. A preliminary report. *N. Engl. J. Med.*, 319: 1676–1680.

Rosenberg, S.A., Lotze, M.T., Yang, J.C., Aebersold, P.M., Linehan, W.M., Seipp, C.A. and White, D.E. (1989) Experience with the use of high-dose interleukin-2 in the treatment of 652 cancer patients. *Ann. Surg.*, 210: 474–484.

Roth, J.A., Morton, D.L. and Holmes, E.D. (1978) Rejection of dinitrochlorobenzene-sensitized guinea pigs. *J. Surg. Res.*, 25: 1–7.

Ryan, J.L. and Shinitzky, M. (1979) Possible role for glycosphingolipids in the control of immune responses. *Eur. J. Immunol.*, 9: 171–175.

Sastry, P.S. (1985) Lipids in nervous tissue: composition and metabolism. *Prog. Lipid Res.*, 24: 69–176.

Senn, H., Orth, M., Fitzke, E., Wieland, H. and Gerok, W. (1989) Gangliosides in normal human serum. Concentration, pattern and transport by lipoproteins. *Eur. J. Biochem.*, 181: 657–662.

Seyfried, T.N., Ando, S. and Yu, R.K. (1978) Isolation and characterization of human liver hematoside. *J. Lipid Res.*, 19: 538–543.

Shibata, M., Hoon, D.S.B., Okun, E. and Morton, D.L. (1992) Modulation of histamine type II receptors on CD8+ T cells by interleukin-2 and cimetidine. *Int. Arch. Allergy Appl. Immunol.*, 97: 8–16.

Sidell, N., Irie, R.F. and Morton, D.L. (1979a) Immune cytolysis of human malignant melanoma by antibody to oncofetal antigen I (OFA-I). I. Complement-dependent cytotoxcity. *Cancer Immunol. Immunother.*, 7: 151–155.

Sidell, N., Irie, R.F. and Morton, D.L. (1979b) Oncofetal antigen I: a target for immune cytolysis of human cancer. *Br. J. Cancer*, 40: 950–953.

Sidell, N., Irie, R.F. and Morton, D.L. (1980) Immune cytolysis of human malignant melanoma by antibody to oncofetal antigen-I (OFA-I). II. Antibody dependent cell mediated cytotoxicity. *Cancer Immunol. Immunother.*, 9: 49–54.

Storm, F.K., Sparks, F.C. and Morton, D.L. (1979) Treatment for melanoma of the lower extremity with intralesional injection of bacille Calmette-Guérin and hyperthermic perfusion. *Surg. Gynecol. Obstet.*, 149: 17–21.

Sunder-Plassman, M. and Bernheimer, H. (1974) Ganglioside in Meningiomen and Heinhauten. *Acta Neuropathol.* (Berl), 27: 289–294.

Svennerholm, L. (1963) Isolation of the major ganglioside of human spleen. *Acta Chem. Scand.*, 17: 860–862.

Tai, T., Paulson, J.C., Cahan, C.D. and Irie, R.F. (1983) Ganglioside GM2 as a human tumor antigen (OFA-I-1). *Proc. Natl. Acad. Sci. USA*, 80: 5392–5396.

Tai, T., Cahan, L.D., Paulson, J.C., Saxton, R.E. and Irie R.F. (1984) Human monoclonal antibody against ganglioside GD2: use in development of anti-GD2 in cancer patients. *J. Natl. Cancer Inst.*, 73: 627–633.

Tai, T., Cahan, L.D., Tsuchida, T., Morton, D.L. and Irie, R.F. (1985) Immunogenicity of melanoma-associated gangliosides in cancer patients. *Int. J. Cancer*, 35: 607–612.

Takahashi, K., Ono, K., Hirabayashi, Y. and Taniguchi, M. (1988) Escape mechanisms of melanoma from immune system by soluble melanoma antigen. *J. Immunol.*, 140: 3244–3248.

Takamizawa, K., Iwamori, M., Mutai, M. and Nagai, Y. (1986) Selective changes in gangliosides of human milk during lactation: a molecular indicator for the period of lactation. *Biochim. Biophys. Acta*, 879: 73–77.

Traylor, D.L. and Hogan, E.L. (1980) Gangliosides of human cerebral astrocytomas. *J. Neurochem.*, 34: 126–131.

Tsuchida, T., Otsuka, H., Niimura, M., Inoue, M., Kukita, A., Hashimoto, Y., Seyama, Y. and Yamakawa, T. (1984) Biochemical study on gangliosides in neurofibromas and neurofibrosarcomas of Recklinghausen's disease. *J. Dermatol.*, 11: 129–136.

Tsuchida, T., Ravindranath, M., Saxton, R.E. and Irie, R.F. (1987a) Gangliosides of human melanoma: altered expression in vivo and in vitro. *Cancer Res.*, 47: 1278–1281.

Tsuchida, T., Saxton, R.E. and Irie, R.F. (1987b) Gangliosides of human melanoma: GM2 and tumorigenicity. *J. Natl. Cancer Inst.*, 78: 55–60.

Tsuchida, T., Saxton, R.E., Morton, D.L. and Irie, R.F. (1987c) Gangliosides of human melanoma. *J. Natl. Cancer Inst.*, 78: 45–54.

Tsuchida, T., Saxton, R.E., Morton, D.L. and Irie, R.F. (1989) Gangliosides of human melanoma II. *Cancer*, 63: 1166–1174.

Vadhan-Raj, S., Cordon-Cardo, C., Carswell, E., Mintzer, D., Dantis, L., Duteau, C., Templeton, M.A., Oettgen, H.F., Old, L.J. and Houghton, A.N. (1988) Phase I trial of a mouse monoclonal antibody against GD3 ganglioside in patients with melanoma; induction of inflammatory responses at tumor sites. *J. Clin. Oncol.*, 6: 1636–1648.

Wallack, M.K., McNally, K.R., Leftheriotis, E., Seigler, H., Balch, C., Wanebo, H., Bartolucci, A.A. and Bash, J.A. (1986) A Southeastern Cancer Study Group phase I/II trial with vaccinia melanoma oncolysates. *Cancer*, 57: 649–655.

Whisler, R.L. and Yates, A.J. (1980) Regulation of lymphocyte responses by human gangliosides. *J. Immunol.*, 125: 2106–2111.

Yamamoto, S., Yamamoto, T., Saxton, R.E., Hoon, D.S.B. and Irie, R.F. (1990) Anti-idiotype monoclonal antibody carrying the internal image of ganglioside GM3. *J. Natl. Cancer Inst.*, 82: 1757–1760.

SECTION V

Gangliosides and Peripheral Neuropathies

L. Svennerholm, A.K. Asbury, R.A. Reisfeld, K. Sandhoff, K. Suzuki, G. Tettamanti and G. Toffano (Eds.)
Progress in Brain Research, Vol. 101

CHAPTER 20

Gangliosides and peripheral neuropathies: an overview

Arthur K. Asbury

University of Pennsylvania School of Medicine, Office of the Vice Dean for Research, 290 John Morgan Building, 36th and Hamilton Walk, Philadelphia, PA 19104-6055, U.S.A.

Introduction

The purpose of this contribution is to provide background information for the succeeding papers on the subject of gangliosides and peripheral neuropathies. It has been known for two decades that peripheral nerve contains gangliosides (Svennerholm et al., 1972; Li et al., 1973). Although gangliosides constitute only a fraction of 1% of total lipids in peripheral nerves or roots (Svennerholm and Fredman, 1990) at least a dozen gangliosides are present in peripheral nerve, and in a pattern that differs from the central nervous system. In addition to gangliosides, other glycoconjugates are found in peripheral nerve, including neutral glycolipids and important myelin- and Schwann cell-associated glycoproteins such as myelin-associated glycoprotein, P_0 and the recently described PMP-22 (Snipes et al, 1992). Several of the glycolipids and glycoproteins found in peripheral nerve tissue are sulfated, of which the most common is the myelin lipid, sulfatide (sulfated galactosylceramide). Virtually all of these glycoconjugates have been implicated in the past 10–12 years as being candidate antigens in instances of peripheral neuropathy in which antibody titers to one or more of these compounds may be measured. Whether antibodies that bind to glycoconjugates are of pathogenetic importance in the neuropathies with which they are associated has stirred lively debate, and is the subject of the following papers on gangliosides and peripheral neuropathies.

There are other issues to address, too. These include whether exogenously administered mixtures of ganglioside can cure or perhaps even cause peripheral neuropathies, and also to comment on some basic features that characterize peripheral neuropathies in general .

Peripheral neuropathies: an overview

Anatomical considerations

The peripheral nervous system is generally considered to begin at the nerve root entry and exit zones of the spinal cord (and at the Obersteiner-Redlich junctions for the cranial nerves) and to extend distally to the peripheral motor or sensory endings. Structures also included in the peripheral nervous system are portions of the pre-ganglionic autonomic fibers that lie outside of the spinal cord and brain stem, the autonomic ganglia and the post-ganglionic fibers. As another way of describing the peripheral nervous system, it is co-terminus with the distribution of Schwann cells. Nevertheless, the primary neural elements that compose the peripheral nervous system also lie in part within the central nervous system, specifically the intraspinal portion of the central processes of primary sensory neurons, the cell bodies of the motor neurons and the proximal portions of their axons and pre-ganglionic autonomic neurons. Therefore, portions of these neural elements are protected by the blood brain barrier, and other portions lie behind the more permeable blood nerve barrier and finally, the dorsal root ganglia themselves are barrier free. These anatomic facts have major implications for distribution of disease and selective vulnerabilities of neural elements that make up the peripheral nervous system.

Neuropathological aspects

The three basic processes by which peripheral nerve fibers are affected are Wallerian degeneration, segmental demyelination and axonal degeneration.

Wallerian degeneration. This results when nerve fibers are transected. Motor and sensory loss in the distribution of the interrupted fibers are apparent immediately, both the axon and the myelin disintegrate distal to the site of transection, conduction fails, and regeneration and recovery through neurite outgrowth from the proximal stump is both slow and generally incomplete. Denervation atrophy occurs in muscles to which the motor nerve fibers have been interrupted. Although Wallerian degeneration is usually caused by physical interruption of nerve fibers, most commonly as a result of trauma, it also occurs during the course of focal ischemia or inflammation of a given nerve trunk. Experimental Wallerian degeneration and regeneration have been studied and re-studied many times, usually with each new technical advance. The application of recombinant technology to Wallerian degeneration has resulted in a marked increase in understanding of the sequence of events and role of growth, adhesion and trophic factors. A recent review of this subject has been made by Scherer and Asbury (1992).

Axonal degeneration. In contrast to Wallerian degeneration, axonal degeneration usually signifies dying back of motor and sensory fibers in a length-dependent manner from the distalmost periphery in the fingers and feet. This axonal process is most frequently encountered with polyneuropathies associated with exogenous toxins or internal disturbances of metabolism. Examples are arsenical neuropathy or uremic neuropathy. Distinction between Waller-ian degeneration and axonal degeneration histologically in a distal nerve may not always be evident, but the processes are quite different. Wallerian degeneration implies transection of otherwise healthy nerve fibers, but axonal degeneration results from dysmetabolic events affecting the entire lower motor neuron or primary sensory neuron. The sites of greatest vulnerability appear to be those furthest from the nerve cell body. Electrodiagnostic features of axonal degeneration are progressive reduction of sensory nerve action potentials and compound muscle action potentials until the nerve becomes inexcitable. Only minor degrees of slowing of conduction velocity are encountered. Recovery from axonal degeneration is variable, but may be surprisingly complete. Remarkably little is known about the neural events leading to axonal degeneration.

Segmental demyelination. Primary breakdown of myelin with preservation of axonal continuity is the hallmark of segmental demyelination. The process usually begins at a node of Ranvier by heminodal involvement. The process may be restricted to the paranode or the entire internode may break down. The physiological consequence of segmental demyelination is conduction block; block can occur with a lesion restricted to a single node of Ranvier (LaFontaine et al, 1982). Other electrodiagnostic features of demyelinating polyneuropathies are slowing of nerve conduction velocities, temporal dispersion and desynchronization of compound evoked potentials, and prolonged distal latencies and late responses. Recovery from segmental demyelination can be rapid and dramatic, because all that is required is remyelination of those sites that are demyelinated and blocked. Commonest examples of primarily demyelinating polyneuropathies are acute Guillain-Barré syndrome and chronic inflammatory demyelinating polyradiculoneuropathy, both thought to be autoimmune in origin, and some of the polyneuropathies associated with dysproteinemias. In some neuropathies mixtures of axonal degeneration and segmental demyelination occur, for instance, in diabetic neuropathies.

Pathophysiological features

The pathophysiology of peripheral neuropathies is most clearly shown by electrophysiological (electrodiagnostic)measures. In nerve fibers undergoing Wallerian degeneration, conduction fails distal to the site of transection in the days, up to 2 weeks, following transection. Similarly, electrodiagnostic features of axonal degeneration are progressive reduction of sensory nerve action potentials and compound muscle action potentials until the nerve becomes inexcitable. Only minor degrees of slowing of conduction velocity are encountered. In contrast, the physiological consequence of segmental demyelination is conduction block: block can occur with a lesion restricted to a single node of Ranvier (LaFontaine et al., 1982). Other electrodiagnostic features of demyelinating polyneuropathies are slowing of nerve conduction velocities, temporal dispersion and desynchronization of compound evoked potentials, and prolonged distal latencies and late responses.

Conduction block is said to be present when a relatively well preserved evoked compound potential is obtained when a given nerve is stimulated distally, but a significantly smaller potential is evoked when the same nerve is stimulated more proximally. The explanation is that some or many of the fibers between the two stimulating electrodes are blocked. The good amplitude of response obtained via the recording electrodes when the stimulus is delivered distally attests to the continuity of axons. There are a number of technical requirements that must be satisfied and pitfalls to be avoided before one can be confident that conduction block has been demonstrated (Cornblath et al, 1991). Conduction block is generally regarded as the hallmark and primary physiological consequence of demyelination, and is not seen in axonal disorders or primary lesions of nerve cell bodies. Clinical recognition of conduction block and its significance has evolved just in the past decade (Lewis et al, 1982; Brown and Feasby, 1984).

Neuropathic symptoms and signs

The manifestations of peripheral neuropathies are usually, but not always, a mixture of sensory, motor and autonomic disturbances. Sensory dysfunctions can be subdivided into those arising from small sensory fiber involvement and those arising from large sensory fiber involvement. They frequently co-exist. Neuropathies affecting small fibers are marked by symptoms of burning raw dysesthesias usually most pronounced in the distal extremities and often painful. Findings on sensory examination are usually hypoesthesia (reduced sensation) to pin prick and thermal stimuli. Light touch is variably affected, but overall the findings are of cutaneous sensory deficit. In contrast, large fiber neuropathies are marked by sensory ataxia. Other major large fiber symptoms are buzzing, tingling, pins and needles type dysesthesia which, while unpleasant, are usually not described as painful. On examination, prominent features are reduced vibration and position sense, areflexia, and lack of balance, particularly in the dark or with eyes closed. Cutaneous sensation is spared. Neuropathies are often pan-sensory, and thus both large and small fiber sensory modalities are impaired.

Damage to motor nerve fibers either by axonal degeneration or demyelination results in weakness of the affected muscles, both as the primary symptom and main finding on examination. Cramps may be a symptom of motor neuropathy, and muscle atrophy attends motor neuron dysfunction. Tendon reflexes are reduced in proportion to the weakness in muscles sub-serving that reflex arc. If large fiber sensory dysfunction co-exists, reflexes disappear completely and early in the course of illness. Dysautonomia, which may be a prominent feature in some neuropathies, is a minor element in those neuropathies in which gangliosides and other glycoconjugates have been implicated, and are not considered further here.

Some correlations between symptoms and physiological events may be drawn. Dysesthesias are considered to result from increased excitability to sensory fibers, often via ectopic impulse generation. In contrast, sensory deficit (hypesthesia) results from axonal degeneration or conduction block in sensory fibers, the common feature being complete interruption of conduction. Motor weakness results from similar events in motor fibers. Conduction slowing, as opposed to interruption of conduction, may have no symptomatic consequences, and can be observed in otherwise normally functioning individuals. Conduction slowing is also seen in symptomatic polyneuropathies, usually demyelinating in nature, but the slowing itself is not the basis for the symptoms or signs.

In the past 5 years, increasing numbers of patients have been identified with asymmetrical multifocal motor neuropathies and conduction block as the evident basis for weakness and atrophy. Some of the patients have had high titers of anti-GM1 antibodies. The complexities and possible interpretation of these observations are discussed in more detail below.

Peripheral neuropathies and exogenous gangliosides

Experimentally, gangliosides have shown considerable potential for promoting enhancement of neurite outgrowth and regeneration both in vivo and in vitro (for review see Gorio and Vitadello, 1987; Ledeen et al., 1984). Nevertheless, controlled trials in a range of neuromuscular disorders have not to date produced definite evidence of efficacy (Bradley et al., 1988; Horowitz, 1989). Mixtures of purified bovine gangliosides were administered in doses ranging up to 100 mg intramuscularly on a daily basis for periods of over a year in dominantly inherited Charcot-Marie-Tooth polyneuropathy, spinocerebellar degeneration with sensory neuropathy, idiopathic polyneuropathies, and amyotrophic lateral sclerosis. Initial optimism concerning usefulness of gangliosides in diabetic neuropathy have not been borne out in subsequent reports (Horowitz, 1986; Hallett et al., 1987).

For a number of years intramuscular doses of mixtures of purified bovine brain gangliosides have been used therapeutically throughout Europe and other parts of the world. It is widely believed that these protein-free glycosphingolipid mixtures are not antigenic and that ganglioside mixtures administered parenterally do not elicit immune-mediated side effects.

Gangliosides and Guillain-Barré syndrome

Recently six cases of purported Guillain-Barré syndrome were brought to the attention of authorities in Germany because the cases were said to have followed upon the administration of exogenous gangliosides. The details were reviewed by Professor K.V. Toyka of Würzburg, who was unfortunately unable to attend this meeting. In a personal communication, Toyka relates that only one of the six cases he reviewed qualified as a case of Guillain-Barré syndrome following administration of gangliosides. He thought that a chance association was likely. Reasons for questioning the other cases were wrong diagnosis and gangliosides not administered until after the onset of symptoms. In addition, no change in overall incidence of GBS is discernible in Italy from 1980 to1990, a period of increasing therapeutic use of gangliosides in that country, according to data published by Granieri et al. (1991). These observations lead one to the tentative conclusion that exogenous gangliosides do not cause appreciable numbers of cases of Guillain-Barré syndrome, at least not in numbers appreciable by these techniques.

In the past year, several cases of GBS-like disorder with high titers of anti-ganglioside antibodies have been reported in which the onset of the polyneuropathy followed closely upon the administration of parenteral gangliosides (Yuki et al., 1991; Latov et al., 1991; Nobile-Orazio et al., 1992). Motor signs predominated in these cases. These cases do not establish a cause and effect relationship between ganglioside administration and either polyneuropathy or the observed titers of anto-ganglioside antibodies, but they do raise the possibility that gangliosides administered parenterally can on occasion trigger polyneuropathy.

The question has also been raised whether gangliosides can exacerbate demyelinating polyneuropathy, a point that is most readily addressed experimentally. Although there have been some conflicting suggestions on this point in the past, three recent studies (Ledeen et al., 1990; Ponzin et al., 1991; Zielasek et al., 1992; Toyka K.V. et al., unpublished data) failed to show that gangliosides enhance experimental autoimmune demyelination in the peripheral nervous system.

Peripheral neuropathies and anti-glycoconjugate antibodies

Background

Several converging threads serve as background. In 1980, Latov et al. (1980) reported a patient with severe demyelinating neuropathy and a high level of monoclonal IgM kappa that bound to peripheral nerve. Immunosuppressive therapy was associated with clinical improvement and marked reduction of the level of circulating monoclonal immunoglobulin (Latov et al., 1980). The constituent to which the IgM bound was later shown to be myelin-associated glycoprotein (MAG) (Braun et al., 1982), and many more such patients have since been identified. Soon after, in patients with dysproteinemic neuropathy it was observed that the IgM that bound to MAG might also bind to gangliosides (Ilyas et al., 1984) and that monoclonal IgM from a patient with a long-standing chronically progressive sensory polyneuropathy bound to polysialosylgangliosides but did not bind to MAG (Ilyas et al., 1985). Thus the issue was raised that gangliosides could be potential autoantigens or at least act as tissue specific binding sites for monoclonal proteins. Moreover, gangliosides could play a role in some of the otherwise idiopathic polyneuropathies. In the following year, Freddo et al. (1986) reported a 64-year-old woman with an IgM lambda dysproteinemia diagnosed as Wäldenstrom's macroglobulinemia and a progressive fatal areflexic motor syndrome with fasciculations and muscle atrophy. The monoclonal immunoglobulin bound avidly to GM1 and GD1b and asialo-GM1.

Motor syndromes

The report of Freddo et al. (1986) and similar ones (Latov et al., 1988; Nardelli et al., 1988; Nobile-Orazio et al., 1990) have triggered a number of surveys of motor neuron disease (Pestronk et al., 1988; Shy et al., 1989; Salazar-Grueso et al., 1990; Sadiq et

al., 1990; Lange et al., 1992). A small proportion of patients exhibit antibody binding to one or more gangliosides. In most who have positive titers, the titers are low and the antibodies are polyclonal. The significance of this observation is uncertain. A few patients with motor neuron syndromes, indeed some with the clinical features and course of classical amyotrophic lateral sclerosis, have had high titers of antibodies (usually polyclonal) to an array of gangliosides. The meaning of these observations, while intriguing, is also unclear.

Concurrently, patients were being noticed who had slowly progressive multifocal lower motor syndromes that resembled motor neuron disease by clinical examination but who had persistent foci of motor conduction block as the basis for their weakness and atrophy (Chad et al., 1986; Parry and Clarke, 1988; Van den Bergh et al., 1989; Krarup et al., 1990; Kaji et al., 1992a,b). Some of these patients had high titers of anti-GM1 antibodies (Pestronk et al., 1988; Baba et al., 1989; Shy et al., 1990), but some did not (Pestronk et al., 1990; Sadiq et al., 1990). Most of these patients were males between the age of 25 and 50, and most have been responsive either to courses of intravenous cyclophosphamide (Feldman et al., 1991) or intravenous immunoglobulin (Kaji et al., 1992a), as discussed further by Griffin in a subsequent chapter. These patients are clearly distinct from classical motor neuron disease, in terms of both their pathophysiology and their response to treatment. In contrast, at least one patient has appeared to follow the course of classical motor neuron disease including development of upper motor neuron findings, but has also had evidence of focal conduction block and high titers to gangliosides (Santoro et al., 1990, 1992). This interesting observation compounds an already vexed problem.

Other syndromes

Antibody binding to gangliosides has been examined in a number of other neuropathic disorders (Pestronk et al., 1990; Sadiq et al., 1990; Pestronk et al., 1991). Sera from patients in the acute phases of Guillain-Barré syndrome (GBS) have been surveyed to look for clues as to the putative autoantigen or autoantigens involved (Ilyas et al., 1988; Svennerholm and Fredman, 1990; Fredman et al., 1991; Ilyas et al., 1992; Nobile-Orazio et al., 1992). Up to one-half or more of GBS patients in these series have measurable titers to gangliosides, but the immunoglobulin sub-types and the gangliosides to which they bind are quite variable. Titers are high in only a few, and frequently sera from normal subjects and other neurological and systemic inflammatory diseases also display low titers to gangliosides. Similar observations have been made in chronic inflammatory demyelinating neuropathy (Fredman et al., 1991; Ilyas et al., 1992). Returning to GBS, Koski et al. (1989) have shown the presence of IgM antibodies that bind to neutral glycolipid. Serial studies demonstrate falling titers concurrent with recovery clinically. Yuki and co-workers reported GBS with extensive axonal damage and high titers of IgG antibody to GM1 following *Campylobacter jejuni* enteritis (1990). Sulfatides have also been implicated in GBS (Fredman et al.,1991). Finally, a recent report from Japan (Chiba et al., 1992) points out that six consecutive patients with Miller Fisher syndrome (a variant of GBS marked by ataxia, areflexia and ophthalmoplegia) demonstrated elevated IgG antibody to GQ1b, that fell later in the course of the illness. Controls including other GBS cases were negative. Taken together, these observations suggest the possibility that gylcolipids, or at least glycoconjugates, could play a role in the immunopathogenesis of autoimmune neuropathies.

Other types of neuropathies deserve brief mention. In about one-half of all neuropathies associated with IgM monclonal gammopathies of undetermined significance, the IgM can be shown to bind to myelin-associated glycoprotein (MAG) (reviewed by Gosselin et al., 1991). The phenotype of these neuropathies tends to be large fiber sensory neuropathies with ataxia and demyelinating features by electrodiagnostic examination. Occasionally neuropathies of this clinical phenotype are associated with antibodies that bind to sulfatide (Pestronk et al., 1991) or to gangliosides with complex polysialosyl groups (Ilyas et al., 1985; Daune et al., 1992; Yuki et al., 1992). More commonly encountered chronic motor-sensory polyneuropathy may on occasion exhibit anti-ganglioside antibodies (Ben Younes-Chennoufi et al., 1992).

Other considerations

The subject of polyneuropathies associated with immunoglobulins that bind to glycoconjugates has produced considerable controversy and a cascade of publications. Nevertheless, progress in understanding has been slow. Part of the problem resides in the unreliability of antibody measurement methods used by different investigators (Marcus et al., 1989) and the continuing inability to compare results across laboratories. Standards used in common by all investigators to allow cross-comparability are seriously needed and are now being prepared by Zielasek, Toyka and collaborators.

Other aspects of the problem are less easily solved. Peripheral nerve has only so many ways it can react to disease (see Overview above), and literally hundreds of agents and conditions are known to cause peripheral neuropathy. It is therefore not surprising that neuropathies of diverse etiology may appear similar clinically. This raises a dilemma. Taking as an example the neuropathies with IgM monoclonal gammopathy, approximately half show binding to MAG and the others do not. Nevertheless, the clinical features of both groups are the same (Gosselin et al., 1991). Does this mean that MAG binding is an epiphenomenon and is extraneous to the pathogenesis of the neuropathy? There are cogent arguments on both sides of this question.

To compound matters, the association of some neurological disorders is with polyclonal antibodies in relatively low titer to various glycoconjugates, as with some cases of GBS, motor neuron disease and chronic inflammatory demyelinating polyneuropathy. Other cases of the same diseases do not have any measurable antibody titers to glycoconjugates. How are these observations to be interpreted? Are the antibody responses merely secondary to neural tissue breakdown by other mechanisms? Or are they indicators of primary immune attack? The evidence and criteria for pathogenicity of antibodies that bind to glycoconjugates are explored in the articles to follow.

References

Baba, H., Daune, G.C., Ilyas, A.A., Pestronk, A., Cornblath, D.R., Chaudhry, V., Griffin, J.W. and Quarles, R. H. (1989) Anti-GM1 ganglioside antibodies with differing fine specificities in patients with multifocal motor neuropathy. *J. Neuroimmunol.,* 25: 143–150.

Ben Younes-Chennoufi, A., Léger, J.-M., Hauw, J.-J., Preud'homme, J.-L., Bouche, P., Aucouturier, P., Ratinahirana, H., Lubetzki, C., Lyon-Caen, O. and Baumann, N. (1992) Ganglioside GD1b is the target antigen for a biclonal IgM in a case of sensory-motor axonal polyneuropathy: Involvement of *N*-acetylneuraminic acid in the epitope *Ann. Neurol.,* 32: 18–23.

Bradley, W.G., Badger, G.J., Tandan, R, Fillyaw, M.J., Young, J., Fries, T.J., Krusinski, P.B., Witarsa, M., Boerman, J. and Blair, C.J. (1988) Double-blind controlled trials of Cronassial in chronic neuromuscular diseases and ataxia. *Neurology,* 38: 1731-1739.

Braun, P.E., Frail, D.E. and Latov, N.(1982) Myelin associated glycoprotein is the antigen for a monoclonal IgM in polyneuropathy. *J. Neurochem.,* 39: 1261–1265.

Brown, W.F. and Feasby, T.E. (1984) Conduction block and denervation in Guillain-Barré polyneuropathy. *Brain,* 107: 219–239.

Chad, D.A., Hammer, K. and Sargent, J. (1986) Slow resolution of multifocal weakness and fasciculation: A reversible motor neuron syndrome. *Neurology,* 36: 1260–1263.

Chiba, A., Kusunoki, S., Shimizu, T. and Kanazawa, I. (1992) Serum IgG antibody to ganglioside GQ1b is a possible marker of Miller Fisher syndrome. *Ann. Neurol.,* 31:677–679.

Cornblath, D.R., Sumner, A.J., Daube, J., Gilliatt, R.W., Brown, W.F., Parry, G.J., Albers, J.W., Miller, R.G. and Petajan, J. (1991) Conduction block in clinical practice. *Muscle Nerve,* 14: 869–871.

Daune, G.C., Farrer, R.G., Dalakas, M.C. and Quarles, P.H. (1992) Sensory neuropathy associated with monoclonal IgM to GD1b ganglioside. *Ann. Neurol.,* 31: 683–685.

Feldman, E.L., Bromberg, M.B., Albers, J.W. and Pestronk, A. (1991) Immunosuppressive treatment in multifocal motor neuropathy. *Ann. Neurol.,* 30: 397–401.

Freddo, L., Yu, R.K., Latov, N., Donofrio, P.D., Hays, A.P., Greenberg, H.S., Albers, J.W., Allessi, A.G. and Keren, D. (1986) Gangliosides GM1 and GD1b are antigens for IgM M-protein in a patient with motor neuron disease. *Neurology,* 36: 454–458.

Fredmann, P., Vedeler, C.A., Nyland, H., Aarli, J.A. and Svennerholm, L. (1991) Antibodies in sera from patients with inflammatory demyelinating polyradiculoneuropathy react with ganglioside LM1 and sulphatide of peripheral nerve myelin. *J. Neurol.,* 238: 75–79.

Gorio, A. and Vitadello, M. (1987) Ganglioside prevention of neuronal functional decay. In: F.J. Seil, E. Herbert and B.M. Carlson, (Eds). *Progress in Brain Research. Vol. 71,* Amsterdam, Elsevier, pp. 203–208.

Gosselin, S., Kyle, R.A. and Dyck, P.J. (1991) Neuropathy associated with monoclonal gammopathies of undetermined significance. *Ann. Neurol.,* 30: 54–61.

Granieri, E., Casetta, I., Govoni, V., Tola, M.R., Paolino, E. and Rocca, W.A. (1991) Ganglioside therapy and Guillain-Barré syndrome. A historical cohort study in Ferra, Italy, fails to demonstrate an association. *Neuroepidemiology,* 10: 161–169.

Hallett, M., Flood, T., Slater, N. and Dambrosia, J. (1987) Trial of ganglioside therapy for diabetic neuropathy. *Muscle Nerve,* 10: 822–825.

Horowitz, S.H. (1986) Ganglioside therapy in diabetic neuropathy. *Muscle Nerve,* 9: 531–536.

Horowitz, S.H. (1989) Therapeutic strategies in promoting peripheral nerve regeneration. *Muscle Nerve,* 12: 314–322.

Ilyas, A.A., Quarles R.H., MacIntosh, T.D., Dobersen, M.J., Trapp, B.D., Dalakas, M.C. and Brady, R.O. (1984) IgM in a human neuropathy related to paraproteinemia binds to a carbohydrate determinant in the myelin-associated glycoprotein and to a ganglioside *Proc. Natl. Acad. Sci.,* 81: 1225–1229.

Ilyas, A.A., Quarles, R.H., Dalakas, M.C., Fishman, P.H. and Brady, R.O. (1985) Monoclonal IgM in a patient with paraproteinemic polyneuropathy binds to gangliosides containing disialosyl groups. *Ann. Neurol.,* 18: 655–659.

Ilyas, A.A., Willison, H.J., Quarles, R.H., Jungalwala, F.B., Cornblath, D.R., Trapp, B.D., Griffin, D.E., Griffin, J.W. and McKhann, G.M. (1988) Serum antibodies to ganglioside in Guillain-Barré syndrome. Ann. Neurol., 23: 440–447.

Ilyas A.A., Mithen, F.A., Dalakas, M.C., Chen, Z.W. and Cook, S.D. (1992) Antibodies to acidic glycolipids in Guillain-Barré syndrome and chronic inflammatory demyelinating polyneuropathy. *J. Neurol. Sci.,* 107: 111–121.

Kaji, R., Shibasaki, H. and Kimura, J.(1992a) Multifocal demyelinating motor neuropathy: Cranial nerve involvement and immunoglobulin therapy. *Neurology,* 42: 506–509.

Kaji, R., Oka, N., Tsuji, J.T., Mezaki, T., Nishio, T., Akiguchi, I. and Kimura, J.(1992b) Pathological findings at the site of conduction block in multifocal motor neuropathy. *Ann. Neurol.,* in press.

Koski, C.L., Chou, D.K.H. and Jungalwala, F.B. (1989) Anti-peripheral nerve myelin antibodies in Guillain-Barré syndrome bind a neutral glycolipid of peripheral myelin and cross reacts with Forssman antigen. *J.Clin.Invest.,* 84: 280–287.

Krarup, C., Stewart, J.D., Sumner, A.J., Pestronk, A. and Lipton, S.A. (1990) A syndrome of asymmetric limb weakness with motor conduction block. *Neurology,* 40: 118–127.

La Fontaine, S., Rasminsky, M., Saida, T. and Sumner, A.J. (1982) Conduction block in rat myelinated nerves following acute exposure to anti-galactocerebroside serum. *J. Physiol.,* 323: 287–306.

Lange, D.J., Trojaborg, W., Latov, N., Hays, A.P., Younger, D.S., Uncini, A., Blake, D.M.,Hirano, M., Burns, S.M., Lovelace, R.E. and Rowland, L.P. (1992). Multifocal motor neruopathy with conduction block: Is it a distinct clinical entity?*Neurology,* 42: 497–505.

Latov, N., Sherman, W.H., Nemni, R., Galassi, G., Shyong, J.S.,

Penn, A.S., Chess, L., Olarte, M.R., Rowland, L.P. and Osserman, E.F. (1980) Plasma cell dyscrasia and peripheral neuropathy with a monoclonal antibody to peripheral myelin. *N. Engl. J. Med.*, 303: 618–621.

Latov, N., Hays, A.P., Donofrio, P.D., Liao, J., Ito, H., McGinnis, S., Manoussos, K., Freddo, L., Shy, M.E. Sherman, W.H., Chang, H.W., Greenberg, H.S., Albers, J.W., Alessi, A.G., Keren, D., Yu, R.K., Rowland, L.P. and Kabat, E.A. (1988) Monoclonal IgM with unique specificity to gangliosides GM1 and GD1b and to lact- *N*-tetraose associated with human motor neuron disease. *Neurology*, 38: 763–768.

Latov, N., Koski, C.L. and Walicke, P.A. (1991) Guillain-Barré syndrome and parenteral gangliosides. *Lancet*, 338: 757.

Ledeen, R.W., Yu, R.K. and Rapport, M.M. (1984) *Ganglioside Structure, Function and Biomedical Function*, New York, Plenum Press.

Ledeen, R.W., Oderfeld-Norwak, B., Brosnan, C.F. and Ervone, A. (1990) Gangliosides offer partial protection in experimental allergic neuritis. *Ann. Neurol.*, 27(Suppl): S69–S74.

Lewis, R.A., Sumner, A.J., Brown, M.J. and Asbury, A.K. (1982) multifocal demyelinating neuropathy with persistent conduction block. *Neurology*, 32: 958–964.

Li, Y.-T., Månsson, J.-E., Vanier, M.T. and Svennerholm, L. (1973) Structure of the major glucosamine-containing ganglioside of human tissues. *J. Biol. Chem.*, 248: 2634–2636.

Marcus, D.M., Latov, N., Hsi, B.P., Gillard, B.K. and participating laboratories (1989) Measurement and significance of antibodies against GM1 ganglioside. *J. Neuroimmunol.*, 25: 255–259.

Nardelli, E., Steck, A.J., Barkas, T., Schluep, M. and Jerusalem, F. (1988) Motor neuron syndrome and monoclonal IgM with antibody activity against gangliosides GM1 and GD1b. *Ann. Neurol.*, 23: 524–528.

Nobile-Orazio, E., Legname, G., Daverio, R., Carpo, M., Giuliani, A., Sonnino, S. and Scarlato, G. (1990) Motor neuron disease in a patient with a monoclonal IgMk directed against GM1, GD1b, and high-molecular-weight neural-specific glycoproteins. *Ann. Neurol.*, 28: 190-194.

Nobile-Orazio, E., Carpo, M., Meucci, N., Grassi, M.P., Capitani, E., Sciacco, M., Mangoni, A. and Scarlato, G. (1992) Guillain-Barré syndrome associated with high titers and anti-GM1, antibodies. *J. Neurol. Sci.*, 109: 200–206.

Parry, G.J. and Clarke, S. (1988) Multifocal acquired demyelinating neuropathy masquerading as motor neuron disease. *Muscle Nerve*, 11; 103–107.

Pestronk, A., Adams, R.N., Clawson, L., Cornblath, D., Kuncl, R.W., Griffin, D. and Drachman, D.B. (1988) Serum antibodies to GM1 gangliosides in amyotrophic lateral sclerosis. *Neurology*, 38: 1457–1461.

Pestronk, A., Cornblath, D.R., Ilyas, A.A., Baba, H., Quarles, R.H., Griffin, J.W., Alderson, K. and Adams, R.N. (1988) A treatable multifocal motor neuropathy with antibodies to GM1 ganglioside. *Ann. Neurol.*, 24: 73–78.

Pestronk, A., Chaudry, V., Feldmann, E.L., Griffin, J.W., Cornblath, D.R., Denys, E.H., Glasberg, M., Kuncl, R.W.,

Olney, R.K. and Yee, W.C. (1990) Lower motor neuron syndromes defined by patterns of weakness, nerve conduction abnormalities, and high titers of antiglycolipid antibodies. *Ann. Neurol.*, 27: 316–326.

Pestronk, A., Li, F., Griffin, J., Feldman, E.L., Cornblath, D., Trotter, J., Zhu, S., Yee, W.C., Phillips, D., Peeples, D.M. and Winslow, B.(1991) Polyneuropathy syndromes associated with serum antibodes to sufatide and myelin-associated glycoprotein. *Neurology*, 41: 357–362.

Ponzin, J.D., Menegus, A.M., Kirschner, G., Nunzi, M.G., Fiori, M.G. and Raine, C.S. (1991) Effects of gangliosides on the expression of autoimmune demyelination in the peripheral nervous system. *Ann. Neurol.*, 30: 678–685.

Sadiq, S.A., Thomas, F.P., Kilidireas, K., Protopsaltis, S., Hays, A.P., Lee, K.W., Romas, S.N., Kumar, N., Van den Berg, L., Santoro, M., Lange, D.J., Younger, D.S., Lovelace, R.E., Trojaborg, W., Sherman, W.H., Miller, J.R., Minuk, J., Fehr, M.A., Roelofs, R.I., Hollander, D., Nichols, F.T., Mitsumoto, H., Kelly, J.J., Swift, T.R., Munsat, T.L. and Latov, N. (1990) The spectrum of neurologic disease associated with anti-GM1 antibodies. *Neurology*, 40: 1067–1072.

Salazar-Grueso, E.F., Routbort, M.J., Martin, J., Dawson, G. and Roos, R.P. (1990) Polyclonal IgM anti-GM1 ganglioside antibody in patients with motor neuron disease and variants. *Ann. Neurol.*, 27: 558–563.

Santoro, M., Thomas, F.P., Fink, M.E., Lange, D.J., Uncini, A, Wadia, N.H., Latov, N. and Hays, A.P. (1990) IgM deposits at nodes of Ranvier in a patient with amyotrophic lateral sclerosis, anti-GM1 antibodies, and multifocal motor conduction block. *Ann. Neurol.*, 28: 373–377.

Santoro, M., Uncini, A., Corbo, M., Staugaitis, S.M., Thomas, F.P., Hays, A.P. and Latov, N. (1992) Experimental conduction block induced by serum from a patient with anti-GM1 antibodies. *Ann. Neurol.*, 31: 385–390.

Scherer, S.S. and Asbury, A.K. (1992) inherited axonal neuropathies. In: R.M. Rosenburg, S.B. Prusiner, S.DiMauro, R.L. Barchi and L.M., Kunkel, (Eds), *The Molecular and Genetic Basis of Neurological Disease*. Butterworth Publishers, Boston, pp. 899–922.

Shy, M.E., Evans, V.A.,Lublin, F.D., Knobler, R.L., Heiman-Patterson, T., Tahmoush, A.J., Parry, G., Schick, P. and DeRyk, T.G.(1989) Antibodies to GM1 and GD1b in patients with motor neuron disease without plasma cell dyscrasia. *Ann. Neurol.*, 25: 511–513.

Shy, M.E., Heiman-Patterson, T., Parry, G.J., Tahmoush, A., Evans, V.A. and Schick, P.K. (1990) Lower motor neuron disease in a patient with auto-antibodies against Gal (β1-3) GalNAc in gangliosides GM1and GD1b: Improvement following immunotherapy. *Neurology*, 40: 842–844.

Snipes, G.J., Suter, U., Welcher, A.A., Shooter, E.M. (1992) Characterization of a novel peripheral nervous system myelin protein (PMP-22/SR13). *J. Cell Biol.*, 117: 225–238.

Svennerholm, L. and Fredman, P. (1990) Antibody detection in Guillain-Barré syndrome. *Ann. Neurol.*, 27 (Suppl.): S36–S40.

Svennerholm, L., Bruce, A., Månsson, J.-E., Rynmark, B.-M. and Vanier, M.-T. (1972) Sphingolipids of human skeletal muscle. *Biochim. Biophys. Acta,* 280: 626–636.

Van den Berg, P., Logigian, E.L. and Kelly, J.J. (1989) Motor neuropathy with multifocal conduction block. *Muscle Nerve,* 11: 26–31.

Yuki, N., Yoshino, H., Sato, S. and Miyatake, T. (1990) Acute axonal polyneuropathy associated with anti-GM1 antibodies following *Campylobacter enteritis. Neurology,* 40: 1900–1902.

Yuki, N., Sato, S., Miyatake, T., Sugiyama, K., Katagiri, T., Sasaki, H. (1991) Motorneuron disease-like disorder after ganglioside therapy. *Lancet,* 337: 1109–1110.

Yuki, N., Miyatani, N., Sato, S., Hirabayashi, Y., Yamazaki, M., Yoshimura, N., Hayashi, Y. and Miyatake, T.(1992) Acute relapsing sensory neuropathy associated with IgM antibody against B-series gangliosides containing a GalNAcβ1-4 (Gal3-2αNeuAc8-2αNeuAc) β1 configuration. *Neurology,* 42: 686–689.

Zielasek, J., Jung, S., Schmidt, B., Ritter, G. and Hartung, HP. (1992) Effects of ganglioside administration on experimental autoimmune neuritis induced by peripheral nerve myelin or P-specific T cell lines. J. *Neuroimmunol,* (in press).

L. Svennerholm, A.K. Asbury, R.A. Reisfeld, K. Sandhoff, K. Suzuki, G. Tettamanti and G. Toffano (Eds.)
Progress in Brain Research, Vol. 101

CHAPTER 21

The structure of human anti-ganglioside antibodies

Donald M. Marcus[1,2] and Nanping Weng[1]

[1]*Departments of Medicine, and* [2]*Microbiology and Immunology, Baylor College of Medicine, Houston, TX 77030, U.S.A.*

Introduction

A subset of patients with peripheral neuropathy has monoclonal immunoglobulins in their sera, most commonly IgM. Many of these IgM antibodies bind to the myelin-associated glycoprotein (MAG) determinant, and a smaller number bind to ganglioside epitopes, especially to GM1 and to asialo GM1 (GA1) (Table I) (Quarles, 1989; Latov, 1990; Steck and Adams, 1991). Several lines of evidence, which will be addressed by other participants in the symposium, indicate that these monoclonal antibodies can cause neuropathies, but some individuals who have monoclonal gammopathies and similar quantities of anti-ganglioside antibodies exhibit no neurological symptoms (Marcus et al., 1989). The specificity of anti-ganglioside antibodies is complex (Ito and Latov, 1988; Baba et al., 1989; Young et al., 1991).

TABLE I

Structures of glycosphingolipids and oligosaccharide-protein conjugates[a]

GM1	Gal(β1-3)GalNAc(β1-4)Gal(β1-4)Glc-Cer
	| (α2-3)
	NeuAc
Asialo GM1 (GA1)	Gal(β1-3)GalNAc(β1-4)Gal(β1-4)Glc-Cer
GalGalNAc-BSA	Gal(β1-3)GalNAc-BSA

[a]Abbreviations: Gal: D-Galactose; Glc: D-Glucose; GalNAc: Acetyl-D-galactosamine; NeuAc: *N*-acetyl-neuraminic acid; Cer: ceramide, *N*-acylsphingosine.

Most bind strongly to both GM1 ganglioside and GA1, but rare antibodies bind preferentially to GM1 and weakly to GA1, and vice versa. Since little is known about the genes encoding these autoantibodies, the structural basis of their fine specificity is unclear.

Recently, elevated titers of polyclonal anti-GM1 and anti-GA1 autoantibodies have been detected in sera of patients with a variety of neurological disorders, including multiple sclerosis, systemic lupus erythematosus (SLE), motor neuron diseases, neuropathies and Guillain-Barré syndrome (Endo et al., 1984; Pestronk et al., 1990; Sadiq et al., 1990; Salazar-Grueso et al., 1990; Ilyas et al., 1992). Most of these sera contain relatively low titers of antibody. In view of the technical problems and lack of standardization associated with measurement of these polyclonal antibodies, and the presence of low titers of anti-ganglioside antibodies in most normal individuals, the role of these autoantibodies in the pathogenesis of neurological diseases is not clear (Marcus et al., 1989; Marcus, 1990).

Results

We have undertaken a study of the structures of anti-ganglioside autoantibodies in order to ascertain their genetic origins, whether there is a requirement for somatic mutation, and to identify the structural basis of their specificity. As a source of autoantibodies, we have immortalized peripheral blood B lymphocytes from two normal individuals and from patients with neuropathy who have elevated titers of anti-GM1 antibodies. Cells producing antibodies against GM1 or GA1 were cloned by limiting dilution, and cDNA encoding VH and VL genes was prepared and sequenced. The first VH sequence that we obtained was from an IgM antibody that was encoded by a VH4 germline gene, and the sequence exhibited only 91% similarity to VH4 germline genes identified previously. Since this gene family was thought to be small, approximately ten genes, and non-polymorphic (Sanz et al., 1989), it seemed likely that the antibody was encoded by a new germline gene. This antibody had been obtained from a pool of lymphocytes from two normal individuals, and we amplified all the VH4 germline genes from these two individuals by the polymerase chain reaction (PCR). We sequenced 75 genes (Table II), including ten unique (non-repetitive) sequences from individual A and eleven unique sequences from individual B (Weng et al., 1992a). Many of these sequences had not been identified previously, and using a very conservative criterion for designating sequences as new, a difference of at least five nucleotides from known germline genes, nine new germline genes were identified. Eight of the nine genes exhibited only 89–96% similarity to the nucleotide sequences of genes identified previously.

Evidence for polymorphism of human immunoglobulin VH genes had been obtained previously by restriction fragment length polymorphism (RFLP) analysis, and by identification of insertion/deletion polymorphisms (Turnbull et al., 1987; Berman et al., 1988; Walter et al., 1991; Willems van Dijk et al., 1991). Our data provide the first information about the extent of this polymorphism, and provide strong evidence that human immunoglobulin gene families are larger and more polymorphic than appreciated previously. These findings have two important implications for the study of autoimmunity. First, unless sequences of expressed antibodies are identical to a germline gene, it is not possible to be confident that the genetic origin of an autoantibody has been identified. Second, polymorphism of immunoglobulin genes probably makes a more significant contribution to the genetic predisposition to autoimmunity, and to hereditary variations in the immune response, than appreciated previously.

Antibody specificity

A summary of the specificities and isotypes of ten autoantibodies produced by cloned B lymphoblastoid cell lines is presented in Table III (Weng et al., 1992b). Four of the antibodies were IgM, six were IgG, seven had kappa light chains and two had lamda light chains. All of the antibodies bound to either GA1 or to a GalGalNAc-BSA conjugate that contains the terminal non-reducing disaccharide of the GA1 structure. Although anti-GM1 antibodies could be detected in supernatants from wells containing 50–100 cells, only one clone, (G3D10) secreted antibodies that bound to GM1. In view of previous data demonstrating that many autoantibodies are polyreactive, we tested our antibodies for binding to single-stranded DNA and to human IgG (rheumatoid factor activity), and these tests were negative.

Sequences of VH and VL genes

A summary of genes encoding the VH and VL domains is presented in Table IV. No restriction of light or heavy chains was noted. Genes from four VH families, three V_K families and two V lamda genes were used to encode these antibodies. Most of the V region genes, 7/9 heavy chains and 6/8 light chains, exhibited less than 97% similarity to known germline

TABLE II

Summary of sequencing data

	Individuals		
	A	B	A + B
Total colonies sequenced	34	41	75
Unique sequences	10	11	15[a]
New genes	5	6	9
Genes reported previously	5	5	6

[a]Individuals A and B have six genes in common, including three new genes and three genes reported previously.

TABLE III

Specificities of antibodies from supernatants of EBV-immortalized B cells

Patients	Clones	Isotype	Antigens		
			GM1	GA1	GalGalNAc-BSA
1[a]	G3D10	IgG/κ	+[b]	+	++
2	H20C3	IgM/κ	−	−	+++
2	HG2B10	IgG/κ	−	+	−
3	K1B12	IgG/λ	−	−	++
4	R1C8	IgG/κ	−	+	−
4	R2C5	IgG/[c]	−	+	−
4	R5A3	IgM/κ	−	+	−
4	R6B8	IgG/κ	−	+	−
N	B5G10	IgM/κ	−	−	++
N	9F2	IgM/λ	ND[d]	ND	+

[a]1–4 are patients and N are normal individuals.
[b]The criteria for a positive reaction are: 500–1000 fluorescence units = +, 1000–1500 = ++, and >1500 = +++.
[c]This light chain was not sequenced.
[d]Not determined.
Reproduced from Weng et al. (1992b) *J. Immunol.*, 149: 2518–2529.

genes. Three of the VH sequences exhibited only 84–87% similarity to known germline genes, and the genes that encode these antibodies have probably not been identified. Omitting these three sequences, the frequency of nucleotide substitutions, and the ratio of replacement/silent substitutions, are summarized in Table V. The higher frequency of nucleotide substitutions, and of replacement substitutions, in the CDR than in the framework regions is strongly suggestive of antigen-driven somatic mutation.

TABLE IV

Summary of V gene usage

Clones	H chain		Potential N-linked	Length of	L chain	
	V_H	J_H	glycosylation site	V_H CDR3	V_L	J_L
G3D10	V_H1	J_H3	—	21	VκI	Jκ1
H20C3	V_H1	J_H5	NPS (CDR2,52[a])	17	VκI	Jκ3
HG2C12	V_H3	J_H4	—	10	VκIII	Jκ5
K1B12	V_H1	J_H6	NYS (CDR3)	23	VλI	Jλ2
R1C8	V_H5	J_H4	NPS (CDR2,60)	16	VκII	Jκ4
R2C5	V_H1	J_H6	NPT (CDR2,57)	19		
R5A3	V_H4	J_H3	NPS (CDR2,60)	12	VκII	Jκ4
R6B8	V_H4	J_H4	NPS (CDR2,60)	14	VκIII	Jκ1
B5G10	V_H4	J_H4	—	9	VκIIIb	Jκ1
9F2	V_H3	J_H4	NYS (CDR2,52)	22	VλI	Jλ2

[a]Codon of asparagine residue.
Reproduced from Weng et al. (1992b) *J. Immunol.*, 149: 2518–2529.

TABLE V

Nucleotide substitution in framework (FW) and complementary determining regions of anti-GA1 autoantibodies

	% Nucleotide substitution			Replacement/silent ratio		
	CDR	FW	CDR/FW	CDR	FW	CDR/FW
H chain	9.0	3.7	2.4	3.5	1.7	2.1
L chain	9.6	3.4	2.7	3.4	0.7	4.9

Reproduced from Weng et al. (1992b) *J. Immunol.*, 149: 2518–2529.

Seven of the nine VH genes had potential N-linked glycosylation sites, but five of these were NPS/T sequences which are rarely glycosylated. Recent observations by Morrison and colleagues (Wallick et al., 1988; Wright et al., 1991) have demonstrated that glycosylation of the variable domains of heavy chains can have a profound effect on antigen binding, ranging from a tenfold increase in affinity to total abolition of binding. We are currently analyzing our antibodies for the presence of carbohydrate.

Heavy chain CDR3 segments

The average length of the CDR3 segment of the anti-GA1 heavy chains was 16 amino acids, and in three antibodies this segment contained more than 20 amino acids. The average length of 260 human CDR3 segments compiled recently is 13 (Kabat et al., 1991). In contrast to murine antibodies, in which a germline D segment that encodes part of CDR3 can usually be identified, the genetic origin of many human CDR3 segments is unclear. Only in four of nine segments was it possible to identify a germline D gene that contributed a core sequence of nucleotides without gaps or mismatches. Partial homologies could be identified by introducing gaps, allowing mismatches, and searching for D-D inversions. This information, and other recent studies of human antibodies, indicate that human heavy chain CDR3 sequences are much larger than their murine counterparts, and make a much larger contribution to antibody diversity and specificity (Sanz, 1991).

Acknowledgement

This work was supported by grants from the National Institutes of Health AI17712 and from the Muscular Dystrophy Association. We wish to thank Drs. James G. Snyder and Li-yuan Yu-Lee for their suggestions and support, and Ms. Charlene Shackelford for her secretarial assistance.

References

Baba, H., Duane G.C., Ilyas, A.A., Pestronk, A., Cornblath, D.R., Chaudhry, V., Griffin, J.W. and Quarles, R.H. (1989) Anti-GM1 ganglioside antibodies with differing fine specificities in patients with multifocal motor neuropathy. *J. Neuroimmunol.*, 25: 143–150.

Berman, J.E., Mellis, S.J., Pollock, R., Smith, C.L., Suh, H., Heinke, B., Kowal, C., Sirti, U., Chess, L., Cantor, C.R. and Alt, F.W. (1988) Content and organization of the human Ig VH locus: defined of three new VH families and linkage to the Ig CH locus. *EMBO J.*, 7: 727–738.

Endo, T., Scott, D.D., Stewart, S.S., Kundu, S.K. and Marcus, D.M. (1984) Antibodies to glycosphingolipids in patients with multiple sclerosis and SLE. *J. Immunol.*, 132: 1793–1797.

Ilyas, A.A., Willison, H.J., Quarles, R.H., Jungalwala, F.B., Cornblath, D.R., Trapp, B.D., Griffin, D.E., Griffin, J.W. and McKhann, G.M. (1992) Serum antibodies to gangliosides in Guillain-Barré syndrome. *Ann. Neurol.*, 23: 440–447.

Ito, H. and Latov, N. (1988) Monoclonal IgM in two patients with motor neuron disease bind to the carbohydrate antigens Gal(β1-3)GalNAc and Gal(β1-3)GlcNAc. *J. Neuroimmunol.*, 19: 245–253.

Kabat, E.A., Wu, T.T., Perry, H.M., Gottesman, K.S. and Foeller, C. (1991) Sequences of proteins of immunological interest. U.S. Department of Health and Human Services.

Latov, N. (1990) Antibodies to glycoconjugates in neurologic disease. *Clin. Aspects Autoimmun.*, 4: 18–29.

Marcus, D.M., (1990) Measurement and clinical importance of antibodies to glycosphingolipids. *Ann. Neurol.*, 27: S53.

Marcus, D.M., Latov, N., Hsi, B.P. and Gillard, B.K. (1989) Measurement and significance of antibodies against GM1. *J. Neuroimmunol.*, 25: 255–259.

Marcus, D.M., Perry, L., Gilbert, S., Preud'homme, J.L. and Kyle, R. (1989) Human IgM monoclonal proteins that bind 3-fucosyl-lactosamine, asialo GM1 and GM1. *J. Immunol.*, 143: 2929–2932.

Pestronk, A., Chaudhry, V., Feldman, E.L., Griffin, J.W., Cornblath, D.R., Denys, E.H., Glasberg, M., Kuncl, R.W., Olney, R.K. and Yee, W.C. (1990) Lower motor neuron syndromes defined by patterns of weakness, nerve conduction abnormalities, and high titers of antiglycolipid antibodies. *Ann. Neurol.*, 27: 316–326.

Quarles, R.H. (1989) Human monoclonal antibodies associated with neuropathy. *Methods Enzymol.*, 179: 291–299.

Sadiq, S.A., Thomas, F.P., Kilidireas, K., Protopsaltis, S., Hays, A.P., Lee, K.-W., Romas, S.N., Kumnar, N., Van den Berg, L., Santoro, M., Lange, D.J., Younger, D.S., Lovelace, R.E., Trojaborg, W., Sherman, W.H., Miller, J.R., Minuk, J., Fehr, M.A., Roelofs, R.I., Hollander, D., Nichols, F.T., Mitsumoto, H. Jr., Kelley, J.J., Swift, T.R., Munsat, T.L. and Latov, N. (1990) The spectrum of neurologic disease associated with anti-GM1 antibodies. *Neurology* 40: 1067–1072.

Salazar-Grueso, E.F., Routbort, M.J., Martin, J., Dawson, G. and Roos, R.P. (1990) Polyclonal IgM anti-GM1 ganglioside antibody in patients with motor neuron disease and variants. *Ann. Neurol.*, 27: 558–563.

Sanz, I., Kelly, P., Williams, C., Scholl, S., Tucker, P. and Capra, J.D. (1989) The smaller human V_H gene families display remarkably little polymorphism. *EMBO J.*, 8: 3741–3748.

Steck, A.J. and Adams, D. (1991) Motor neuron syndrome and monoclonal IgM antibodies to gangliosides. In: Lewis P.Rowland, (Ed.), *Amyotrophic Lateral Sclerosis and Other Motor Neuron Diseases*, Raven Press, pp. 421–425.

Turnbull, I.F., Bernard, O., Sriprakash, K.S. and Mathews, J.D. (1987) Human immunoglobulin variable region genes: a new V_H sequence used to detect polymorphism. *Immunogenetics*, 25: 184–192.

Wallick, S.C., Kabat, E.A. and Morrison, S.L. (1988) Glycosylation of a V_H residue of a monoclonal antibody against $\alpha(1\text{-}6)$ dextran increases its affinity for antigen. *J. Exp. Med.*, 168: 1099–1110.

Walter, M.A., Dosch, H.M. and Cox, D.W. (1991) A deletion map of the human immunoglobulin heavy chain variable region. *J. Exp. Med.*, 174: 335–349.

Weng, N.-P., Snyder, J.G., Yu-Lee, L.-Y., and Marcus, D.M. (1992a) Polymorphism of human immunoglobulin V_H4 germline genes. *Eur. J. Immunol.*, 22: 1075–1082.

Weng, N.-P., Yu-Lee, L.-Y., Sanz, I., Patten, B.M. and Marcus, D.M. (1992b) Structure and specificities of anti-ganglioside autoantibodies associated with motor neuropathies. *J. Immunol.*, 149: 2518–2529.

Willems van Dijk, K., Sasso, E.H. and Milner, C.B. (1991) Polymorphism of the human Ig V_H4 gene family. *J. Immunol.*, 146: 3646–

Wright, A., Tao, M.-H., Kabat, E.A. and Morrison, S.L. (1991) Antibody variable region glycosylation: position effects on antigen binding and carbohydrate structure. *EMBO J.*, 10: 2717–2723.

Young, K.B., Thomas, P.K., King, R.H.M., Waddy, H., Will, R.G., Hughes, A.C., Gregson, N.A. and Leibowitz, S. (1991) The clinical spectrum of peripheral neuropathies associated with benign monoclonal IgM, IgG and IgA paraproteinaemia. *J. Neurology*, 238: 383–391.

L. Svennerholm, A.K. Asbury, R.A. Reisfeld, K. Sandhoff, K. Suzuki, G. Tettamanti and G. Toffano (Eds.)
Progress in Brain Research, Vol. 101

CHAPTER 22

Antibodies to glycoconjugates in neuropathy and motor neuron disease

Norman Latov

Department of Neurology, Columbia University, Black Bldg Rm 3-323, 650 W. 168th Street, New York, NY 10032, U.S.A.

Introduction

Increased titers of antibodies that recognize carbohydrate determinants of glycolipids or glycoproteins (glycoconjugates) are associated with distinct neuropathic syndromes. In one of these syndromes, the demyelinating neuropathy associated with anti-MAG antibodies, there is good evidence that the antibodies cause the neuropathic disease. In several other syndromes, there is a close correlation between the clinical syndrome and the specificity of the associated autoantibody, suggesting that the antibody is related to the disease. This paper will describe the syndromes that are associated with anti-glycoconjugate antibodies, and discuss what is known about the specificity of the antibodies and the mechanisms which might be responsible for their development.

Clinical syndromes associated with anti-glycoconjugate antibodies

Anti-MAG antibodies and demyelinating neuropathy

In approximately 50% of patients with neuropathy and IgM monoclonal gammopathy, the IgM antibodies react with a carbohydrate determinant that is shared by the myelin-associated glycoprotein (MAG), the P_0 glycoprotein, and the glycolipid-sulfated glucuronyl paragloboside (SGPG) and sulfated glucuronyl lactosaminylparagloboside (SGLPG) (reviewed in Latov et al., 1988a; Vital et al., 1989). Patients with anti-MAG antibodies frequently present

with a slowly progressive sensorimotor demyelinating neuropathy which improves following therapeutic reduction of the autoantibody concentrations. The anti-MAG antibodies in most patients occur as nonmalignant monoclonal gammopathies, but they may also be associated with macroglobulinemia or chronic lymphocytic leukemia, or occur without gammopathies.

There is considerable evidence that the anti-MAG antibodies cause the neuropathy. Pathological studies of patients' nerves show demyelination, consistent with the specificity of the autoantibodies, and there are deposits of the anti-MAG antibodies and complement on affected myelin sheaths (Monaco et al., 1990). Intraneural injection of patients' serum into cat sciatic nerves induces demyelination (Hays et al., 1987; Willison et al., 1988; Trojaborg et al., 1989), and systemic administration of anti-MAG antibodies to chickens causes neuropathy and demyelination with separation of the myelin lamellae at the minor dense line, as is seen in the human disease (Tatum, 1993).

Anti-GM1 IgM antibodies and motor neuropathy or motor neuron disease

IgM anti-GM1 antibodies were first reported in patients with IgM monoclonal gammopathy and motor neuron disease or predominantly motor neuropathy (Freddo et al., 1986a; Latov et al., 1988a; Ilyas et al., 1988a; Nardelli et al., 1988; Kusunoki et al., 1989; Nobile-Orazio et al., 1990; Jauberteau et al.,

1990; Sadiq et al., 1990). Increased IgM anti-GM1 antibody titers were also detected in the absence of monoclonal gammopathy (Shy et al., 1989; Salazar-Grueso et al., 1990) and some of the patients were found to have multifocal motor conduction block (Pestronk et al., 1988, 1990; Baba et al., 1989; Sadiq et al., 1990; Kaji et al., 1992). Chemotherapy (Latov et al., 1988; Pestronk et al., 1988; Shy et al., 1990; Feldman et al., 1991), or infusion of human IgG (Kaji et al., 1992) have been associated with clinical improvement.

In most patients the IgM antibodies bind to the Gal(1-3)GalNAc epitope which is shared by GM1, GD1b and several glycoproteins (Ito et al., 1988; Thomas et al., 1989a), but some of the antibodies are highly specific for GM1 (Kusunoki et al., 1989; Sadiq et al., 1990; Corbo et al., 1992) or cross-react with GM2 (Ilyas et al., 1988a; Baba et al., 1989).

It is not known whether the anti-GM1 antibodies cause or contribute to the disease, or whether they are only an associated abnormality; the pathology is not yet known. Although GM1 and other Gal(1-3)GalNAc-bearing glycoconjugates are highly concentrated and widely distributed in the central and peripheral nervous systems (Thomas et al., 1989b), they are mostly cryptic and unavailable to the antibodies. However, anti-GM1 antibodies bind to GM1 at the surface of motor neurons (Thomas et al., 1990; Corbo et al., 1993), and both GM1 and cross reactive glycoproteins are present at the nodes of Ranvier in peripheral nerve (Santoro et al., 1990; Gregson et al., 1991; Corbo et al., 1993; Apostolski et al., 1993), so that the antibodies might exert their effects at these sites. In one study, rabbits immunized with GM1 or Gal(1-3)GalNAc-BSA developed conduction abnormalities with immunoglobulin deposits at the nodes of Ranvier (Thomas et al., 1991), and in another, serum from a patient with increased titers of anti-GM1 antibodies and IgM deposits at the nodes of Ranvier produced demyelination and conduction block when injected into rat sciatic nerve (Santoro et al., 1992). In neither study, however, was it definitively demonstrated that it was the anti-GM1 antibody activity which caused the physiological abnormalities. The reason for the predominant motor

involvement in this syndrome is unknown, but there is an enrichment of GM1 in myelin sheaths of motor nerves in comparison to sensory nerves, possibly making them more susceptible to the antibodies' effects (Ogawa-Goto et al., 1992). Another possibility is that antibody binding to motor neurons might result in upregulation or increased exposure of the reactive epitopes along the motor fibers, increasing their exposure to the autoantibodies (Corbo et al., 1993).

Anti-sulfatide antibodies in sensory neuropathy

Increased titers of monoclonal or polyclonal IgM anti-sulfatide antibodies have been reported in association with predominantly sensory neuropathy (Pestronk et al., 1991; Quattrini et al., 1992). Several of the patients appeared to have a small fiber sensory neuropathy or ganglioneuritis, with normal electrophysiological and nerve biopsy studies. Immunocytochemical studies revealed that the antibodies bound to the surface of rat dorsal root ganglia neurons (Quattrini et al., 1992). Sulfatide is also highly concentrated in peripheral nerve myelin, but it appears unavailable for antibody binding on the surface of the myelin sheath. Some anti-sulfatide antibodies cross-react with MAG and are associated with demyelination (Ilyas et al., 1992a).

Anti-GD1b antibodies and predominantly sensory demyelinating neuropathy

Four patients with IgM monoclonal anti-GD1b antibodies and predominantly sensory neuropathy have been described (Ilyas et al., 1985; Daune et al., 1992; Yuki et al., 1992; Younes-Chennouf et al., 1992). Several of the antibodies cross-reacted with other disialosyl-bearing gangliosides including GD2, GD3, GT1b and GQ1b. In one of the patients the neuropathy began acutely as in the Guillain-Barré syndrome, but following initial improvement, the disease progressed in a stepwise fashion (Yuki et al., 1992).

Anti-chondroitin sulfate antibodies in neuropathy

Three patients with monoclonal IgM anti-chondroitin sulfate antibodies and axonal neuropathy with endoneurial deposits of the monoclonal IgMs have been described (Sherman et al., 1983; Freddo et al., 1985, 1986c; Kabat et al., 1984; Yee et al., 1989). A fourth patient had monoclonal IgM anti-chondroitin C antibodies associated with a predominantly sensory axonal neuropathy, and the IgM bound to the Schmidt-Lanterman incisures (Quattrini et al., 1991).

Antibodies to other glycoconjugates in neuropathy

Isolated cases of neuropathy associated with monoclonal IgM antibodies to other glycolipids have been reported. One patient had a monoclonal IgM that bound to sialosyllactosaminyl paragloboside (Baba et al., 1985; Miyatani et al., 1987). Two patients had IgM M-proteins specific for GM2, GM1b-GalNAc, and GD1a-GalNAc (Ilyas et al., 1988b). One patient had a motor neuropathy with antibodies to GD1a (Bollensen et al., 1989). Another monoclonal IgM from a patient with chronic lymphocytic leukemia and neuropathy bound to myelin and cross-reacted with denatured DNA and with a conformational epitope of phosphatidic acid and gangliosides (Freddo et al., 1986b; Spatz et al., 1990).

Antibodies to glycoconjugates in the Guillain-Barré syndrome (GBS)

Increased titers of antibodies to several glycolipids have been described in GBS. Several patients with acute demyelinating neuropathy or typical GBS, had antibodies to the gangliosides LM1, GD1b, GD1a, or GT1b, (Ilyas et al., 1988c, 1991), to a Forssman-like glycolipid (Koski et al., 1989) or to sulfatide or LM1 (Svennerholm and Fredman, 1990; Fredman et al., 1991). Antibodies to GQ1b were reported in the Miller Fisher variant of GBS (Chiba et al., 1992). Several studies reported increased titers of IgG or IgA anti-GM1 antibodies in GBS, particularly in patients with axonal disease and following infection with *Campylobacter jejuni* (Yuki et al., 1990; Walsh et al., 1991; Gregson et al., 1991; van den Berg et al., 1992; Nobile-Orazio et al., 1992; Ilyas et al., 1992b; Garcia-Guijo et al., 1992; Schonffer, 1992). Acute motor neuropathy (Figeuras et al., 1992) with increased antibodies to GM1 (Latov et al., 1991;

Nobile-Orazio et al., 1992), or to GM2 (Yuki et al., 1991) was also reported to occur following parenteral administration of gangliosides.

Development of anti-glycoconjugate antibodies

Anti-glycoconjugate antibodies share several important characteristics. They are typically directed against carbohydrate determinants, they are frequently IgMs and occur as monoclonal gammopathies, and they are present in low titers in many normal individuals. B-cells capable of secreting anti-carbohydrate antibodies such as anti-MAG or anti-GM1 antibodies are common constituents of the normal human immune repetoire, but are normally suppressed or rendered anergic early in development (Nossal, 1989; Schwartz, 1989). The reason for their activation in later life in some patients is unknown, but elucidation of the mechanisms regulating their expression could help in the development of new and more effective therapies.

Carbohydrate antigens recognized by anti-glycoconjugate antibodies

Carbohydrates differ from peptides in their interactions with the immune system. Carbohydrate epitopes are typically recognized by B-cells only, whereas peptide epitopes are also recognized by T-cells in association with self MHC molecules. Carbohydrate epitopes function as haptens, which by themselves have no immunogenic properties, but when associated with a carrier they can induce an antibody response. T-dependent carbohydrate antigens are carbohydrates which are associated with peptide carriers and which induce a T-cell-dependent antibody response to the carbohydrate determinants. This occurs when B-cells that bear antibodies specific for the carbohydrate determinants bind and internalize the complex and present the associated peptides to T-cells. The reactive T-cells in turn activate the carbohydrate specific B-cells. Other carbohydrate-carrier conjugates activate B-cells in the absence of direct interaction with T-cells, and these are called T-independent (TI) antigens. These activate B-cells, in part, by cross-linking their surface immunoglobulins, and are not entirely T-cell-independent, as they may require the presence of T-cell-derived lymphokines, or that the responding B-cells be previously activated. T-I antigens are further divided into those associated with carriers which can, by themselves, polyclonally activate B-cells (TI-1 antigens), and those which have no polyclonal activating properties (TI-2 antigens) (Mond et al., 1987). T-dependent antigens typically induce high affinity IgG antibodies which contain somatic mutations and are highly specific for the immunizing antigen, whereas T-independent antigens induce antibodies which are predominantly IgMs and have lower affinities. The type of antibody response elicited by a particular carbohydrate epitope is therefore dependent on its associated carrier and on the presence of B- or T-cells capable of responding to the glycoconjugate.

Glycolipids, which are like haptens, do not by themselves induce an antibody response. However, when presented in association with other molecules as in cell membranes, coated on the surface of bacteria, incorporated into liposomes, or with BCG in Freund's adjuvant, they can induce antibodies to their carbohydrate determinants (Galanos et al., 1971; Kinsky et al., 1977; Lai et al., 1985; Livingston et al., 1987; Rapport et al., 1980). Glycolipids in cell membranes might also become incorporated into the envelopes of budding viruses and induce an immune response (Pathak et al., 1990). In these systems, the glycolipids are probably associated with molecules which serve as carriers. Accordingly, in patients with the Guillain-Barré syndrome, the anti-GM1 antibody response following *Campylobacter* enteritis might be induced by the *Campylobacter* lipopolysaccharide which has been shown to contain GM1-like carbohydrate determinants (Aspinall et al., 1992), or by complexes containing GM1 and the *Campylobacter* enterotoxin which binds to GM1 (Klipstein and Engert, 1985). The response following parenteral administration of gangliosides might be induced by complexes of GM1 with tissue antigens with which it became associated. T-cell responses to gangliosides such as delayed type hypersensitivity have also been described (Offner et al., 1985; Knorr et al., 1986), and these reactions could have been directed against

gangliosides associated with protein antigens, or possibly against the gangliosides themselves by MHC non-restricted mechanisms (Siliciano et al., 1985). Immunological cross-reactivity with ganglioside epitopes in peripheral nerve might have triggered the Guillain-Barré syndrome.

Characteristics of B-cells that secrete IgM anti-glycoconjugate antibodies

Low titers of IgM anti-carbohydrate antibodies, such as those directed at MAG or GM1, are found in many normal individuals (McGinnis et al., 1988; Sadiq et al., 1990), and lymphocytes from newborn umbilical cord blood can be activated by Epstein-Barr virus (EBV) to secrete anti-GM1 or MAG antibodies (Lee et al., 1990). These B-cells belong to the CD5+ subpopulation (Lee et al., 1991; Graves et al., 1992), which has been implicated in the secretion of IgM antibodies with autoreactive specificities and in the development of IgM monoclonal gammopathies and chronic lymphocytic leukemia in both humans and mice (Stall et al., 1988; Casali and Notkins, 1989; Stoheger et al., 1989). The B-cells are normally rendered anergic in early development, but they might be activated at a later date by a sufficiently strong stimulus. This stimulus might be provided by a cross-reactive antigenic determinant in an infectious agent or tumor, by idiotypic interactions resulting from an immune response to unrelated antigens (Zanetti, 1986; Painter et al., 1986), or by the carbohydrate antigens in association with a carrier. Since most human anti-carbohydrate antibodies are IgMs, they resemble antibodies responding to T-independent antigens. However, some carbohydrate determinants induce predominantly IgM responses regardless of whether they are associated with T-dependent or T-independent carriers (Pawlita et al., 1982; Matsuda et al., 1989), and it may require a particularly strong stimulus to activate anergic B-cells (Pike et al., 1987; Goodnow et al., 1991). Antibodies secreted by CD5+ B-cells are frequently encoded by a restricted repetoire of immunoglobulin heavy or light chain variable region genes without somatic mutations, and although most anti-MAG antibody heavy chains are encoded by variable region genes belonging to the

Vh3 family, their light chains are more diverse, and it is not yet known whether they are encoded by germline genes or whether they contain somatic mutations (Brouet et al., 1989; Spatz et al., 1992). In the Guillain-Barré syndrome, anti-ganglioside antibodies are frequently IgGs or IgAs, suggesting that the immune reaction is T-cell driven, but it is not known whether these antibodies are derived from the same B-cells that secrete the IgM antibodies or from a separate population (Linton et al., 1989).

Many of the anti-carbohydrate antibodies occur as IgM monoclonal gammopathies, suggesting that they are derived from single B-cell clones which undergo transformation. The transformation could occur spontaneously, or as a consequence of chronic antigenic stimulation, chemical mutation (Gelman and Dennis, 1981), or viral infection (Hanto et al.,1982; Ahmed and Oldstone, 1984) B-cells expressing antibodies that bind to carbohydrate antigens on the surface of a transforming virus might be particularly susceptible to infection and transformation. Genetic predisposition may be an additional factor in the development of monoclonal gammopathies (Radl et al., 1986; Jensen et al., 1988).

Little is known about the mechanisms that regulate CD5+ B-cells or that activate the polyclonal or monoclonal B-cells which secrete anti-carbohydrate autoantibodies. In one study, monoclonal B-cells that secrete anti-MAG antibodies were shown to be stimulated by activated T-cells (Latov et al., 1988), and in other studies, B-cells that secrete anti-carbohydrate antibodies were suppressed or deleted following exposure to the carbohydrate antigens (Dintzis et al., 1992; Murakami at al., 1992). It is not known how the B-cells that secrete anti-glycoconjugate antibodies in patients with neuropathy become stimulated or whether they respond to the same regulatory mechanisms as normal B-cells.

References

Ahmed, R. and Oldstone, M.B.A. (1984) Mechanisms and biological implications of virus induced polyclonal B-cell activation. In: A.L., Notkins and M.B.A., Oldstone (Eds), *Concepts in Viral Pathogenesis*. Springer-Verlag, New York, pp. 231–240.

Apostolski, S., Sadiq, S.A., Hays, A.P. and Latov, N. (1993) Hyaluronectin and OMgp are autoantigens in peripheral nerve. (abstract) *Trans. Am. Soc. Neurochem.*, 24: 128.

Aspinall, G.O., McDonald, A.G., Raju, T.S., Pang, H., Mills, S.D., Kurjanczyk, L.A. and Penner, J.L. (1992) Serological diversity and chemical structure of *Campylobacter jejuni* low-molecular weight lipopolysaccharides. *J. Bacteriol*, 174: 1324–1332.

Baba, H., Miyatani, N., Sato, S., Yuasa, T. and Miyatake, T. (1985) Antibody to glycolipid in a patient with IgM paraproteinemia and polyradioculoneuropathy. *Acta. Neurol. Scand.*, 72: 218–221.

Baba, H., Daune, G.C., Ilyas, A.A., Pestronk, A., Cornblath, D., Chaundhry, V., Griffin, J. and Quarles, R. (1989) Anti-GM1 ganglioside antibodies with differing specificities in patients with multifocal motor neuropathy. *J Neuroimmunol.*, 25: 143–150.

Bollensen, E., Schipper, H.I. and Steck, A.J. (1989) Motor neuropathy with activity of monoclonal IgM antibody to GD1a ganglioside. *J. Neurol.*, 236: 353–355.

Brouet, J.-C., Dellagi, K., Gendron, M.C., Chevalier, A., Schmitt, C. and Mihaesco, E. (1989) Expression of a public idiotype by human monoclonal IgM directed to myelin associated glycoprotein and characterization of the variable subgroup of their heavy and light chains. *J. Exp. Med.*, 170: 1551–1558.

Casali, P. and Notkins, A.L. (1989) CD5+ B lymphocytes, polyreactive antibodies and the human B-cell repetoire. *Immunol. Today*, 9: 364–368.

Chiba, A., Kusunoki, S., Shimizu, T. and Kanazawa, I. (1992) Serum IgG antibody to ganglioside GQ1b is a possible marker of Miller Fisher syndrome. *Ann. Neurol.*, 31: 677-679.

Cook, D., Dalakas, M., Galdi, A., Biondi, D. and Porter, H. (1990) High-dose intravenous immunoglobulin in the treatment of demyelinating neuropathy associated with monoclonal gammopathy. *Neurology*, 40: 212–214.

Corbo, M., Quattrini, A., Lugaresi, A., Santoro, M., Latov, N. and Hays, A.P. (1992) Patterns of reactivity of human anti-GM1 antibodies with spinal cord and motor neurons. *Ann. Neurol.*, 32: 487–493.

Corbo, M., Quattrini, A., Latov., N. and Hays, A.P. (1993) Localization of GM1 and Gal (β1-3) GalNAc antigenic determinants at the nodes of Ranvier in peripheral nerve. *Neurology*, 43: 809–816.

Daune, G.C., Farrer, R.G., Dalakas, M.C. and Quarles, R.H. (1992) Sensory neuropathy associated with immunoglobulin M to GD1b ganglioside. *Ann. Neurol.*, 31: 683–685.

Dennis, R.D., Antonicek, H., Wiegandt, H. and Schachner, M. (1988) Detection of the L2/HNK-1 carbohydrate epitope on glycoproteins and acidic glycolipids of the insect *Calliphora vicina*. *J. Neurochem.*, 51: 1490–1496.

Dintzis, H.M. and Dintizis, R. (1992) Profound specific suppression by antigen of persistent IgM, IgG and IgE antibody production. *Proc. Natl. Acad. Sci. USA*, 89:1113–1117.

Feldman, E.L., Bromberg, M.B., Albers, J.W. and Pestronk, A. (1991) Immunosuppressive treatment of multifocal motor neuropathy. *Ann. Neurol.*, 30:397–401.

Figeuras, A., Morales-Olivas, F.J., Capella, D., Palop, V. and Laporte, J.-R. (1992) Bovine ganglioside and acute motor polyneuropathy. Br. Med. J., 305: 1330–1331.

Freddo, L., Hays, A.P., Sherman, W.H. and Latov, N. (1985) Axonal neuropathy in a patient with IgM M-protein reactive with nerve endoneurium. *Neurology*, 35: 1321–1325.

Freddo, L., Yu, R.K., Latov, N., Donofrio, P.D., Hays, A.P., Greenberg, H.S., Albers, J.W., Allessi, A.G. and Keren, D. (1986a) Gangliosides GM1 and GD1b are antigens for IgM M-proteins in a patient with motor neuron disease. *Neurology*, 36: 454–458.

Freddo, L., Hays, A.P., Nickerson, K.G., Spatz, L., McGinnis, S., Lieberson, G., Vedeler, C.A., Shy, M.E., Autilo-Gambetti, L., Grauss, F.C., Petito, F., Chess, L. and Latov, N. (1986b) Monoclonal anti-DNA IgMk in neuropathy binds to myelin and to a conformational epitope formed by phosphatidic acid and gangliosides. *J. Immunol.*, 137: 3821–3825.

Freddo, L., Sherman, W.H. and Latov, N. (1986c) Glycosaminoglycan antigens in peripheral nerve; studies with antibodies from a patient with neuropathy and monoclonal gammopathy. *J. Neuroimmunol.*, 12: 57–64.

Fredman, P., Vedeler, C.A., Nyland, H., Aarli, J.A. and Svennerholm, L. (1991) Antibodies in sera from patients with inflammatory demyelinating polyradiculoneuropathy react with ganglioside LM1 and sulfatide of peripheral nerve myelin. *J. Neurol.*, 238: 75–79.

Galanos, C., Luderitz, O. and Westphal, O. (1971) Preparation and properties of antisera against the lipid A component of bacterial lipopolysaccharides. *Eur. J. Biochem.*, 24: 116–122.

Garcia-Guijo, C., Garcia-Merino, A., Rubio, G., Guerrero, A., Martinez, A.C. and Arpa, J. (1992) IgG anti-ganglioside antibodies and their subclass distribution in two patients with acute and chronic motor neuropathy. *J. Neuroimmunol.*, 37: 141–148.

Gelman, E.P. and Dennis, L.H. (1981) Plasma cell dyscrasia after alkylating agent therapy for Hodgkins disease (Letter). *N. Engl. J. Med.*, 35: 135.

Goodnow, C.C., Brink, R. and Adams, E. (1991) Breakdown of self tolerance in anergic B lymphocytes. *Nature*, 352: 532–536.

Graves, M.C., Ravindranath, R.M.H. and Ravindranath, M.H. (1992) CD5+ B cells synthesize anti-GM1 and ansialo-GM1 antibodies in lower motor neuron disease. (abstract) *Neurology*, 42 (suppl. 3): 1992.

Gregson, N.A., Jones, D., Thomas, P.K. and Willison, H.J. (1991) Acute motor neuropathy with antibodies to GM1 ganglioside. *J. Neurol.*, 238: 447–451.

Hanto, D., Frizzera, G., Gajl-Peczalska, Sakamoto, K., Purtilok, Balfour, Jr. H.H., Simmons, R.L. and Najarian, J.S. (1983) Epstein-Barr virus induced B-cell lymphoma after renal transplantation. Acyclovir therapy and transition from polyclonal to monoclonal B-cell proliferation. *N. Engl. J. Med.*, 306: 913–918.

Hass, D.C. and Tatum, A.H. (1988) Plasmapheresis alleviates neuropathy accompanying IgM anti-myelin associated glycoprotein paraproteinemia. *Ann. Neurol.*, 23: 394–396.

Hays, A.P., Latov, N., Takatsu, M. and Sherman, W.H. (1987) Experimental demyelination of nerve induced by serum of patients with neuropathy and anti-MAG proteins. *Neurology*, 37: 242–256.

Ilyas, A.A., Quarles, R.H., Dalakas, M.C., Fishman, P.H. and Brady, R.O. (1985) Monoclonal IgM in a patient with paraproteinemic polyneuropathy binds to gangliosides containing disialosyl groups. *Ann. Neurol.*, 18: 655–659.

Ilyas, A.A., Willison, H.J., Dalakas, M., Whitaker, J.N. and Quarles, R.H. (1988a) Identification and characterization of gangliosides reacting with IgM paraproteins in three patients with neuropathy and biclonal gammopathy. *J. Neurochem.*, 51: 851–858.

Ilyas, A.A., Li, S.-C., Chou, D.K.H., Li, Y.-T., Jungalwala, F.B., Dalakas, M.C. and Quarles, R.H. (1988b) Gangliosides GM2, IVGalNAcGM1b, and IVGalNAcGD1a as antigens for monoclonal immunoglobulin M in neuropathy associated with gammopathy. *J. Biol. Chem.*, 263: 4369–4373.

Ilyas, A.A., Willison, H.J., Quarles, R.H., Jungalwala, F.B., Cornblath, D.R., Trapp, B.D., Griffin, J.W. and McKhann, G.M. (1988c) Serum antibodies to gangliosides in Guillain-Barre syndrome. *Ann. Neurol.*, 23: 440–447.

Ilyas, A.A., Mithen, F.A., Chen, Z.W. and Cook, S.D. (1991) Search for antibodies to neutral glycolipids in sera of patients with Guillain-Barré syndrome. *J. Neurol. Sci.*, 102: 67–75.

Ilyas, A., Cook, S.D., Dalakas, M.C. and Mithen, F.A. (1992a) Anti-IgM paraproteins from some patients with polyneuropathy associated with IgM paraproteinemia also react with sulfatide. *J. Neuroimmunol.*, 37: 85–92.

Ilyas, A.A., Mithen, F.A., Chen, Z.W. and Cook, S.D. (1992b) Anti-GM1 antibodies in Guillain-Barré syndrome. *J. Neuroimmunol.*, 36: 69–76.

Ito, H. and Latov, N. (1988) Monoclonal IgM in two patients with motor neuron disease bind to the carbohydrate antigens Gal(β1-3)GalNAc and Gal(β1-3)GlcNAc. *J. Neuroimmunol.*, 19: 245–253.

Jauberteau, M.O., Gualde, N., Freud'homme, J.L., Rigaud, M., Gil, R., Vallat, J.M. and Baumann, N. (1990) Human monoclonal IgM with autoantibody activity against two gangliosides GM1 and GD1b in a patient with motor neuron syndrome. *Clin. Exp. Immunol.*, 80: 186–191.

Jensen, T.S., Schroder, H.D., Ionsson, V., Ernerudh, J., Stigsby, B., Kamieniecka, Z., Hippe, E. and Trojaborg, W. (1988) IgM monoclonal gammopathy and neuropathy in two siblings. *J. Neurol. Neurosurg. Psychiatry*, 51: 1308–1315.

Kabat, K.A., Liao, J., Sherman, W.H. and Osserman, E.F. (1984) Immunological characterization of the specificities of two human monoclonal IgMs reacting with chondroitin sulfates. *Carbohydrate Res.*, 130: 289–298.

Kaji, R., Shibasaki, H. and Kimura, J. (1992) Multifocal demyelinating motor neuropathy: Cranial nerve involvement and immunoglobulin therapy. *Neurology*, 42: 506–509.

Kinsky, S.C. and Nicoletti, R.A. (1977) Immunological properties of model membranes. *Ann. Rev. Biochem.*, 46:49–67.

Klipstein, F.A. and Engert, R.F. (1985) Immunological relationship of the B subunits of *Campylobacter jejuni* and *Escherichia coli* heat-labile enterotoxin. *Infect. Immunol.*, 48: 629–633.

Knorr-Held, S., Brendel, W., Kiefer, H. et al. (1986) Sensitization against brain ganglioside after therapeutic swine brain implantation in a multiple sclerosis patient. *J. Neurol.*, 233: 54–56.

Koski, C.L., Chou, D.K.H. and Jungalwala, F.B. (1989) Anti-peripheral nerve myelin antibodies in Guillain-Barré syndrome bind a neutral glycolipid of peripheral myelin and cross reacts with Forssman antigen. *J. Clin. Invest.*, 84: 280–287.

Kusunoki, S., Shimizu, T., Matsumura, K., Maemura, K. and Mannen, T. (1989) Motor dominant neuropathy and IgM paraproteinemia: the IgM M-protein binds to specific gangliosides. *J. Neuroimmunol.*, 21: 177–181.

Lai, E., Kabat, E.A. and Mobraaten, L. (1985) Genetic and nongenetic control of the immune response of mice to a synthetic glycolipid, stearylisomaltotetraose. *Cell. Immunol.*, 92: 172–183.

Latov, N., Hays, A.P. and Sherman, W.H. (1988a) Peripheral neuropathy and anti-MAG antibodies. *CRC Crit. Rev. Neurobiol.*, 3: 301–332.

Latov, N., Hays, A.P., Donofrio, P.D., Liao, J., Ito, H., McGinnis, S., Manoussos, K., Freddo, L., Shy, M.E., Sherman, W.H., Chang, H.W., Greenberg, H.S., Albers, J.W., Allessi, A.G., Keren, D., Yu, R.K., Rowland, L.P. and Kabat, E.A. (1988b) Monoclonal IgM with unique reactivity to gangliosides GM1 and GD1b and to lacto-N-tetraose in two patients with motor neuron disease. *Neurology,* 38: 763–768.

Latov, N., Koski, C.L. and Walicke, P.A. (1991) Guillain-Barré syndrome and parenteral gangliosides (letter). *Lancet*, 338: 757.

Lee, K.W., Inghirami, G., Sadig, S.A., Thomas, F.P., Spatz, L., Knowles, D.M. and Latov, N. (1990) B-cells that secrete anti-MAG or anti-GM1 antibodies are present at birth and anti-MAG antibody B-cells are CD5+. (abstract) *Neurology*, 40 (suppl. 1): 367.

Lee, K.W., Inghirami, G., Spatz, L., Knowles, D.M. and Latov, N. (1991) The B-cells that express anti-MAG antibodies is neuropathy and non-malignant IgM monoclonal gammopathy belong to the CD5 population. *J. Neuroimmunol.*, 31: 83–88.

Linton, P.-J., Decker, D.J. and Klinman, M.R. (1989) Primary antibody-forming cells and secondary B-cells are generated from separate precursor cell subpopulations. *Cell*, 59: 1049–1059.

Livingston, P.O., Calves, M.J. and Natoll, Jr. E.J. (1987) Approaches to augmenting the immunogenicity of the ganglioside GM2 in mice: purified GM2 is superior to whole cells. *J. Immunol.*, 138: 1524–1529.

Matsuda, T. and Kabat, E.A. (1989) Variable region cDNA sequences and antigen binding specificity of mouse monoclonal antibodies to isomaltosyl oligosaccharides coupled to proteins. *J. Immunol.*, 142: 863–870.

McGinnis, S., Kohriyama, T., Yu, R.K., Pesce, M.A. and Latov, N. (1988) Antibodies to sulfated glucuronic acid containing glycosphingolipids in neuropathy associated with anti-MAG antibodies and in normal subjects. *J. Neuroimmunol.*, 17: 119–126.

Miyatani, N., Baba, H., Sato, S., Nakamura, K., Yuasa, T. and Miyatake, T. (1987) Antibody to sialosyllactosaminylparagloboside in patient with IgM paraproteinemia and polyradiculoneuropathy. *J. Neuroimmunol.*, 14: 189–196.

Monaco, S., Bonetti, B., Ferrari, S. et al. (1990) Complement dependent demyelination in patients with IgM monoclonal gammopathy and polyneuropathy. *N. Engl. J. Med.*, 322: 649–652.

Mond, J.J. and Brunswick, M. (1987) A role for IFN- and NK cells in immune responses to T-cell regulated antigens type 1 and 2. *Immunol. Rev.*, 99: 105–118.

Murakami, M., Tsubata, T., Okamoto, M., Shimizu, A., Kumagai, S., Imura, H. and Honjo, T. (1992) Antigen-induced apoptotic death of Ly-1 B-cells responsible for autoimmune disease in transgenic mice. *Nature*, 357: 77–80.

Nardelli, E., Steck, A.J., Barkas, T., Schluep, M. and Jerusalem, F. (1988) Motor neuron disease and monoclonal IgM with antibody activity against gangliosides GM1 and GD1b. *Ann. Neurol.*, 23: 524–528.

Nobile-Orazio, E., Baldini, L., Barbieri, S., Marmirolli, P., Spagnol, G., Francomano, E. and Scarlato, G. (1988) Treatment of patients with neuropathy and anti-MAG IgM M-proteins. *Ann. Neurol.*, 24: 93–97.

Nobile-Orazio, E., Legname, G., Deverio, R., Carpo, M., Giuliana, A., Sonnino, S. and Scarlato, G. (1990) Motor neuron disease in a patient with a monoclonal IgMk directed against GM1, GD1b, and high molecular weight neural-specific glycoproteins. *Ann. Neurol.*, 28: 190–194.

Nobile-Orazio, E., Carpo, M., Meucci, N., Grassi, M.P., Capitani, E., Sciacco, M., Mangoni, A. and Scarlato, G. (1992) Guillain-Barré syndrome associated with high titers of anti-GM1 antibodies. *J. Neurol. Sci.*, 109: 200–206.

Nossal, G.J.V. (1989) Immunologic tolerance: collaboration between antigen and lymphokines. *Science*, 245: 147–153.

Offner, H., Standage, B.A., Burger, D.R. and Vandenbark, A.A. (1985) Delayed type hypersensitivity to gangliosides in the Lewis rat. *J. Neuroimmunol.*, 9: 145–157.

Ogawa-Goto, K., Funamoto, N., Ohta, Y., Abe, T. and Nagashima, K. (1992) Myelin gangliosides of human peripheral nervous system: an enrichment of GM1 in the motor nerve myelin isolated from cauda equina. *J. Neurochem.*, 59: 1844–1848.

Painter, C., Monestier, M., Bonin, B. and Bonna, C.A. (1986) Functional and molecular studies of V genes expressed in autoantibodies. *Immunol. Rev.*, 94: 75–95.

Pathak, S., Illavia, S.J., Khalili-Shirazi, A. and Webb, H.E. (1990) Immunoelectron microscopic labeling of a glycolipid in the envelopes of brain cell-derived budding viruses, Semliki Forest, influenza, and measles, using a monoclonal antibody directed chiefly against galactocerebroside resulting from Semliki Forest virus infection. *J. Neurosci.*, 96: 293–302.

Pawlita, M., Potter, M. and Rudikoff, S. (1982) Kappa chain restriction in anti-galactan antibodies. *J. Immunol.*, 129: 615–618.

Pestronk, A., Cornblath, D.R., Ilyas, A.A., Baba, H., Quarles, R.H., Griffin, J.W., Alderson, K. and Adams, R.N. (1988) A treatable multifocal motor neuropathy with antibodies to GM1 ganglioside. *Ann. Neurol.*, 24: 73–78.

Pestronk, A., Chaudhry, V., Feldman, E.L. et al. (1990) Lower motor neuron syndromes defined by patterns of weakness, nerve conduction abnormalities and high titers of antiglycolopid antibodies. *Ann. Neurol.*, 27: 316–326.

Pestronk, A., Li, F., Griffin, J., Feldman, E.L., Cornblath, D., Trotter, J., Zhu, S., Yee, W.C., Phillips, D., Peeples, D.M. and Winslow, B. (1991) Polyneuropathy syndromes associated with serum antibodies to sulfatide and myelin associated glycoprotein. *Neurology*, 41: 357–362.

Pike, B.L., Alderson, M.R. and Nossal, G.J.V. (1987) T-independent activation of single B-cells: an orderly analysis of overlapping stages in the activation pathway. *Immunol. Rev.*, 99: 177–152.

Quattrini, A., Nemni, R., Fazio, R., Iannaccone, S., Lorenzetti, I., Grassi, F. and Canal, N. (1991) Axonal neuropathy in a patient with monoclonal IgM kappa reactive with Schmidt-Lanterman incisures. *J. Neuroimmunol.*, 33: 73–79.

Quattrini, A., Corbo, M., Dhaliwal, S.K., Sadig, S.A., Lugaresi, A., Oliveira, A., Uncini, A., Abouzahr, K., Miller, J.R., Lewis, L., Estes, D., Cardo, L., Hays, A.P. and Latov, N. (1992). Anti-sulfatide antibodies in neurological disease; binding to rat dorsal root ganglia neurons. *J. Neurol. Sci.*, 112: 152–159.

Radl, J., DeGlopper, E., van den Berg, P. and Wanzwieten, M.J. (1986) Idiopathic paraproteinemia III. Increased frequency of paraproteinemia in thymectomized, aging C57GL/Kahukig and CBA/Brarij mice. *J. Immunol.*, 125: 31–35.

Rapport, M.M., Graf, L., Huang, Y.L., Brunner, W. and Yu, R.K. (1980) Antibodies to brain gangliosides: titer and specificity of antisera. In: L. Svennerholm, P. Mandel, H. Dreyfus, and P.F. Urban, (Eds), *Structure and Function of Gangliosides*. Plenum Press, New York, pp. 327–334.

Sadig, S.A., Thomas, F.P., Kilidireas, K., Protopsaltis, S., Hays, A.P., Lee, K.W., Romas, S.N., Kumar, N., van den Berg, L., Santoro, M., Lange, D.J., Younger, D.S., Lovelace, R.E., Trojaborg, W., Sherman, W.H., Miller, J.R., Minuk, J., Fehr, M.A., Roelofs, R.E., Hollander, D., Nichols, III, Mitsumoto, H., Kelley, Jr. J.J., Swift, T.R., Mansat, T.L. and Latov, N. (1990) The spectrum of neurological disease associated with anti-GM1 antibodies. *Neurology*, 40: 1067–1072.

Salazar-Grueso, E.F., Routbort, M.J., Martin, J., Dawson, G. and Roos, R.P. (1990) Polyclonal IgM anti-GM1 ganglioside antibody in patients with motor neuron disease and variants. *Ann. Neurol.*, 27: 558–563.

Santoro, M., Thomas, F.P., Fink, M.E., Lange, D.J., Uncini, A., Wadia, N.H., Latov, N. and Hays, A.P. (1990) IgM deposits at the nodes of Ranvier in a patient with amyotrophic lateral sclerosis, anti-GM1 antibodies, and multifocal motor conduction block. *Ann. Neurol.*, 28: 373–377.

Santoro, M., Uncini, A., Corbo, M., Staugaitis, S.M., Thomas, F.P., Hays, A.P. and Latov, N. (1992) Experimental conduction block induced by serum from a patient with anti-GM1 antibodies. *Ann. Neurol.*, 31: 385–390.

Schonhoffer, P.S. (1992) GM1 ganglioside for spinal cord injury (letter) . *N. Engl. J. Med.*, 326: 493.

Schwartz, R.H. (1989) Acquisition of immunologic self tolerance. *Cell*, 57: 1073–1081.

Sherman, W.H., Latov, N., Hays, A.P., Takatsu, M., Nemni, R., Galassi, G. and Osserman, E.F. (1983) Monoclonal IgMk antibody precipitating with chondroitin sulfate C from patients with axonal polyneuropathy and epidermolysis. *Neurology*, 33: 192–201.

Shy, M.E., Evans, V.A., Lublin, F.D., Knobler, R.L., Heiman-Patterson, T., Tamoush, A.J., Parry, G., Schick, P. and DeRyk T.G. (1989) Antibodies to GM1 and GD1b in patients with motor neuron disease without plasma cell dyscrasia. *Ann. Neurol.*, 25: 511–513.

Shy, M.E., Heiman-Patterson, T., Parry, G.J. et al. (1990) Lower motor neuron disease in a patient with antibodies against Gal (β1-3) GalNAc in gangliosides GM1 and GD1b: improvement following immunotherapy. *Neurology*, 40: 842–844.

Siliciano, R.F., Keegan, A.D., Dintzis, R.Z., Dintzis H.M., and Shin, H.S. (1985) The interaction of nominal antigen with T-cell antigen receptors. *J. Immunol.*, 135: 906–914.

Spatz, L.A., Wong, K.K., Williams, M., Desai, R., Golier, J., Berman, J.E., Alt, F.W. and Latov, N. (1990) Cloning and sequence analysis of the variable heavy and light chain regions of an anti-myelin/DNA antibody from a patient with peripheral neuropathy and chronic lymphocytic leukemia. *J. Immunol.*, 144: 2821–2828.

Spatz, L.A., Williams, M., Brender, B., Desai, R. and Latov, N. (1992) DNA sequence analysis and comparison of the variable heavy and light chain regions of the two IgM monoclonal anti-MAG antibodies. *J. Neuroimmunol.*, 36: 29–39.

Stall, A.M., Farinas, C., Tarlington, D.M., Lalor, P.A., Herzenberg, L.A., Strober, S. and Herzenberg, L.A. (1988) Ly-1 B-cell clones similar to human chronic lymphocytic leukemias routinely develop in older normal mice and young autoimmune (New Zealand Black-related) animals. *Proc. Natl. Acad. Sci.*, 85: 7312–7316.

Sthoeger, Z.M., Wakai, M., Tse, D.B., Viciguerra, V.P., Allen, S.I., Budman, D.R., Lichtman, S.M., Schulman, P., Weiselberg, L.R. and Chiorazzi, N. (1989) Production of autoantibodies by CD5 expressing B-lymphocytes from patients with chronic lymphocytic leukemia. *J. Exp. Med.*, 169: 255–268.

Svennerholm, L. and Fredman, P. (1990) Antibody detection in the Guillain-Barré syndrome. *Ann. Neurol.*, 27 (suppl): 36–40.

Tatum, A.H. (1993) Experimental paraprotein neuropathy; demyelation by passive transfer of human Igm anti-myelin associated glycoprotein. *Ann Neurol.*, 33: 502–506.

Thomas, F.P., Lee, A.M., Romas, S. and Latov, N. (1989a) Monoclonal IgMs with anti-Gal (β1-3) GalNAc activity in lower motor neuron disease; identification of glycoprotein antigens in neural tissue and cross reactivity with serum immunoglobulins. *J. Neuroimmunol.*, 23: 167–174.

Thomas, F.P., Adapon, P.H., Goldberg, G.P., Latov, N. and Hays, A.P. (1989b) Localization of neural epitopes that bind to IgM monoclonal autoantibodies (M-proteins) from two patients with motor neuron disease. *J. Neuroimmunol.*, 21: 31–39.

Thomas, F.P., Thomas, J.E., Sadig, S.A., van den Berg, L.H., Latov, N. and Hays, A.P. (1990) Human monoclonal anti-Gal (β1-3) GalNAc autoantibodies bind to the surface of bovine spinal motorneurons. *J. Neuropathol. Exp. Neurol.*, 49: 89–95.

Thomas, F.P., Trojaborg, W., Nagy, C., Vallejos, U., Santoro, M., Sadig, S.A., Latov, N. and Hays, A.P. (1991) Experimental autoimmune neuropathy with anti-GM1 antibodies and immunoglobulin deposits at the nodes of Ranvier. *Acta Neuropathol.*, 82: 378–383.

Trojaborg, W., Galassi, G., Hays, A.P., Lovelace, R.E., Alkaitis, M. and Latov, N. (1989) Electrophysiologic study of experimental demyelination induced by serum of patients with IgM M-proteins and neuropathy. *Neurology,* 39: 1581–1586.

van den Berg, L.H., Marrink, J., de Jager, A.E.J., de Jong, H.J., van Imhoff, G.W., Latov, N. and Sadiq, S.A. (1992) Anti-GM1 antibodies in patients with Guillain-Barré syndrome. *J. Neurol. Neurosurg. Psychiatry*, 55: 6–11.

Vital, A., Vital, C., Julien, J., Baquey, A. and Steck, A.J. (1989) Polyneuropathy associated with IgM monoclonal gammopathy; immunological and pathological study in 31 patients. *Acta. Neuropathol.*, 79: 160–167.

Walsh, F.S., Cronin, M., Koblar, S., Doherty, P., Winer, J., Leon, A. and Hughes, R.A.C. (1991) Association between glycoconjugate antibodies and Campylobacter infection in patients with Guillain-Barré syndrome. *J. Neuroimmunol.*, 34: 43–51.

Willison, H.J., Trapp, B.D., Bacher, J.D., Dalakas, M.C., Griffin, J.W. and Quarles, R.H. (1988) Demyelination induced by intraneural injection of human anti-myelin associated glycoprotein antibodies. Muscle and Nerve, 11: 1169–1176.

Yee, W.C., Hahn, A.F., Hearn, S.A. and Rupar, A.R. (1989) Neuropathy in IgM paraproteinemia; immunoreactivity to neural proteins and chondroitin sulfate. *Acta. Neuropathol.*, 78: 57–64.

Younes-Chennoufi, A.B., Leger, J.M., Hauw, J.J., Preud'homme, J.L., Bouche, P., Aucouturier, P., Ratinahirana, H., Lubetzki, C., Lyon-Caen, O. and Baumann, N. (1992) Ganglioside GD1b is the target antigen for a biclonal IgM in a case of sensory-motor axonal neuropathy: *Ann. Neurol.*, 32: 18–23.

Yuki, N., Yoshino, H., Sato, S. and Miyatake, T. (1990) Acute axonal polyneuropathy associated with anti-GM1 antibodies following Campylobacter enteritis. *Neurology*, 40: 1900–1902.

Yuki, N., Sato, S., Miyatake, T., Sugiyama, K., Katagiri, T. and Sasaki, H. (1991) Motoneuron disease like disorder after ganglioside therapy. *Lancet*, 337: 1109–1110.

Yuki, N., Miyatani, N., Sato., S., Hirabayashi, Y., Yamazaki, M., Yoshimura, N., Hayashi, Y. and Miyatake, T. (1992) Acute relapsing sensory neuropathy associated with IgM antibody against B-series gangliosides. *Neurology*, 42: 686–689.

Zanetti, M. (1986) Idiotypic regulation of autoantibody production. *Crit. Rev. Immunol.*, 6: 151–183.

L. Svennerholm, A.K. Asbury, R.A. Reisfeld, K. Sandhoff, K. Suzuki, G. Tettamanti and G. Toffano (Eds.)
Progress in Brain Research, Vol. 101
© 1994 Elsevier Science BV. All rights reserved.

CHAPTER 23

Gangliosides and related glycoconjugates in myelin: relationship to peripheral neuropathies

A.J. Steck, D. Burger, S. Picasso, T. Kuntzer, E. Nardelli[1] and M. Schluep

Department of Neurology and Laboratory of Neurobiology, Centre Hospitalier Universitaire Vaudois, CH-1011 Lausanne, Switzerland and [1]Department of Neurology, University of Verona, Italy

Introduction

The biology of glycoconjugates and glycoproteins and immunopathological responses to them in reference to diseases of the nervous system is an area of intensive research. Several carbohydrate structures that are target determinants in peripheral nerve diseases have been isolated: the myelin-associated glycoprotein (MAG) and P_0, two myelin glycoproteins and the two glycolipids SGPG and SGLPG bear the L2/HNK-1 carbohydrate epitope that is the target antigen in a human demyelinating neuropathy. Other anti-carbohydrate antibodies, such as antibodies to gangliosides, mostly GM1, are present in patients with lower motor neuron disease or motor neuropathy.

We give here an update on our current work on the oligosaccharide structures of MAG and P_0, discuss the development of an animal model for neuropathy by immunization of chickens with SGPG and SGLPG and review the clinical significance of anti-GM1 antibodies.

Carbohydrates as antigenic determinants in peripheral nerve diseases

MAG and P_0 are two myelin glycoproteins that are both unique to the nervous system. MAG (M_r 100,000) is the major myelin glycoprotein of the central nervous system (CNS), whereas it is a minor pro-

tein constituent of myelin sheaths in both the CNS and the peripheral nervous system (PNS) (for review, see Brady and Quarles, 1988). Human MAG possesses 9 potential N-glycosylation sites (Sato et al., 1989; Spagnol et al., 1989) and contains 30% of carbohydrate by weight (Brady and Quarles, 1988). Because of its localization in periaxonal membranes in PNS and CNS myelin MAG is thought to be involved in glia-neuron and in glia-glia interactions (Brady and Quarles, 1988). P_0 (M_r 28,000) is the major protein component of PNS myelin accounting for 50-60% of the myelin proteins (Greenfield et al., 1973). Human P_0 is a glycoprotein containing 1 unique N-glycosylation site according to cDNA sequencing (Hayasaka et al., 1991) in agreement with the fact that it possesses only 6–8% of carbohydrate by weight (Bollensen et al., 1988). By homophilic interaction of its extracellular domains, P_0 stabilizes the intraperiod line of compact PNS myelin (D'Urso et al., 1990; Filbin et al., 1990; Schneider-Schaulies et al., 1990). MAG and P_0 bear the L2/HNK-1 carbohydrate epitope that is also present in other neural cell adhesion molecules belonging to the immunoglobulin superfamily such as L1 and N-CAM (Schachner, 1989). It has been suggested that the L2/HNK-1 epitope may be involved in cell-cell interactions (Keilhauer et al., 1985; Künemund et al., 1988). Interestingly, this L2/HNK-1 epitope is similar or identical to that recognized by the human M-IgM found in patients with peripheral demyelinating neuropathy associated with gammopathy (McGarry et al., 1983; Schuller-Petrovic et al., 1983; Steck et al., 1983). The epitope for patients' M-IgM and HNK-1 is also present on two PNS 3-sulfoglucuronyl glycosphingolipids referred to as SGPG and SGLPG (Ariga et al., 1987; Ilyas et al., 1990). It has been shown that the presence of the sulfated glucuronic acid was required for the high affinity binding of both patients' and mouse M-IgM to SGPG (Ilyas et al., 1990). Current evidences strongly suggest a causative role for patients' M-IgM directed against the L2/HNK-1 epitope in the neuropathy: (i) a characteristic ultrastructural alteration consisting of a widening of the myelin lamellae in peripheral nerve with a 23-nm spacing between the separated leaflets of the intermediate line (King and Thomas, 1984); (ii) deposits of the corresponding class of heavy and light chain type in the affected nerve (Mendell et al., 1985); (iii) deposits of terminal-complement complex corresponding to the sites of myelin lamellae widening (Monaco et al., 1990); (iv) intraneural injection of M-IgM in the appropriate species causes focal demyelination (Trojaborg et al., 1989); and (v) the production of the characteristic myelin lamellae widening by systemic transfusion of patient's M-IgM in chickens (Tatum, 1989).

Gangliosides are glycosphingolipids which are more abundant in the CNS and PNS (Fishman and Brady, 1976; Ledeen and Yu, 1982) than in other organs. Recent studies indicate that exogenous gangliosides promote extension of neuronal fibres of normal nerves both in vitro (Ledeen, 1984) and in vivo (Gorio and Camignoto, 1984), and the neurite extension of neuronal cell lines in vitro (Ledeen, 1984). On the other hand, exogenous gangliosides seem to alter the morphology and growth of astrocytes in culture, suggesting that gangliosides differently affect neuronal and glial events (Hefti et al., 1985). The ganglioside GM1 is the basic structural unit of mammalian brain gangliosides. The other three major gangliosides of mammalian brain which are GDla, GDlb and GTlb contain additional sialic acid residues attached to the GM1 structure (for review see Ledeen and Yu, 1982). Antibodies reacting with carbohydrate epitopes on GM1 are present in patients with lower motor neuron disease, sensorimotor neuropa-

thy, or motor neuropathy with or without conduction block. Therapeutic reduction of antibody concentrations can result in clinical improvement, suggesting that these antibodies may be pathogenic (for review see Latov, 1990).

Oligosaccharide structures of MAG and P_0

The epitope recognized by patients' M-IgM on MAG, P_0 and glycolipids is also recognized by the mouse monoclonal IgM HNK-1 (McGarry et al., 1983; Sato et al., 1983) that was raised against human natural killer cells (Abo and Blach, 1981). The results obtained by Chou et al. (1985) together with the fact that human M-IgM binds other glycoproteins that are not recognized by HNK-1 (Dennis et al., 1987) suggest that the epitopes recognized by HNK-1 and human M-IgM are closely related but may be different. Other monoclonal antibodies from rat have also been raised against the same epitope which is now referred to as the L2/HNK-1 epitope (Schachner, 1989). The structural characterization and chemical modifications of the carbohydrate moiety of SGPG allowed the identification of the sulfated glucuronic acid as an important determinant for binding of the antibodies belonging to the L2/HNK-1 family (Chou et al., 1985; Ariga et al., 1987; Ilyas et al., 1990). The presence of sulfate groups in the carbohydrate moiety of both MAG and P_0 was demonstrated a long time ago (Matthieu et al., 1975a,b), but the existence of glucuronic acid residues in the oligosaccharide structures of glycoproteins bearing the L2/HNK-1 epitope was demonstrated only in the ependymins of fish brain (Shashoua et al., 1986) although the later study did not directly demonstrate the presence of sulfate on the glucuronic acid residue. Others have failed to demonstrate the presence of a sulfated glucuronic acid residue in N-linked oligosaccharides of bovine P_0 (Uyemura and Kitamura, 1991), human P_0 (Field et al., 1992) or whole murine brains (Wing et al., 1989). The removal of sialic acid and sulfate groups by neuraminidase and methanolysis neutralizes >95% of L2/HNK-1 positive oligosaccharide structures of human P_0 suggesting that glucuronic acid is absent from these structures (Field et al., 1992). It seems therefore that the L2/HNK-1 epitope in glycoproteins is definitely different from that in glycolipids, since it may lack glucuronic acid residue. During the last few years, our attention was focused on the oligosaccharide structure that bears the L2/HNK-1 epitope on human MAG and P_0. In a first study (Burger et al., 1990) we have shown that the epitope for HNK-1 and patient's M-IgM is borne by complex and hybrid-types of N-linked oligosaccharide structures on MAG and P_0. The N-linked oligosaccharides possess a common pentasaccharide core with the structure Man$^-$1-3(Man$^-$1-6)Manβ1-4GlcNacβ1-GlcNAcβ1-N-Asn. As a function of the sugar residues added to this core structure three different types of oligosaccharides can occur: high mannose type structures that contain additional mannose residues exclusively, complex type structures containing additional residues such as N-acetylglucosamine, galactose, fucose and sialic acid, and hybrid types of these two structures. Taking advantage of the fact that N-linked oligosaccharide structures display differential affinities for lectins, we have used serial lectin affinity chromatography to characterize the N-linked oligosaccharide structures of human MAG and P_0 (Burger et al., 1992a) and to identify those structures bearing the epitope for human M-IgM and HNK-1 (Burger et al., 1992b). Both these studies have shown that the L2/HNK-1 epitope is mainly or even exclusively borne by oligosaccharide structures with a $^-$(1-6) fucose residue in the core in MAG and in P_0. This result has been confirmed by those of Field et al. (1992) showing a high degree of fucosylation of L2/HNK-1 positive N-oligosaccharides of human P_0. Both MAG and P_0 bear a number of various N-oligosaccharide structures since no less than 16 and 15 glycopeptide fractions were isolated as a function of their oligosaccharide structure, respectively (Burger et al., 1992a). By comparing the oligosaccharide structures of human MAG and P_0, we have shown that despite a large structural heterogeneity, these glycoproteins display oligosaccharide content homologies that may reflect similarities in their function in the myelination process and maintenance of the myelin structure. A comparative analysis of bovine and human P_0 has shown that both glycoproteins bear very similar

oligosaccharide structures (D. Burger and K. Uyemura, unpublished results), i.e. mainly hybrid type structures as demonstrated by mass spectrometry (Uyemura and Kitamura, 1991). These results, together with those recently published by Field et al. (1992), indicate that the major oligosaccharide structures present on P_0 are of the hybrid type and that these structures may be multisulfated.

In order to identify which of the 9 potential N-glycosylation sites of MAG were glycosylated, purified human MAG was digested with V8 protease and the glycopeptides separated by lectin affinity chromatography on Con A-Sepharose and RCAI-agarose as a function of their carbohydrate moiety. The glycopeptides were further separated by RP-HPLC and the amino acid sequence of each glycopeptide determined. From the 9 glycosylation sites of MAG, 7 were found to be glycosylated, the Asn$_{332}$ was free of carbohydrate and the Asn$_{106}$ was partially glycosylated (D. Burger and A.J. Steck, submitted).

In the nervous system, a number of glycoproteins that are involved in cellular recognition and/or adhesion share common carbohydrate structures including the L2/HNK-1 epitope (Schachner, 1989). These carbohydrate epitopes are well conserved during evolution (Bajt et al., 1990), suggesting that these carbohydrate structures play an important functional role in the development and maintenance of the nervous system tissues. The L2/HNK-1 carbohydrate epitope is present mainly, if not exclusively in molecules involved in adhesion such as MAG, P_0, L1, NILE, Ng-CAM, N-CAM, J1, hexabrachion, tenascin, cytotactin, F11, cytotactin receptor, and integrin (Schachner, 1989). Recent evidence from our laboratory supports the idea that the L2/HNK-1 epitope is also present on MOG (N. Kerlero de Rosbo et al., in press) a new member of the immunoglobulin gene superfamily (M. Gardinier et al., 1992).

Pathogenesis of demyelinating polyneuropathy with anti-MAG antibodies

The pathogenesis of this polyneuropathy raises 3 main questions: which factors or conditions produce the proliferation of the anti-MAG M-IgM secreting B cell clone? how does the M-IgM cross the blood nerve barrier (BNB) to reach their PNS targets? and how are these M-IgM producing such singular abnormalities of myelin structure as the 23-nm spacing of the lamellae at the intermediate line?

We have injected chickens intramuscularly with purified SGPG and SGLPG (Ariga et al., 1987), since these antigens enable to generate polyclonal antibodies that display the same specificity as patients' M-IgM or HNK-1 (Kohriyama et al., 1988). Injected animals showed elevated levels of anti-SGPG antibodies that cross-reacted with MAG as assessed by immunostaining of Western blot of myelin proteins. Currently, the animals are observed for the appearance of leg weakness and nerve conduction abnormalities. The development of this animal model should help characterize the role of autoantibodies in this demyelinating neuropathy.

Clinical relevance of anti-GMl antibodies

Many publications have discussed the presence and role of antibodies to ganglioside GM1 in the serum of patients with motor neuron disorders and peripheral neuropathies (for a review see Pestronk, 1991). Antibodies to GM1 can occur in monoclonal and polyclonal form. Some of the IgM paraproteins that do not react with MAG and SGPG from neuropathy patients exhibit reactivity against ganglioside antigens (Quarles, 1989). Though antibodies to gangliosides often show considerable cross-reactivity and numerous specificities have been reported, two major binding patterns have been recognised: (1) several patients presenting with a lower motor neuron syndrome were identified in which the monoclonal IgM bound to GM1, asialo-GM1 and GDlb (Steck and Adams, 1991). The basis for this cross-reactivity is the terminal disaccharide Galβ1-3GalNAc that is found in all three of these gangliosides; (2) another example of human IgM antibodies from sensory neuropathy patients with monoclonal reactivity against gangliosides includes one that reacts with gangliosides containing a disialosyl configuration (Ilyas et al., 1985). This last antibody has a murine equivalent, clone A2B5, that binds specifically to ganglioside

GQ1b (Kasai and Yu, 1983) as well as other gangliosides containing disialosyl groups. It may not be uninteresting to note that patients who received a murine anti-GD2 monoclonal antibody for melanoma treatment developed a peripheral neuropathy (Dropcho et al., 1992).

Recent publications have analysed the role of polyclonal anti-GM1 antibodies (Adams et al., 1991; Santoro et al., 1992). Attention has been focused on a subgroup of patients presenting with syndromes of motor involvement of the peripheral nervous system in association with high titers of polyclonal anti-GM1 antibodies. In some of these patients electrophysiological studies have revealed multifocal motor conduction blocks. We found in a cohort of patients with motor neuropathy (all except one had multifocal conduction blocks) antibodies to GM1 in 9 out of 10 patients (Adams et al 1991). In contrast, however, Moulonguet et al. (1991) reported a cohort of 18 patients with a pure motor neuropathy and conduction blocks; from these patients, 17 were tested for the presence of anti-GM1 antibodies and only 7 patients were positive. In a more recent study (Table I) we found anti-GM1 antibodies in 47% of the cases. Therefore motor neuropathy may be a disease of different pathogenesis or may even represent distinct conditions with different prognosis. The latter is suggested by the following two observations. First, anti-GM1 antibodies are found frequently in patients presenting with lower motor neuron syndromes, that may be clinically similar to multifocal motor neuropathy but lack conduction block on electrophysiological studies. Our data show that 33% of these patients have anti-GM1 antibodies. Such patients with extensive weakness and atrophy especially late in the course of the disease may be difficult to distinguish from a motor neuron disease. Second, it has been suggested that the motor neuropathy with multifocal persistent conduction blocks may be fatal in a manner similar to motor neuron disease (Magistris and Roth, 1992). Many of the conditions associated with anti-GM1 antibodies share in common the features of an insidious purely motor deficit with fasciculations and sometimes conduction blocks. Whether these clinical syndromes are separate entities or rep-

TABLE I

Results of 289 sera tested for anti-GM1 antibodies by ELISA. Clinical data, including (a) neurological diagnosis(ses); (b) associated non-neurological disease(s); (c) electrophysiological tests; and (d) neuroimaging or nerve biopsy were obtained from the various centers from which sera were addressed for testing. This table summarizes the percentages of positive and negative anti-GM1 titers obtained in different neurological disorders. ALS: amyotrophic lateral sclerosis; LMNS: lower motor neuron syndrome; S-M PN: sensory-motor polyneuropathy; GBS: Guillain-Barré Syndrome; OND: other neurological disease

Dignosis	Positive anti-GM1 antibody titer		Negative anti-GM1 antibody titer	
ALS	31/113	(27.4%)	82/113	(72.6%)
LMNS	18/54	(33.3%)	36/54	(66.7%)
Motor Neuropathy	17/36	(47.2%)	19/36	(52.8%)
S-M PN	13/41	(31.7%)	28/41	(68.3%)
Acute GBS	4/8		4/8	
Chronic GBS	8/15	(53.3%)	7/15	(46.7%)
Sensory PN	2/7		5/7	
OND	5/23	(21.7%)	18/23	(78.3%)

resent both ends of the same condition remains to be shown.

Very few patients with anti-GM1 antibodies had post-mortem studies performed (Steck and Adams, 1991). We recently studied a patient with a progressive lower motor neuron syndrome with conduction blocks and high titers of anti-GM1 antibodies (Adams et al., in press). Neuropathological findings included a predominantly proximal motor neuropathy with multifocal IgM deposits on myelinated nerve fibers associated with a secondary loss of spinal motor neurons. The exact localization of the Ig deposits could not be ascertained but the pattern of deposits was compatible with the localisation at the

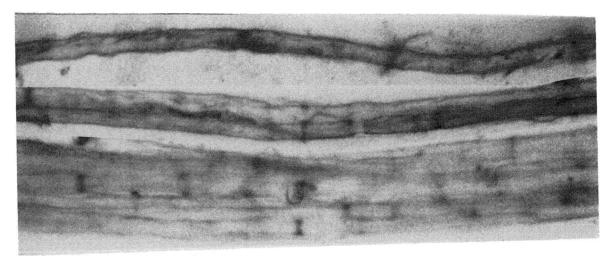

Fig. 1. Teased nerve fibers from rabbit sciatic nerve showing immunostaining of the nodes of Ranvier after incubation with a serum from a patient with multifocal motor neuropathy and anti-GM1 antibodies (1:200, ABC technique)

nodes of Ranvier as described in experimental models (Thomas et al., 1991). These findings support a probable autoimmune origin of this lower motor neuron syndrome with retrograde degeneration of spinal motor neurons.

Gangliosides are important components of nerve membrane in the peripheral and central nervous system, concentrated at the node of Ranvier and at synaptic terminals. It is interesting to note that IgM deposits have been localised at the nodes of Ranvier in a patient presenting with anti-GM1 antibodies and multifocal motor conduction block (Santoro et al., 1991). Furthermore, sera from some patients with anti-GM1 antibodies bind preferentially on peripheral nerve to the node of Ranvier as shown by indirect immunocytochemistry (Fig. 1). It is, however, not clear under which conditions the antibodies gain access to their targets and why there is a preferential vulnerability of motor nerves. Evidence of interaction with the motor system was provided by the demonstration that anti-GM1 antibodies bind to nerve terminals (Schluep and Steck, 1988). Further pathological studies are necessary in patients with high titers of anti-GM1 antibodies to further identify the targets and role of anti-GM1 antibodies.

Conclusion

Most of the antibodies that may be implicated in the pathogenesis of autoimmune neuropathies are directed against carbohydrate epitopes of glycoproteins or glycolipids. The immune system displays distinct reactivities against carbohydrate and proteins. It is assumed that carbohydrates are mostly recognised by B cells, whereas proteins are usually recognised by both T and B cells. The study of the B lymphocytes from normal subjects has shown that a discrete (CD5+)B-cell subset is responsible for the production of what has been described as polyreactive or natural antibodies (Casali and Notkins, 1989). Using the EBV technology to study the B-cell repertoire of patients with autoimmune disease has revealed that in contrast to normal subjects, a significantly higher frequency of B lymphocytes was committed to the production of monoreactive autoantibodies relevant to

the particular disease. These studies should lead to a better definition of the B cell repertoire in patients with anti-carbohydrate antibodies and help elucidate the origin of the auto-antibody producing B-cells.

Acknowledgements

The experiments reported here are carried out with the support of the Swiss National Science Foundation, the Swiss Multiple Sclerosis Society and the Roche Research Foundation.

References

Abo, T. and Blach, C.M. (1981) A differentiation antigen of human NK and K cells identified by a monoclonal antibody (HNK-1). J. Immunol., 127: 1024–1029.

Adams, D., Kuntzer, T., Burger, D., Chofflon, M., Magistris, M.R., Regli, F. and Steck, AJ. (1991) Predictive value of anti-GM1 ganglioside antibodies in neuromuscular diseases: a study of 180 sera. J. Neuroimmunol., 32: 223–230.

Adams, D., Kuntzer, T., Steck, A.J., Lobrinus, A., Janzer, R. and Regli, F. (in press) Motor conduction block and high titer of anti-GM1 ganglioside antibodies: pathological evidence of a motor neuropathy. J. Neurol. Neurosurg. Psychiatry.

Ariga, T., Kohriyama, T., Freddo, L., Latov, N., Saito, M., Kon, K., Ando, S., Suzuki, M., Hemling, M.E., Rinehart, K.L., Kusunoki, S. and Yu, R.K. (1987) Characterization of sulfated glucuronic acid containing glycolipids reacting with IgM M-proteins in patients with neuropathy. J. Biol. Chem., 262: 646–653.

Bajt, M.L., Schmitz, M., Schachner, M. and Zipser, B. (1990) Carbohydrate epitopes involved in neural cell recognition are conserved between vertebrates and leech. J. Neurosci. Res.,27: 276–285.

Bollensen, E., Steck, A.J. and Schachner, M. (1988) Reactivity with the peripheral myelin glycoprotein P_0 in sera from patients with IgM monoclonal gammopathy and polyneuropathy. Neurology, 38: 1266–1270.

Brady, R.O. and Quarles, R.H. (1988) Developmental and pathophysiological aspects of the myelin-associated glycoprotein. Cell. Mol. Neurobiol.,8: 139–148.

Burger, D. and Steck, A.J. (1992) N-oligosaccharide structures and N-glycosylated sites of MAG and P_0, the two major human myelin glycoproteins. Neurochem. Int. (in press).

Burger, D., Simon, M., Perruisseau, G. and Steck, A.J. (1990) The epitope(s) recognized by HNK-1 antibody and IgM paraprotein in neuropathy is present on several N-linked oligosaccharide structures on human P_0 and myelin-associated glycoprotein. J. Neurochem., 54: 1569–1575.

Burger, D., Perruisseau, G., Simon, M. and Steck, A.J. (1992a) Comparison of the N-linked oligosaccharide structures of the two major human myelin glycoproteins MAG and P_0:

Assessment and relative occurrence of oligosaccharide structures by serial lectin affinity chromatography of [^{14}C]glycopeptides. *J. Neurochem.*, 58: 845–853.

Burger, D., Perruisseau, G., Simon, M. and Steck, A.J. (1992b) Comparison of the N-linked oligosaccharide structures of the two major human myelin glycoproteins MAG and P$_0$: Assessment of the structures bearing the epitope for HNK-1 and human monoclonal immunoglobulin M found in demyelinating neuropathy. *J. Neurochem.*, 58: 854–861.

Casali, P. and Notkins, A.B. (1989) Probing the human B-cell repertoire with EBV: polyreactive antibodies and CD5+ B lymphocytes. *Annu. Rev. Immunol.*, 7: 513–535.

Chou, D.K.H, Ilyas, A.A., Evans, J.E., Quarles, R.H. and Jungalwala, F.B. (1985) Structure of a glycolipid reacting with monoclonal IgM in neuropathy and with HNK-1. *Biochem. Biophys.Res. Commun.*, 128: 383–388.

Dennis, J.W., Laferte, S., Waghorne, C., Breitman, M.L. and Kerbel, R.S. (1987) Beta 1-6 branching of Asn-linked oligosaccharides is directly associated with metastasis. *Science,*236: 582–585.

Dropcho, E.J., Saleh, M.N., Grizzle, W.E. and Oh, S.J. (1992) Peripheral neuropathy following treatment of melanoma with a murine anti-GD2 monoclonal antibody (MoAb). *Neurology,* 42 (Suppl 3): 457.

D'Urso, D., Brophy, P.J., Staugaitis, S.M., Gillespie, S., Frey, A.B., Stempak, J.G. and Colman, D.R. (1990) Protein zero of peripheral nerve myelin: biosynthesis, membrane insertion, and evidence for homotypic interaction. *Neuron,* 2: 449–460.

Field, M.C., Wing, D.R., Dwek, R.A., Rademacher, T.W., Schmitz, B., Bollensen, E. and Schachner, M. (1992) Detection of multisulphated N-linked glycans in the L2/HNK-1 carbohydrate epitope expressing neural adhesion molecule P$_0$. *J. Neurochem.*, 58: 993–1000.

Filbin, M.T., Walsh, F.S., Trapp, B.D., Pizzey, J.A. and Tennekoon, G.I. (1990) Role of myelin P$_0$ protein as a homophilic adhesion molecule. *Nature,* 344: 871–872.

Fishman, P.H. and Brady, R.O. (1976) Biosynthesis and function of gangliosides. *Science,* 194: 906–915.

Gardinier, M.V., Amoguet, P., Linington, C. and Matthieu, J.M. (1992) Myelin/oligodendrocyte glycoprotein is a unique member of the immunoglobulin superfamily. *J. Neurosci. Res.*, 33: 177–187.

Gorio, A. and Camignoto, G. (1984) Enhancing reinnervation of muscle by ganglioside treatment. In: G. Serratrice, D. Cros, C. Desnuelle, J.L. Gastaut, J.F. Pellissier, J. Pouget, and A. Sachino (Eds), *Neuromuscular Diseases,* Raven Press, NY, pp. 287–292.

Greenfield, S., Brostoff, S., Eylar, E.H. and Morell, P. (1973) Protein composition of myelin of the peripheral nervous system. *J. Neurochem.*, 20: 1207–1216.

Hayasaka, K., Nanao, K., Tahara, M., Sato, W., Takada, G., Miura, M. and Uyemura, K. (1991) Isolation and sequence determination of cDNA encoding the major structural protein of human peripheral myelin. *Biochem. Biophys. Res.Commun.*, 180: 515–518.

Hefti, F., Hartikka, J. and Frick, W. (1985) Gangliosides alter morphology and growth of astrocytes and increase the activity of choline acetyltranferase in cultures of dissociated septal cells. *J. Neurosci.*, 5: 2086–2094.

Ilyas, A.A., Quarles, R.H., Dalakas, M.C., Fishmann, P.H. and Brady, R.O. (1985) Monoclonal IgM in a patient with paraproteinemic polyneuropathy binds to gangliosides containing disialosyl groups. *Ann Neurol.*, 18: 655–659.

Ilyas, A.A., Chou, D.K.H., Jungalwala, F.B., Costello, C. and Quarles, R.H. (1990) Variability in the structural requirements for binding of human monoclonal anti-myelin-associated glycoprotein immunoglobulin M antibodies and HNK-1 to sphingolipid antigens. *J. Neurochem.*, 55: 594–601.

Kasai, N. and Yu, R.K. (1983) The monoclonal antibody A2B5 is specific to ganglioside GQ1c. *Brain Res.*, 277: 155–158.

Keilhauer, G., Faissner, A. and Schachner, M. (1985) Differential inhibition of neurone-neurone, neurone-astrocyte and astrocyte-astrocyte adhesion by L1, L2, and N-CAM antibodies. *Nature,* 316: 728–730.

King, R.H.M. and Thomas, P.K. (1984) The occurrence and significance of myelin with unusually large periodicity. *Acta Neuropathol.*, 63: 319–329.

Kohriyama, T., Ariga, T. and Yu, R.K. (1988) Preparation and characterization of antibodies against a sulfated glucuronic acid-containing glycosphingolipid. *J. Neurochem.*, 51, 869–877.

Künemund, V., Jungalwala, F.B., Fischer, G., Chou, D.K.H., Keilhauer, G. and Schachner, M. (1988) The L2/HNK-1 carbohydrate of neural cell adhesion molecules is involved in cell interactions. *J. Cell Biol.*, 106: 213–223.

Latov, N. (1990) Neuropathy and anti-GM1 antibodies. *Ann. Neurol.*, 27 (suppl), S41–S43.

Ledeen, R.W. (1984) Biology of gangliosides: neuritogenic and neuronotrophic properties. *J. Neurosci. Res.*, 12: 147–159.

Ledeen, R.W. and Yu, R.K. (1982) Gangliosides: structure, isolation, and analysis. *Methods Enzymol.*, 83: 139–191.

Magistris, M. and Roth, G. Motor neuropathy with multifocal persistent conduction blocks. *Muscle Nerve,* 15: 1056–1057.

Matthieu, J.M., Everly, J.L., Brady, R.O. and Quarles, R.H. (1975a) [^{35}S]sulfate incorporation into myelin glycoproteins. II. Peripheral nervous system. *Biochim. Biophys. Acta,* 392: 167–174.

Matthieu, J.M., Quarles, R.H., Poduslo, J.F. and Brady, R.O. (1975b) [^{35}S]sulfate incorporation into myelin glycoproteins. I. Central nervous system. *Biochim. Biophys. Acta,* 392: 159–166.

McGarry, R.C., Helfand, S.L., Quarles, R.H. and Roder, J.C. (1983) Recognition of myelin-associated glycoprotein by the monoclonal antibody HNK-1. *Nature,* 306: 376–378.

Mendell, J.R., Sahenk, Z., Whitaker, J.N., Trapp, B.D., Yates, A.J., Griggs, R.C. and Quarles, R.H. (1985) Polyneuropathy and IgM monoclonal gammopathy: studies on the pathogenic role of anti-myelin-associated glycoprotein antibody. *Ann. Neurol.* 17: 243–254.

Monaco, S., Bonetti, B., Ferrari, S., Moretto, G., Nardelli, E., Tedesco, F., Mollnes, T.E., Nobile-Orazio, E., Manfredini, E.,

Bonazzi, L. and Rizzuto, N. (1990) Complement-mediated demyelination in patients with IgM monoclonal gammopathy and polyneuropathy. *N. Engl. J. Med.* 322: 649–652.

Moulonguet, A., Bouche, P., Adams, D., Ropert, A., Leger, J.M., Said, G. and Meininger, V. (1991) Patterns of conduction blocks motor neuropathy: a review of 18 patients. *Neurology,* 41 (suppl): 260.

Pestronk, A. (1991) Motor neuropathies, motor neuron disorders, and antiglycolipid antibodies. *Muscle Nerve,* 14: 927–936.

Quarles, R.H. (1989) Human monoclonal antibodies associated with neuropathy. *Methods Enzymol.,* 179: 927–936.

Santoro, M., Thomas, F.P., Fink, M.E., Matthew, E., Lange, D.J., Latov, N. and Hays, A.P. (1991) IgM deposits at nodes of Ranvier in a patient with amyotrophic lateral sclerosis, anti-GM1 antibodies and multifocal motor conduction block. *Ann. Neurol.,* 28: 373–377.

Santoro, M., Uncini, A., Corbo, M., Staugaitis, S.M., Thomas, F.P., Hays, A.P. and Latov, N. (1992) Experimental conduction block induced by serum from a patient with anti-GM1 antibodies. *Ann. Neurol.,* 31: 385–390.

Sato, S., Baba, H., Tanaka, M., Yanagisawa, K. and Miyatake, T. (1983) Antigenic determinant shared between myelin-associated glycoprotein from human brain and natural killer cells. *Biomed. Res.,* 4: 489–494.

Sato, S., Fujita, N., Kurihara, T., Kuwano, R., Sakimura, K., Takahashi, Y. and Miyatake, T. (1989) cDNA cloning and amino acid sequence for human myelin-associated glycoprotein. *Biochem. Biophys. Res. Commun.,* 163: 1473–1480.

Schachner, M. (1989) Families of neural adhesion molecules. In: *Carbohydrate Recognition in Cellular Function. Ciba foundation symposium.* John Wiley & Son, Chichester, pp. 156–168.

Schluep, M. and Steck, A.J. (1988) Immunostaining of motor nerve terminals by IgM M protein with activity against gangliosides GM1 and GDlb from a patient with motor neuron disease. *Neurology,* 38: 1890–1892.

Schneider-Schaulies, J., von Brunn, A. and Schachner, M. (1990) Recombinant peripheral myelin protein P_0 confers both adhesion

and neurite outgrowth-promoting properties. *J. Neurosci. Res.,* 27: 286–297.

Schuller-Petrovic, S., Gebhart, N., Lassmann, H., Rumpold, H. and Kraft, D. (1983) A shared antigenic determinant between natural killer cells and nervous tissue. *Nature,* 306: 179–181.

Shashoua, V.E., Daniel, P.F., Moore, M.E. and Jungalwala, F.B. (1986) Demonstration of glucuronic acid on brain glycoproteins which react with HNK-1 antibody. *Biochem. Biophys. Res. Commun.,* 183: 902–909.

Spagnol, G., Williams, M., Srinivasan, J., Golier, J., Bauer, D., Lebo, R.V. and Latov, N. (1989) Molecular cloning of human myelin-associated glycoprotein. *J. Neurosci. Res.,* 24: 137–142.

Steck, A.J. and Adams, D. (1991) Motor neuron syndromes and monoclonal IgM antibodies to gangliosides. *Adv. Neurol.,* 56: 421–425.

Steck, A.J., Murray, N., Meier, C., Page, N. and Perruisseau, G. (1983) Demyelinating neuropathy and monoclonal IgM antibody to myelin-associated glycoprotein. *Neurology,* 33, 19–23.

Tatum, A.H. (1989) Experimental IgM anti-myelin paraprotein demyelinating neuropathy: ultrastructural characterization. *Ann. Neurol,* 26: 298 (abstract).

Thomas, F.P., Trojaborg, W., Nagy, C., Santoro, M., Sadiq, S.A., Latov, N. and Hays, A.P. (1991) Experimental autoimmune neuropathy with anti-GM1 antibodies and immunoglobulin deposits at the nodes of Ranvier. *Acta Neuropathol.,* 82: 378–383.

Trojaborg, W., Galassi, G., Hays, A.P., Lovelace, R.E., Alkaitis, M. and Latov, N. (1989) Electrophysiologic study of experimental demyelination induced by serum of patients with IgM M-proteins and neuropathy. *Neurology,* 39: 1581–1586.

Uyemura, K. and Kitamura, K. (1991) Comparative studies on myelin proteins in mammalian peripheral nerve. *Comp. Biochem. Physiol.,* 98C: 63–72.

Wing, D.R., Field, M.C., Schmitz, B., Schachner, M., Dweck, R.A. and Rademacher, T.W. (1989) Use of whole brain oligosaccharides as probe for a functional neural epitope. *Proceeding of the Xth International Symposium on Glycoconjugates,* Jerusalem, Israel. pp 374–375.

L. Svennerholm, A.K. Asbury, R.A. Reisfeld, K. Sandhoff, K. Suzuki, G. Tettamanti and G. Toffano (Eds.)
Progress in Brain Research, Vol. 101

CHAPTER 24

Antiglycolipid antibodies and peripheral neuropathies: links to pathogenesis

John W. Griffin

Johns Hopkins University, School of Medicine, Baltimore, MD, U.S.A.

Introduction

My charge in this Symposium can be simply summarized: to assess critically the pathogenetic role of antiglycolipid antibodies in producing neuropathies. For most of the now sizable number of neuropathies in which antibodies against glycoconjugates have been reported (see Asbury, 1993) there are no data to support or exclude a pathogenetic role for the antibodies. This presentation will examine in detail two classes of antibodies in which there is more substantial pathogenetic information — the so-called 'anti-MAG' antibodies and the 'anti-GM1' antibodies. I will conclude that in the former instance there is good evidence that the antiglycoconjugate antibody does cause the neuropathy, but that the nature of the epitope remains unresolved. For the anti-GM1 antibodies there is increasing evidence that at least one of the disorders associated with the antibody is immune-mediated, but it is premature to conclude that the antibody causes the disease.

Neuropathy associated with 'anti-MAG' antibodies

Twelve years ago Latov et al. (1980) identified a patient with a peripheral neuropathy and a monoclonal IgM paraprotein that bound to myelin. Subsequently, the monoclonal IgM antibodies in many such patients were shown to recognize a specific glycoprotein, the myelin-associated glycoprotein (MAG) (Braun et al., 1982; Ilyas et al., 1984; Gregson and Leibowitz, 1985; Mendell et al., 1985a). Of all the patients with IgM paraproteins and peripheral neuropathy, as many as 50% may have evidence of such 'anti-MAG' antibodies. The clinical manifestations associated with 'anti-MAG' antibody usually entail a slowly progressive neuropathy, with physiologic evidence of demyelination as well as variable degree of axonal loss, and with a clinical picture that is often dominated by large-fiber sensory dysfunction and gait ataxia. The pathologic picture includes demyelination (Fig. 1A) and remyelination with a variable number of onion bulbs — whorls of supernumerary Schwann cells surrounding nerve fibers and reflecting the Schwann cell proliferation that accompanies demyelination (Fig. 1B). In addition, a distinctive feature is 'wide spacing' of myelin lamellae — in some fibers the distance between several neighboring lamellae of the myelin sheath increases from about 12 to about 24 nm (Figs. 1C and D).

MAG, initially identified by Quarles et al. (1973), is a component of peripheral nerve Schwann cells, as well as oligodendrocytes in the CNS. It is distinguished among the myelin proteins by its extensive glycosylation, which contributes 30% of MAGs total molecular weight. Steck (1993) will review the extracellular consensus sites for N-linked glycosylation on MAG, and the sulfation of these MAG oligosaccha-

314

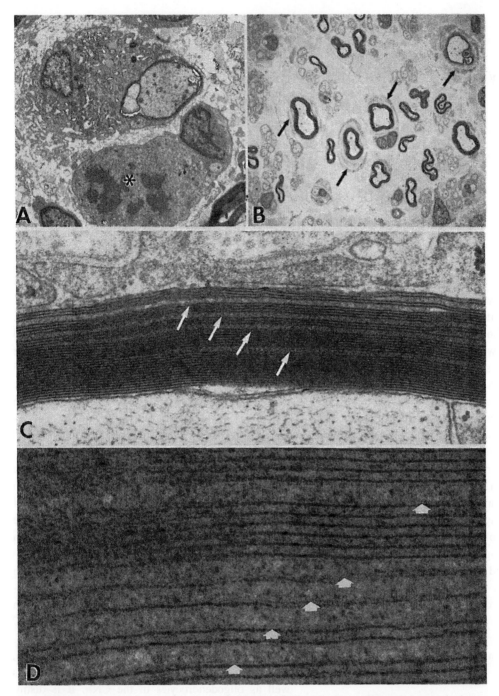

Fig. 1. Pathologic changes in neuropathy associated with 'anti-MAG' antibodies. *A*. Electron micrograph of a transverse section from a sural nerve biopsy. This fiber is undergoing demyelination. Note that a neighboring Schwann cell (asterisk) contains a mitotic figure, reflecting the associated Schwann cell proliferation. *B*. Several fibers are surrounded by supernumerary Schwann cells forming minor onion bulbs. These structures reflect earlier Schwann cell proliferation associated with demyelination. *C*. This remyelinated fiber contains severely widely spaced myelin lamellae (arrows). *D*. At high magnification, another fiber has several major dense lines (arrows) separated by interlamellar distances that are approximately twice the normal periodicity.

rides (Matthieu et al., 1975; Noronha et al., 1986). For the present purposes the essential point is that the epitope seen by the human IgM 'anti-MAG' antibodies contains sulfated oligosaccharides and is also recognized by a series of murine monoclonal antibodies including HNK-1 and (usually) L2 (Steck, 1993). Shortly after the identification of the MAG reactivity of the human IgM paraproteins, their cross-reactivity with some lower molecular weight nerve glycoproteins (O'Shannessy et al., 1986) and with two nerve glycolipids (Chou et al., 1985; Quarles et al., 1986; Ilyas et al., 1986) was also recognized. These glycolipids include sulfate-3-glucuronylneolactotetraosyl-ceramide (here abbreviated SGNLC and often termed 'sulfate-3-glucuronyl paragloboside' [SGPG]) and sulfate-3-glucuronyl-N-acetylactosaminyl paragloboside (SGLPG).

Do these antibodies cause neuropathy? And if so, what is the molecular target and the immunopathogenesis? I will review first new data on passive transfer of neuropathy to experimental animals. It will be useful to summarize the cellular biology of the myelin-associated glycoprotein itself, before considering whether MAG, other glycoproteins, or glycolipids might be the pathogenetically important molecular targets of the antibodies.

IgM monoclonal 'anti-MAG' antibodies can transfer the essential features of the human neuropathy to experimental animals

Are 'anti-MAG' antibodies pathogenic? The demonstration of antibodies directed against a Schwann cell constituent in patients with demyelinating nerve disease suggested the obvious inference: these paraproteins cause nerve damage directly. An important piece of supporting information was the demonstration of IgM within the regions of myelin wide-spacing in the peripheral nerves of patients with 'anti-MAG' antibodies (Mendell et al., 1985b) (Fig. 2). However, attempts to transfer the disease passively have, until recently, been unsatisfying, and the possibility that these antibodies might occur as a consequence of long-standing neuropathy, rather than as the cause of it,

has been raised (Mendell et al., 1985b; Quarles et al., 1992).

The initial passive transfer studies utilizing systemic administration of 'anti-MAG' paraproteins were unsuccessful, perhaps reflecting the difficulty in passage of IgM through the blood-nerve barrier. Direct injection of these paraproteins into the nerve produced demyelination in the cat (Hays et al., 1987; Willison et al., 1988). However, the resulting lesions were inflammatory and their production was complement-dependent. Both of these features differed from the indolent, non-inflammatory demyelination found in man.

Recent data from Tatum (1992) have removed any doubt about the capacity of these antibodies to produce demyelination and the characteristic wide-spacing of myelin. Tatum chose the young chick as the experimental animal, reasoning that, similar to man, the chick has a high concentration of the HNK-1 epitope on its MAG molecules and on the lower molecular weight myelin glycoproteins. IgM from a patient with a high-titer IgM 'anti-MAG' paraprotein was administered intraperitoneally into chicks daily, beginning on the second day after hatching. In vitro studies documented reduced velocities of nerve conduction, a characteristic feature of demyelination, within 8 weeks. The histopathologic findings were dramatic. A high proportion of axons developed myelin wide-spacing, particularly in their outer lamellae. Clear-cut demyelination with macrophage-mediated myelin removal, but without T cell infiltration, was prominent with higher doses, and minor onion bulbs were generated.

The human IgM selectively localized to the regions of the Schwann cells that normally contain MAG. The demyelination in these young chicks was more severe than in older animals, suggesting that periods of myelin formation, presumably including development, remyelination, and regeneration, may be times of particular susceptibility to 'anti-MAG' antibodies. Whether the antibodies inhibit remyelination remains uncertain, but the experiments of Tatum dramatically demonstrate the potential of antibodies directed against the HNK-1/L2 epitope to produce demyelination.

316

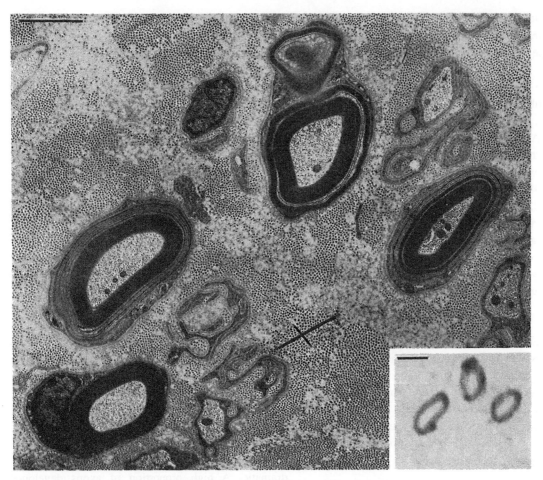

Fig. 2. This sural nerve biopsy from a patient with 'anti-MAG' antibodies illustrates the deposition of IgM in sites of myelin wide-spacing. IgM is visualized immunocytochemically within the myelin sheaths of the fibers identified by arrows in the inset (scale marker = 10 μm). The electron micrograph (marker = 5 μm) shows that these fibers have widely spaced myelin. (Modified from Mendell et al. [1985a], courtesy of Dr. Bruce Trapp. Reprinted by permission of the *Annals of Neurology*.)

What is the target molecule responsible for demyelination by 'anti-MAG' antibodies?

The HNK-1/L2 epitope. The sulfated oligosaccharides detected by these antibodies are common to a variety of cell adhesion molecules in addition to MAG (as described below), and they have an extensive distribution outside as well as within the nervous system. MAG from man, cats, and chickens expresses abundant HNK-1/L2 antigen, whereas MAG from rodents contains relatively little (O'Shannessy et al.,

1985). Even though normally present at low levels, the epitope can have biologic importance in rodents. For example, the HNK-1/L2 epitope is present particularly within ventral roots and motor fibers (Schachner, 1990; Martini et al., 1992). Recent evidence suggests that this epitope may selectively facilitate outgrowth of motor axons during development and nerve regeneration (Martini et al., 1988; Martini et al., 1992). These species differences have undoubtedly confounded the elucidation of the pathogenetic significance and pathogenetic mechanisms of the 'anti-MAG' antibodies. In addition, there are variations in HNK-1- and L2-reactive epitopes, as summarized by Steck (1993).

Another source of initial confusion has been the diffuse staining of myelin sheaths by the 'anti-MAG' antibodies; the initial observations were made at a time when there was controversy about whether MAG might be a constituent of compact myelin (Webster et al., 1983; Favilla et al., 1984). The distinctive localizations of the MAG polypeptide within the Schwann cell (Trapp and Quarles, 1984; Trapp et al., 1984; Trapp et al., 1989b), described below, have now been abundantly confirmed. The myelin staining by the 'anti-MAG' paraproteins is now presumed to represent, at least in part, the presence of SGNLC and SGLPG and probably the lower molecular weight glycoproteins (Ilyas et al., 1986; Welcher et al., 1991; Snipes et al., 1992) in compact myelin.

The cell biology of MAG. MAG is an approximately 100-kDa glycoprotein that is found in oligodendrocytes in the central nervous system (CNS) as well as in Schwann cells of the peripheral nervous system (PNS). Its amino acid sequence, as deduced from cDNA clones (Arquint et al., 1987; Salzer et al., 1987; Lai et al., 1988), indicates that it is an intrinsic membrane glycoprotein and a member of the immunoglobulin gene (IgG) superfamily, a family that includes such adhesion molecules as the nerve cell adhesion molecule (NCAM), the neuronal-glial cell adhesion molecule (NG-CAM), and the L1 protein (Mirsky et al., 1986; Martini and Schachner, 1986; Trapp et al., 1989b; Burgoon et al., 1991) (see Edelman, 1993). MAG contains the Arg-Gly-Asp (RGD) sequence, suggesting that MAG may mediate interactions with other transmembrane or extracellular proteins. Two different cytoplasmic domains for MAG are generated by alternative splicing, resulting in a 72-kDa polypeptide (L-MAG) and a 67-kDa polypeptide (S-MAG). In the peripheral nervous system, at least 95% of the total MAG is S-MAG. In addition to its glycosylation, MAG undergoes a number of other post-translational modifications, including phosphorylation, acylation, and post-translational palmitylation (Pedraza et al., 1990).

MAG is found in myelin preparations, but it is not a component of compact myelin. Rather, the distribution of the myelin-associated glycoprotein corre-

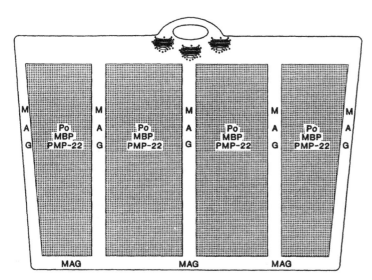

Fig. 3. Schematic diagram of an 'unrolled' Schwann cell, illustrating the distribution of the proteins of compact myelin (stippled), including P$_0$, myelin basic protein (MBP), and the lower molecular weight glycoprotein PMP-22, and of MAG within the noncompacted regions containing cytoplasmic channels. These sites include the adaxonal plasmalemma, the paranodal loops, and the Schmidt-Lanterman incisures. (Courtesy of Dr. Bruce Trapp.)

sponds to the periaxonal membranes and the regions of non-compacted myelin (Trapp and Quarles, 1984; Trapp et al., 1984; Trapp et al., 1989a,b). In a peripheral internode, the distribution of MAG is as illustrated in Fig. 3. MAG is found in the adaxonal Schwann cell membrane, in the paranodal loops, and in the cytoplasmic extensions of the Schwann cell cytoplasm that extend through the myelin sheath and are termed Schmidt-Lanterman incisures. In addition, MAG is present in the internal and external mesaxons (Trapp et al., 1989a). There is no evidence that MAG is present on the abaxonal Schwann cell plasmalemma, except at the mesaxons.

This pattern of localization within the peripheral nervous system has led to the suggestion that MAG functions as a membrane spacer, preventing the apposition of membranes more closely than 12–14 nm, and in particular preventing compaction of myelin and formation of major dense lines where MAG is present (for review see Quarles et al., 1992). Within the compact myelin, MAG is apparently replaced by the P_0 glycoprotein. The localization of MAG in Schmidt-Lanterman incisures, paranodal loops, and mesaxons correlates closely with the distribution of F-actin in myelin-forming Schwann cells, suggesting that the C-terminal cytoplasmic domain of MAG may interact with actin microfilaments, and may be involved in the mechanism of cellular motility required for myelination (Trapp et al., 1989b).

Finally, the location of MAG in the periaxonal membrane and its specificity for myelin-forming Schwann cells (compared to non-myelinating Schwann cells) have led to the suggestion that it is a necessary adhesion molecule for myelination. The expression of recombinant MAG in primary Schwann cells has been shown to promote the initial investment of axons by Schwann cells (Owens et al., 1990). Anti-sense techniques have been used to reduce expression of MAG mRNA in cultured Schwann cells (Owens and Bunge, 1991). Reduced MAG levels were associated with failure of myelination and failure of segregation of axons into the characteristic 1:1 relationship with Schwann cells. These data strongly support the presumption that MAG is essential for Schwann cell-axolemmal interactions that result in sequestration, ensheathment, and myelination. The nature of the axolemmal 'receptor' for MAG is the subject of current investigations.

Other glycoproteins expressing the HNK-1/L2 epitope. The reactivity of the human 'anti-MAG' antibodies with a group of lower-molecular-weight glycoproteins has received less attention, but is of potential importance. O'Shannessy et al. (1986) demonstrated that several bands in the range of 19–28 kDa were prominently stained by these monoclonal antibodies. Unlike MAG itself, these bands are found only within the PNS. It is noteworthy that a recently defined glycoprotein, PMP-22 (Welcher et al., 1991; Snipes et al., 1992), is within this molecular weight range. The presence of PMP-22 appears to be required for normal PNS myelination; its absence in the Trembler mouse strain is associated with hypomyelination (Kohriyama et al., 1988), and abnormalities of PMP-22 may underlie a prevalent heritable demyelinating neuropathy of man, Charcot-Marie-Tooth disease type I. This possibility is suggested by the fact that in man PMP-22 is coded by the region of chromosome 17 that is known to contain a microduplication in many of the CMT-1 families.

Lipid antigens expressing the HNK-1 epitope. Whether or not the SGNLC or other glycolipids are pathogenetically important targets in the human neuropathy remains unresolved. Injection into rat nerves of antibodies raised against SGNLC has been reported to produce demyelination (Maeda et al., 1991a). Rats sensitized with SGNLC develop axonal degeneration in the dorsal columns (Maeda et al., 1991b). Because cerebral microvessels express this marker, the possibility of vascular damage contributing to the nervous system injury has been raised (Miyatani et al., 1990). Rabbits immunized with SGNLC develop weakness and electrophysiologic changes in the sciatic nerves but no pathologic assessment is available (Kohriyama et al., 1988). At the present time, the evidence for antibodies against SGNLC and SGLPG as pathogenetic agents in demyelinating neuropathy remains fragmentary.

Conclusions

The localization of the IgM in the regions of myelin wide-spacing in the human nerves and the passive transfer of both demyelinating activity and myelin wide-spacing by injection of 'anti-MAG' antibodies into animals provide convincing evidence that these monoclonal antibodies are directly responsible for the nerve damage. The nature of the responsible target antigen is less certain. Recently, the affinity of a number of IgM paraproteins for MAG and for SGNLC has been examined (Brouet et al., 1992). All of the human monoclones had a 10–100-fold higher affinity for MAG than for SGNLC. However, it is uncertain how such in vitro assays might translate into the in vivo setting, in which the gangliosides are present within membranes. In addition, a small amount of antibody might produce disease by binding to a critical epitope, whereas abundant antibody might have no pathogenetic effect when bound to other epitopes. MAG, the lower molecular weight glycoproteins, and the glycolipids all remain viable candidates as the target antigen. The data summarized above suggest a series of possibilities by which antibodies reactive to epitopes on MAG, other PNS glycoproteins, or glycolipids might interfere with the process of myelination, with myelin maintenance, or with axon-Schwann cell interactions, but the precise immunopathogenesis remains to be established.

Multifocal motor neuropathy

Antibodies against GM1 and related glycolipids have been identified in several disorders that involve the motor neuron or motor nerve fiber (Sadiq et al., 1990; see Latov, 1993), including some cases of amyotrophic lateral sclerosis (ALS), a progressive and usually fatal degeneration of motor neurons. In many of these syndromes the clinical classification is controversial and immunotherapy has been without benefit. One disorder, multifocal motor neuropathy, stands apart because of its relatively stereotyped clinical expression and its frequently dramatic response to therapy. As reviewed by Latov, (1993), antiglycolipid antibodies are found in a high proportion of patients with this disorder, although they are neither specific nor essential for its diagnosis. This presentation will present the view that multifocal motor neuropathy is a clinically definable and discrete entity in which the treatment responses support an immunopathogenic mechanism. However, the data remain insufficient to conclude that the identified antiganglioside antibodies are responsible for the nerve damage.

Clinical features

The short history of multifocal motor neuropathy, recognized as a distinct entity only within the last 6 years (Chad et al., 1986; Parry and Clarke, 1988; Pestronk et al., 1988) is recounted by Asbury (1993). Pestronk et al. (1988) found elevated anti-GM1 antibody titers in 2 patients who had had almost exclusive involvement of motor nerves and evidence of multifocal conduction block. This general picture has now been confirmed (Chad et al., 1986; Roth et al., 1986; Parry and Clarke, 1988; Pestronk et al., 1988; Van den Bergh et al., 1989; Krarup et al., 1990; Lange et al., 1990; Pestronk et al., 1990; Feldman et al., 1991; Chaudhry et al., 1992). At present the suggested diagnostic criteria include progressive lower motor neuron weakness; multifocal motor demyelination with partial motor conduction block (PMCB), and normal sensory evoked responses on electrodiagnostic testing. Two useful clinical features are the asymmetry of the weakness and the loss of tendon reflexes in limbs with relatively strong muscles. The presence of elevated antiglycolipid antibodies are not an essential feature of the diagnosis, but they are present in about 50% of cases.

Multifocal motor neuropathy was previously confused with chronic inflammatory demyelinating neuropathy (CIDP), an immune-mediated demyelinating neuropathy, on the one hand and with motor neuron disease on the other. Some cases that are otherwise typical of CIDP can be strikingly patchy and asymmetrical, as recognized by Lewis et al. (1982). However, CIDP almost always has physiologic and clinical footprints of sensory fiber involvement. The distinction of multifocal motor neuropathy from most cases of ALS is made by the relative symmetry of weakness in ALS, the loss of tendon reflexes only late and in very weak muscles in ALS, and the paucity of significant conduction block on electrodiagnostic testing. Although rare individuals may manifest convincing features of multifocal motor neuropathy that evolve into ALS (Steck, 1993), most cases of multifocal motor neuropathy are easily distinguished and have a much less ominous prognosis.

Data on the pathology of multifocal motor neuropathy are only fragmentary at this time. Kaji et al. (1992a) biopsied the musculocutaneous nerve of a patient with a severe focal lesion and identified inflammatory demyelination. We have examined 9 sural nerve biopsies from patients with multifocal motor neuropathy (A.M. Corse, V. Chaudhry and J.W. Griffin, unpublished). Consistent with the normal clinical and physiologic findings, these sensory nerves are only mildly abnormal, but all showed some evidence of remyelination and several had small onion bulbs, both features reflecting previous demyelination. The basis for the prominent motor fiber loss and consequent muscle atrophy is not explained by simple demyelination, and the persistence of conduction block at the same site, in some cases lasting many years, is unique among acquired demyelinating conditions. The disorder behaves as if motor fibers were selectively demyelinated at focal sites, with the demyelination evolving into Wallerian-like degeneration of motor fibers distal to the sites in some nerves, and with effective remyelination prevented in others. This sequence would require a distinctive pathogenetic mechanism. If GM1 is the responsible antigen, its localization in the axon at the nodes of Ranvier (Thomas et al., 1989) and its relative enrichment in motor fibers (Svennerholm, pers. commun.) may contribute to the distinctive features.

Treatment of multifocal motor neuropathy

Multifocal motor neuropathy was initially found to improve following immunosuppression by cyclophosphamide (Pestronk et al., 1988; Pestronk et al., 1990; Feldman et al., 1991), but not with prednisone or plasmapheresis (Pestronk et al., 1988; Feldman et al., 1991). Kaji (1992b) reported improvement with intravenous immunoglobulin infu-

sions in two patients, and two other brief reports have appeared (Azulay et al., 1992; Nobile-Orazio et al., 1992). Chaudhry et al. (1992) at Johns Hopkins treated 9 patients with intravenous human immunoglobulin (HIG), an immunoreactive therapy reviewed by Dwyer (1992). All were diagnosed as having MMN based on the presence of chronic, progressive, asymmetric, predominantly distal limb weakness of a lower motor neuron nature and electrophysiologic evidence of multifocal motor demyelination with PMCB. None had prominent sensory symptoms or significant sensory abnormalities, either clinically or electrophysiologically. Four had high titers of anti-GM1 antibody. No patients had upper motor neuron findings, and none met the criteria for the diagnosis of CIDP (Cornblath et al., 1991).

All had some objective improvement and some improved rapidly and dramatically. The degree of functional improvement varied: 5 patients improved from being disabled to near-normal function, whereas in the other 4, their functional status improved only modestly. Improvement began within hours in 2 patients, and within 7 days in the other 7, with the peak of improvement at 14 days. The duration of improvement lasted from 6 to 15 weeks, and patients again improved with retreatment.

The major features of MMN and its response to HIG are illustrated by the following case history.

This 44-year-old man developed, over 17 years, a series of individual nerve palsies that cumulated in severe functional deficits in both arms. His initial symptom in 1974 was left wrist drop developing over a few days and persisting. In 1978 he developed, again over a few days, inability to extend his right fingers. He then had progressive weakness of the right arm over the next 10 years. Since 1987 he had not been able to flex his right arm at the elbow, or to elevate the right arm at the shoulder, and more recently he had lost use of his fingers to the point that he was unable to button buttons and was having difficulty using a keyboard. By 1991 he was at risk for being unable to continue as a diplomat in Africa.

His neurological examination in 1991 demonstrated strikingly asymmetrical weakness in the arms. For example, his left biceps was only mildly weak whereas his right was just able to flex the elbow against gravity. Similarly, he was not able to overcome gravity with the right deltoid. His sensory examination was entirely normal. In the arms, tendon reflexes were elicitable only at the left biceps; they were normal in the legs.

His electrodiagnostic examination confirmed the presence of marked slowing of conduction velocities in some nerves and provided evidence consistent with conduction block. He was treated with an intravenous infusion of HIG (2.4 g/kg over 3 days). He noted subjective improvement in strength in his arms within 24 h and within 3 days was able to elevate his right arm over his head and to flex powerfully at the right elbow. His improvement continued over at least 10 days and persisted for approximately 12 weeks with dramatic increases in quantitative measures of strength. He was re-treated when his strength began to decline and had a second profound response. In this individual the expensive and temporary therapy has allowed continued productive work.

Conclusions

Within the wide spectrum of disorders associated with antiganglioside antibodies are some that are unlikely to have an immune-mediated pathogenesis. In many settings the antibody may be a consequence of nervous system disease or may be only indirectly related. Multifocal motor neuropathy stands out as a disorder that can respond dramatically to immunotherapies, supporting the likelihood of an immune pathogenetic mechanism. The relationship of the anti-GM1 antibody to the pathogenesis remains a matter of speculation, but the anatomic and cellular distributions of GM1 in the PNS make it a candidate as the molecular target.

Acknowledgements

Dr. Bruce Trapp provided helpful discussion. Studies from this laboratory were supported by NIH grant PO1-22849.

References

Arquint, M., Roder, J., Chia, L.-S., Down, J., Wilkinson, O., Bayley, H., Braun, P. and Dunn, R. (1987) Molecular cloning and primary structure of myelin-associated glycoproteins. *Proc. Natl. Acad. Sci. U.S.A.*, 84: 600–604.

Azulay, J.-P., Blin, O., Bille, F., Gerard, C., Boucraut, J., Pouget, J. and Serratrice, G. (1992) High-dose intravenous human immunoglobulins are effective in the treatment of lower motor neuron syndromes associated with elevated serum anti-GM1 antibody titers: A double-blind, placebo-controlled study. *Neurology*, 42 (Suppl 3): 334. (Abstract)

Braun, P.E., Frail, D.E. and Latov, N. (1982) Myelin-associated glycoprotein is the antigen for a monoclonal IgM in polyneuropathy. *J. Neurochem.*, 39: 1261-1265.

Brouet, J.C., Mariette, X., Chevalier, A. and Hauttecoeur, B. (1992) Determination of the affinity of monoclonal human IgM for myelin-associated glycoprotein and sulfated glucuronic paragloboside. *J. Neuroimmunol.*, 36: 209–215.

Burgoon, M.P., Grumet, M., Mauro, V., Edelman, G.M. and Cunningham, B.A. (1991) Structure of the chicken neuron-glia cell adhesion molecule, Ng-CAM: Origin of the polypeptides and relation to the Ig superfamily. *J. Cell Biol.*, 112: 1017–1029.

Chad, D.A., Hammer, K. and Sargent, J. (1986) Slow resolution of multifocal weakness and fasciculations: a reversible motor neuron syndrome. *Neurology*, 36: 1260–1263.

Chaudhry, V., Corse, A.M., Cornblath, D.R., Kuncl, R.W., Drachman, D.B., Freimer, M.L., Miller, R.G. and Griffin, J.W. (1992) Multifocal motor neuropathy: Response to human immune globulin. Submitted.

Chou, K.H., Ilyas, A.A., Evans, J.E., Quarles, R.H. and Jungalwala, F.B. (1985) Structure of a glycolipid reacting with monoclonal IgM in neuropathy and with HNK-1. *Biochem. Biophys. Res. Commun.*, 128: 383–388.

Cornblath, D.R., Asbury, A.K., Albers, J.W., Feasby, T.E., Hahn, A.F., McLeod, J.G., Mendell, J.R., Parry, G.J., Pollard, J.D. and Thomas, P.K. (1991) Research criteria for diagnosis of chronic inflammatory demyelinating polyneuropathy (CIDP). *Neurology*, 41: 617–618.

Dwyer, J.M. (1992) Manipulating the immune system with immune globulin. *N. Engl. J. Med.*, 326: 107–116.

Favilla, J.T., Frail, D.E., Palkovits, C.G., Stoner, G.L., Braun, P.E. and Webster, H.deF. (1984) Myelin-associated glycoprotein (MAG) distribution in human central nervous tissue studied immunocytochemically with monoclonal antibody. *J. Neuroimmunol.*, 6: 19–30.

Feldman, E.L., Bromberg, M.B., Albers, J.W. and Pestronk, A. (1991) Immunosuppressive treatment in multifocal motor neuropathy. *Ann. Neurol.*, 30: 397–401.

Gregson, N.A. and Leibowitz, S. (1985) IgM paraproteinaemia, polyneuropathy and myelin-associated glycoprotein (MAG). *Neuropathol. Appl. Neurobiol.*, 11: 329–347.

Hays, A.P., Latov, N., Takatsu, M. and Sherman, W.H. (1987)

322

Experimental demyelination of nerve induced by serum of patients with neuropathy and an anti-MAG IgM M-protein. *Neurology*, 37: 242–256.

Ilyas, A.A., Quarles, R.H., MacIntosh, T.D., Dobersen, M.J., Trapp, B.D., Dalakas, M.C. and Brady, R.O. (1984) IgM in a human neuropathy related to paraproteinemia binds to a carbohydrate determinant in the myelin-associated glycoprotein and to a ganglioside. *Proc. Natl. Acad. Sci. U.S.A.*, 81: 1225–1229.

Ilyas, A.A., Dalakas, M.C., Brady, R.O. and Quarles, R.H. (1986) Sulfated glucuronyl glycolipids reacting with anti-myelin-associated-glycoprotein monoclonal antibodies including IgM paraproteins in neuropathy: Species distribution and partial characterization of epitopes. *Brain Res.*, 385: 1–9.

Kaji, R., Oka, N., Tsuji, J.T., Mezaki, T., Nishio, T., Akiguchi, I. and Kimura, J. (1992a) Pathological findings at the site of conduction block in multifocal motor neuropathy. *Ann. Neurol.*, (in press).

Kaji, R., Shibasaki, H. and Kimura, J. (1992b) Multifocal demyelinating motor neuropathy: Cranial nerve involvement and immunoglobulin therapy. *Neurology*, 42: 506–509.

Kohriyama, T., Ariga, T. and Yu, R.K. (1988) Preparation and characterization of antibodies against a sulfated glucuronic acid-containing glycosphingolipid. *J. Neurochem.*, 51: 869–877.

Krarup, C., Stewart, J.D., Sumner, A.J., Pestronk, A. and Lipton, S.A. (1990) A syndrome of asymmetrical limb weakness and motor conduction block. *Neurology*, 40: 118–127.

Lai, C., Watson, J.B., Bloom, F.L., Sutcliffe, J.G. and Milner, R.J. (1988) Neural protein 1B236/myelin-associated glycoprotein (MAG) defines a subgroup of the immunoglobulin superfamily. *Immunol. Rev.*, 100: 127–149.

Lange, D.J., Blake, D.M., Hirano, M., Burns, S.M., Latov, N. and Trojaborg, W. (1990) Multifocal conduction block motor neuropathy: Diagnostic value of stimulating nerve roots. *Neurology*, 40 (Suppl): 182.

Latov, N., Sherman, W.H., Nemni, R., Galassi, G., Shyong, J.S., Penn, A.S., Chess, L, Olarte, M.R., Rowland, L.P. and Osserman, E.F. (1980) Plasma-cell dyscrasia and peripheral neuropathy with a monoclonal antibody to peripheral-nerve myelin. *N. Engl. J. Med.*, 303: 618–621.

Lewis, R.A., Sumner, A.J., Brown, M.J. and Asbury, A.K. (1982) Multifocal demyelinating neuropathy with persistent conduction block. *Neurology*, 32: 958–964.

Maeda, Y., Bigbee, J.W., Maeda, R., Miyatani, N., Kalb, R.G. and Yu, R.F. (1991a) Induction of demyelination by intraneural injection of antibodies against sulfoglucuronyl paragloboside. *Exp. Neurol.*, 113: 221–225.

Maeda, Y., Brosnan, C.F., Miyatani, N. and Yu, R.K. (1991b) Preliminary studies on sensitization of Lewis rats with sulfated glucuronyl paragloboside. *Brain Res.*, 541: 257–264.

Martini, R., Bollensen, E. and Schachner, M. (1988) Immunocytochemical localization of the major peripheral nervous system glycoprotein P0 and the L2/HNK-1 and L3 carbohydrate structures in developing and adult mouse sciatic nerve. *Dev. Biol.*, 129: 330–338.

Martini, R., Xin, Y., Schmitz, B. and Schachner, M. (1992) The L2/HNK-1 carbohydrate epitope is involved in the preferential outgrowth of motor neurons on ventral roots and motor nerves. (in press)

Martini, R. and Schachner, M. (1986) Immunoelectron microscopic localization of neural cell adhesion molecules (L1, N-CAM, and MAG) and their shared carbohydrate epitope and myelin basic protein in developing sciatic nerve. *J. Cell Biol.*, 103: 2439–2448.

Matthieu, J.-M., Quarles, R.H., Poduslo, J.F. and Brady, R.O. (1975) ^{35}S-Sulfate incorporation into myelin glycoproteins: I. Central nervous system. *Biochim. Biophys. Acta*, 392: 159–166.

Mendell, J.R., Sahenk, Z., Whitaker, J.N., Trapp, B.D., Yates, A.J., Griggs, R.C. and Quarles, R.H. (1985a) Polyneuropathy and IgM monoclonal gammopathy: Studies on the pathogenetic role of anti-myelin-associated glycoprotein antibody. *Ann. Neurol.*, 17: 243–254.

Mendell, J.R., Sahenk, Z., Whitaker, J.N., Trapp, B.D., Yates, A.J., Griggs, R.C. and Quarles, R.H. (1985b) Polyneuropathy and IgM monoclonal gammopathy. Studies on the pathogenic role of anti-MAG antibody. *Ann. Neurol.*, 17: 243–254.

Mirsky, R., Jessen, K.R., Schachner, M. and Goridis, C. (1986) Distribution of the adhesion molecules N-CAM and L1 on peripheral neurons and glia in adult rats. *J. Neurocytol.*, 15: 799–815.

Miyatani, N., Kohriyama, T., Maeda, Y. and Yu, R.K. (1990) Sulfated glucuronyl paragloboside in rat brain microvessels. *J. Neurochem.*, 55: 577–582.

Nobile-Orazio, E., Meucci, N., Barbieri, S., Corbetta, G., Carpo, M. and Scarlato, G. (1992) Intravenous immunoglobulin therapy in multifocal motor neuropathy. *Neurology*, 42 (Suppl 3): 178–179. (Abstract)

Noronha, A.B., Ilyas, A., Antonicek, H., Schachner, M. and Quarles, R.H. (1986) Molecular specificity of L2 monoclonal antibodies that bind to carbohydrate determinants of neural cell adhesion molecules and their resemblance to other monoclonal antibodies recognizing the myelin-associated glycoprotein. *Brain Res.*, 385: 237–244.

O'Shannessy, D.J., Willison, H.J., Inuzuka, T., Dobersen, M.J. and Quarles, R.H. (1985) The species distribution of nervous system antigens that react with anti-myelin-associated glycoprotein antibodies. *J. Neuroimmunol.*, 9: 255–268.

O'Shannessy, D.J., Ilyas, A.A., Dalakas, M.C., Mendell, J.R. and Quarles, R.H. (1986) Specificity of human IgM monoclonal antibodies from patients with peripheral neuropathy. *J. Neuroimmunol.*, 11: 131–136.

Owens, G.C., Boyd, C.J., Bunge, R.P. and Salzer, J.L. (1990) Expression of recombinant myelin-associated glycoprotein in primary Schwann cells promotes the initial investment of axons by myelinating Schwann cells. *J. Cell Biol.*, 111: 1171–1182.

Owens, G.C. and Bunge, R.P. (1991) Schwann cells infected with a recombinant retrovirus expressing myelin-associated glycoprotein antisense RNA do not form myelin. *Neuron*, 7: 565–575.

Parry, G.J. and Clarke, S. (1988) Multifocal acquired neuropathy masquerading as motor neuron disease. *Muscle Nerve*, 11: 103–107.

Pedraza, L., Owens, G.C., Green, L.A.D. and Salzer, J.L. (1990) The myelin-associated glycoproteins: Membrane disposition, evidence of a novel disulfide linkage between immuno-globulin-like domains, and posttranslational palmitylation. *J. Cell Biol.*, 111: 2651–2661.

Pestronk, A., Cornblath, D.R., Ilyas, A.A., Baba, H., Quarles, R.H., Griffin, J.W., Alderson, K. and Adams, R.N. (1988) A treatable multifocal motor neuropathy with antibodies to GM1 ganglioside. *Ann. Neurol.*, 24: 73–78.

Pestronk, A., Chaudhry, V., Feldman, E.L., Griffin, J.W., Cornblath, D.R., Denys, E.H., Glasberg, M., Kuncl, R.W., Olney, R.K. and Yee, W.C. (1990) Lower motor neuron syndromes defined by patterns of weakness, nerve conduction abnormalities, and high titers of antiglycolipid antibodies. *Ann. Neurol.*, 27: 316–326.

Quarles, R.H., Everly, J.L. and Brady, R.O. (1973) Evidence for the close association of a glycoprotein with myelin. *J. Neurochem.*, 21: 1177–1191.

Quarles, R.H., Ilyas, A.A. and Willison, H.J. (1986) Antibodies to glycolipids in demyelinating diseases of the human peripheral nervous system. *Chem. Phys. Lipids*, 42: 235–248.

Quarles, R.H., Colman, D.R., Salzer, J.L. and Trapp, B.D. (1992) Myelin-associated glycoprotein: Structure-function relationships and involvement in neurological diseases. In: R.E. Martenson (Ed.), *Myelin: Biology and Chemistry*, CRC Press, Boca Raton, pp. 413–448.

Roth, R.G., Rohr, J., Magistris, M.R. and Ochsner, F. (1986) Motor neuropathy with proximal multifocal persistent conduction block, fasciculations, and myokymia: Evolution to tetraplegia. *Eur. Neurol.*, 25: 416–423.

Sadiq, S.A., Thomas, F.P., Kilidireas, K., Protopsaltis, S., Hays, A.P., Lee, K.W., Romas, S.N., Kumar, N., Van den Berg, L., Santoro, M., Lange, D.J., Younger, D.S., Lovelace, R.E., Trojaborg, W., Sherman, W.H., Miller, J.R., Minuk, J., Fehr, M.A., Roelofs, R.I., Hollander, D., Nichols, F.T., Mitsumoto, H., Kelly, J.J., Swift, T.R., Munsat, T.L. and Latov, N. (1990) The spectum of neurologic disease associated with anti-GM1 antibodies. *Neurology*, 40: 1067–1072.

Salzer, J.L., Holmes, W.P. and Colman, D.R. (1987) The amino acid sequences of the myelin-associated glycoproteins: homology to the immunoglobulin gene superfamily. *J. Cell Biol.*, 104: 957–965.

Schachner, M. (1990) Novel functional implications of glial recognition molecules. *Semin. Neurosci.*, 2: 497–507.

Snipes, G.J., Suter, U., Welcher, A.A. and Shooter, E.M. (1992) Characterization of a novel peripheral nervous system myelin protein (PMP-22/SR13). *J. Cell Biol.*, 117: 225–238.

Tatum, A.H. (1992) Experimental paraprotein neuropathy: Demyelination by passive transfer of human IgM anti-MAG. Submitted.

Thomas, F.P., Adapon, H.P., Goldberg, G.P., Latov, N. and Hays, A.P. (1989) Localization of neural epitopes that bind to IgM monoclonal autoantibodies (M-proteins) from two patients with motor neuron disease. *J. Neuroimmunol.*, 21: 31–39.

Trapp, B.D. and Quarles, R.H. (1984) Immunocytochemical localization of the myelin-associated glycoprotein: Fact or artifact? *J. Neuroimmunol.*, 6: 231–249.

Trapp, B.D., Quarles, R.H. and Griffin, J.W. (1984) Myelin-associated glycoprotein and myelinating Schwann cell-axon interaction in chronic beta,beta'-iminodipropionitrile neuropathy. *J. Cell Biol.*, 98: 1272–1278.

Trapp, B.D., Andrews, S.B., Cootauco, C. and Quarles, R.H. (1989a) The myelin-associated glycoprotein is enriched in multivesicular bodies and periaxonal membranes of actively myelinating oligodendrocytes. *J. Cell Biol.*, 109: 2417–2426.

Trapp, B.D., Andrews, S.B., Wong, A., O'Connell, M. and Griffin, J.W. (1989b) Co-localization of the myelin-associated glycoprotein and the microfilament components f-actin and spectrin in Schwann cells of myelinated fibers. *J. Neurocytol.*, 18: 47–60.

Van den Bergh, P., Logigian, E.L. and Kelly, J.J. Jr, (1989) Motor neuropathy with multifocal conduction blocks. *Muscle Nerve*, 12: 26–31.

Webster, H.deF., Palkovits, C.G., Stoner, G.L., Favilla, J.T., Frail, D.E. and Braun, P.E. (1983) Myelin-associated glycoprotein: Electron microscopic immunocytochemical localization in compact developing and adult central nervous system myelin. *J. Neurochem.*, 41: 1469–1479.

Welcher, A.A., Suter, U., De Leon, M., Snipes, G.J. and Shooter, E.M. (1991) A myelin protein is encoded by the homologue of a growth arrest-specific gene. *Proc. Natl. Acad. Sci. U.S.A.*, 88: 7195–7199.

Willison, H.J., Trapp, B.D., Bacher, J.D., Dalakas, M.C., Griffin, J.W. and Quarles, R.H. (1988) Demyelination induced by intraneural injection of human anti-myelin-associated glycoprotein antibodies. *Muscle Nerve*, 11: 1169–1176.

SECTION VI

Gangliosides and Functional Recovery of Injured Nervous System

L. Svennerholm, A.K. Asbury, R.A. Reisfeld, K. Sandhoff, K. Suzuki, G. Tettamanti and G. Toffano (Eds.)
Progress in Brain Research, Vol. 101

CHAPTER 25

CNS Pharmacology of gangliosides

Giancarlo Pepeu[1], Barbara Oderfeld-Nowak[2] and Fiorella Casamenti[1]

[1]*Department of Preclinical and Clinical Pharmacology, University of Florence, Viale Morgagni 65, 50134 Florence, Italy and* [2]*Nencki Institute of Experimental Biology, Polish Academy of Sciences, 3 Pasteura, Warszawa, Poland*

Introduction

If pharmacology 'embraces the knowledge of history, source, physical and chemical properties, compounding, biochemical and physiological effects, the mechanisms of action, absorption, distribution, biotransformation and excretion, and therapeutic uses of drugs' (Benet et al., 1990), even with the limitation to the central nervous system, it would be very difficult to include in a single chapter all information on ganglioside pharmacology presently available. However, the history, structure, physico-chemical properties and many aspects of the mechanism of action of the gangliosides are described in other chapters of this book, and have been the subject of recent extensive reviews (Skaper et al., 1989; Samson 1990; Cuello, 1990; Rahmann, 1992). We can therefore limit the scope of this chapter to describing the actions of gangliosides on the CNS 'in vivo', the effective doses, and their penetration into the brain. Considerations on the mechanism of action and therapeutic applications will stem from the analysis of the actions. Most of the experiments on CNS have been carried out with the monosialoganglioside GM1. In some cases a mixture of gangliosides including GM1 (21%), GD1a (39.7%), GD1b (16%) and GTI (19%) has been used and it will be referred to in the text as MIX. Recently the GM1 inner ester AGF2 has also been used.

Pharmacological actions

Pharmacological actions of gangliosides have been demonstrated on neural cells and tissues 'in vitro', and 'in vivo' on intact animals but mostly on animals with brains lesions of a different nature. The purpose of some of the brain lesions was to mimick pathological conditions occurring in man. Examples are MPTP administration for Parkinson's disease (Schneider et al., 1992), lesions of the nucleus basalis for the cholinergic hypofunction associated with Alzheimer's disease (Pepeu et al., 1986), and brain vessel occlusions for stroke.

Actions on intact animals

The effects brought about by GM1 administration in intact rodents are summarized in Table I. From the few data available it appears that GM1 in intact animals is only active during development and aging. It facilitates neuronal development and consequently the acquisition of learned behaviors in newborn and young animals, and improves the age-impaired cholinergic mechanisms, Conversely, no effect of GM1 on the cholinergic system of intact adult

---

Abbreviations: ACh, Acetylcholine; ChAT, choline acetyl-transferase; GAD, glutamate decarboxylase; HACU, high affinity choline uptake, DA, dopamine; MIX, ganglioside mixture (see text); MPTP, 1-methyl1-4-phenyl1-1,2,3,6-tetrahydropyridine; NA, noradrenaline; NGF, nerve growth factor; 5,7-OH-HT, 5,7-dihydroxytryptamine; 6-OH-DA, 6-hydroxy-dopamine; NB, nucleus basalis magnocellularis

TABLE I

Effect of GM1 administration in intact animals

Animal species	Dose (mg/kg)	Duration (days)	Effects observed	Ref.
Newborn rats	5 each s.c.	10	Enhancement of CNS maturation	1
1-month-old mice	30 i.p.	21	Improved passive avoidance retention	2
1-month-old mice	40 i.p.	21	Facilitate adaptive reactions	3
7-week-old rats	20 i.p.	2	Retardation or facilitation of memory formation according to time of administration	4
	40×2 i.p.	1		
24-month-old rats	30 i.p.	30	Increase striatal ACh, ChAT, and HACU	5
	5 mg day i.c.v.	30		

Ref. 1: Mahadik and Karpiak, 1986; 2: Fagioli et al., 1990; 3: Fagioli et al., 1991; 4: Haselhorst et al., 1990; 5: Hadjiconstantinou et al., 1992.

rodents has been ever demonstrated at doses effective in lesioned animals, and in senescent rats (Hadjicostantinou et al., 1992).

Only the serotoninergic system has been shown to be affected by GM1 administration in intact adult rats. Three-day administration of GM1 (30 mg/kg i.p) brought about a small but consistent increase in number and affinity of the cortical spiperone-binding 5HT receptors (Agnati et al., 1983a), while a 30-day treatment with the same dose of GM1 increased 5HT metabolism in the spinal cord (Hadjicostantinou and Neff, 1986).

In young rats, the effect of GM1 on acquisition and retention of conditioned responses could depend on the improvement of synaptic transmission, observed in the dentate gyrus of 6–9-week old rats (Ramirez et al. 1990).

Effect of gangliosides on animals with brain lesions

After the first demonstration by Wojcik et al. (1982) that the administration of a ganglioside mixture was able to facilitate the recovery of the cholinergic system in the hippocampus of rats with an electrolytic lesion of the septum, many investigators

demonstrated that exogenous gangliosides could reduce the injury and facilitate the recovery in animals in which different CNS areas were damaged by mechanical and electrolytic lesions, neurotoxin administration, and ischemia. Examples of the protection and/or recovery brought about by gangliosides are given in Tables II, III, IV and V.

Mechanical and electrolytic lesions

Table II shows that ganglioside administration, starting immediately after lesioning, or 2 days later (Toffano et al., 1983), attenuates the damage induced to cholinergic, dopaminergic and serotoninergic systems by electrolytic and mechanical lesions. The dose used was generally 30 mg/kg i.p. but even 5 mg/kg i.p. were effective in some cases. The recovery of functions such as HACU (Pedata et al., 1984) and DA uptake (Raiteri et al., 1985) was detected already after 3-4 days of GM1 treatment while the recovery of enzymes such as ChAT or tyrosine hydroxylase required longer periods of treatment.

Gangliosides do not always protect from mechanical lesions. For instance, no facilitation of behavioral recovery was observed by Butler et al. (1987) after

TABLE II

Effects of Ganglioside administration on mechanical and electrolytic lesions of the brain

Animal species	Region	Lesion	Drug dose (mg/kg)	Duration (days)	Effect	Ref
Rat	septum	electrolyt	Mix 50 i.p.	3,5,18,50	ChAT,AChE recovery in the hippocampus	1
Rat	NB	electrolyt	GM1 30 i.p.	4	cortical HACU recovery	2
Rat	NB	electrolyt	GM1 30 i.p.	20	cortical ChAT recovery	3
Rat	mesence- phalon	electrolyt	GM1 30 i.p.	6,14	prevention of hippocampal 5HT decrease	4
Rat	entorhinal cortex	suction	Mix 50 i.m.	16	recovery of alternation	5
Rat	sensory- motor cortex	suction	GM1 30 i.p.	10	recovery striatal gluta- matergic synapses	6
Rat	caudate	radio frequency	GM1 30 i.p.	30	recovery of spatial learning	7
Rat	nigrostriatal hemitransection		GM1 5 i.p. 30 i.p.	8–76	regenerations of DA neurons	8
Rat	nigrostriatal hemitransection		GM1 10 i.p.	56	protection and recovery of DA function	9
Rat	nigrostriatal hemitransection		AGF 30 i.p.	5	protection of DA function	10
Rat	spinal cord hemitransection		GM1 30 i.p.	30	stimulation of 5HT and DA spinal cord metabolism	11

1: Wojcik et al., 1982; 2: Pedata et al., 1984; 3: Casamenti et al., 1985; 4: Lombardi et al., 1988; 5: Karpiak 1983; 6: Shifman, 1991; 7: Sabel et al., 1984; 8: Toffano et al., 1983; 9: Agnati et al., 1983b; 10: Raiteri et al., 1992; 11: Hadjicostantinou and Neff, 1986.

GM1 administration to rats with bilateral visual abla- tion. Ganglioside effect seems to depend on the site of the lesion but mainly on its severity (Ramirez and Sabel, 1990). Gradkowska et al. (1986) demonstrated that the enhancement of AChE and ChAT recovery induced by GM1 in the hippocampus was directly proportional to the degree of fiber degeneration. Lombardi et al. (1988) were unable to demonstrate a facilitation of spontaneous recovery of the hippocam- pal 5-HT system, damaged by severing the dorsal hippocampal afferents, with the same administration schedule by which GM1 prevented the loss of 5-HT neurons induced by an electrolytic lesion placed in the mesencephalon. Finally, although gangliosides generally exert no effect in unlesioned adult animals, they may stimulate the cholinergic system contralat- eral to the lesioned region, as shown in the cerebral

cortex after a unilateral NB lesion by Pedata et al. (1984), Casamenti et al., (1985), Cuello et al. (1989).

Neurotoxic lesion

Examples of the protection exerted by GM1 and its inner ester AGF2 on neurochemical and behavioral alterations induced by local administration of neuro- toxins are reported in Table III. Prolonged treatments, sometimes initiated before the neurotoxin administra- tion (Kojima et al., 1984; Lombardi et al., 1989; Emerich and Walsh, 1991), attenuate the damage induced by a heterogeneous group of neurotoxins, including 6-OH-DA, which specifically damages adrenergic neurons, excitatory amino acids acting through glutamatergic receptors, and alkaloids dis- rupting axoplasmic transport.

Neurotoxic effects brought about by systemically

TABLE III

Effects of GM1 ganglioside administration on brain lesions induced by local neurotoxin administraition

Animal species	Region	Agent	GM1 dose (mg/kg)	Duration days	Effect	Ref.
Rat	cortex	6-OH-DA	30 i.p.	3+14	Enhanced NA recovery	1
Rat	NB	Ibotenic acid	30 i.p.	21	Protection of NB cholinergic neurons and behavioral recovery	2
Rat	NB	Ibotenic acid	30 i.p.	21	Recovery of cortical HACU and ChAT activity	3
Rat	striatum	Ibotenic acid	30 i.p.	24	ChAT and GAD recovery in the striatum	4
Rat	striatum	Quinolinic acid	30 i.p.	3+12	Recovery of striatal ChAT and GAD	5
Rat	i.c.v.	Vincristine	60 i.p., s.c.	11	Complete hippocampal ChAT, HACU, and behavioral recovery	6
Rat	i.c.v.	Colchicine	AGF 10 i.m.	3+14	Prevention loss cholinergic neurons and cognitive deficit	7

1: Kojima et al., 1984; 2: Casamenti et al., 1989; 3: Di Patre et al., 1989a; 4: Contestabile et al., 1990; 5: Lombardi et al., 1989; 6: Di Patre et al., 1989b; 7: Emerich and Walsh, 1991.

administered agents can also be reduced by ganglioside administration as shown by the examples reported in Table IV.

Gangliosides prevent the degeneration of serotoninergic and dopaminergic neurons induced by specific toxins, and the unspecific toxic effects of

TABLE IV

Effects of ganglioside administration on brain lesions induced by systemic neurotoxin administration

Animal species	Agent	Drug	Dose (mg/kg)	Duration (days)	Effect	Ref.
Newborn rat	5,7-OH-HT	GM1	30 s.c.	4	Prevention of cortical 5-HT degeneration	1
Mouse	MPTP	GM1	30 i.p.	23	Recovery of striatal DA level	2
Mouse	MPTP	GM1	30 i.p.	35	Partial recovery of striatal DA system	3
Monkey	MPTP	GM1	15–30 i.m.	42,56	Recovery of striatal DA system and parkinsonian-like symptoms	4
Mouse	MPTP	Mix	0.4 i.c.v.	1	Prevention DA neuron loss	5
Mouse	ethanol	GM1 AGF	20 i.p. 20	1	Reduced mortality, reduced sleeping time, reduced ATPase decrease	6

1: Jonsson et al., 1984; 2: Hadjicostantinou et al., 1986; 3: Date et al., 1989; 4: Schneider et al., 1992; 5: Gupta et al., 1990; 6: Hungund et al., 1990.

TABLE V

Effects of ganglioside administration on ischemic lesions of the brain

Animal species	Region	Lesion	Drug dose (mg/kg)	Duration (days)	Effect	Ref.
Rat	Cortex	Devascularization	GM1 5 i.c.v.	7	Prevention of NB retrograde degeneration	1
Rat	Cortex	Devascularization	GM1 30 i.p.	30	Prevention of NB retrograde degeneration	2
Rat	Cortex	Artery occlusion	GM1 10 i.m.	14	Reduction of functional deficits	3
Rat	Cortex	Artery occlusion	GM1 10 i.m.	20	Reduction of cognitive deficits	4
Gerbil	Hippocampus	Transient carotid occlusion	GM1 30 i.p.	3	Decrease of excitatory amino acid output	5
Rat	Striatum	Transient carotid occlusion	AGF 5 i.p.	7	Reduction of DARPP32 immunoreactivity disappearance	6

1: Cuello et al., 1989; 2: Stephens et al., 1989; 3: Barucha et al., 1991; 4: Ortiz et al., 1990; 5: Lombardi and Moroni, 1992; 6: Zoli et al., 1989.

ethanol. It is remarkable that both ethanol-induced damage (Hungund et al., 1990) and MPTP-induced loss of DA neurons (Gupta et al., 1990) are attenuated by one or two administrations of ganglioside prior to the neurotoxin injections. However, in primates with experimental Parkinsonism induced by MPTP, recovery was obtained by treating the animals with GM1 for 6–8 weeks post lesion (Schneider et al., 1992). Furthermore, according to Weihmuller et al. (1989), continuous administration of GM1 (30 mg/kg i.p) is necessary to maintain the biochemical and behavioral recovery of MPTP-treated mice followed for 30 days after the last MPTP injection.

While it is difficult to find a common link between the neurotoxic mechanisms of these agents and the effects obtained with short preventive and long therapeutic treatments with gangliosides, it must be pointed out that GM1 administration is not always able to reduce the injury induced by neurotoxins. Stephens et al. (1988) demonstrated that GM1, at the same doses which protected against the effect of devascularization, was inactive against the retrograde degeneration

of the basal forebrain cholinergic neurons induced by cortical application of kainic acid. According to Cadete-Leite et al. (1991) GM1 (35 mg/kg s.c.) administered for 4 weeks after withdrawal to rats treated with ethanol for 6 months did not reduce the degeneration of the hippocampal neurons.

Ischemic lesions

The experiments demonstrating a reduction in ischemic injury brought about by gangliosides, and the possible mechanisms involved, have recently been the object of an extensive review (Karpiak et al., 1990). Therefore, in Table V only a small number of examples are presented. They demonstrate firstly that GM1 injected either i.c.v. or i.p. prevents retrograde neuronal degeneration and the cognitive deficit resulting from partial chronic ischemia induced by cortical devascularization (Cuello et al., 1989; Stephens et al., 1988; Elliott et al., 1989). Secondly, that it reduces the functional and cognitive deficits following focal cortical ischemia caused in the rat by left common carotid artery ligation and medial cerebral artery occlusion (Ortiz et al., 1990; Barucha et al., 1991).

The reduction in the functional and cognitive deficits results from the decrease in edema formation in the primary infarct area and peri-infarct zones, the protection of plasma membrane Na, K-ATPase, and membrane fatty acids brought about by ganglioside treatment (see refs in Karpiak et al., 1990). These effects may in turn be related to a reduction in glutamate toxicity (Costa et al., this volume), and to the decrease of further excitatory amino acid release (Lombardi and Moroni, 1992)

Ganglioside pharmacokinetics

All ganglioside actions on the brain reported in the Tables were obtained by intramuscular, intraperitoneal and subcutaneous administrations. Some of the actions were reproduced by i.c.v. injection of smaller doses (Cuello et al., 1989; Hadjicostantinou et al., 1992). In no case were gangliosides only active when given by i.c.v. route. It may therefore be assumed that the gangliosides distribute themselves through the body and cross the blood brain barrier in sufficient amounts to exert their effects. In the few studies (Lang, 1981; Ghidoni et al., 1989; Bellato et al., 1992) in which distribution of radioactive gangliosides in the brain has been measured, a level of radioactivity from 5 to 10 times smaller than in the liver was found in the brain. The radioactivity persisted in the brain for at least 48h after administration, 90% represented by GM1, when GM1 lactone was administered, and was larger in the hemisphere with a nigrostriatal hemitransection than in the intact hemisphere (Bellato et al., 1992). The latter finding indicates that lesions may facilitate ganglioside penetration in the brain.

Treatment doses and duration

The data reported in the Tables show that many different doses and treatment schedules have been used to demonstrate the effects of gangliosides in the CNS, sometimes without an apparent rationale. The effects are dose-dependent and the effective doses range between 5 and 60 mg/kg. Dose-dependency has been specifically investigated by Di Patre et al. (1989a), studying GM1 protection against the hippocampal damage induced by i.c.v. vincristine, and by Borszeix et al. (1989) in studying the late consequences of forebrain ischemia.

A few studies have compared different treatment schedules. According to the type of lesion placed in the brain, and of the enzyme, function and behavior investigated, improvement and recovery have been observed after a treatment as short as 2–4 days (Pedata et al., 1984; Raiteri et al., 1992) or as long as 76 days (Toffano et al., 1983). In most experiments a treatment of 2–3 weeks seems necessary in order to achieve a clear cut recovery. The administration is generally started immediately after lesioning, but recovery has sometimes been obtained even if the administration was delayed by 5–7 days. With longer delays no effect is seen (Stephens et al., 1987). Pretreatments have frequently been used. Evidence that a post treatment alone was ineffective and recovery was obtained by pretreatment followed by post treatment has been given by Kojima et al. (1984), Lombardi et al, (1989).

Mechanism of action

In the final parpagraph an attempt will be made to outline the possible mechanisms of action responsible for the protection and recovery exerted by gangliosides in the large number of damaging conditions reported in this review.

The first question that can be asked is whether gangliosides protect the neurons from the injury or facilitate recovery through a trophic effect. Evidence can be presented to support both types of actions. The finding that GM1 given immediately after ibotenic acid injection in the NB reduces the loss of cholinergic neurons (Casamenti et al., 1989) indicates that GM1 protects the neurons from ibotenic acid neurotoxicity. Similar conclusions have been drawn by Mahadik et al. (1988), and Emerich and Walsh (1991) after neurotoxic lesion, and by Mahadik et al. (1989) after ischemic lesions. Conversely, the demonstrations that beneficial effects can be obtained even after delayed administration of gangliosides (Stephens et al., 1987; Di Patre et al., 1989a), and that in some cases longer treatments results in better recovery (Toffano et al., 1983; Contestabile et al., 1990) indicate that gangliosides facilitate recovery.

The next question is how ganglioside exert their protective action. Since a common final pathway for all conditions leading to cellular death is now proposed, gangliosides may prevent injuries by acting on some step of this pathway. It has been demonstrated that gangliosides reduce excitatory amino acid toxicity and break the pathogenetic chain of events that results in altered intracellular Ca^{2+}homeostasis (Guidotti et al., 1991; Costa et al., this volume). Other mechanisms may be involved, such as membrane protection, reduction of free radical formation, modulation of protein kinases (Cuello, 1990; Karpiak et al., 1990).

Facilitation of recovery may result from several actions for which there is experimental evidence. First, gangliosides facilitate sprouting 'in vitro' (Roisen et al., 1981) and 'in vivo' (Agnati et al., 1983b). Second, gangliosides potentiate recovery induced by NGF 'in vivo' (Cuello et al., 1989; Di Patre et al., 1989b), a finding suggesting that they may interact with endogenous trophic factors released by the lesions. Third, a direct trophic effect of gangliosides on synaptic formation and on cholinergic neurons has been demonstrated in developing and aging intact rodents (see Table I). Fourth, gangliosides may facilitate neuronal recovery by reducing the glial responses to lesion (Lescaudron et al., 1992). Finally, the observations that gangliosides have no effect if the lesion is very large and a transection complete (Ramirez and Sabel, 1990), that they are effective if administered during a time window after the lesion, and that they are less active in aging than in young animals (Stephens et al., 1987) all support the hypothesis that gangliosides facilitate ongoing spontaneous recovery of surviving neurons.

The potentially therapeutic benefits of gangliosides seem therefore to result from a dual action, protection from injuries, whatever their nature, and facilitation of recovery.

Acknowledgement

During the preparation of this article Dr. Barbara Oderfeld-Nowak was supported by a EC TEMPUS fellowship.

References

Agnati, L.F., Benfenati, F., Battistini, N., Cavicchioli, L., Fuxe, K. and Toffano, G. (1983a) Selective modulation of ^{3}H-spiperone labeled 5-HT receptors by subchronic treatment with the ganglioside GM1 in the rat. *Acta Physiol. Scand.*, 117: 311–314.

Agnati, L.F., Fuxe, K., Calza, L., Benfenati, F., Cavicchioli, L., Toffano, G. and Goldstein, M. (1983b) Gangliosides increase the survival of lesioned nigral dopamine neurons and favour recovery of dopaminergic synaptic function in striatum of rats by collateral sprouting, *Acta Physiol. Scand.*, 119: 347–363.

Barucha, V.A., Wakade, C.G., Mahadik, S.P. and Karpiak, S.E. (1991) GM1 ganglioside treatment reduces functional deficits associated with cortical focal ischemia. *Exp. Neurol.*, 114: 136–139.

Bellato, P., Milan, F., Facchinetti, E. and Toffano, G. (1992) Disposition of exogenous tritium-labelled GM1 lactone in the rat, *Neurochem. Int.*, 20: 359–369.

Benet, L.Z., Mitchell, J.R. and Sheiner, L.B. (1990) General principles. In: A. Goodman Gilman, T.W. Rall, A.S. Nies and P.Taylor (Eds), *The Pharmacological Basis of Therapeutics, 8th ed.*, Pergamon, New York, pp. 1–32.

Borszeix, M.G., Cahn, R. and Cahn, J. (1989) Effect of brain gangliosides on early and late consequences of a transient incomplete forebrain ischemia in the rat. *Pharmacology*, 38: 167–176.

Butler, W.M., Griesback, E., Labbe, R. and Sterin, D. (1987) Gangliosides fail to enhance behavioral recovery after bilateral ablation of visual cortex. *J. Neurochem. Res.*, 17: 404–409.

Cadete-Leite, A., Brandao, F., Madeira, M.D. and Paula-Barbosa, M.M. (1991) Effects of GM1 ganglioside upon neuronal degeneration during withdrawal from alcohol. *Alcohol*, 8: 417–423.

Casamenti, F., Bracco, L., Bartolini, L. and Pepeu, G. (1985) Effects of ganglioside treatment in rats with a lesion of the cholinergic forebrain nuclei, *Brain Res.*, 338: 45–52.

Casamenti, F., Di Patre, P.L., Milan, F., Petrelli, L. and Pepeu, G. (1989) Effects of nerve growth factor and GM1 ganglioside on the number and size of cholinergic neurons in rats with unilateral lesion of the nucleus basalis. *Neurosci. Lett.*, 103: 87–91.

Contestabile, A., Virgili, M., Migani, P. and Barnabei, O. (1990) Effect of short- and long-term ganglioside treatment on the recovery of neurochemical markers in the ibotenic acid-lesioned rat striatum. *J, Neurosci. Res.*, 26: 483–487.

Cuello, A.C. (1990) Glycosphingolipids that can regulate nerve growth and repair. *Adv. Pharmacol.*, 21: 1–50.

Cuello, A.C., Garofalo, L., Kenigsberg, R.L. and Maysinger, D. (1989) Gangliosides potentiate in vivo and in vitro effects of

334

nerve growth factor on central cholinergic neurons. *Proc. Natl. Acad. Sci. USA*, 86: 2056–2060.

Date, I., Felten, S.Y. and Felten, D.L. (1989) Exogenous GM1 gangliosides induce partial recovery of the nigrostriatal dopaminergic system in MPTP-treated young mice but not in aging mice. *Neurosci. Lett.*, 106: 282–286.

Di Patre, P.L., Abbamondi, A., Bartolini, L. and Pepeu, G. (1989a) GM1 ganglioside counteracts cholinergic and behavioral deficits induced in the rat by intracerebral injection of vincristine. *Eur. J. Pharmacol.*, 162: 43–50.

Di Patre, P.L., Casamenti, F., Cenni, A. and Pepeu, G. (1989b) Interaction between growth factor and GM1 monosialoganglioside in preventing cortical choline acetyltransferase and high affinity choline uptake decrease after lesion of the nucleus basalis. *Brain Res.*, 480: 219–224.

Elliott, P.J., Garofalo, L. and Cuello, A.C. (1989) Limited neocortical devascularizing lesions causing deficits in memory retention and choline acetyltransferase activity. Effects of the monosialoganglioside GM1. Neuroscience, 31: 63–76.

Emerich, D.F. and Walsh, T.J. (1991) Ganglioside AGF2 prevents cognitive impairments and cholinergic cell loss following intraventricular colchicine *Exp. Neurol.*, 112: 328–337.

Fagioli, S., Castellano, C., Oliverio, A. and Toffano, G. (1990) Effect of chronic GM1 ganglioside administration on passive avoidance retention in mice. *Neurosci. Lett.*, 109: 212–216.

Fagioli, S., Rossi-Arnaud, C. and Amassari-Teule, M. (1991) Open field behaviours and spatial learning performance in C57BL/6 mice: early stage effects of chronic GM1 ganglioside administration. *Psychopharmacology*, 105: 209–212.

Ghidoni, R., Fiorilli, A., Trinchera, M., Venerando, B., Chigorno, V. and Tettamanti, G. (1989) Uptake, cell penetration and metabolic processing of exogenously administered GM1 ganglioside in rat brain. *Neurochem. Int.*, 15: 455–465.

Gradowska, M., Skup, M., Kiedrowski, L., Calzolari, S. and Oderfeld-Nowak, B. (1986) The effect of GM1 ganglioside on cholinergic and serotoninergic systems in the rat hippocampus following partial denervation is dependent on the degree of fiber degeneration. *Brain Res.*, 375: 217–422.

Guidotti, A., de Erasquin, G., Brooker, G., Favaron, M., Manev, H. and Costa, E. (1991) Receptor-abuse-dependent antagonism. A new strategy in drug targeting for excitatory amino acid-induced neurotoxicity. In: B.S. Meldrum, F. Moroni, R.P. Simon and J.H. Wood (Eds), *Excitatory amino acids*, Raven Press, New York, pp.635–646.

Gupta, M., Schwarz, J., Chen, X.L. and Roisen, D.J. (1990) Gangliosides prevent MPTP toxicity in mice - an immunocytochemical study. *Brain Res.*, 527: 330–334.

Hadjiconstantinou, M. and Neff, N.H. (1986) Treatment with GM1 ganglioside increase rat spinal cord indole content. *Brain Res.*, 366: 343–345.

Hadjiconstantinou, M., Rossetti, Z.L., Paxton, R.C. and Neff, N.H. (1986) Administration of GM1 ganglioside restores the dopamine content in striatum after chronic treatment with MPTP. *Neuropharmacology*, 25: 1075–1077.

Hadjiconstantinou, M., Karadsheh, N.S., Rattan, A.K., Tejwani, G.A., Fitkin J.C. and Neff, N.H. (1992) GM1 ganglioside enhances cholinergic parameters in the brain of senescent rats, *Neuroscience*, 46: 681–686.

Haselhorst, U., Krusche, A., Schenk, H., Hantke, H, and Grecksch, G. (1990) Effect of ganglioside on memory formation of a conditioned avoidance response (CAR). *Biomed. Biochim. Acta*, 49: 523–526.

Hungund. B.L., Reddy, M.V., Barucha, V.A. and Mahadik, S.P. (1990) Monosialogangliosides (GM1 and AGF2) reduce acute ethanol intoxication: sleep time, mortality, and cerebral cortical Na^+,K^+-ATPase. *Drug. Dev. Res.* 19: 443–451.

Jonsson, G., Gorio, A., Hallman, H., Janigro, D., Kojima, H. and Zanoni, R. (1984) Effect of GM1 ganglioside on neonatally neurotoxin induced degeneration of serotonin neurons in the rat brain. *Dev. Brain Res.*, 16: 171–180.

Karpiak, S.E. (1983) Ganglioside treatment improves recovery of alternation behavior after unilateral entorhinal cortex lesion *Exp. Neurol.*, 81: 330–339.

Karpiak S.E., Mahadik, S.P. and Wakade C.G. (1990) Ganglioside reduction of ischemic injury. *Crit. Rev. Neurobiol.*, 5: 221–237.

Kojima, H., Gorio, A., Janigro, D. and Jonsson, G. (1984) GM1 ganglioside enhances regrowth of noradrenaline nerve terminals in rat cerebral cortex lesioned by the neurotoxin 6-hydroxy-dopamine. Neuroscience, 13: 1011–1022.

Lang, W. (1981) Pharmacokinetic studies with 3H-labeled exogenous gangliosides injected intramuscularly into rats. In: M.M. Rapport and A. Gorio (Eds), *Gangliosides in neurological and neuromuscular function, development and repair*, Raven Press, New York, pp. 241–251.

Lescaudron, L., Bitran, B.S. and Stein, D.G. (1992) GM1 ganglioside effects on astroglial response in the rat nucleus basalis magnocellularis and its cortical projection areas after electrolytic or ibotenic lesion. *Exp. Neurol.*, 116: 85–95.

Lombardi, G. and Moroni, F. (1992) GM1 ganglioside reduces ischemia-induced excitatory amino acid output: a microdialysis study in the gerbil hippocampus. *Neurosci. lett.*, 134: 171–174.

Lombardi, G., Beni, M., Consolazione, A. and Moroni, F. (1988) Lesioning and recovery of the serotoninergic afferents: differential effects of GM1 ganglioside. *Neuropharmacology*, 27: 1085–1088.

Lombardi, G., Zanoni, R. and Moroni, F. (1989) Systemic treatments with GM1 ganglioside reduce quinolinic acid-induced striatal lesions in the rat. *Eur. J. Pharmacol.*, 174: 123–125.

Mahadik, S.P. and Karpiak, S.E. (1986) GM1 ganglioside enhances neonatal cortical development *Neurotoxicology*, 7: 486–497.

Mahadik, S.P., Vilim, F., Korenowsky, A. and Karpiak, S.E. (1988) GM1 ganglioside protects nucleus basalis from excitotoxin damage: reduced cortical cholinergic losses and animal mortality. *J. Neurosci. Res.*, 20: 479–483.

Mahadik, S.P., Hawyer, D.B., Hungund, B.L., Li, Y.S. and Karpiak, S.E. (1989) GM1 ganglioside treatment after global ischemia protects changes in membrane fatty acid and properties of Na⁺,K⁺-ATPase and Mg²⁺-ATPase. *J. Neurosci. Res.*, 24: 402–419.

Ortiz, A., MacDonnall, J.S., Wakade, C.G. and Karpiak, S.E. (1990) GM1 ganglioside reduces cognitive dysfunction after focal cortical ischemia. *Pharmacol. Biochem. Behav.*, 37: 679–684.

Pedata, F., Giovannelli, L. and Pepeu, G. (1984) GM1 ganglioside facilitates the recovery of high affinity choline uptake in the cerebral cortex of rats with a lesion of the nucleus basalis magnocellularis. J. Neurosci. Res., 12: 421–427.

Pepeu, G., Casamenti, F., Pedata, F., Cosi. C. and Marconcini Pepeu, I. (1986) Are the neurochemical changes induced by lesion of the nucleus basalis in the rat a model of Alzheimer's disease? *Prog. Neuro-Psychopharmacol. Biol. Psychol.*, 10: 541–551.

Rahmann, H. (1992) Calcium-ganglioside interactions and modulation of neuronal function. *Curr. Aspects Neurosci.*, 4: 87–125.

Raiteri, M., Versace, P. and Marchi, M. (1985) GM1 monosialiganglioside inner ester induces early recovery of striatal dopamine uptake in rats with unilateral nigrostriatal lesion. *Eur. J. Pharmacol.*, 118: 347–350.

Raiteri, M., Marchi, M., Bonanno, G., Fedele, E. and Versace, P. (1992) Dopa mine release and dopaminergic inhibition of acetylcholine release in rat striatal slices after nigro-striatal hemitransection and parenteral ganglioside administration. *Eur. J. Pharmacol.*, 213: 17–24.

Ramirez, O.A. and Sabel, B.A. (1990) Toward a unified theory of ganglioside-mediated functional restoration after brain injury: lesion size, not lesion site, is the primary factor determining efficacy. *Acta Neurobiol. Exp.*, 50: 415–438.

Ramirez, O.A., Gomez, R.A. and Carrer, H.F. (1990) Gangliosides improve synaptic transmission in dentate gyrus of hippocampal rat slices. *Brain Res.*, 506: 291–293.

Roisen, F.J., Bartfeld, H., Nagele, R. and Yorke, G. (1981) Ganglioside stimulation of axonal sprouting in vitro. *Science*, 214: 135–150.

Sabel, B.A., Slavin, M.D. and Stein, D.G. (1984) GM1 Ganglioside treatment facilitate behavioral recovery from bilateral brain damage. *Science*, 225: 340–342.

Samson, J.C. (1990) GM1 Ganglioside treatment of central nervous system injury: clinical evidence for improved recovery. *Drug Dev. Res.*, 19: 209–224.

Schneider, J.S., Pope, A., Simpson, K., Taggart, J., Smith, M.G. and DiStefano, L. (1992) Recovery from experimental parkinsonism in primates with GM1 ganglioside treatment. *Science*, 256: 843–846.

Shifmann, M. (1991) The effect of gangliosides upon recovery of aspartate/glutamatergic synapses in striatum after lesions of the rat sensorimotor cortex. *Brain Res.*, 568: 323–324.

Skaper, S.D., Leon, A. and Toffano, G. (1989) Ganglioside function in the development and repair of the nervous system. *Mol. Neurobiol.*, 3: 173–199.

Stephens, P.H., Tagari, P.C. Garofalo, L., Maysinger, D., Piotte, M. and Cuello, A.C. (1987) Neural plasticity of basal forebrain cholinergic neurons: effects of gangliosides. *Neurosci. Lett.*, 80: 80–84.

Stephens, P.H., Tagari, P.C. and Cuello, A.C. (1988) Retrograde degeneration of basal forebrain cholinergic neurons after neurotoxin lesion of the neocortex: application of ganglioside GM1. *Neurochem. Int.*, 12: 475–481.

Toffano, G., Savoini, G., Moroni, F., Lombardi, G., Calzà, L. and Agnati, L.F. (1983) GM1 ganglioside stimulates the regeneration of dopaminergic neurons on the central nervous system. *Brain Res.*, 261: 163–166.

Weihmullar, F.B., Hadyicostantinou, M., Bruno, J.P. and Neff, N.H. (1989) Continued administration of GM1 ganglioside is required to maintain recovery from neuroleptic-induced sensorimotor deficits in MPTP-treated mice. *Life Sci.*, 45: 2495–2502.

Wojcik, M., Ulas, J. and Oderfeld-Nowak, B. (1982) The stimulating effect of ganglioside injections on the recovery of choline acetyltransferase and acetylcholinesterase activities in the hippocampus of the rat after septal lesion. *Neuroscience*, 7: 495–499.

Zoli, M., Grimaldi, R., Agnati, L.F., Zini, I., Merlo Pich, E., Toffano, G. and Fuxe, K. (1989) Neurohistochemical studies on striatal lesions induced by transient forebrain ischemia. Evidence for protective effects of the ganglioside analogue AGF2. *Neurosci. Res. Commun*, 4: 153–158.

L. Svennerholm, A.K. Asbury, R.A. Reisfeld, K. Sandhoff, K. Suzuki, G. Tettamanti and G. Toffano (Eds.)
Progress in Brain Research, Vol. 101
© 1994 Elsevier Science BV. All rights reserved.

CHAPTER 26

Cooperative effects of gangliosides on trophic factor-induced neuronal cell recovery and synaptogenesis: studies in rodents and subhuman primates

A. Claudio Cuello, Lorella Garofalo, Paolo Liberini and Dusica Maysinger

McGill University, Department of Pharmacology and Therapeutics, McIntyre Medical Building, 3655 Drummond Street, Suite 1325, Montreal, Quebec, H3G 1Y6 Canada

Introduction

Gangliosides are normal constituents of the outer plasma membrane of most mammalian cells and are particularly conspicuous in the central nervous system. They are expressed differentially in a temporospatial manner during development. In the mature central nervous system (CNS), their role is not fully understood. Despite our limited understanding of the biology of gangliosides in neural cells, much interest has been paid to their potential role as agents which can facilitate or accelerate the repair of the nervous systems (for reviews, see Ledeen, 1985; Skaper et al., 1989; Cuello, 1990a).

The notion that gangliosides could induce neuritogenesis emerged from the anatomopathological study of a biopsic sample from a child suffering from GM2 gangliosidosis. In this material Purpura and Suzuki (1976) demonstrated enlargements in processes of cortical pyramidal neurons which were named 'meganeurites'. Later, post-mortem ultrastructural observations made from the same case revealed even more bizarre meganeurites with abundant ramifications bearing multiple synaptic contacts (Purpura, 1978). Purpura and Suzuki (1976) proposed that these formations were due to neuritogenic properties of gangliosides. This idea provoked a great deal of interest and indeed this effect has been confirmed in a variety of in vitro systems (for reviews, see Ledeen,

1985; Skaper et al., 1989; Cuello, 1990a). The prospect that such properties of gangliosides could be exploited for therapeutical purposes was rapidly assessed in the peripheral nervous system. Ceccarelli et al. (1976) found that systemic administration of gangliosides facilitated the reinnervation of sympathetically denervated nictitating membranes. This was followed by an extensive series of studies by Gorio et al. (1983) examining the effects of gangliosides on the rate of skeletal muscle reinnervation. In the CNS, it was shown that gangliosides facilitated cholinergic neurochemical recovery following anterograde lesions of the septal nucleus (Wojcik et al., 1982), partial transections of the fimbria-fornix, (Oderfeld-Nowak et al., 1984) or lesions of the nucleus basalis magnocellularis (nbm) (Pedata et al., 1984) and enhanced survival of dopaminergic cells of the substantia nigra after cerebral hemitransections (Toffano et al., 1983; Agnati et al., 1983).

Our group has been particularly interested in testing whether exogenously applied gangliosides could prevent the retrograde (Sofroniew et al., 1983) and anterograde atrophy (Garofalo et al., 1992a) of CNS cholinergic neurons of the nbm or cortex secondary to cortical damage (infarction). It is important to stress that this experimental model, in contrast to the disconnection of the cholinergic septal-hippocampal pathway, is not an axotomy lesion model. The unilateral infarction is produced by causing a desvacular-

338

ization of a portion of the neocortex (mostly parietal 1 and neighbouring cortices). The ensuing infarction involves the regional terminal network of cholinergic fibers (see Fig. 1 for schematic representation of infarcted cortical regions). This type of lesion produces well defined anatomical and biochemical deficits in cholinergic neurons of the nbm (Sofroniew et al., 1983; Stephens et al., 1985). In this model we have noted that the systemic administration of large doses of GM1 [30 mg/kg/ intraperitoneally (i.p.) per day], from the time of onset of the lesion and continuing for 30 days, resulted in total protection of neurons from retrograde shrinkage. This was determined by quantifying the cross sectional area of ChAT immunoreactive neurons and the ChAT enzymatic activity of microdissected nbm samples (Stephens et al., 1985; Cuello et al., 1986). In more recent investigations we have shown that lower doses of GM1 applied for shorter periods (usually 7 days) can also effectively protect cholinergic neurons (Cuello et al., 1989; Maysinger et al., 1989). However, this treatment requires the early administration of GM1 [either

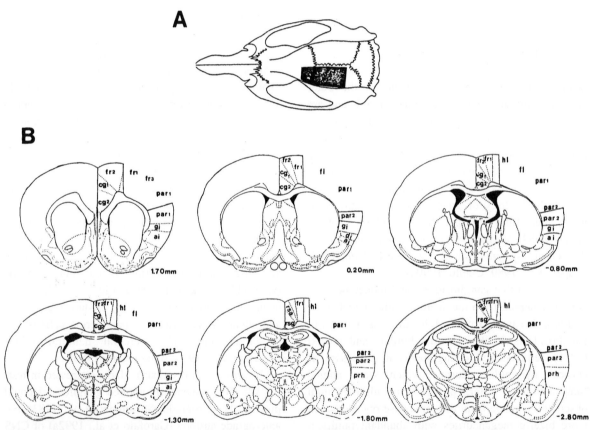

Fig. 1. (A) Dorsal view of a rat skull. The shaded area indicates the amount of skull bone removed prior to cortical lesioning. (B) Typical rostral to caudal extent of the unilateral devascularizing lesion. Numbers represent millimeters form bregma according to Paxinos and Watson atlas (1986). Note that the lesion encroaches the Frontal 1 and 3, hindlimb, forelimb and parietal 1 and 2 areas of the neocortex. *Abbreviations*: Cg 1,2 cingulate cortex areas 1 and 2; Fr 1,3, frontal cortex areas 1 and 3; Par 1,2, parietal cortex areas 1 and 2; gi, granular insular cortex; di, dysgranular insular cortex; ai, agranular insular cortex; prh, perirhinal cortex; rSA, retrosplenial agranular cortex; rsg, retrosplenial granular cortex; HL, hindlimb area of cortex; FL, forelimb area of cortex (from Elliott et al., 1989; with the publisher's permission).

i.p. or intracerebroventricularly (i.c.v.)] as delays beyond 48 h. from the time of lesion, renders the ganglioside treatment ineffective (Stephens et al., 1987; Garofalo and Cuello, in preparation). Interestingly, the uptake of [³H]-choline into synaptosomes obtained from the remnant neocortex of the lesioned side is enhanced by the ganglioside treatment (Garofalo and Cuello, 1990). Furthermore, cortical acetylcholine release, induced by high molarity potassium in the presence of physostigmine in vivo, is enhanced in lesioned animals treated with GM1 i.c.v. This has been demonstrated, using microdialysis procedures, by Maysinger et al. (1988). Using this cholinergic lesion model, it was shown, as also demonstrated by Casamenti et al. (1985), for nbm lesioned rats, that ganglioside administration improves the behavioral performance of lesioned rats (Elliott et al., 1989).

The recent compelling evidence that nerve growth factor (NGF) is an effective agent which can rescue axotomized CNS cholinergic neurons of the septal-hippocampal pathway (Hefti, 1986; Williams et al., 1986; Kromer, 1987) prompted us to examine its effects on the atrophy of the basolo-cortical cholinergic system and to compare them with those of gangliosides. Using this lesion model we found that both agents were equally efficacious in protecting the morphological and biochemical deficits occurring in cholinergic nbm neurons following injury (Cuello et al., 1989). However, a detailed dose/response analysis demonstrated that the potency of NGF to increase ChAT activity was severalfold greater (Garofalo and Cuello, in preparation) than that of GM1. These responses are illustrated in Fig. 2. Another important finding of these studies is that gangliosides, when administered together with NGF, appear to act cooperatively with this neurotrophin (Cuello et al., 1987, 1989). This cooperativity has been most clearly observed in the cerebral cortex where the combined application of NGF and GM1 can raise ChAT enzymatic activity over 200% of control levels in the remnant cortex of the infarcted side. This observation prompted us to enquire whether the infarction and neurotrophic treatment provoked changes in the cholinergic terminal network of the remaining cortex.

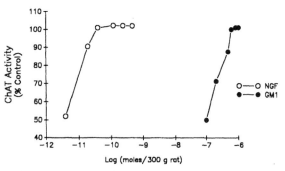

NBM DOSE–RESPONSE

Fig. 2. Dose-response curves for GM1 or NGF effects on nbM ChAT activity in unilaterally decorticated rats. Rats were lesioned as indicated in the text and were immediately treated, i.c.v. via minipump, with either GM1 (dosage range: 150–1500 µg/day, for 7 days) or NGF (dosage range: 0.5–12 µg/day, for 7 days). Animals were sacrificed 30 days post-lesion (i.e., 23 days following termination of treatment) and ChAT activity was measured in microdissected nbM samples according to previous methods (Fonnum, F., 1975). Drug concentrations are plotted as moles of administered compound per day.

To address this question, we used high resolution immunocytochemistry combined with quantitative image analysis. The results obtained up to present indicate that the lesion per se produces an anterograde retraction and shrinkage of the cholinergic network and that a substantial fiber and synaptic plasticity can be induced by exogenous NGF. Some of these effects can be further modulated by the co-administration of the monosialoganglioside GM1 (Garofalo et al., 1992b and Garofalo et al., 1993 in press). These findings and the behavioral consequences of these treatments will be discussed in greater detail below.

In order to better estimate the potential therapeutic value of gangliosides or neurotrophins in CNS trauma or degenerative processes, we reproduced the rodent experimental model, involving loss of a portion of the cortical cholinergic network of nbm projections in subhuman primates. A similar cholinergic atrophic reaction was elicited in primates (Cercopithecus aethiops) and the neurotrophic therapy was equally effective in such animals. The details of these investigations is also discussed in this chapter.

Exogenously administered gangliosides act cooperatively with NGF remodelling cortical cholinergic innervation. Evidence for trophic factor induced presynaptic hypertrophy and synaptogenesis

Our studies applying putative neurotrophic agents (NGF , gangliosides) in the cholinergic basolo-cortical lesion model suggested that these substances might induce an important re-arrangement of the cholinergic innervation (see Introduction). This possibility has been thoroughly assessed by examining the pattern of ChAT immunoreactive fibers in the remaining neocortex of lesioned animals (treated and untreated) and the equivalent region in unlesioned animals (NGF-treated and untreated). The factor treatment consisted of the i.c.v. application of 12 μg/day, for 7 days, of 2.5S NGF, 5 mg/kg/day for 7 days of GM1 or the simultaneous administration of the two agents. The i.c.v. delivery was assured by stereotaxically inserting a permanent cannulae, with its tip positioned in the lateral ventricle, which was connected to an osmotic minipump by coiled flexible polyethylene tubing (Vahlsing et al., 1989) containing either the neurotrophic agents diluted in artificial cerebrospinal fluid (c.s.f.) or c.s.f. alone.

Thirty days after the partial, unilateral, cortical infarction and beginning of treatments the animals

were processed for ChAT immunoreactivity, for both light and electron microscopy. The length of ChAT immunoreactive fibers was determined using an image analysis system. The program involved the automatic detection, from background, skeletonization of immunoreactive elements, and reduction to single pixel strings. The total number of pixels quantified per field therefore, represents total fiber length. In each animal, two focal planes and 12 adjacent fields were measured encompassing layers I–VI. This revealed that although lesions do not cause detectable changes in ChAT enzymatic activity a marked reduction of the cortical ChAT-IR fiber network is observed in the adjacent cortex of vehicle-treated lesioned animals (Garofalo et al., 1992a) (see Table I). Interestingly, the immediate administration of either NGF or GM1 is sufficient to prevent the apparent contraction of the cortical cholinergic network and furthermore, an augmented extension of detected ChAT immunoreactive fibers is observed when these two agents are given in combination (see Table I).

In these animals we have also investigated the incidence of ChAT-IR varicosities as these are sites for storage of acetylcholine as well as the enzyme involved in its synthesis, and are where physiological release of the neurotransmitter occurs. This study was accomplished by using a similar image analysis system combined with immunohistochemistry, except that 24 fields per animal were scanned at higher mag-

TABLE I

Cortical cholinergic innervation

Group	Total network of ChAT-IR fibers (%)	Number of ChAT-IR varicosities (%)	Mean cross-sectional area of ChAT-IR boutons (%)
Control	100	100	100
Lesion + Vehicle	70*	75*	64*
Lesion + GM1	87	107	78*
Lesion + NGF	106	130*	135*
Lesion + NGF/GM1	140*	146*§	169*§

Animals were unilaterally decorticated and immediately treated, via minipump, with either: vechile (artifical CSF + 0.1% BSA), GM1 (5mg/kg per day for 7 days), NGF (12 μg/day for 7 days) or both NGF and GM1 at the mentioned doses. Thirty days post-lesion the animals were processed for ChAT immunocytochemistry for light and electron microscopic analysis. Values are expressed as per cent control.
*$P<0.01$ from control unoperated group.
§ $P<0.01$ from lesion + NGF treated group. Anova post-hoc Newman-Keuls' test.

nification. The program employed was designed to recognize immunoreactive elements, falling within the size range of nerve varicosities, along fibers. These studies (Garofalo et al., 1992a, b) revealed a 25% diminution in the number of detectable varicosities in lesioned-untreated animals while the treatment with ganglioside i.c.v. was able to maintain varicosity number at control levels (see Table I). The application of NGF increased varicosity number above control levels and this effect was further enhanced by the simultaneous administration of the ganglioside GM1 (See Table I). We have obtained electron microscopic

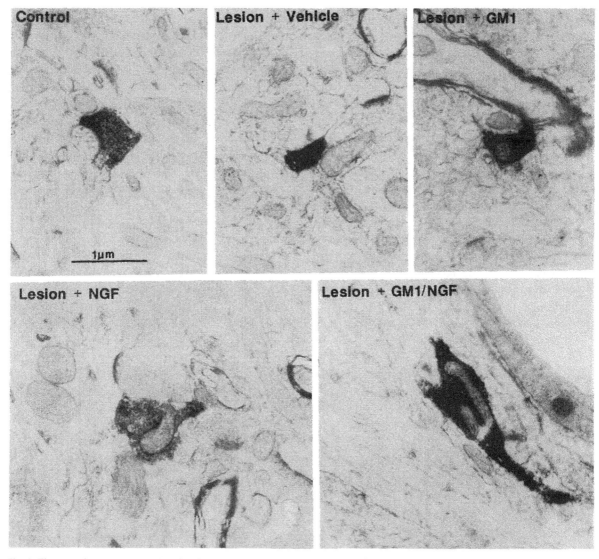

Fig. 3. Electron micrographs of cortical ChAT immunoreactive boutons in layer V of rat somatosensory cortex. Animals were lesioned and treated with either GM1 (1.5 mg/day, 7 days), NGF (12 μg/day, 7 days) or GM1 + NGF (at the mentioned doses) as described in text. Unoperated rats served as controls. Thirty days post-surgery (i.e., 23 days following termination of treatment) animals were perfused through the ascending aorta and processed for ChAT immunocytochemical electron microscopic quantitative analysis as previously described (Garofalo et al., 1992a). Representative profiles of cholinergic presynaptic terminals boutons from each case are shown.

data (Garofalo et al., 1992a and Garofalo, et al., 1993 in press) indicating that these changes in varicos number represent an increment in synaptic sites. Furthermore, 3-D reconstruction of the cortical cholinergic boutons of NGF plus GM1 treated rats showed that the areas of synaptic contacts may be notably expanded (Garofalo et al., 1993 in press). We have in these experiments also analyzed the cross sectional area of the cortical cholinergic boutons. For this, the first 20 ChAT immunoreactive boutons observed at the E.M. level were recorded on video tape and were subsequently transferred to the image analysis system for further processing. This study revealed that the cortical lesion, in addition to other changes previously reported, also causes a shrinkage of cholinergic boutons. This parallels the reported and repeatedly confirmed reduction in size of nbm cholinergic cell bodies (Sofroniew et al., 1983; Stephens et al., 1987; Cuello et al., 1986, 1989; Garofalo et al., 1992a). Gangliosides alone did not significantly modify the boutons however, they clearly potentiated the NGF-induced hypertrophy of cortical cholinergic presynaptic sites (see Table I). Typical examples of the ultrastructural changes occurring in cholinergic boutons following cortical lesions and treatment with putative neurotrophic agents is illustrated in Fig.3.

Ganglioside and NGF effects on the behavior of animals bearing unilateral cortical infarctions

In this chapter we will also discuss the behavioral effects of maximal doses of exogenous NGF or GM1 when given to adult rats with unilateral cortical lesions, alone or in combination. Prior to any surgical manipulation, adult male Wistar rats were pretrained in two behavioral tasks, passive avoidance and the Morris water maze. Only rats displaying analogous levels of performance in the two tasks were used in these studies. Following acquisition of these two tasks animals were cortically lesioned as indicated in the Introduction. Immediately following the lesion rats received i.c.v., via minipumps, either vehicle (CSF + 0.1% bovine serum albumin (BSA), GM1 (750 μg/day), NGF (6 μg /day) or both GM1 and NGF at these doses for 2 weeks. The rats were subsequently retested in the tasks 30 days post-lesion (i.e., 2 weeks after the end drug administration).

Passive avoidance

Prior to surgery all rats were habituated for 3 days to the passive avoidance box. After 2 days of free access to both compartments rats entering the dark side received scrambled footshocks. All rats were trained (15-min trial interval) to remain in the light

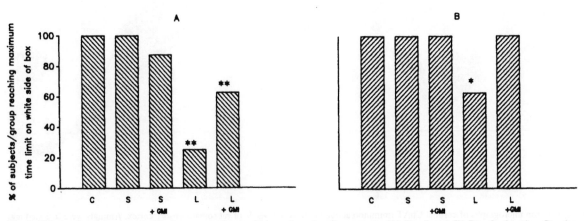

Fig. 4. Effects of unilateral cortical lesions on passive avoidance, 4 weeks after surgery. (A) Day 1 and (B) Day 2 of testing. Results are shown as percentage of animals/group reaching the cut-off time (300 sec) in the white side of the box (n = 8–11). C = control, S = sham operated and L = lesioned group. * and ** represent $P<0.05$ and $P<0.01$, respectively, compared to control (from Elliott et al., 1989, with permission).

compartment using a 300-sec latency criteria. Retention of the learnt response was tested 24 h. later. Thirty days post-surgery (i.e., 2 weeks after the end of drug infusions) animals were retested in this behavioral task. Latency to enter the dark side of the passive avoidance box was recorded on 2 consecutive days (test days 1 and 2). Animals which reentered the box on test day 1 were reshocked and were retested 24 h. later (test day 2).

The comparison of the latency time means demonstrated that lesion vehicle and lesion GM1-treated rats show retention deficits in this task and that exogenous GM1 can facilitate task reacquisition (Elliott et al., 1989; Garofalo and Cuello, 1990). By contrast, NGF treated rats showed substantial retention of the learnt task on test day 1 similar to that noted for control animals. On test day 2 the GM1, NGF and control groups were indistinguishable while

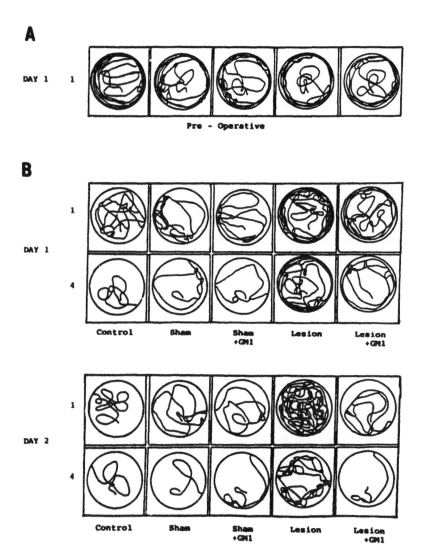

Fig. 5. Typical swimming patterns as recorded by video camera and traced with the aid of an image analysis system of: (A) all groups of animals prior to any surgical manipulations and drug treatments on the first trial of the first day of training and (B) all groups 1 month following surgical manipulations and 2 weeks following respective drug treatment on the first and fourth trials of the first testing day (Day 1) and on the first and fourth trials of the second testing day (Day 2) (from Elliott et al., 1989, with permission).

the vehicle treated animals still displayed a deficit. Lesioned rats which received both NGF and GM1 did not differ from NGF alone-treated animals on either test day 1 or 2.

Morris water maze

In these studies (Elliott et al., 1989; Garofalo and Cuello, 1990; and in preparation), we used a training paradigm consisting of one block of four trials, spaced 15 min apart, per day for 4 consecutive days. Start positions within each block varied between trials. As previously reported (Elliott et al., 1989), unoperated rats quickly learn the water maze task in such way that by the fourth trial on Day 2 of training, the platform was found by an average time of 10–20 sec. After the end of training, animals were operated, treated with neurotrophic agents and 30 days post-lesion (i.e., 2 weeks after end of drug infusions) were retested using the same experimental paradigm as described for training. These experiments revealed that on test day 1 both lesion vehicle and lesion GM1-treated animals showed significantly longer latencies to find the platform than all other groups. By contrast, escape latency times for NGF or NGF/GM1-treated lesioned animals did not differ significantly from control animals. Although the latency to escape the water was similar for lesion vehicle or GM1-treated rats on the first day of testing, analysis of the swim profiles showed that the lesion vehicle treated animals spent more time circling the perimeter of the pool than did rats from other groups. On test day 2, only lesion vehicle treated rats still showed significantly greater escape latency times when compared to control rats. Thus, GM1-treated lesioned rats quickly reacquired the task whereas lesioned rats which received vehicle required more training trials and showed a more random search strategy. Figure 5 illustrates the typical swimming patterns of lesioned and unlesioned vehicle or ganglioside treated rats. In summary, although ganglioside treatment does not maintain retention of the task it does, however, facilitate task reacquisition in lesioned animals as compared to their vehicle treated counterparts.

Primate nbM cholinergic neurons are protected from retrograde degeneration by gangliosides

Previous studies have shown that the monkey cytoarchitecture of the nbM and the organization of its projections to the neocortex are analogous to those observed in humans (Mesulam et al., 1983, 1984, 1986; Kitt et al., 1987). These neurons are highly dependent on the integrity of their cortical connections. In Nissl-stained preparations the equivalent of the nucleus magnocellularis of Meynert (nbM) (the Ch4 group being a modern nomenclature for its cholinergic component) practically disappear following extensive neocortical infarctions (Pearson et al., 1983). This closely resembles the neuropathological findings observed in human neurodegenerative diseases involving the nbM, such as AD and suggests that the degenerative changes noted in the nucleus basalis of Meynert in brains with Alzheimer's disease most likely arise as a secondary consequence of a primary cortical disturbance (Whitehouse et al., 1982). These similarities prompted us to reproduce, in non-human primates (*Cercopithecus aethiops*), the cortical devascularizing lesion model to investigate whether degenerative changes affecting the neocortex surrounding the lesion site and in the nbM ipsilateral to the lesion side are prevented by administration of trophic agents. So far, in this lesion model, we have investigated the effects of the monosialoganglioside GM1 and of recombinant human NGF (rhNGF) applied alone or in combination. One of the main goals of the present study was to examine if a cooperativity between the two trophic agents, previously described in the rat (Cuello et al., 1989), also occurs in the monkey.

Fifty monkeys (*Cercopithecus aethiops*) randomly divided in five groups (sham-operated; lesioned vehicle-treated; lesioned rhNGF-treated; lesioned GM1-treated; lesioned GM1/rhNGF-treated) underwent surgery and trophic factor treatment. The surgical procedures used for this experimental model are briefly described in Fig 6. The pharmacological treatment consisted of the placement of gelatin films containing the active molecules (specifically: 2.8 mg rhNGF/animal; 175 mg GM1/animal; 175 mg GM1 +

2.8 mg rhNGF/animal) onto the devascularized area of neocortex. In contrast, lesioned vehicle-treated monkeys received empty gelatin films. Following a 6-month survival the monkeys were processed either for biochemistry (ChAT assay) or for immunocyto-chemistry (ICC) of ChAT and NGFr antigenic sites.

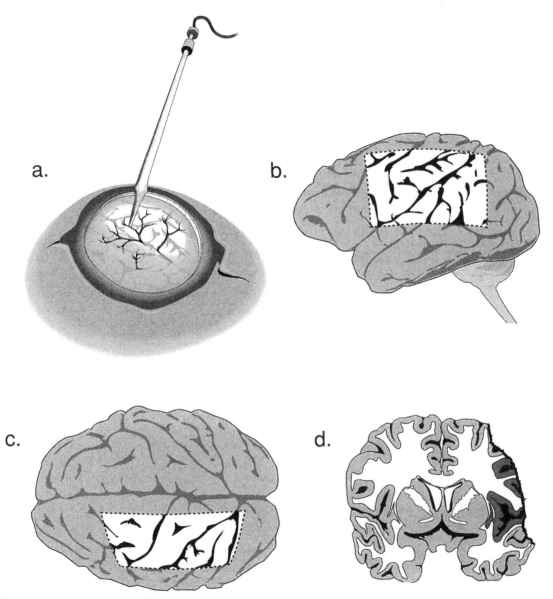

Fig. 6. Schematic representation of the surgical procedures used in this study. In lesioned primates, after removing a portion of the fronto-parietal-temporal skull bone (3.5 cm × 5.0 cm), pial blood vessels supplying the exposed neocortex were coagulated with a bipolar cautery device (a). Gelatin films containing vehicle or active molecules (2.8 mg rhNGF/gel; 175 mg GM1/gel; 175 mg GM1 + 2.8 mg rhNGF/gel) were placed onto the devascularized area of neocortex illustrated in panels b and c. Following a 6-month survival the monkeys were processed either for biochemistry or for immunocytochemistry. The extent of the atrophy affecting the neocortex in lesioned animals is shown schematically in (d) which illustrates a typical coronal monkey brain section from a lesioned monkey.

The unilateral devascularizing lesion of the *C. aethiops* neocortex induced a retrograde degeneration

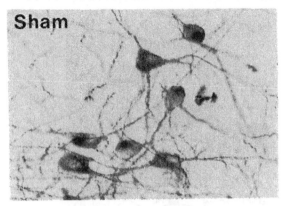

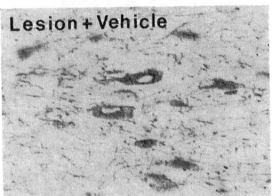

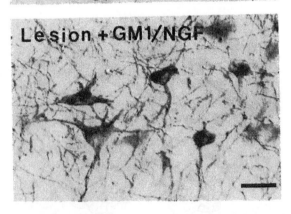

Fig. 7. Effect of the lesion and treatments on the morphology of NGFr immunoreactive neurons in the intermediate nbM of *Cercopithecus aethiops*. Note the cell shrinkage, the eccentric displacement of the nuclei and the loss of neuritic processes in lesioned vehicle-treated animals. The combined administration of GM1 with rhNGF maintains the normal cell phenotype. Scale bar = 15 μm.

of the nbM which is reflected in specific biochemical and morphological alterations (Pioro et al., 1993; Liberini et al., 1993). In these studies we have found that the enzymatic activity of ChAT in the nbM ipsilateral to the devascularized cortex was reduced 30% from sham operated values. The neocortical areas closely surrounding the lesion site showed a similar loss of ChAT activity. The reported reduction of ChAT activity in both these brain areas were fully prevented by the administration of rhNGF, GM1 or GM1 + rhNGF. However, animals which received the combined treatment showed a trend toward an increase in cortical ChAT activity above sham-operated values (118 ± 4.2%). This is in accordance with previous investigations in the cortically devascularized rat indicating a synergistic interaction between β-NGF and GM1 on the up regulation of ChAT activity in the neocortex of lesioned animals (Cuello et al., 1989).

The morphological changes affecting ChAT-IR neurons within the monkey nbM ipsilateral to the lesion consisted of cell body shrinkage to about 60% of sham-operated cell size, eccentric displacement of the nuclei and a marked loss of neuritic processes. The quantitative analysis of other forebrain areas revealed that such damage affected only the intermediate nbM (Ch4id and Ch4iv) and not its anterior (Ch4a) or posterior (Ch4p) regions or the basal nucleus contralateral to the cortical lesion (Pioro et al., 1993; Liberini et al., 1993). This selective localization of retrograde degeneration is probably related to the topographical organization of projections from the nbM to neocortex. As previously described (Mesulam et al., 1983, 1984; Kitt et al., 1987), the intermediate nbM subdivision provides a major cholinergic input to the cortical areas corresponding to the surgically induced infarction. It is interesting to note that either rhNGF, GM1 or GM1+rhNGF were able to prevent the shrinkage of ChAT and NGFr-IR neurons within the intermediate region of the nbM ipsilateral to the lesion side. However, a complete recovery of the neuritic processes was observed only in the double-treated monkeys (Liberini et al., 1993). These cellular changes observed in the primate nbM are illustrated in Fig. 7.

Significance of ganglioside-neurotrophin effects on basalocortical cholinergic neurons

Trophic agents are defined as biological substances that cause an increase of metabolism, neurite growth and cell survival or differentiation (Nieto-Sampedro et al., 1983; Freed et al., 1985). In the last few years, a variety of molecules showing trophic activity for neurons have been described. A decrease, or lack, of these substances has been proposed to be responsible for several neurodegenerative pathologies, like Alzheimer's disease, Parkinson's syndrome and amyotrophic lateral sclerosis (Appel, 1981; Arendt et al., 1983; Hefti and Weiner, 1986).

Nerve growth factor (NGF) is the best characterized neurotrophic factor and serves as a paradigm for more recently discovered trophic agents (Leibrock et al., 1989; Maisonpierre et al., 1990; Ernfors et al., 1990; Berkemier et al., 1991). This molecule has been shown to exert neurotrophic effects during development of rat basal forebrain cholinergic neurons and in the maintenance of normal function of these cells during adult life (Whittemore et al., 1987). The recent identification of nerve growth factor receptors (NGFr) in cholinergic neurons of rat (Dawbarn et al., 1988; Kiss et al., 1988 and Pioro and Cuello, 1990) and primate basal forebrain (Kordower et al., 1988; Schatteman et al., 1988, Mufson et al., 1989) raised the possibility that this cell population could respond to exogenous trophic agents. Investigations carried out using rats or non-human primates (*Macaca fascicularis*) with unilateral transection of the fimbria fornix showed that NGF infusion prevents degenerative changes in axotomized cholinergic neurons of the medial septal nucleus in rats and monkeys (Hefti, 1986; Williams et al., 1986; Kromer, 1987; Koliatsos et al., 1990, 1991; Tuszynski et al., 1990, 1991).

The reported findings indirectly suggest that β-NGF is normally supplied to basal forebrain cholinergic neurons by their target cells. Destruction of their terminal fields results in loss of retrograde transport of β-NGF and, hence, in neuronal degeneration. The administration of exogenous rhNGF, that can bind to NGFr located on nbM cell bodies or surviving processes, can reduce the neuron shrinkage. In addition to NGF, the administration of the mono-sialo-ganglioside GM1 in primates provokes distinctive neuroprotective effects on degenerating cholinergic neurons. Whether the cellular changes detected with these treatments including the application of gangliosides are due to prevention of neurite loss or represents a compensatory sprouting of nbM cholinergic neurons is not known. A sprouting reaction induced by trophic agents following neuronal loss might serve to build up lost connections and keep circuits operative. In the course of neurodegenerative pathologies, this process is probably beneficial and may reflect an ongoing turnover capacity to form synapses. Nevertheless, we can not exclude that a sprouting reaction could become misdirected and lead to the development of abnormal or dysfunctional connections, thus worsening the basic pathologic process (Geddes et al., 1986). However, the administration of drugs in the current experimental design, does not result in apparent behavioral impairments in treated monkeys. However, this aspect remains to be fully investigated for both monkeys and rats. In lesioned rats, our observations (Elliott et al., 1989; and Garofalo and Cuello, 1990; and in preparation) would indicate that NGF alone definitively protects the animals from 'forgetting' learnt tasks after infarctions of a large area of the neocortex involving mostly parietal regions. However, whereas the gangliosides alone produce rather dramatic effects on neurochemical markers, cell size and the extent of the cholinergic fiber network which are comparable to those caused by exogenous NGF, distinct behavioral effects arise from these treatments. Some of the above discussed behavioral performance of cortically lesioned factor-treated rats appears to correlate better with changes observed in axonal varicosities and on the ultrastructure of these cortical cholinergic presynaptic elements (see Fig 3 and Table I). Nevertheless, no further significant improvement of behavioral performance was noted in NGF + GM1-treated animals which showed the greatest elevated ChAT activity (Cuello et al., 1989), cholinergic fiber network and hypertrophic synapses (Garofalo et al., 1992b; and in preparation). It is possible that more extensive and complex analysis of behavior might detect subtler differences that the presently used tests are unable to demonstrate. It is important to note that the increased cortical cholinergic innervation brought about by both NGF and the ganglioside GM1 find their functional counterpart in the enhanced uptake of [³H]-choline by synaptosomes (Garofalo and Cuello, 1990) and the in vivo release of endogenous acetylcholine, as shown using microdialysis techniques (Maysinger et al., 1989, 1992).

In the cortically devascularized monkeys, the time-course of the experiments described here indicates that GM1 and rhNGF, applied alone or in combination, induce a long-term protection of nbM cholinergic neurons (6-month survival). Previous experiments performed in monkeys which underwent fimbria fornix lesions and rhNGF treatment were examined at shorter survival times (2–3 months). With that regimen, a hypertrophy of ChAT-IR cell bodies in the medial septal nucleus either on the lesion side or contralateral to it, was reported (Koliatsos et al., 1991). It is possible that comparable responses might have also occurred in our model following early postinjury stages.

It is not clear in which manner the gangliosides act in a co-operative fashion with NGF in these in vivo models (rodent and primate). The sequence of events that follow receptor-ligand interactions may include gene activation, ion channel modulation, allosteric or postranslational modifications of proteins (Greenberg et al., 1985; Kruijer et al., 1985; Milbrant, 1986; Leonard et al., 1987) and induction of specific second messenger systems, involved in trophic activities (Cremins et al., 1986; Hama et al., 1986). The interaction of gangliosides with β-NGF in the central cholinergic system may occur at any or many of these levels (Cuello, 1990). Some of the possible sites of action which could explain the co-operativity of gangliosides with NGF are schematically presented in Fig. 8.

It is very likely that interactions between NGF and gangliosides occur as a consequence of their incorporation into cell membranes. Indeed, it has been shown that exogenously applied GM1 is incorporated in the plasma membrane (Toffano et al., 1980), where endogenous gangliosides are normally found (Hakamori, 1986, 1989). Membrane-anchored gangliosides could assist in the binding of NGF to trkA, its natural high affinity receptor (Kaplan et al., 1991; Klein et al., 1991; Hempstead et al., 1991). It has been reported that immobilized GM1 is capable of binding β-NGF with low affinity (Schwartz and Spirman, 1982). Therefore, it is possible that gangliosides may act by providing additional binding sites for growth factors or by modulating their efficacy (Doherty et al., 1985; Leon et al., 1988; Ferreira et al., 1990). A similar mode of action has been attributed to gangliosides for their co-operativity with TSH for the activation of the TSH receptor (Lacetti et al., 1983) and for GM1 and the activation of adenylate cyclase of cholera toxin. In this last case the membrane ganglioside binds the beta-subunit of the toxin favouring the binding of the alpha subunit to the adenylate cyclase (Fishman, 1982). Although the biological significance of the complexing of low (p75) and high affinity (p140, trkA) NGF receptor remains controversial (for review, see Bothwell, 1991) there is a possibility that exogenous gangliosides might contribute to the formation of complexes between the dimeric ligand and the two types of receptor molecules. Gangliosides might more directly change the affinity of receptors (e.g., inducing allosteric changes) within the membrane milieu. There is already some experimental evidence for such a change with the vitronectin receptor in the presence of the ganglioside GD2 (Cheresh and Klier, 1986; Cheresh et al., 1987). Although no experimental evidence is available, the known changes induced by gangliosides on membrane fluidity and mobility of proteins might produce interactions betwen trophic factor receptors and other membrane bound or soluble proteins. The same biophysical changes could elicit trophic factor receptor capping or internalization, a process which the low affinity NGFr has been shown to undergo in cholinergic neurons of the rat nbM (Pioro et al., 1990).

In non-neural cells gangliosides can apparently affect trophic factor biological effects by modulating the degree of phosphorylation of their corresponding receptor molecules (Bremer, et al., 1984, 1986).

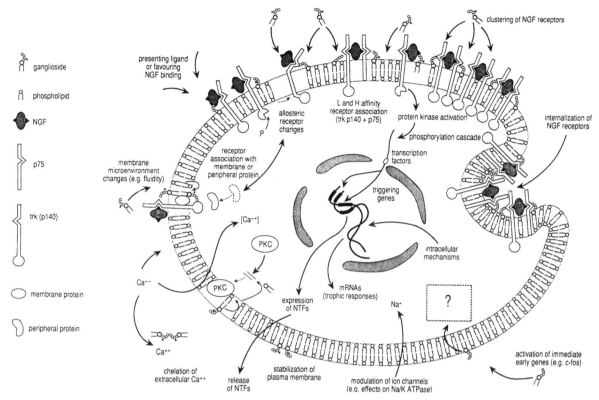

Fig. 8. Summary of possible participation of exogenously administered gangliosides in trophic responses elicited by nerve growth factor (NGF). See text.

Another possible site of action of exogenous gangliosides might reside in the NGF signal transduction cascade. Recently, Ferrari et al. (1992) have shown that gangliosides prevent the inhibition by k-252a, a general kinase inhibitior, of NGF induced C-Fos induction and neurite regeneration in PC12 cells. Furthermore, Hilbush and Levine (1991) have shown that exogenous gangliosides upon incorporation into the cell membrane stimulated a Ca^{2+}-dependent protein kinase only in the presence of NGF. This results in the phosphorylation of peptide T2 of the catecholamine biosynthetic enzyme tyrosyne hydroxylase (T-OH) in PC12 cells, in vitro. This site is a known substrate for Ca^{2+}/calmodulin-dependent protein kinase II and is not normally phosphorylated in response to NGF (Haycock, 1990). The activation of Ca^{2+}/calmodulin kinase II is thought to require an increase of intracellular Ca^{2+} and the formation of

Ca^{2+}/calmodulin complexes which in turn binds and activates the enzyme. Increased concentrations of GM1 at the cell surface might modulate Ca^{2+} channel function as has been shown for Na^+ channels (Spiegel et al., 1986). It is also possible that GM1, or its intracellular metabolites, interact with Ca^{2+}/calmodulin kinase increasing its affinity for Ca^{2+} and the formation of Ca^{2+}-calmodulin complexes. In both circumstances the incorporation of gangliosides would facilitate trophic responses by modulating Ca^{2+}-dependent signal pathways in the presence of NGF and, probably, other neurotrophins. On the other hand, exogenously applied gangliosides might hamper the neurotoxic effects resulting from the excessive accumulation of intracellular Ca^{2+} as discussed below.

Recent reports have provided further biochemical evidence to suggest possible mechanisms by which gangliosides could act as neuroprotective agents in our experimental model. During ischemic damage of the CNS, the glutamate concentration at synapses surrounding the ischemic area is increased for sustained periods of time, resulting in 'abusive' stimulation of glutamate receptors (Engelsen, 1982; Benveniste et al., 1989). The abuse of glutamate receptors results in the amplification of several second-messenger responses that normally occur after receptor stimulation under physiological conditions (Manev et al., 1990). It has been suggested that glutamate-induced neurotoxicity may be related to an abnormal amplification of several signals. Among them, the increase of intracellular levels of Ca^{2+} appears to be important in causing neuronal death. It is remarkable that the destabilization of Ca^{2+} metabolism, induced also by the translocation of protein kinase C (PKC) from the cytosol to the neuronal membrane, persists even after glutamate withdrawal (Manev et al., 1990), and is inevitably followed by neuronal death. The persistent increase of PKC translocation and intracellular Ca^{2+} may trigger the activation of such Ca^{2+}-dependent protease and lipase, which may ultimately be responsible for the glutamate-induced neuronal death (Choi, 1988). The paroxysmal stimulation of glutamate receptors is reduced in vitro and in vivo by GM1 treatment (Vaccarino et al., 1987; Manev et al., 1989; Manev et al., 1990). We can speculate therefore, that also in our ischemic injury model ganglioside administration could reduce secondary neural damage in the neocortex surrounding the lesion, thus, protecting areas potentially involved in NGF synthesis and release. In this way, GM1 could exert dual effects after ischemic brain injury, antagonizing acute excitatory amino acid-related toxicity, thus, reducing the extent of brain damage, and potentiating endogenously occurring neurotrophic factors (Ferrari et al. 1983, Leon et al. 1990).

Recent investigations have shown that, in non-human primates, the early administration of the monosialoganglioside GM1 ameliorates Parkinson-like symptoms acutely induced by the neurotoxin 1-methyl-4-phenyl-1,2,3,6-tetrahydropyridine (MPTP) (Schneider et al. 1992). In this experiment, it was shown that striatal levels of dopamine were increased in lesioned GM1-treated animals, indicating that the ganglioside administration is able to induce a neurochemical recovery in a well-established model of Parkinson's disease. Moreover, tyrosine hydroxylase (TH) immunohistochemistry revealed increased TH-positive staining within the striatal regions in GM1-treated animals. This may be the result of sprouting, a GM1-induced stabilization of MPTP-damaged neurons or a combination of these factors (Schneider et al. 1992). The GM1 treatment was initiated within 60 h after the MPTP exposure, when active degenerative processes were still taking place. Neurons damaged by MPTP but, still viable, may have been rescued by GM1 treatment and further stimulated to undergo neuritic sprouting (Schneider et al., 1992). This view is supported by in vitro findings showing that after the exposure to MPP$^+$, the oxidation product of the MPTP toxin, the survival of fetal dopaminergic neurons can be stimulated to recover a normal cell phenotype in the presence of GM1 (Stull et al., 1991).

In vivo studies (Stephens et al., 1987; Karpiak et al., 1991, Schneider et al., 1992) as well as pilot clinical trials (Geisler et al., 1991) have indicated the early administration of gangliosides is a 'permissive condition' (Cuello, 1990; Cuello et al., 1990) for any neuroprotective effect. By contrast, β-NGF can be administered after a certain delay (Hagg et al., 1988). However, the greatest protection is also obtained after an early administration (Hagg et al., 1988). Consequently, in our experimental design we chose to administer these agents immediately after lesioning. Previous in vivo analysis showed that the biodegradable polymers used to incorporate the drugs degrade within 21 days in the rat (Tamargo et al., 1989). On this basis, we assume that in our monkey model rhNGF and GM1 were present on the cortical surface for at least 3 weeks. The protection of nbM neurons, whose cell bodies are not in close proximity to its site of administration, suggests either a circulation of the GM1 through the ventricular system into regions adjacent to the nbM or its effect via the nbM cortically-projecting axons whose injured terminals would be subadjacent to the gelatin film. In either case, the efficacy of our therapy applied at a site distant to the responsive nbM cells bodies is noteworthy and its site of action may differ from that of the intraventricular infusion of GM1 in the rat (Cuello et al., 1989) or NGF in monkeys (Tuszynsky et al., 1990; Koliatsos et al., 1990; Tuszynsky et al., 1991; Koliatsos et al., 1991). This latter method of drug delivery has the disadvantage that the compounds are dispersed into interstitial fluids throughout the brain (Vahlsing et al., 1989). Because in primates the distance over which the compounds need to diffuse in order to reach the nbM is greater than that for the septal nuclei (Szabo and Cowan, 1984), larger doses could perhaps have been needed to obtain effective concentrations. In fact, it was reported that, after fimbria fornix lesions, the areas most affected by the NGF therapy were toward the ventricular side of the monkey septum in which the drug concentration was maximal (Koliatsos et al., 1991). Thus, the effectiveness of intracerebroventricular administration of trophic factors appears to be conditioned by the distance between the ventricular system and the target (Vahlsing et al., 1989; Koliatsos et al., 1991). By contrast, with the system of delivery adopted in our experimental paradigm (Otto, et al.,1989), the compound released by the gelatin film is assumed to be taken up and directly transported to cholinergic cell bodies within the nbM. Nevertheless, it is not possible to exclude that the molecule released by the gelatin film can be largely metabolized at a local level. This could explain the lack of full prevention of nbM retrograde changes in animals treated with only rhNGF. In such a case, one can assume that the simultaneous administration of GM1 may potentiate a subthreshold concentration of rhNGF thus producing a maximal response (Cuello et al., 1989). The effectiveness of the treatment applied in our experimental paradigm, suggests that other chronic delivery systems could also be used to provide the active molecules at a local level [e.g., genetically modified cells (Rosenberg et al., 1988; Gage et al., 1991) or microspheres (Maysinger et al., 1989)].

In conclusion, the monosialoganglioside GM1 and rhNGF, applied alone or in combination, are able to prevent the degeneration of injured cholinergic neurons in the primate nbM for at least 6 months after neocortical lesioning and up to 2 months in the rat.

Other primate studies have shown an NGF-protective effect upon cholinergic septal neurons 1 month following fimbria fornix lesion (Tuszynsky et al., 1990, 1991; Koliatsos et al., 1990; Koliatsos et al., 1991). Therefore, our results suggest that a relatively short period of treatment with GM1, rhNGF, or rhNGF + GM1 may be sufficient to protect nbM cholinergic neurons after a cortical infarction.

The successful demonstration of GM1-rhNGF efficacy in an animal model of neurodegeneration should encourage clinical investigators to consider the possibility of applying this therapy to a variety of neurological disorders involving NGF-responsive cell groups. Potential candidates would include primary cortical pathologies (degenerative, vascular or traumatic) inducing retrograde degeneration of basal forebrain cholinergic neurons. Moreover, these data in conjunction with those obtained in a recent study about the effect of GM1 in MPTP-treated monkeys (Schneider et al., 1992), indicate that the combined administration of GM1 and rhNGF might be beneficial in preventing progression or alleviating symptoms of selected cases of Parkinson's disease. Potential therapeutic effects of this approach should also be investigated in a variety of pathologies involving the peripheral nerve, such as certain types of diabetic neuropathy and neuropathies induced by toxic or antineoplastic agents.

As we have proposed previously (Cuello et al., 1990) a hypothetical neurotrophic-agent therapy in Alzheimer's disease (or other degenerative conditions involving CNS cholinergic systems) would be most effective if applied early in the course of neurodegenerative processes. That is to say when a larger population of responsive neurons are still available for the trophic action. Early intervention will also depend upon development of diagnostic markers for these diseases.

Acknowledgements

The authors are grateful for the financial assistance provided by the Medical Research Council (MRC) and the Centres of Excellence Network for Neuronal Regeneration and Functional Recovery. We would also like to thank Mr. Alan Forster for photographic expertise, Mr. David Rolling for the illustrations and Marsha Warmuth for editorial assistance.

References

Agnati, L.F., Fuxe, K., Calza, L., Benfenati, F., Caviccholi, L., Toffano, G. and Goldstein, M. (1983) Gangliosides increase the survival of lesioned nigral dopamine neurons and favour the recovery of dopaminergic synaptic function in striatum of rat by collateral sprouting. *Acta Physiol. Scand.*, 119, 347–363.

Appel, S.H. (1981) A unifying hypothesis for the cause of amyotrophic lateral sclerosis, Parkinsonism, and Alzheimer's disease. *Ann. Neurol.* 10: 499–505.

Arendt, T., Bigl, V., Arendt, A. and Tennstedt, A. (1983) Loss of neurons in the nucleus basalis of Meynert in Alzheimer's disease, paralysis agitans and Korsakoff's disease. *Acta Neuropathol. Berl.*, 61: 101–108.

Benveniste, H., Jorgensen, M.B., Sandberg, M., Christensen, T., Hagberg, H. and Diemer, N.H. (1989) Ischemic damage in hippocampal CA1 is dependent on glutamate release and intact innervation from CA3. *J. Cereb. Blood Flow Metab.*, 9: 629–639.

Berkemier, L.R., Winslow, J.W., Kaplan, D.R., Nikolics, K., Goeddel, D.V. and Rosenthal, A. (1991). Neurotrophin-5: a novel neurotrophic factor that activates *trk* and *trkB*. *Neuron*, 7: 857–866.

Bothwell, M. (1991) Keeping Track of Neurotrophin receptors. *Cell*, 65: 915–918.

Bremer, E.G., Hakomori, S.I., Bowen-Pope, D.F., Rains, E. and Ross, R. (1984) Ganglioside mediated modulation of cell growth, growth factor binding and receptor phophorylation. *J. Biol. Chem.*, 259: 6818–6825.

Bremer, E.G., Schlessinger, J. and Hakomori, S.I. (1986) Ganglioside-mediated modulation of cell growth. Specific effects of GM3 on tyrosine phosphorylation of the epidermal growth factor receptor. *J. Biol. Chem.*, 261(5): 2434–2440.

Casamenti, F., Bracco, L., Bartolini, L. and Pepeu, G. (1985) Effects of ganglioside treatment in rats with a lesion of the cholinergic forebrain. *Brain Res.* 338: 45–52.

Ceccarelli, B., Aporti, F. and Finesso, M. (1976) Effect of brain ganglioside on functional recovery in experimental regeneration and reinnervation. *Adv. Exp. Med. Biol.*, 71: 275–293.

Cheresh, D.A. and Klier, F.G. (1986) Disialoganglioside GD2 distributes preferentially into substrate-associated microprocesses on human melanoma cells during their attachment to fibronection. *J. Cell Biol.*, 102: 1887–1897.

Cheresh, D.A., Pytela, R., Pierschbacher, D., Klier, F.G., Ruoslahti, E. and Reisfeld, R.A. (1987) An Arg-Gly-Asp-directed receptor on the surface of human melanoma cells exist in a divalent cation dependent functional complex with the disialoganglioside GD2. *J. Cell. Biol.*, 103: 1163–1173.

Choi, D.W. (1988) Calcium mediated neurotoxicity: Relationship to specific channel types and role in ischemic damage. *Trends Neurosci.*, 11: 465–469.

Cremins, J., Wagner, J.A. and Halegouia, S. (1986) Nerve growth factor action is mediated by cyclic AMP and Ca^{2+}/phospholipid dependent protein kinases. *J. Cell. Biol.*, 103: 887–893.

Cuello, A.C. (1990) Glycosphingolipids that can regulate nerve growth and repair. *Adv. Pharmacol.*, 21: 1–50.

Cuello, A.C., Stephens, P.H., Tagari, P.C., Sofroniew, M.V.and Pearson, R.C.A. (1986) Retrograde changes in the nucleus basalis of rat, caused by cortical damage, are prevented by exogenous ganglioside GM1. *Brain Res.*, 376: 373–377.

Cuello, A.C., Maysinger, D., Garofalo, L., Tagari, P., Stephens, P.H., Pioro, E. and Piotte M. (1987) Influence of gangliosides on plasticity of forebrain cholinergic neurons. In: K Fuxe and L.F. Agnati, (Eds), *Receptor-receptor Interactions*, McMillan Press, London, pp. 62–77.

Cuello, A.C., Garofalo, L., Kenisberg, R.L. and Maysinger, D. (1989) Gangliosides potentiate in vivo and in vitro effects of nerve growth factor on central cholinergic neurons. *Proc. Natl. Acad. Sci. USA*, 86: 2056–2060.

Cuello, A.C., Garofalo, L., Kenigsberg, R.L., Maysinger, D., Pioro, E.P. and Ribeiro-da-Silva, A. (1990) Degeneration and regeneration of basal forebrain cholinergic neurons. In: L.A. Harrocks et al. (Eds.), *Trophic Factors and the Nervous System*, Raven Press, Ltd., New York, pp. 307–326.

Dawbarn, D., Allen, S.J. and Semenenko, F.M. (1988) Coexistence of choline acetyltransferase and nerve growth factor receptors in the rat basal forebrain. *Neurosci. Lett.*, 94: 138–144.

Doherty, P., Dickson, J.G., Flanigan T.P. and Walsh F.S. (1985) Ganglioside GM1 does not initiate but enhances neurite regeneration of nerve growth factor-dependent sensory neurons. *J. Neurochem.* 44: 1259–1265.

Elliott, P.J., Garofalo, L. and Cuello, A.C. (1989) Limited neocortical devascularizing lesions causing deficits in memory retention and choline acetyltransferase activity — Effects of the monosialoganglioside GM1. *Neuroscience*, 31: 63–76.

Engelsen, B. (1982) Neurotransmitter glutamate: its clinical importance. *Acta Neurol. Scand.*, 74: 337–355.

Ernfors, P., Ibanes, C.F., Ebendal, T., Olson, L. and Persson, H. (1990). Molecular cloning and neurotrophic activities of a protein with similarities to nerve growth factor: developmental and topographical expression in brain. *Proc. Natl. Acad. Sci. USA*, 87: 5454–5485.

Ferrari, G., Fabris, M. and Gorio, A. (1983) Gangliosides enhance neurite outgrowth in PC12 cells. *Dev. Brain Res.*, 8: 215–221.

Ferrari, G., Fabris, M., Fiori, M.G., Gabellini, N. and Volonté, C. (1992) Gangliosides prevent the inhibition by K-252a of NGF responses in PC12 cells. *Dev. Brain Res.*, 65: 35-42.

Ferreira, A., Busciglio, J., Landa, C. and Caceres, A. (1990) Ganglioside enhanced neurite growth: evidence for a selective induction of high-molecular-weight MAP-2. *J. Neurosci*, 10: 293–302.

Fishman, P.H. (1982) Role of membrane gangliosides in the binding and action of bacterial toxins. *J. Membr. Biol.,* 69, 85–97.

Freed, W. J., deMedinacelli, L. and Wyatt, R.J. (1985) Promoting functional plasticity in damaged nervous system. *Science,* 277: 1544–1552.

Fonnum, F. (1975) A rapid radiochemical method for the determination of choline acetyltransferase. *J. Neurochem,* 24: 407–409.

Gage, F.H., Kawaja, D. and Fisher, L.J. (1991) Genetically modified cells: applications for intracerebral grafting. *TINS,* 14: 328–333.

Garofalo, L. and Cuello, A.C. (1990) Nerve growth factor and the monosialoganglioside GM1 modulate cholinergic markers and affect behavior of decorticated adult rats. *Euro. J. Pharmacol.* 183: 934–935.

Garofalo, L., Ribeiro-da-Silva, A. and Cuello, A.C. (1992a) Nerve growth factor induced synaptogenesis and hypertrophy of cortical cholinergic terminals. *Proc. Natl. Acad. Sci. USA,* 38: 2639–2643.

Garofalo, L., Ribeiro-da-Silva, A. and Cuello, A.C. (1992b) Ultrastructural three dimensional reconstruction of cortical ChAT immunoreactive varicosities: Effects of lesion, GM1 and/or NGF treatment. *Society for Neuroscience,* 22nd Annual Meeting, 1992, Session: 329.3, Abstract A-36.

Garofalo, L., Ribeiro-da-Silva, A. and Cuello, A.C. (1993) Potentiation of nerve growth factor-induced alterations in cholinergic fibre length and presynaptic terminal size in cortex of lesioned rats by the monosialoganglioside GM1. *Neuroscience,* in press.

Geddes, J.W., Anderson, K.J. and Cotman, C.W. (1986) Senile plaques as abberant sprot-stimulating structures. *Exp. Neurol.,* 94: 767–776.

Geisler, F.H., Dorsey, F.C. and Coleman, W.P. (1991) Recovery of motor function after spinal-cord injury — A randomized, placebo-controlled trial with GM1 ganglioside. *N. Engl. J. Med.,* **324**: 1829–1838.

Gorio, A., Marini, P. and Zanoni, R. (1983) Muscle reinnervation. III. Motoneuron sprouting capacity, enhancement by exogenous gangliosides. *Neuroscience,* **8**: 417–429.

Greenberg, M.E., Greene, L.A. and Ziff, E.B. (1985) Nerve growth factor and epidermal growth factor induce rapid transient changes in proto-oncogene transcription in PC12 cells. *J. Biol. Chem.,* 260: 14101–14110.

Hagg, T., Manthorpe, M., Vahlsing, H.L and Varon, S. (1988) Delayed treatment with nerve growth factor reverses the apparent loss of cholinergic neurons after acute brain damage. *Exp. Neurol.,* 101: 303–312.

Hakomori, S. (1986) Glycosphingolipids. *Sci. Am.,* 254: 44–53.

Hakomori, S. (1989) Glycosphingolipids as modulators of growth factor receptors. In: *Trophic Factors and the Nervous System,* Satellite to the 20th Annual Meeting of the American Society for Neurochemistry, Abstr., p. 11.

Hama, T., Huang, P. and Guroff, G. (1986) Protein kinase C as a component of nerve growth factor sensitive phosphorylation system in PC12 cells. *Proc. Natl. Acad. Sci. USA,* 83: 2352–2357.

Haycock, J.W. (1990) Phosphorylation of tyrosine hydroxylase *in Situ* at serine 8, 19, 31 and 40. *J. Biol. Chem.,* 265: 11682–11691.

Hefti, F. (1986) Nerve growth factor promotes survival of septal cholinergic neurons after fimbrial transections. *J. Neurosci.,* 6, 2155–2162.

Hefti, F. and Weiner, W.J. (1986) Nerve growth factor and Alzheimer's disease. *Ann. Neurol.,* 20: 275–281.

Hempstead, B.L., Martin-Zanca, D., Kaplan, D.R., Prada, L.F. and Chao, M.V. (1991) High affinity NGF binding requires co-expression of the *trk* proto-oncogene and the low affinity NGF receptor. *Nature,* 350, 678–683.

Hilbush, B.S. and Levine, M. (1991) Stimulation of a Ca^{2+}-dependent protein kinase by GM1 ganglioside in nerve growth factor-treated PC12 cells. *Proc. Natl. Acad. Sci.,* 88: 5616–5620.

Kaplan, D.R., Hempstead, B.L., Martin-Zanca, D., Chao, M.V. and Prada, L.P. (1991) The *trk* proto-oncogene product: A signal transducer receptor for NGF. *Science,* 252: 554–558.

Karpiak, S.E., Wakade, C.G., Tagliavia, A. and Mahadic, S.P. (1991) Temporal changes in edema, Na^+, K^+, and Ca^{2+} in focal cortical stroke: GM1 ganglioside reduces ischemic injury. *J. Neurosci. Res.,* 30: 512–520.

Kiss, J., McGovern, J. and Patel, A.J.U. (1988) Immunohistochemical localization of cells containing nerve growth receptors in the different regions of the adult rat forebrain. *Neuroscience,* 27: 731–748.

Kitt, C.A., Mitchell, S.J., DeLong, M.R., Wainer, B. and Price, D.L. (1987) Fiber pathways of the basal forebrain in monkeys. *Brain Res.,* 406: 192–206.

Klein, R., Jing, S., Nanduri, V., O'Rourke, E. and Barbacid, M. (1991) The trk proto-oncogene encodes a receptor for nerve growth factor. *Cell.,* 65: 189–197.

Koliatsos, V.E., Nauta, H.J.W., Clatterbuck, R.E., Holtzman, D.M., Mobley, W.C. and Price, D.L. (1990) Mouse nerve growth factor prevents degeneration of axotomized basal forebrain cholinergic neurons in the monkey. *J. Neurosci.,* 10: 3801–3813.

Koliatsos, V.E., Clatterbuck, R.E., Nauta, H.J.W., Knüsel, B., Burton, L.E., Hefti, F.F., Mobley, W.C., and Price, D.L. (1991) Human nerve growth factor prevents degeneration of basal forebrain cholinergic neurons in primates. *Ann. Neurol.,* 30: 831–840.

Kordower, J.H., Bartus, R.T., Bothwell, M., Schatteman, and Gash D.M. (1988) Nerve growth factor receptor immunoreactivity in nonhuman primate (*Cebus apella*): distribution, morphology, and colocalization with cholinergic enzymes *J. Comp. Neurol.,* 277: 465–486.

Kromer, L.F. (1987) Nerve growth factor treatment after brain injury prevents neuronal death. *Science,* 235: 214–216.

Kruijer, W., Schubert, D. and Verma, I. (1985) Induction of the proto-oncogene *fos* by nerve growth factor. *Proc. Natl. Acad. Sci. USA,* 82: 7330–7334.

Lacetti, P., Grollman, E.F., Aloj, S.M. and Kohn, L.D. (1983) Ganglioside dependent return of TSH receptor function in a rat

354

thyroid tumor with a TSH receptor defect. *Biochem. Biophys. Res. Commun.* 10: 772–778.

Ledeen, R.W. (1985) Gangliosides of the Neuron. *Trends Neurosci.* 8: 169–174.

Leibrock, J., Lottspeich, F., Hohn, A., Hofer, M., Hengerer, B., Masiakowski, P., Thoenen, H. and Barde, Y-A. (1989) Molecular cloning and expression of brain-derived neurotrophic factor. *Nature,* 341: 149–152

Leon, A., Del Toso, R., Presti, D., Benevegnu, D., Facci, L., Kirschner, G., Tettamanti, G. and Toffano, G. (1988) Development and survival of neurons in dissociated fetal mesencephalic serum-free cell cultures: II. Modulatory effects of gangliosides. *J. Neurosci.,* 8: 746–753.

Leon, A., Lipartiti, M., Seren, M.S., Lazzaro, A., Mazzari, S., Koga, T., Toffano, G. and Skaper, S.D. (1990) Hypoxic-ischemic damage and the neuroprotective effects of GM1 Ganglioside. *Stroke,* 21: 95–97.

Leonard, D.G.B., Ziff, E.B. and Green, L.A. (1987) Identification and characterization of mRNAs regulated by nerve growth factor in PC12 cells. *Mol. Cell. Biol.,* 7: 3156–3167.

Liberini, P., Pioro, E.P., Maysinger, D., Ervin, F.R., and Cuello, A.C. (1993) Long-term protective effect of human recombinanant growth factor and monosialoganglioside GM1 treatment on primate nucleus basalis cholinergic neurons after neocortical infraction. *Neuroscience,* 53 (3): 625–637..

Maisonpierre, P.C., Belluscio, L, Squinto, S., Ip, N.Y., Furth, M.E., Lindsay, R.M. and Yancopoulous, G.D. (1990) Neurotrophin-3: a neurotrophic factor related to NGF and BDNF. *Science,* 247: 1446–1451.

Manev, H., Favaron, M., Vicini, S., Guidotti, A. and Costa, E. (1989) Glutamate-induced neuronal death in primary cultures of cerebellar granule cells: protection by synthetic derivatives of endogenous sphingolipids. *J. Pharm. Exp. Ther.,* 252: 419–428.

Manev, H., Costa, E., Wroblewsky, J.T. and Guidotti, A. (1990) Abusive stimulation of excitatory amino acid receptors: a strategy to limit neurotoxicity. *FASEB J.,* 4: 2789–2797.

Maysinger, D., Herrera-Marschitz, M., Carlsson, A., Garofalo, L., Cuello, A.C. and Ungerstedt, U. (1988) Striatal and cortical acetylcholine release in vivo in rats with unilaterial decortication: Effects of treatment with monosialoganglioside GM1. *Brain Res.,* 461: 355–360.

Maysinger, D., Garofalo, L., Jalsenjak, I. and Cuello, A.C. (1989) Effects of microencapsulated monosialoganglioside GM1 on cholinergic neurons. *Brain Res.,* 496: 165–172.

Maysinger, D., Herrera-Marschitz, M., Goiny, M., Ungerstedt, U. and Cuello, A.C. (1992) Effects of nerve growth factor on cortical and striatal acetylcholine and dopamine release in rats with cortical devascularising lesions. *Brain Res.,* 577: 300–305

Mesulam, M.M., Mufson, E.J. , Levey, A.Y. and Wainer, B.H. (1983) Cholinergic innervation of cortex by basal forebrain : cytochemistry and cortical connections of the septal area, diagonal band nuclei, nucleus basalis (substantia innominata) and hypothalamus in the rhesus monkey. *J. Comp. Neurol.,* 214: 170–197.

Mesulam, M.M., Mufson, E.J., Levey, A.I. and Wainer, B.H. (1984) Atlas of cholinergic neurons in the forebrain and upper brainstem of the macaque based on monoclonal choline acetyltransferase immunohistochemistry and acetylcholinesterase histochemistry. *Neuroscience* 12: 669–686.

Mesulam, M.M., Mufson, E.J. and Wainer, B.H. (1986) Three-dimensional representation and cortical projection topography of the nucleus basalis (Ch4) in the macaque: concurrent demonstration of choline acetyltransferase and retrograde transport with a stabilized tetramethylbenzidine method for horseradish peroxidase. *Brain. Res.,* 367: 301–308.

Mildbrant, J. (1986) Nerve growth factor rapidly induces c-*fos* mRNA in PC12 rat pheochromocytoma cells. *Proc. Natl. Acad. Sci. USA,* 83: 4789–4793.

Mufson, E.J., Bothwell, M., Hersh, L.B. and Kordower, J.H. (1989) Nerve growth factor receptor immunoreactive profiles in the normal, aged human basal forebrain: colocalization with cholinergic neurons. *J. Comp. Neurol.,* 285: 196–217.

Nieto-Sampedro, M., Manthorpe, M., Barbin, G., Varon, S. and Cotman, C.W. (1983) Injury-induced neuronotrophic activity in adult rat brain: Correlation with survival of delayed implants in the wound cavity. *J. Neurosci.,* 3: 2219–2229.

Oderfeld-Nowak, B., Skup, M., Ulas, J., Jezierska, M., Gradknowska, R and Zaremba, M. (1984) Effect of GM1 ganglioside treatment on postlesion responses of cholinergic neurons in rat hippocampus after various partial deafferentations. *J. Neurosci. Res.,* 12: 409–420.

Otto, D., Frotsher, M. and Unsicker, K. (1989) Basic fibroblast growth factor and nerve growth factor administered in gel foam rescue medial septal neurons after fimbria fornix transection. *J. Neurosci. Res.,* 22: 83–91.

Pearson, R.C.A., Gatter, K.C. and Powell, T.P. (1983) Retrograde cell degeneration in the basal nucleus of monkey and man. *Brain. Res.,* 261: 321–326.

Pedata, F., Giovanelli, L, and Pepeu, G. (1984) GM1 ganglioside facilitates the recovery of high-affinity choline uptake in the cerebral cortex of rats with a lesion of the nucleus basalis magnocellularis. *J. Neurosci.,* 12: 421–427.

Pioro, E. and Cuello, A.C. (1990) Distribution of nerve growth factor receptor-like in the adult rat central nervous system. Effect of colchicine and correlation with the cholinergic system-I. Forebrain. *Neuroscience,* 34: 57–87.

Pioro, E.P., Ribeiro-da-Silva, A. and Cuello A.C. (1990) Immunoelectron microscopic evidence of nerve growth factor receptor metabolism and internalisation in rat nucleus basalis neurons. *Brain Res.,* 527: 109–115.

Pioro, E.P., Maysinger, D., Ervin, F.R., Desypris, G. and Cuello, A.C. (1993) Primate nucleus basalis of Meynert p75^{ngfr}-containing cholinergic neurons are protected from retrograde degeneration by the ganglioside GM$_1$. *Neuroscience,* 53 (1): 49–56.

Purpura, D.P. (1978) Ectopic dendritic growth in mature pyramidal neurones in human ganglioside storage disease. Nature (Lond.) 276: 520–521.

Purpura, D.P. and Suzuki, K. (1976) Distortion of neuronal geometry and formation of aberrant synapses in neuronal storage disease. *Brain Res.,* 116: 1–21.

Rosenberg, M.B., Friedmann, T., Robertson, R.C., Tuszynsky, M., Wolfe, J.A , Breakfield, X.O. and Gage, F.H. (1988) Grafting genetically modified cells to the damaged brain: restorative effects of NGF expression. *Science,* 242: 1575–1578.

Schatteman, G.C., Gibbs, L., Lanahan, A.A., Claude, P. and Bothwell, M. (1988) Expression of NGF receptors in developing and adult primate central nervous system. *J. Neurosci.,* 8: 860–873.

Schwartz, M. and Spirman, N. (1982) Sprouting from chicken embryo dorsal root ganglia induced by nerve growth factor is specifically inhibited by affinity purified antiganglioside antibodies. *Proc. Natl. Acad. Sci. USA,* 79: 6080–6083.

Schneider, J.S., Pope, A., Simpson, K., Taggart, J., Smith, M.G. and DiStefano, L. (1992) Recovery from experimental parkinsonism in primates with GM1 ganglioside treatment. *Science,* 256: 843–846.

Skaper, S.D., Leon, A. and Toffano, G. (1989) Ganglioside function in the development and repair of the nervous system. *Mol. Neurobiol.,* 3: 173–198.

Sofroniew, M.V., Pearson, R.C.A., Eckenstein, F., Cuello, A.C. and Powell, T.P.S. (1983) Retrograde changes in cholinergic neurons in the basal forebrain of rat following cortical damage. *Brain Res.,* 289: 370–374.

Spiegel, S., Handler, J.S. and Fishman, P.H. (1986) Gangliosides modulate sodium transport in cultured toad kidney epithelia. *J. Biol. Chem.,* 261(33): 15755–15760.

Stephens, P.H. , Cuello, A.C., Sofroniew, M.V., Pearson, R.C.A. and Tagari, P. (1985) The effects of unilateral decortication upon choline acetyltransferase and glutamate decarboxylase activities in the nucleus basalis and other areas of the rat brain. *J. Neurochem.,* 45: 1021–1026.

Stephens, P.H., Tagari, P.C., Garofalo, L., Maysinger, D., Piotte, M. and Cuello, A.C. (1987) Neuronal plasticity of basal forebrain cholinergic neurons: effects of gangliosides. *Neurosci. Lett.* 80: 80–84.

Stull, N., Iacovitti, L., DiStefano, L. and Shneider, J.S. (1991) GM1 ganglioside treatment promotes recovery but does not protect fetal dopaminergic neurons from MPP$^+$-induced damage. *Soc. Neurosci. Abstr.* 17: 702.

Szabo, J. and Cowan, W.M. (1984) A stereotaxic atlas of the brain of the cynomolgous monkey (*Macaca fascicularis*). *J. Comp. Neurol.,* 222: 265–300.

Tamargo, R.J., Epstein, J.I., Reinhard, C.S., Chasin, M and Brem, H. (1989) Brain biocompatibility of a biodegradable, controlled-release polymer in rats. *J. Biomed. Mat. Res.,* 23: 253–266.

Toffano, G., Benvegnu, A., Bonetti, A., Facci, L., Leon, A., Orlando, F., Ghidoni, R. and Tettamanti, G. (1980) Interaction of GM1 ganglioside with crude rat brain neuronal membranes. *J. Neurochem.,* 35: 861–866.

Toffano, G., Savoini, G., Moroni, F., Lombardi, G., Calza, L and Agnati, L.F. (1983) GM1 ganglioside stiumaltes the regeneration of dopaminergic neurons in the central nervous system. *Brain Res.,* 261: 163-166.

Tuszynski, M.H., Sang, U H., Amaral, D.G. and Gage, F. (1990) Nerve growth factor infusion in the primate brain reduces lesion–induced neural degeneration. *J. Neurosci.,* 10: 3604–3614.

Tuszynski, M.H., Sang, U.H., Yoshida, K. and Gage, F.H. (1991) Recombinant human growth factor infusions prevent cholinergic neural degeneration in the adult primate brain. *Ann.Neurol.,* 30: 625–636.

Vaccarino, F., Guidotti, A. and Costa, E. (1987) Ganglioside inhibition of glutamate-mediated protein kinase C translocation in primary cultures of cerebellar neurons. *Proc. Natl. Acad. Sci. USA* 84: 8707–8711.

Vahlsing, H.L., Varon, S., Hagg, T., Fass-Holmes, B., Dekker, A., Manley, M. and Manthorpe, M. (1989) An improved device for continuous intraventricular infusions prevents the introduction of pump-derived toxins and increases the effectiveness of NGF treatments. *Exp. Neurol.,* 105: 233–243.

Whitehouse, P.J., Price, D.L., Strubble, R.G., Clark, A.W., Coyle, J.T. and Delong, M.R. (1982) Alzheimer's disease and senile dementia loss of neurones in the basal forebrain. *Science,* 215, 1237–1239.

Whittemore, S.R. and Seiger, A. (1987) The expression, localization and functional significance of β nerve growth factor in the central nervous system. *Brain Res. Rev.,* 12: 439–464.

Williams, L.R., Varon, S., Peterson, G.M. Wictorin, K., Fischer, W., Bjorklund, A. and Gage, F.H. (1986) Continuous infusion of nerve growth factor prevents basal forebrain neuronal death after fimbria-fornix transection. *Proc. Natl. Acad. Sci. USA,* 83: 9231–9235.

Wojcik, M., Ulas, J. and Oderfeld-Nowak, B. (1982) The stimulating effect of ganglioside injections on the recovery of choline acetyltransferase and acetylcholinesterase activities in the hippocampus of the rat after septal lesions. *Neuroscience,* 7: 495–499.

L. Svennerholm, A.K. Asbury, R.A. Reisfeld, K. Sandhoff, K. Suzuki, G. Tettamanti and G. Toffano (Eds.)
Progress in Brain Research, Vol. 101

357

CHAPTER 27

Gangliosides in the protection against glutamate excitotoxicity

E. Costa, D.M. Armstrong, A. Guidotti, A. Kharlamov, L. Kiedrowski, H. Manev[1], A. Polo and J.T. Wroblewski

Fidia-Georgetown Institute for the Neurosciences, Georgetown University Medical School, 3900 Reservoir Rd., N.W., Washington, D.C. 20007, U.S.A. and [1]Fidia S.p.A., Via Ponte della Fabbrica, 3/a, 35031 Abano Terme, Padova, Italy

Introduction

In brain, the trans-synaptic activation of ionotropic transmitter receptors located in the postsynaptic membranes mediates a rapid communication between neurons and produces postsynaptic potentials lasting from a fraction of a millisecond to several milliseconds. Two kinds of elementary events modulate this communication: the presynaptic quantal release of transmitter and the postsynaptic random opening of the single channels that are associated with postsynaptic ionotropic receptors. Each channel opening is triggered by the binding of the transmitter to specific recognition sites located in the extracellular domain of the receptor and the postsynaptic neuronal response is a synchronous flux of ions leading to a minute depolarization in excitatory synapses or a minute hyperpolarization in inhibitory synapses. The profile of the neuronal response depends on the size, the duration and the nature of this ion flux. When receptor-operated Ca^{2+} channels are part of the receptor mosaic located at postsynaptic membranes the ensuing intraneuronal influx of cations includes a small amount of Ca^{2+}. Since the free ionic Ca^{2+} $[Ca^{2+}]_i$ in cytosol is about 50 nM, this influx of Ca^{2+} changes the $[Ca^{2+}]_i$ significantly. This change acts as a second messenger, by activating various Ca^{2+}-dependent intraneuronal enzymes. Hence, the change in the catalytic activity of the enzymes determined by the rapid $[Ca^{2+}]_i$ oscillation becomes part of a 'per se' electri-

cally silent response of the postsynaptic neuron. Among various ionotropic receptors present in brain, those responsive to glutamate often allow a simultaneous influx of Na^+ and Ca^{2+} (Sakmann, 1992). Hence, a Ca^{2+} messenger response can be associated to the Na^+ induced depolarization elicited by glutamate. The Ca^{2+} participation in the glutamate signal transduction depends on the polymorphism of the ionotropic glutamate receptors that is represented in many clusters of glutamate ionotropic receptors located vis a vis the glutamatergic nerve terminal (Sommer et al., 1990; Hollmann et al., 1991; Sommer et al., 1991). The notion that there is a great variability in the structural molecular determinants (receptor subunits) of ionotropic receptors gated by glutamate is becoming an interesting reality (Sakmann, 1992; Sommer and Seeburg, 1992); so far more than a dozen different genes encode for structurally different subunits of these ionotropic receptors (Sakmann, 1992; Sommer et al., 1991) (Table I). There are parts of these ionotropic receptor subunits (transmembrane domain) which have a greater degree of intra-family homology than others (extracellular and intracellular receptor domains). Moreover within a given subunit the structural heterogeneity may be generated by alternative splicing or by single amino acid substitution at the ion filtering ring of the ionic channel (Table I). During the last 10 years the progress of our understanding of glutamate receptor polymorphism has increased in proportion with the depth of our

TABLE I

Ionotropic glutamate receptors

Subunit gene	kDA approx.	Homology to Glur-1 (%)	Agonists	Alternative splicing	Homomeric function	Specific aminoacid ion filtering	Cation selective channel
Glur-1				Flip-flop; α, β		Q	Ca^{2+}; Mg^{2+}
Glur-2	100	68–73	AMPA	Flip-flop	YES	R	Na^+
Glur-3				Flip-flop		Q	Ca^{2+}
Glur-4				Flip-flop; α, β		Q	Ca^{2+}
Glur-5					NO		None
Glur-6	100	38–40	Kainate	?	YES	Q/R	Na^+, Ca^{2+}, Mg^{2+}
Glur-7					NO		None
KA 1	100	35	Kainate	?	NO	Q	None
KA 2							
NMDA R_1	160	22		Z1-1, Z1-2	Yes*		Ca^{2+}, no Mg^{2+}
NMDA R_{2A}	160						None
NMDA R_{2B}	160	12–18	NMDA	?	NO	N	
NMDA R_{2C}	133						

*Different from native NMDA receptor function.

knowledge on this receptor subunit structure (Table I). It all began with the assumption that glutamate receptors were only ionotropic, and empirically three receptor types were distinguished according to their susceptibility to the agonistic efficacy of synthetic structural analogues of glutamate: NMDA (*N*-methyl-D-asparate), kainate and AMPA (α-amino-3-hydroxy-5-methyl-4-isoxazole). We now know that each of the three receptor families includes a number of structurally different subunits, which are approximately 70% homologous. However, a lower homology exists between the subunits of different receptor families (Table I). We also know that homomeric glutamate ionotropic receptors are not always functional and therefore native receptors are probably heterooligomeric (Sakmann, 1992; Sommer and Seeburg, 1992). The degree of subunit heterogeneity and stoichiometry operative in glutamate receptor

assembly are still unknown. However, the number of different gene encoding for these subunits suggests that a high degree of structural variability may characterize the mosaic of glutamate receptor in various brain subsynaptic membranes.

In 1986, we reported (Nicoletti et al., 1986a,b) that in brain are expressed also glutamate metabotropic receptors with signal transduction operated via a G protein linked to a specific phospholipase C that releases IP_3 (inositol triphosphate) from membrane inositide phospholipid (Table II). The existence of such a metabrotropic receptor was reported also by others (Sladeczek et al., 1988), but the acceptance of this novelty was not prompt. Recently, five structurally different subtypes of this receptor were cloned (Masu et al., 1991; Tanabe et al., 1992) (Table II) and this evidence prompted the undiscussed acceptance of glutamatergic receptors with IP_3 or cyclic-AMP

TABLE II

Metabotropic glutamate receptors

Receptor	kDA	Response	PTX	Location
● mGluR1α	133	● PLC activation	↓↓↓	● Hippocampus: CA2-CA3
● mGluR1β	102	● AC activation	↑↑	● Cerebellum: Purkinje cells
				● Olfactory bulb: tufted & mitral cells
● mGluR2	96	● AC inhibition	↓↓↓	● Cerebellum: Golgi cells
● mGluR3	99	?		● Wilde: cerebral cortex, dentate gyrus
				● Glial cells
● mGluR4	102	?		● Wide
				● Cerebellum: granule cells
● mGluR5	128	● PLC activation	↓	● Cerebral cortex
				● Hippocampus: CA1
				·● Cerebellum: granule cells
				● Olfactory bulb: internal granule cells

Abbreviation: PTX, Pertussis toxin.

(inhibitory) transduction mechanisms (Table II). The signal transduction of two of these receptor (mGluR3 and mGluR4) remains still unknown.

It appears that metabotropic glutamate receptor stimulation has a trophic action in CNS (Conn and Desai, 1991), perhaps via the modulation of the neuronal expression of mRNAs encoding for neurotrophic factors such as BDNF and other neurotrophines (Comelli et al., 1992). Hence, if on one hand, glutamate causes excitotoxicity via destabilization of $[Ca^{+2}]_i$ homeostasis, on the other hand, it prompts the biosynthesis of trophic substances via metabotropic receptor stimulation. A mechanism by which the metabotrophic glutamate receptor stimulation elicits the tropic action is the expression of IEG (immediate early genes) which act as third nuclear messengers stimulating the transcription of DNA encoding for trophic factors (Szekely et al., 1990). The increase of IEG expression derives from Ca^{2+}-dependent mechanisms including phosphorylation of nuclear proteins regulatory for the IEG expression. This phosphorylation relates to the release of Ca^{2+} stored in endoplasmic reticulum by IP_3 derived from the hydrolysis of membrane phosphatidylinositols by phospholipase C.

Glutamate receptor signal transduction in CNS pathologies

During convulsions the time interval between two successive presynaptic events shortens and if the number of glutamate gated channels that are Ca^{2+} permeable and the number of glutamatergic synapses located in the postsynaptic neuron are sufficient the rate of Ca^{2+} influx can overwhelm the efficiency of the $[Ca^{2+}]_i$ extruding mechanisms; consequently, the $[Ca^{2+}]_i$ content in the postsynaptic neuron begins to accumulate after each synaptic event. Such an accumulation leads to a persistent activation of the cytosolic Ca^{2+}-dependent protein kinases. Among them protein kinase C (PKC), when activated, tends to translocate to the neuronal membrane causing phosphorylation of membrane regulatory proteins that may either increase the ionotropic channel open time, reduce the efficacy of the Ca^{2+} extrusion mechanisms, or change both processes simultaneously.

Thus, on reiterated convulsive discharges the $[Ca^{2+}]_i$ content can reach levels (500–700 nM) that activate simultaneously various proteolytic enzymes (Manev et al., 1991) and thereby trigger neuronal cytoskeleton abnormalities leading to unavoidable neuronal death.

During brain ischemia elicited by various mechanisms (cardiovascular, thrombotic, traumatic, etc.) the rates of the elementary presynaptic events leading to glutamate release increase paroxysmically causing this transmitter to accumulate in the interstitial fluid. The concurrent neuronal death of glutamatergic neurons caused by hypoxia and the inherent oxygen free radical formation further increases the glutamate content of interstitial fluids leading to a chain of events characterized by the simultaneous and paroxysmal stimulation of every glutamate receptor in an area surrounding the primary ischemia (area penumbra). The borders of this area remain undefined, for the dynamics of the process extends in relation to the neuronal geometry, the glutamate receptor subtypes that are stimulated and density of the glutamatergic ionotropic receptors located in the membranes of the neurons involved. By such mechanism the neurons in the area penumbra are at risk of excitotoxic death. Thus the primary ischemic focus becomes surrounded by a dynamically expanding area of neurons rapidly degenerating and target of possible excitotoxicity. Glutamate though an important factor is not the only cause of neuronal death in brain ischemias. Formation of free radicals and other excitotoxins contribute to the evolution of the brain ischemic damage. When a patient with brain ischemia reaches a physician the only intervention possible is to act by reducing the death risk of the area penumbra neurons. In fact, the sudden and unexpected nature of brain ischemias virtually eliminates a preventive treatment.

Clearly, a therapeutic strategy to reduce neuronal death which is the end point in the sequelae of neuronal damage of area penumbra is that of blocking ionotropic glutamate receptor function in order to prevent the enlargement of the primary ischemic brain damage; area penumbra appears to enlarge and extend on the basis of the continuous pathological stimulation of glutamate receptors triggered by the accumulation of glutamate in the brain interstitial fluids. Since the neurotoxic action of glutamate is mediated by the stimulation of the same ionotropic receptors that mediate the physiological responses of this transmitter, it is impossible by a pharmacological strategy based on the use of isosteric or allosteric receptor antagonists to selectively block the receptors which are pathologically stimulated and leave unaffected the same receptors in those brain areas that are still functioning. In fact, such a simultaneous pharmacological inhibition of receptors activated by pathological and physiological mechanisms blocks the compensatory mechanisms in those brain structures that neither are affected by the primary ischemia nor by the area penumbra pathology. Since excitotoxicity follows the abusive stimulation of NMDA and non-NMDA selective ionotropic glutamate receptors, it is important to study which steps in the chain of events triggered by abusive stimulation of ionotropic glutamate receptors are specific of the pathological response and important in triggering neuronal death. This study would allow to discover whether it is possible to intervene pharmacologically on some selective step located downstream from the glutamate recognition site and thereby inhibit those events that are specific of the abnormal signal transduction while sparing the physiological glutamatergic communication in those brain structures operative in functional compensatory mechanisms.

Specific casade of events associated with stimulation of various glutamate receptor subtypes

Histochemical studies employing specific labelled ligands for recognition sites of various glutamate ionotropic receptor families, specific antisera directed against peptides included in the subunits assembled in various ionotropic receptor families, or in situ hybridization studies of mRNAs encoding for ionotropic glutamate receptor subunits have revealed that the brain distribution of various glutamate receptor subtypes is uneven (Choi, 1991). In line with such a variable receptor distribution, pharmacotherapeutic interventions on specific receptor families aimed at preventing or protecting from glutamate excitotoxicity have different efficacy in different brain regions. For example, the allosteric NMDA receptor antagonist dizolcipine (MK801) provides a consistent partial protection against cortical infarction after middle cerebral artery occlusion (MCA) while it exerts only a minimal protection against striatal damage after global brain ischemia (Meldrum et al., 1992). Moreover, dizolcipine causes direct neuronal damage in limbic cortex (Olney et al., 1989). In contrast, excitotoxicity in striatum is reduced by NBQX (2,3-dihydroxy-6-nitro-7-sulfamoyl-benzo[F]quinoxaline) an isosteric antagonist for non-NMDA ionotropic receptor families (Meldrum et al., 1992).

As mentioned, early glutamate metabotropic receptors do not mediate excitotoxicity, but actually may participate in triggering natural protective mechanisms against excitotoxicity. A concept that is emerging is that ionotropic receptor structures may modify during function. For instance, the expression of the mRNA encoding for the non-NMDA glutamate receptor subunit GluR2 that is developmentally regulated, appears to be down regulated in area penumbra (Zukin et al., 1992). Since the presence of GluR2 subunit, virtually obliterates Ca^{2+} permeability of ionotropic glutamate receptors selective for AMPA one might infer that a permutation of the subunits

assembled in a given receptor may be triggered as a protective mechanism against area penumbra glutamate excitotoxicity. In fact, as mentioned earlier, such excitotoxicity depends on the destabilization of $[Ca^{2+}]_i$ homeostasis caused by Ca^{2+} influx via glutamate-gated cationic channels and the subsequent translocation of PKC from cytosol to the neuronal membranes.

Currently, pharmacological research directed to reduce excitotoxicity follows two major trends: (1) blockade of the glycine modulatory (cotransmitter?) site located on NMDA selective glutamate ionotropic receptors; and (2) inhibition of critical and selective events in the cascade leading to excitotoxicity.

Let us first consider the cascade known to be operative in the function of metabotropic glutamate receptors following their intermittent physiological stimulation. Though as shown in Table I the molecular nature of this transduction is only known for three (mGluR1, 2, 5) of the five structurally different receptor subtypes, the current understanding of the signal transduction at glutamate metabotropic receptors is depicted in Fig. 1. A crucial step in the activation of various modalities of signal transduction appears to be the receptor coupling to a G protein. This coupling is probably variable and depends on the selection from a G protein pool available in the postsynaptic membrane. As shown in Fig. 1, IP_3

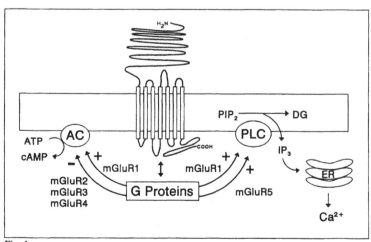

Fig. 1

releases Ca^{2+} stored in the endoplasmic reticulum. This leads to Ca^{2+}-dependent PK activation. When the transduction leads to inhibition of adenylate cyclase multiple actions (Table I) can be expected as a consequence of PKA inhibition. Please note that increase in IP_3 does not promptly lead to $[Ca^{2+}]_i$ homeostasis destabilization because IP_3 can also increase the Ca^{2+} uptake into endoplasmic storage sites thereby attenuating the destabilizing influences of Ca^{2+}-dependent Ca^{2+} release from the endoplasmic reticulum stores. In other words, a series of IP_3-dependent processes regulate intracellular Ca^{2+} dynamic, this occurs with a minimal import of Ca^{2+} from extracellular fluids (Fig. 1). In contrast, the abusive stimulation of Ca^{2+} permeable glutamate ionotropic receptors when protracted leads unescapably to $[Ca^{2+}]_i$ homeostasis destabilization because of a continuous neuronal influx of extracellular Ca^{2+}. The rates of such influx relate to the density of Ca^{2+} permeable ionotropic receptors present in a given neuron (Fig. 2). One should keep in mind that there are intrinsic rates of receptor desensitization following continuous stimulation. They are markedly different for the members of various ionotropic receptor families, their rank order appears to be NMDA < kainate < AMPA.

The neuronal $[Ca^{2+}]_i$ overload that follows persis-tent abusive activation of Ca^{2+} permeable ionotropic glutamate receptor families accelerates several molecular events illustrated in Fig. 2 (Manev et al., 1990b). These include enhanced lipolysis and proteolysis, induction of IEG (Szekely et al., 1990), activation and translocation of protein kinases (Vaccarino et al., 1988), and coordinated expression of genetic programs leading to neuronal structural changes. Probably, these include the alteration in the assembly of different glutamate receptor subunits which was mentioned above. In the case of the stimulation of NMDA selective ionotropic receptors, not only must the NO synthase activation be included (Kiedrowski et al., 1991) but also the consequent activations of hydroxyl radical formation guanylate cyclase (Novelli et al., 1987) and protein ADP ribosylation (Wroblewski, 1991) (Fig. 2).

$[Ca^{2+}]_i$ homeostatic mechanism: destabilization by glutamate receptor stimulation

A complex balance between Ca^{2+} release from endoplasmic reticulum stores, Ca^{2+} influx from extraneuronal fluids and the relationships between Ca^{2+} and Na^{2+} extrusion mechanisms must be considered in order to understand the destabilization of $[Ca^{2+}]_i$ homeostasis elicited by persistent glutamate receptor

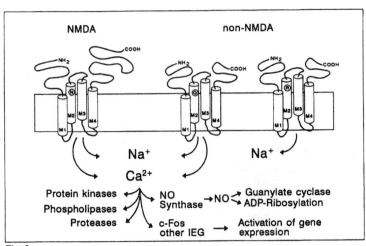

Fig. 2

stimulation. As mentioned earlier, the Ca^{2+} influx occurs via glutamate regulated Ca^{2+} permeable channels and specific voltage or IP_3 regulated Ca^{2+} channels. The latter two sources are always of minor significance because the voltage-dependent channels desensitize promptly and the IP_3 gated channels essentially contribute Ca^{2+} to the dynamics of the endoplasmic reticulum Ca^{2+} stores.

The function of $[Ca^{2+}]_i$ extrusion mechanisms is linked to the cytosolic Na^+ content and its regulation. As can be seen in Fig. 3, following paroxysmal stimulation of ionotropic glutamate receptors the neuronal cytosolic content of Na^+ can increase to 100 mM and more. The increase in cytosolic Na^+ content fails to return to basal levels promptly, following the glutamate withdrawal (Fig. 3). Its return to the basal levels

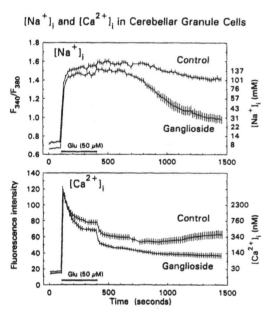

Fig. 3. Effect of ganglioside treatment on intracellular free sodium (top) and calcium (bottom) concentration in cerebellar granule cells. The cells were treated for 2 h with 100 μM ganglioside GT1B or with vehicle (control) and then simultaneously loaded with sodium and calcium fluorescent probes, SBFI (sodium-binding benzofuran isophtalate) and Fluo-3, respectively. $[Na^+]_i$ was calculated from 340 nm/380 nm excitation ratios and $[Ca^{2+}]_i$ from fluorescence signal measured using 480 nm excitation light. Calibration was performed in situ at the end of each experiment. Data are means ± S.E.M. from 25 to 30 cells.

following glutamate withdrawal depends on the function of Na^+/K^+ ATPase, Na^+/Ca^{2+} exchanger and Ca^{2+}ATPase. These enzymes can be regulated by phosphorylation and therefore their catalytic activities can be changed when Ca^{2+}-dependent kinases are activated. In particular, if Na^+ remains elevated the $[Ca^{2+}]_i$ extrusion is less efficacious (Fig. 3). On physiological intermittent receptor stimulation the translocation of cytosolic PKC to neuronal membranes is a dynamic (oscillatory) process. This rapidly reversible PKC translocation probably relates to the $[Ca^{2+}]_i$ oscillation. On persistent glutamate receptor stimulation the rate of $[Ca^{2+}]_i$ return to basal levels is slow (Fig. 3). Also the dissociation of PKC from the neuronal membranes is retarded (Table III) (Manev et al., 1990b). The physicochemical characteristics of PKC interaction with neuronal membranes and the oscillation of $[Ca^{2+}]_i$ appear to depend on the mode of glutamate receptor stimulation. The PKC translocation resulting from abusive stimulation of glutamate ionotropic receptors has slower dissociation kinetics (Table III) (Manev et al., 1990b) whereas that following intermittent physiological stimulation of the glutamate receptor appears to be extremely fast. An understanding of the molecular mechanism underlying this difference is still a matter of speculation. Nevertheless, from some experiments to be discussed later in this paper it appears that the regulation of this kinetics may be an important target in order to obtain drugs that selectively antagonize the neuronal excitotoxicity elicited by paroxysmal stimulation of ionotropic glutamate receptors. Support to this inference comes from experiments in which an activation of protein kinases causes degeneration and death in primary neuronal cultures (Table III) (Manev et al., 1990b). Moreover, several lines of investigation indicate that PKC activation plays a pivotal role in increasing the neuronal death occurring in the neurons of area penumbra that are at risk. In primary cultures of cerebellar granule cells the addition of okadaic acid (OKA), a toxin which increases protein phosphorylation by inhibiting the serine/threonine phosphoprotein phosphatase 1 and 2A, causes neuronal death (Mattson, 1991) (Table III). Protection from OKA neurotoxicity was obtained either by the

TABLE III

Persistent phosphorylation and cell death in neuronal cultures: antagonism by gangliosides

Treatment	Effect on PKC	Effect on neurons	Antagonism of the effect	Reference
Glutamate	Translocation-activation	Not determined	Gangliosides	Vaccarino et al., 1987
Glutamate	Slowly reversible translocation	Neurotoxicity	Gangliosides	Favaron et al., 1988; Manev et al., 1989
Glutamate	Slowly reversible translocation	Neurotoxicity	Gangliosides derivatives	Manev et al., 1990b
Glutamate	Slowly reversible translocation	Neurotoxicity	PKC down-regulation	Favaron et al., 1990
Phorbol ester	Slowly reversible activation	Neurotoxicity	H7	Mattson, 1991
Okadaic acid[a]	Activation	Neurotoxicity	None	Fernandez et al., 1991
Okadaic acid[a]	Activation	Neurotoxicity	H7[b]; PKC down-regulation	Candeo et al., 1992

[a]Inhibits phosphoprotein phosphatase.
[b]Inhibits protein kinase C.

concomitant treatment of the cultures with the protein kinase inhibitor 1-(5-isoquinolinesulfonyl)L-2-methylpiperazine (H7) (Table III) or by down-regulating PKC with phorbol ester derivatives (TPA) prior to OKA administration (Table III). TPA down-regulation 'per se' protects the cultures of cerebellar granule neurons from glutamate excitotoxicity. Also, natural gangliosides and their semisynthetic derivatives can inhibit PKC translocation induced by neurotoxic doses of glutamate and protect from the associated excitotoxicity induced by this transmitter (Table III).

One can surmise that the process linking PKC activation and translocation to excitotoxicity might be a protein operative in the regulation of $[Ca^{2+}]_i$ extrusion via the regulation of Na^+/K^+ ATPase or the Na^+/Ca^+ exchange or both. The Ca^{2+} ATPase pump of neuronal membranes may also be involved, moreover, Ca^{2+} extrusion appears in relation to the Na^+/K^+ ATPase activity. Recently, in isolated trigeminal neurons it was demonstrated that intracellularly applied PKC reduces the Mg^{2+} block of NMDA selective glutamate receptors (Chen and Huang, 1992). It is known that glutamate gated channels may activate elementary currents that fall into several amplitude classes, and that also glutamate channel gating may function at various conductance levels (Sakmann, 1992). Probably the phosphorylation state of PK specific consensi located in the intracellular domain of glutamate ionotropic receptor subunits may contribute together with the expression of a variety of homo or heteroligoemeric assembly of channel isoforms to receptor conductance variability. A possibility is that channel subtype structural variability and phosphorylation of receptors co-localized in a mosaic-like fashion in the postsynaptic membrane contribute to a dynamic variation in channel conductance regulated by receptor subunit phosphorylation. Hence, low levels of protein kinase activity may facilitate the prevailing of low conductance state in many receptor-mosaic constituents, that brings about a condition of hindrance toward $[Ca^{2+}]_i$ homeostasis destabilization induced by abusive receptor stimulation.

Glutamate destabilization of $[Ca^{2+}]_i$ homeostasis: inhibition by gangliosides

The link between glutamate induced destabilization of $[Ca^{2+}]_i$ homeostasis and excitotoxicity has focused attention on therapeutic strategies that prevent the $[Ca^{2+}]_i$ homeostasis destabilization by acting to change the kinetic or to limit the translocation of protein kinase C from cytosol to neuronal membranes (Favaron et al., 1988, Manev et al., 1989; Favaron et al., 1990, Manev et al., 1990b; Mattson, 1991). However, if such therapeutic strategy were possible it would allow to selectively reduce the Ca^{2+}-dependent amplification leading to neurotoxicity. A selective inhibition of mechanisms causing amplification of Ca^{2+} responses during paroxysmal glutamate receptor stimulation has been termed receptor abuse-dependent antagonism (RADA) (Manev et al., 1990b). This strategy implies that the therapeutic agent should spare the initial step in the ionotropic transduction (ion flux though the gated channel) and act on an obligatory step (PKC translocation?) in the cascade of events leading to amplification of Ca^{2+} responses and to unavoidable $[Ca^{2+}]_i$ homeostasis destabilization and neuronal death.

A report by Connor et al. (1988) indicated that sphingosine antagonizes the $[Ca^{2+}]_i$ accumulation elicited by successive NMDA applications, without preventing the initial increase of Ca^{2+} influx, elicited the NMDA application. Unfortunately, sphingosine cannot be used as a therapeutic agent because it disrupts the membranes and causes neurotoxicity. Gangliosides are sphingosine derivatives, they are natural constituents of neuronal membranes, but unlike sphingosine are virtually devoid of intrinsic neurotoxicity. A physiological importance of gangliosides can be inferred from the changes in the profile of gangliosides present in neuronal membranes during the development and maturation of the CNS. Moreover, a hereditary abnormality of membrane ganglioside composition leads to serious neuropathology (Purpura, 1978; Baumann et al., 1980). Finally in newborn infants who died of hypoxia the brain ganglioside content decreases in a manner that can be correlated with the brain damage severity (Qi and Xue, 1991). When gangliosides are added to neuronal cultures they are slowly incorporated in their membranes, thereby participating in the metabolism and turnover of endogenous membrane glycosphingolipids (Ghidoni et al., 1989).

The incorporation of exogenous gangliosides into neuronal membranes fails to change basal electrogenic membrane resting characteristics or the ion flux through cationic ionotropic receptors caused by stimulation of transmitter gated channels (Manev et al., 1990a). This lack of ganglioside induced perturbations of glutamate modulated ion flux can be appreciated from the data reported in Figs. 3 and 4. Neither Na^+ nor Ca^{2+} initial fluxes elicited by excitotoxic glutamate concentrations are changed by the ganglioside insertion into neuronal membranes. The micellar aggregation of gangliosides in solution, limits the concentrations of gangliosides in monomeric configurations which is the only molecular configuration of gangliosides suitable for membrane insertion (Sonnino et al., 1990). This micellar aggregation

$[Ca^{2+}]_i$ in Cerebellar Granule Cells

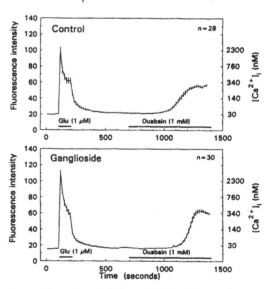

Fig. 4. Effect of the inhibition of Na^+/K^+ ATPase on the intracellular free calcium concentration in cerebellar granule cells treated with vehicle (control) (top) or with ganglioside (GT1B) (bottom). Ganglioside treatment and fluorescent probe loading were the same as described in Fig. 3. Data are means ± S.E.M. from 28 to 30 cells.

slows the ganglioside insertion rates into neuronal membrane. Experimenting with primary neuronal cultures this slow rate of ganglioside insertion imposes long preincubation to obtain a ganglioside protection against glutamate neurotoxicity. Once the free gangliosides are eliminated from the culture media, these sphingolipids protect the culture from the excitotoxity elicited by glutamate in a manner proportional to the quantity of ganglioside inserted in the neuronal membranes (Favaron et al., 1988). Hence, gangliosides fail to change homeostasis and homeostatic control mechanisms of $[Ca^{2+}]_i$, but reduce the destabilization of $[Ca^{2+}]_i$ homeostasis by excitotoxic concentrations of glutamate (Fig. 3). In the presence of a sufficient ganglioside insertion (1–2 nmol/mg protein) into the neuronal membrane (Favaron et al., 1988) the increase of $[Ca^{2+}]_i$ content by an excitotoxic dose of glutamate is not influenced but $[Ca^{2+}]_i$ returns to basal level faster than in neurons that were not treated with gangliosides (de Erausquin et al., 1990) (Fig. 3).

The amount of ganglioside inserted in the neuronal membrane that decreases $[Ca^{2+}]_i$ homeostasis destabilization (Fig. 3) also protects neurons from excitotoxicity (Favaron et al., 1988) (Table III). Thus, gangliosides reduce the crucial pathological amplification of glutamate induced $[Ca^{2+}]_i$ increase without affecting the glutamate signal transduction at ionotropic receptors. Indeed gangliosides appear to follow the action profile expected for a RADA drug (Favaron et al., 1988; Manev et al., 1989; Manev et al., 1990b). Ganglioside protection against $[Ca^{2+}]_i$ homeostasis destabilization, induced by abusive stimulation of glutamate ionotropic receptor permeable to Ca^{2+}, may depend on the reduction of abnormally increased phosphorylation of regulatory proteins operative in Ca^{2+} homeostasis due to the persistent PKC translocation into neuronal membranes (Favaron et al., 1990). The substrate for this phosphorylation has not been studied in detail and one might speculate that it includes modulators or modulatory sites of the Na^+/Ca^{2+} exchanger, Na^+/K^+ ATPase and the consensi of the intracellular domain of certain subunits assembled to form glutamate gated ionotropic receptors. The ganglioside primary target might be the regula-

tion of PKC association to and dissociation from neuronal membranes during the Ca^{2+}-dependent activation and translocation of the PK. The membrane insertion of critical amounts of exogenous gangliosides (1–2 nmol/mg protein), (Favaron et al., 1988), shortens the prolongation of the PKC translocation following glutamate withdrawal that appears to be associated with glutamate excitotoxicity (Favaron et al., 1990). From the data of Fig. 4 it appears clear that gangliosides fail to reduce the increase in $[Ca^{2+}]_i$ elicited by ouabain inhibition of Na^+/K^+ ATPase. Hence the function of ATPase is a condition necessary to observe the ganglioside induced protection against $[Ca^{2+}]_i$ homeostasis destabilization. Moreover, the action of gangliosides in facilitating the decrease of Na^+ increase after glutamate withdrawal (Fig. 3) points out to the possibility of a double site of action of gangliosides: (a) they facilitate Ca^{2+} extrusion by accelerating intracellular Na^+ decrease (Fig. 3) probably via a Na^+/K^+ ATPase activation; and (b) they reduce the amplification of Ca^{2+} response elicited by PKC translocation by shortening the duration of this translocation (Table III).

The natural ganglioside inability to reach a degree of therapeutically valid insertion into neuronal membranes when given orally (Table IV), prompted the synthesis of lysoganglioside derivatives with suitable physicochemical properties to assure an appropriate oral absorption and with an equilibrium in body fluid between monomeric configuration and micellar aggregation of the compound which favors the monomeric configuration, and thereby increases the lysoganglioside insertion in neuronal membranes (Manev et al., 1990c).

Natural gangliosides and glutamate excitotoxicity

In primary cultures of cerebellar granular neurons the stimulation of ionotropic receptors by glutamate or NMDA in the absence of Mg^{2+} or by kainate in the presence of Mg^{2+}, induces a dose-dependent neuronal death (Favaron et al., 1988). The application of appropriate NMDA doses for 15 min or of kainate for 30–60 min induces a delayed neuronal death beginning 1 or 2 h, after the excitotoxin withdrawal and reaches completion in about 12–14 h (Favaron et al., 1988). Natural gangliosides in a dose-dependent manner prevent neuronal death. The physicochemical properties of gangliosides dissolved in water-based solvents (buffers-biological fluids) even in concentrations as low as 1 μM, determine that a very small proportion of the compound dissociates as a monomer, while the bulk of the compound forms micelles which do not insert in neuronal membranes (Sonnino et al., 1990). Since as mentioned earlier the monomeric form is the one that inserts in the neuronal membranes it takes 30–60 min to reach equilibrium between micelles, monomer and membrane inserted monomer (Favaron et al., 1988). Thus the ganglioside must be preincubated for about 60 min to allow for a ganglioside equilibrium. The number of sialic acid molecules present in the chemical structure of the ganglioside favors efficacy (Favaron et al., 1988). Hence the asialo GM1 fails to display excitotoxicity protection while the rank order of efficacy of the natural products tested was GT1b>GD1b>GM1. This structural activity relationship is maintained in in vitro models of anoxia and in hypoglycemia induced neuronal death (Facci et al., 1990).

Ganglioside derivatives and excitotoxicity

Because natural gangliosides fail to reach the brain when given by mouth (Table IV) and because they insert and concentrate in neuronal membrane slowly, the synthesis of ganglioside derivatives having physicochemical properties suitable for oral administration and possessing rapid rates of membrane insertion become an attractive research strategy. Two main approaches were selected: one directed to a modification of the ceramide moiety of the natural ganglioside molecule and the other directed to the carbohydrate moiety including the sialic acid.

Among the derivatives synthesized following the ceramide directed strategy the compounds with the N-acetyl-sphingosine or N-dichloroacetyl-sphingosine residue of ceramide appear to be particularly interesting because of their potency and rapidity of action. These are: LIGA 4(11³ Nen 5Ac Cg Ose₄-

2d-*erythro*-1,3-dihydroxy-2-acetylamide-4-*trans*-octadecene) and LIGA 20 (11³ Nen 5Ac Cg Ose₄-2D-*erythro*-1,3 dihydroxy-2 dichloroacetyl-amide-4-*trans*-octadecene) (Manev et al., 1990c).

The structural differences between LIGAs and natural gangliosides are manifested in different equilibria between monomer and micellar aggregation in the aqueous solution (Sonnino et al., 1990). The micellar concentration of LIGA 4 is lower than that of natural gangliosides. It appears that ganglioside derivatives which form a higher proportion of monomers in aqueous solution tend to insert more rapidly and to a greater extent in the neuronal membranes (Manev et al., 1990c). Moreover, the lower proportion of micellar aggregation, when they are in solution, facilitates their oral absorption. LIGA 4 and LIGA 20, given orally, can be found in brain in higher concentrations than compounds containing the ceramide moiety (Table IV).

In neuronal cultures protection against glutamate excitotoxicity by LIGA 4 and LIGA 20 is rapid, therefore preincubation is not required (Manev et al., 1990c). Moreover, these drugs are one order of magnitude more potent than GM1. The LIGA 20 protection is particularly longer lasting because in vitro it forms an active metabolite (Sonnino et al., 1990). It must be added that like GM1 also LIGA 4 and LIGA 20 do not modify the glutamate gating of ionotropic channels (Manev et al., 1990c). Recently, a Ca^{2+}-dependent ganglioside binding protein (gangliomodulin) was isolated from a soluble fraction of mouse brain and was identified with calmodulin (Higashi et al., 1992). Thus gangliosides may regulate calmodulin-dependent enzymes when they are inserted in neuronal membranes. Since Ca^{2+} ATPase was shown

TABLE IV

Brain and plasma drug content in rats receiving 70 mmol/kg (p.o.)

Drug	Brain (μM)	Plasma (μM)
GM1	0.07	0.05
LIGA 4	0.55	0.69
LIGA 20	0.86	1.30

to be activated by calmodulin in a Ca^{2+}-independent manner, the binding of gangliosides inserted in neuronal membranes to calmodulin or to the enzymes themselves may facilitate the catalytic activity of the Ca^{2+} ATPase and contribute to the ganglioside antagonisms against the $[Ca^{2+}]_i$ homeostasis destabilization induced by glutamate.

In vivo studies of natural gangliosides and their semisynthetic derivatives

We have discussed earlier how a progressive and persistent accumulation of glutamate in interstitial fluids surrounding brain ischemic areas, leads to a paroxysmal stimulation of ionotropic Ca^{2+} permeable channels gated by glutamate and to destabilization of $[Ca^{2+}]_i$ homeostatic processes. Such a destabilization is followed by excitotoxic neuronal death (Choi, 1988). Several experimental models of focal brain ischemia are currently available in various animal species (Watson et al., 1985). In the rat the model of cortical photothrombosis is convenient to evaluate drug efficacy in attenuating the progress of excitotoxicity in area penumbra.

In this model, the skull of the rat is exposed under general anesthesia by an injection of the dye Rose Bengal (80 mg/kg i.v.) and successively the skull is irradiated with a beam of cold white light*. The interaction of the light with the dye releases oxygen singlets that mediate an endothelial cell damage causing a photothrombosis. The end result is a reproducible focal cortical lesion. In our laboratory the area of illumination by a light beam of 3 mm was centered 1.8 mm posterior to the Bregma and 2.8 mm lateral to the midline, the underlying parietal sensory motor cortex is damaged and a highly reproducible infarct is produced within 1 h. This procedure provides a consistent lesion in which the hemodynamic consequences may be demonstrated arteriographically (Dietrich et

al., 1987) and the accompanying edema can be revealed by MRI technology (Pierpaoli, pers. commun.). The hemodynamic consequences associated with the lesion may be demonstrated to reflect those of spontaneous cortical cerebral infarction in humans (Dietrich et al., 1987). Moreover, metabolic alterations and edema can be demonstrated to spread within 6–8 h to the contralateral cortex; decreased cerebral blood flow and glucose utilization occur within 30 min to 4 h in the infarcted area. The surroundings of this primary lesion exhibit congestion and increased glucose metabolism. Within 6–8 h the decrease in cerebral blood flow spreads ipsilaterally to areas distal from the infarct. Alterations in the blood brain barrier and in water and electrolyte content also have been documented in the infarcted area. Vasogenic edema is detected in areas distal to the infarct which may even induce secondary ischemic brain damage.

The infarcted area is small at 30 min, enlarges in 4–12 h and by 24–48 h becomes delimited by a massive infiltration of macrophages and later by a reactive gliosis. Examination of Nissl stained tissue sections revealed, in the area penumbra, neuronal damage and degeneration (Costa et al., 1992). The severity of this lesion decreases in a relation to the distance from the infarct. The role of glutamate in the progression of area penumbra neuronal damage was demonstrated by the ability of a pretreatment with MK801 or NBQX to reduce the size and severity of this neuronal degeneration (Table V). For example examination of Nissl stained tissue sections revealed that in the area penumbra there was a severe reduction of pyramidal neuron density, which resulted in gross distortion of the normal cytoarchitecture (Costa et al., 1992). Also in the cortical-parietal area V the content of AMPA ionotropic receptor subunits (GluR1-GluR2) and of their encoding in mRNAs were shown to be reduced. In contrast, this area evinces a marked increase of heat shoke protein-like immunoreactivity. We do not know yet whether this increase is associated with a change in the respective mRNA. All these observations, taken together, indicate that the cortical photothrombotic lesion with the surrounding area penumbra can serve as a useful model to test for drug

---

*This light source consisted of the power supply (single output rated-15VDC:8.4–10.4 A) fan, dichroic halogen bulb with parabolic reflector (12 V, 100 W, wavelength 400–1200 nm) with peak energy at 1000 nm, 3400 K.

efficacy in limiting the progressive extension and the neuronal damage of area penumbra.

In this context it is important to evaluate the dynamic evolution of the area penumbra where neurons are at risk by measuring [³H]PDBU binding (Fig. 5). Since this compound is a selective high affinity ligand of PKC the increased binding monitors the extent of PKC activation and translocation operative at any given time. Thus this marker allows to follow in a time- and volume-related manner the PKC translocation in various brain areas and drug efficacy on this translocation.

In other words this high affinity [³H]PDBU binding monitors the impending excitotoxicity of glutamate accumulating in interstitial fluids. Within 1 h after the lesion, an increase of [³H]PDBU binding is observed in the immediate surroundings of the infarcted cortical area. Thereafter [³H]PDBU binding begins to extend beyond the experimental infarctual lesion reaching a peak at 3–6 h, yet by 12 h it begins to shrink and returns to normal levels at 24 h. In support of the inference that an increase of high affinity [³H]PDBU binding denotes the perinfarctual area where it operates a paroxysmal abusive stimulation of inotropic Ca^{2+} receptors activated by glutamate is the evidence that the increased [³H]PDBU binding is

abolished by a MK801 pretreatment and partially reduced by NBQX administration (Table V). As mentioned earlier these compounds block NMDA selective (MK801) and AMPA/Kainate selective (NBQX) glutamate receptors, respectively. LIGA 4 and LIGA 20 which fail to act on glutamate channel gating at various ionotropic receptors (Manev et al., 1990c) also can shorten the increase of [³H]PDBU binding (Table V). These results support the inferences made earlier in this paper on the ability of LIGA 4 and LIGA 20 to reduce glutamate excitotoxicity by reducing PKC activation and translocation (Table V). Moreover, that they act via a RADA mechanism is documented by their inability to prevent the increase of FOS expression, an index of lack of action on glutamate channel gating (Table V).

Since a reduction of [³H]PDBU binding is also provided by a pretreatment with GM1 we can conclude that natural and lysogangliosides are a new class of compounds that reduce glutamate excitotoxicity by acting downstream from the receptor ion flux filter regulated by the glutamate channel gating. In fact, as can be shown by Ca^{2+} imaging (Fig. 4) gangliosides do not inhibit the cell fluorescence increase elicited by glutamate but inhibit the persistence of this fluorescence following withdrawal of high doses

TABLE V

Effect of MK-801 and LIGA 20 on rose bengal induced cortical photothrombosis

	PKC activity[a] ([³H]PDBu binding)	FOS[b] expression	Number of neurons[c]
Vehicle	5.8 + 0.82*	△	25 + 1.2*
MK-801 (8.3 μmol/kg i.v.)	6.5 + 1.1*	▲	30 + 1.6*
LIGA 20 (89.2 μmol/kg i.v.)	6.8 + 0.91*	△	28 + 1.2*
Photochemical lesion	21 + 1.6**	▲▲▲	14 + 1.0**
Lesion + MK-801	8.3 + 1.9*	▲	21 + 1.2*
Lesion + LIGA 20	12 + 0.93**,*	▲▲▲	22 + 1.6*

Three hours after the lesion the different parameters were determined. MK-801 and LIGA 20 were injected 1 h before the lesion.
[a][³H]Phorbol 12,13-dibutyrate binding is evaluated as optical density of area penumbra.
[b]FOS immunostaining: △, basal level; ▲, modest increase; ▲▲▲, large increase.
[c]Nissl-positive neurons were counted in 0.125 × 0.125-mm segments of area penumbra (Kharlamov et al. in press).

370

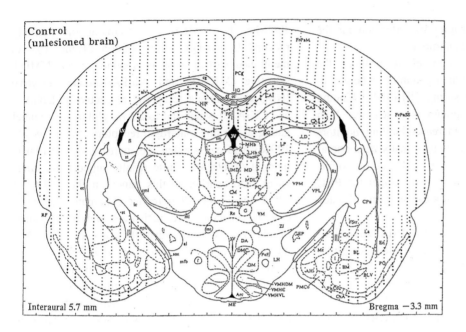

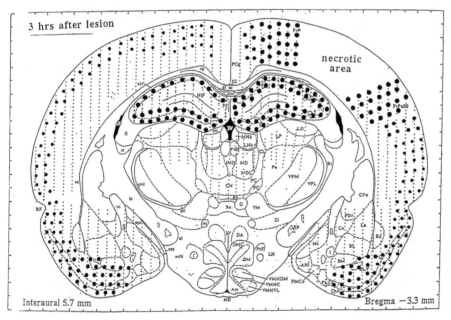

(Kharlamov et al. in press)

Fig. 5. Autoradiographic distribution of protein kinase C after photochemically induced focal brain ischemia in rats (scheme of [³H]PDBU distribution through the coronal plane of rat brain). ..., Basal level of PKC activity; ●●●, Minimum increase of PKC activity; ●●●, Maximum increase in PKC activity.

of glutamate (Fig. 3). This persistence is an index of $[Ca^{2+}]_i$ homeostasis destabilization and predicts excitotoxicity. Moreover, Table V shows that LIGA 20 shortens the time course of PKC translocation and reduces in vivo the neuronal excitotoxicity in area penumbra. This is documented in Table V by the neuronal counts that clearly indicate a sparing action of LIGA 20 on the impending neuronal death of the area penumbra.

Another marker to evaluate the ionotropic receptor stimulation by the glutamate accumulating in brain interstitial fluids is the detection by the in situ hybridization of the increased IEG expression mRNAs and/or proteins. FOS (Table V), JUN, ZIF268 and NUR-77 and their mRNAs are increased in brain sections from animals with photothrombotic lesions. The increase in FOS protein and *fos* mRNA is completely abolished by MK801 (Table V) but not by NBQX, indicating that a stimulation of NMDA selective glutamate receptors is operative on this IEG response (Szekely et al., 1990). However, the blocking affect of MK801 on FOS expression is transient because of the short biological half-life of this drug that contrasts with the longer lasting increase of brain interstitial fluid glutamate accumulation.

Conclusions

Natural gangliosides and LIGA compounds in doses that reduce the increase of [^{3}H]PDBU binding in the area penumbra and fail to block glutamate channel gating are completely ineffective in preventing the expression of the IEG. This suggests that IEG induction is the direct consequence of the Ca^{2+} influx through the glutamate gated ionotropic channel. In fact, stimulation with specific agonists of AMPA receptors that are less permeable to Ca^{2+} than NMDA receptors fail to induce IEG. We can conclude that natural and lysogangliosides also in vivo fail to modify the glutamate gating of ionotropic channels. These are blocked by MK801 which virtually abolishes IEG induction in the area penumbra (Table V). Thus we can infer that the ganglioside block of Ca^{2+} amplification (Fig. 3) occurs downstream from the glutamate channel gating. Since MK801, the natural ganglio-

sides and LIGA 4 and 20 reduce neuronal death in area penumbra (Table V), it is clear that excitotoxicity requires Ca^{2+} amplification mechanisms but not IEG induction. Perhaps IEG induction relates to the neuroplastic response to glutamate. If this inference were to be correct, blockers of NMDA selective glutamate receptors would suppress glutamate mediated neuroplasticity which instead will remain operative during ganglioside or LIGA treatment. With regard to the inhibition of Ca^{2+} amplification by LIGA and natural gangliosides two mechanisms appear to be important. One is the reduction of PKC translocation and activation and the other is Na^+/K^+ ATPase stimulation and the consequent facilitation of Na^+ efflux which in turn appears to indirectly facilitate Ca^{2+} extrusion by the Na^+/Ca^{2+} exchanger. At this time, only the lack of action of gangliosides on $[Ca^{2+}]_i$ increase elicited by ouabain (Fig. 4) supports this contention.

An extra dividend of the photothrombosis model is the possibility to quantitatively evaluate the motor deficit caused by the cortical lesion. For example, when the animal is held by the tail, the anterior and posterior limbs ipsilateral to the lesion are extended while the contralateral limbs are not. Moreover, the paw dexterity decrease induced by the lesion can be rated by the time required to remove a tape placed in one paw. The lesioned rats unlike normal rats cannot stay on an inclined plane nor can they hang on a vertical grid. Both tests can be quantified by measuring hanging time and latency to fall in the inclined plane, respectively. The size of the cortical infarction can be assessed by staining the tissue with triphenyl tetrazolium chloride (TTC). Motor deficit degree evaluated as described above correlates with the size of the lesion. The motor deficit severity peaks at about 48 h after the lesion, and stabilizes at a lower level within 7 days. Natural gangliosides or LIGA 4 or 20 administration reduce the intensity of the deficit caused by the photothrombotic lesion and speed up recovery time. Interestingly this amelioration coincides with the decrease in the number of neurons that are disappearing from the cortical layer V in the area penumbra surrounding the cortical infarction (Costa et al., 1992).

References

Baumann, N., Harpin, M.L. and Jacque, C. (1980) Brain gangliosides in shiverer mouse: Comparison with other dysmyelinated mutants, quaking and jumpy. In: N. Baumann (Ed.), Neurological Mutations Affecting Myelination, INSERM Symposium 14, Elsevier North Holland, Amsterdam, pp 257–262.

Candeo, P., Favaron, M., Lengyel, I., Manev, R.M., Rimland, J.M. and Manev, H. (1992) Pathological phosphorylation causes neuronal death: Effect of okadaic acid in primary culture of cerebellar granule cells. J. Neurochem 59: 1558–1561.

Chen, L. and Huang, L.Y.M. (1992) Protein kinase C reduces Mg^{2+} block of NMDA-receptor channels as a mechanisms of modulation. Nature, 356: 521–523.

Choi, D.W. (1988) Ionic dependence of glutamate neurotoxicity. J. Neurosci., 7: 369–379.

Choi, D.W. (1991) Excitatory amino acid neurotransmitters: Anatomical systems. In: B.S. Meldrum (Ed.), Excitatory Amino Acid Antagonists, Blackwell Scientific Publications, Oxford, pp. 14–38.

Comelli, M.C. Seren, M.S., Guidolin, D., Manev, R.M., Favaron, M., Rimland, J.M., Canella, R., Negro, A. and Manev. H. (1992) Photochemical stroke and brain derived neurotropic factor mRNA expression. Neuroreport, 3: 437–476.

Conn, P.J. and Desai, M.A. (1991) Pharmacology and physiology of metabotropic glutamate receptors in mammalian central nervous system. Drug Dev. Res., 24: 207–229.

Connor, J.A., Wadman, W.J. and Hockberger, P.E. (1988) Sustained dendritic gradients of Ca^{2+} induced by excitatory amino acid in CAI hippocampal neurons. Science, 240: 649–653.

Costa, E., Kharlamov, A., Guidotti, A., Hayes, R. and Armstrong, D. (1992) Sequelae of biochemical events following photochemical injury of rat sensory-motor cortex: Mechanism of ganglioside protection. Physiopathol. Exp. Ther., 14: 17–23.

De Erausquin, G., Manev, H., Guidotti, A., Costa, E. and Brooker, G. (1990) Gangliosides normalize distorted single cell intracellular free Ca^{2+} dynamics after toxic doses of glutamate in cerebellar granule cells. Proc. Natl. Acad. Sci. USA, 87: 8017–8021.

Dietrich, W.D., Watson, B.D., Busto, R., Ginsberg, M.D. and Bethea, J.R. (1987) Photochemically induced cerebral infartion I early microvascular alternations. Acta Neurol., 72: 315–334.

Facci, L., Leon, A. and Skaper, S.D. (1990) Hypoglycemic neurotoxicity in vitro: Involvement of excitatory amino acid receptors and attenuation by monosialoganglioside GM1. Neuroscience. 37: 709–716.

Favaron, M., Manev, H., Alho, H., Bertolino, M., Ferret, B., Guidotti, A. and Costa, E. (1988) Gangliosides prevent glutamate and kainate neurotoxicity in primary neuronal cultures of neonatal rat cerebellum and cortex. Proc. Natl. Acad. Sci., USA, 85: 7351–7355.

Favaron, M., Manev, H., Siman, R., Bertolino, M., Szekely, A.M.,

de Erausquin, G., Guidotti, A. and Costa, E. (1990) Down regulation of protein kinase C protects cerebellar granule neurons in primary culture from glutamate induced neuronal death. Proc. Natl. Acad. Sci., USA, 87: 1983–1987.

Fernandez, M.T., Zitko, V., Gascon, S. and Novelli, A. (1991) The murine Toxin Okadeic acid is a potent neurotoxin for cultured cerebellar neurons. Life Sci., 49: 157–162.

Ghidoni, R., Riboni, L. and Tettamanti, G. (1989) Metabolisms of exogenous gangliosides in cerebellar granular cells differentiated in culture. J. Neurochem., 53: 1567–1574.

Higashi, H., Omozi, A. and Yamagoto, T. (1992) Calmodulin, a ganglioside binding protein. J. Biol. Chem., 267: 9831–9838.

Hollman, M., O'Shea-Greenfield, A., Rogers, W. and Heineman, S. (1991) Ca^{2+} permeability of KA-AMPA-gated glutamate receptor channels depend on subunit composition. Science, 252: 851–853.

Kharlamov, A., Guidotti, A., Costa, E., Hayes, R. and Armstrong, D. (1993) Semisynthetic sphingolipids prevent protein kinase C translocation and neuronal damage in the proyocal area following a photochemically-induced thrombotic brain cortical lesion. J. Neuroscience.

Kiedrowski, L., Manev, H., Costa, E. and Wroblewski, J.T. (1991) Inhibition of glutamate induced neuronal death by sodium nitroprusside is not mediated by nitric oxide. Neuropharmacology, 30: 1241–1243.

Manev, H., Favaron, M., Guidotti, A. and Costa, E. (1989) Delayed increase of Ca^{2+} influx elicited by glutamate: Role in neuronal death. Mol. Pharmacol., 36: 106–112.

Manev, H., Favaron, M., Vicini, S. and Guidotti, A. (1990a) Ganglioside mediated protection from glutamate induced neuronal death. Acta Neurobiol. Exp., 50: 475–488.

Manev, H., Costa, E., Wroblewski, J.T. and Gudiotti, A. (1990b) Abusive stimulation of excitatory amino acid receptors: A strategy to limit neurotoxicity. FASEB J., 4: 2789–2797.

Manev, H., Favaron, M., Vicini, S., Guidotti, A. and Costa, E. (1990c) Glutamate induced neuronal death in primary cultures of cerebellar granule cells: Protection by synthetic derivatives of endogenous sphingolipids. J. Pharmacol. Exp. Ther., 252: 419–427.

Manev, H., Favaron, M., Siman, R., Guidotti, A. and Costa, E. (1991) Glutamate neruotoxicity is independent of calpain 1 inhibition in primary cultures of cerebellar granule cells. J. Neurochem., 57: 1288–1295.

Masu, M., Tanabe, Y., Tsuchida, K., Shigemoto, R. and Nakanishi, S. (1991) Sequence and expression of a metabotropic glutamate receptor. Nature, 349: 760–765.

Mattson, M.P. (1991) Evidence of the involvement of protein kinase C in neurodegenerative changes in cultured human cortical neurons. Exp. Neurol., 111: 95–103.

Meldrum, B.S., Smith, S.E., Le Peillet, E., Moncada, C. and Arvin, B. (1992) Antagonists acting at non-NMDA receptors as cerebroprotective agents in global and focal ischemia. In: R.P. Simon (Ed.), Excitatory Amino Acids, Vol. 9, Thieme Medical Publishers, New York, pp. 235–239.

Nicoletti, F., Iadarola, M.J., Wroblewski, J.T. and Costa, E. (1986) Excitatory amino acid recognition sites coupled and with inosital phospholipid metabolism: Developmental changes and interaction with α_1 adreneceptors. *Proc. Natl. Acad. Sci. USA*, 83: 1931–1935.

Nicoletti, F., Wroblewksi, J.T., Novelli, A., Alho, H., Guidotti, A. and Costa, E. (1986) The activation of isonitol phospholipid metabolism as a signal-transducing system for excitatory amino acids in primary cultures of cerebellar granule cells. *J. Neurosci.*, 6: 1905–1911.

Novelli, A., Nicoletti, F, Wroblewski, J.T., Alho, H., Costa, E. and Guidotti, A. (1987) Excitatory amino acid receptors coupled with granulate cyclase in primary cultures of cerebellar granule cells. *J. Neurosci.*, 7: 547–554.

Olney, J.W., Labruyere, J. and Price, M.T. (1989) Pathological changes induced in cerebrocortical neurons by phencyclidine and related drugs. *Science*, 244: 1360–1362.

Purpura, D. (1978) Ectopic dendritic growth in mature pyromidal neurons in human ganglioside storage disease. *Nature*, 276: 520–521.

Qi, Y. and Xue, Q.M. (1991) Ganglioside levels in hypoic brains from neonatal and premature infants. *Mol. Chem. Neuropathol.*, 16: 87–95.

Sakmann, B. (1992) Elementary steps in synaptic transmission revealed by currents through single ion channels. *Neurons*, 8: 613–629.

Sladeczek, F., Recasens, M. and Bockaert, J. (1988) A new mechanism for glutamate reception action: Phosphoinositide hydrolysis. *Trends Neurosci.*, 11: 545–549.

Sommer, B. and Seeburg, P.H. (1992) Glutamate receptor channels: novel properties and new clones. *TIPS*, 13: 291–296.

Sommer, B., Keinanen, K., Verdon, T.A., Wisden, W., Burnashev, N., Herb, A., Kohler, M., Tagaki, T., Sakmann, B. and Seeburg, P.H. (1990) Flip and Flop: A cell specific functional switch in glutamate-operated channels of CNS. *Science*, 249: 1580–1585.

Sommer, B., Kohler, M., Sprengle, R. and Seeburg, P.H. (1991) RNA editing in brain controls a determinant of ion flow in glutamate gated channel. *Cell*, 67: 11–20.

Sonnino, S., Cantu, L., Corti, M., Acquotti, D., Kirschner, G. and Tettamanti, G. (1990) Aggregation properties of semisynthetic GM_1 ganglioside (11^3 Neu 5-Ac Gg Ose_4 Cer) containing an acetyl group as acylmoiety. *Chem. Phys. Lipids*, 56: 49–57.

Szekely, A.M., Costa, E. and Grayson, D.R. (1990) Transcriptional program coordination by NMDA-sensitive glutamate receptor stimulation in primary cultures of cerebellar neurons. *Mol. Pharmacol.*, 38: 624–633.

Tanabe, Y., Masu, M., Ishis, T., Shigemoto, R. and Nakanishi, S. (1992) A family of metabotropic glutamate receptors. Neuron, 8: 169–179.

Vaccarino, F., Guidotti, A. and Costa, E. (1988) Ganglioside inhibition of glutamate mediated protein kinase C translocation in primary cultures of cerebellar neurons. *Proc Natl. Acad. Sci. USA*, 84: 8707–8710.

Watson, D.B., Dietrich, W.D., Busto, R., Wachtel, M.S., Ginstery, M.D. (1985) Induction of reproducible brain infarction by photochemically initiated thrombosis. *Am. Neurol.*, 17: 497–504.

Wroblewksi, J.T., Raulli, R., Lazarewicz, J.W., Kiedrowski, L., Costa, E. and Wroblewska, B. (1991) Intracellular messengers generated by glutamate receptors in primary cultures of cerebellar neurons. In: E.A. Barnard (Ed.), *Transmitter Amino Acid Receptors: Structures, Transduction and Models of Drug Development, Vol. 6*, Thieme Medical Publishers, New York, pp. 379–393.

Zukin, R.S., Pellegrini-Giampietro, D.E., McGurk, J.F. and Bennett, M.V.L. (1992) Molecular biological studies of glutamate receptors in the developing brain. In: R.P. Simon (Ed.), *Excitatory Amino Acids, Vol. 9*, Thieme Medical Publishers, New York, pp. 47–51.

L. Svennerholm, A.K. Asbury, R.A. Reisfeld, K. Sandhoff, K. Suzuki, G. Tettamanti and G. Toffano (Eds.)
Progress in Brain Research, Vol. 101

CHAPTER 28

Clinical approaches to dementia

Guy M. McKhann

Zanvyl Krieger Mind/Brain Institute and Department of Neurology, The Johns Hopkins University, Baltimore, MD 21218 U.S.A.

Introduction

Dementia is an age-related process. If one studies a general, community-based population, the incidence is only 3% for those between 65 and 74 years, but rises to 47% for those over 84 years (Evans et al., 1989). Similar data has been obtained from studies of a Stockholm population, with an incidence of 60% in those over 95 years (Bengt Winblad, pers. commun.). Thus there is both good news and bad news. The good news is thst if you reach 85–90 years of age, as many of us will, you have a 50% chance of being mentally quite competent. The bad news is that you have an equal chance of developing dementia. What if you lived until 100 years, 110 years, or even 120 years of age which is considered by some to be the maximal extent of the human life-span? Is dementia a natural, and inevitable outcome of aging? I am not sure there is an answer to this question, but including the effects of aging with a presumed disease process, such as Alzheimer disease, makes studies of dementia difficult.

Before going further it is necessary to define the term 'dementia' and to describe its possible causes. The definition is a functional one, with the following criteria (McKhann et al., 1984):

1. decline in memory;
2. dysfunction in two or more areas of cognition;
3. no disturbances of consciousness; and
4. interference with social and occupational functions by cognitive impairment.

Using this definition, it is clear that dementia, like mental retardation, is a symptom, not a disease. In Table I are listed the more common etiologies. The differential diagnosis varies with age. For the age group 50–90 years, Alzheimer disease and vascular dementia predominate; whereas in younger patients, AIDS would lead the list in many countries.

TABLE I

Differential diagnosis of dementia

Alzheimer disease	50–60%
Vascular Dementia	10–20%
Drugs and toxins	1–5%
Intracranial masses	1–5%
Head injury	1–5%
Other neurodegenerative diseases	
Parkinson's disease	1%
Huntington's disease	1%
Pick's disease	
Progressive supranuclear palsy	
Metachromatic leukodystrophy	
Neuronal storage diseases	
Infections	
Prion-Associated diseases	
Creutzfeld-Jakob disease	
AIDS	
Meningitis	
Encephalitis	
Syphilis	
Lyme disease	
Nutritional and metabolic diseases	

What are the criteria for Alzheimer disease?

Several years ago, I chaired a group to develop criteria for Alzheimer disease (McKhann et al., 1984). The reasoning behind this was that a number of studies of mechanisms and therapy for Alzheimer disease were being proposed. For proper evaluation of these studies, it was essential that investigators be dealing with subjects meeting the same criteria. We divided the subjects into 'probable Alzheimer disease' and 'possible Alzheimer disease', based on clinical, imaging, and cognitive testing criteria, and 'definite Alzheimer disease' based on pathologic confirmation.

These criteria have held up in subsequent years, and been widely used, particularly the 'probable' category. The correlation between clinical criteria and pathologic confirmation has been high (Morris et al., 1988; DeKosky et al., 1992). However, the pathological basis of Alzheimer disease, as distinguished from the aging brain, may not be as definitive as we thought in 1984, particularly when based on a biopsy of frontal regions of the brain. As discussed by Terry (1993, this volume), the classical pathological findings in Alzheimer disease are the presence of senile plaques, neurofibrillary tangles and amyloid within the central core of plaques and around blood vessels. These changes, however, do not correlate well with the degree of dementia and are found, to some degree, in the non-demented aged. In addition, about 20% of patients who meet the clinical criteria of Alzheimer disease do not have the characteristic pathology in biopsy or autopsy brain material (DeKosky et al., 1992). As discussed by Terry (1993, this volume), the important correlation may be with the decrease in synapses, as indicated by the synaptic marker, synaptophysin (Masliah et al., 1989, 1991).

These clinical-pathological correlations make it harder to accept the classical pathological findings in Alzheimer disease as the sine qua non for diagnosis beyond a general diagnostic category. The problem may be similar to the issues raised by the pathologic findings of demyelination. A pathologic process is defined, but there may be multiple underlying mechanisms.

Is Alzheimer disease all one disease?

In common with other conditions which are defined by clinical symptoms, such as schizophrenia or depression, there is some question whether what we arbitrarily define as a single clinical entity may not have multiple causes. In trying to dissect Alzheimer disease, one possible discriminating factor is age of onset. The idea that age is a criterion goes back to Alzheimer original description of a 51-year-old woman (Alzheimer, 1907) and the subsequent categorization of dementia into presenile and senile populations. This division was subsequently discarded on the basis of the similarities in pathology in the two groups, leading to the concept of 'Senile Dementia of the Alzheimer Type'. However, clinical observations suggest that younger patients may be distinguished by differing clinical presentations and the likelihood of familial occurrence. For example, in patients below age 70 the incidence in first-degree relatives is 2–3 times the expected incidence, while in those over age 70, the familial incidence is not increased. Some authors have suggested that the early onset, often familial, cases are more likely to have more severe cognitive defects, particularly aphasia and apraxia (Breitner and Folstein, 1984). Markesbery (1992) has found by imaging and pathologic studies that there is more widespread cortical damage in younger patients. Swearer et al., (1992), on the other hand, came to the conclusion that even though early onset, familial Alzheimer disease may have a different etiology than late onset, sporadic Alzheimer disease, the clinical expression is similar.

Alzheimer disease is genetically heterogeneous

There are families with dominant patterns of inheritance of Alzheimer disease, usually with persons with younger ages of onset. Linkages have been observed in specific families with three different chromosomes; chromosome 14 (Schellenberg et al., 1992), chromosome 19 (Roses et al., 1990; Schellenberg et al., 1992a,b) and chromosome 21 (St. George-Hyslop et al., 1987; Goate et al., 1989). The association with chromosome 21 is even more complex, because a few

Table II

Criteria for Clinical Diagnosis of Alzheimer disease

I. The criteria for the clinical diagnosis of PROBABLE Alzheimer disease include:
   dementia established by clinical examination and documented by the Mini-Mental Test, Blessed Dementia Scale, or some similar examination, and confirmed by neuropsychological tests;
   deficits in two or more areas of cognition;
   progressive worsening of memory and other cognitive functions;
   no disturbances of consciousness;
   onset between ages 40 and 90, most often after age 65; and
   absence of systemic disorders or other brain diseased that in and of themselves could account for the progressive deficits in memory and cognition.

II. The diagnosis of PROBABLE Alzheimer disease is supported by:
   progressive deterioration of specific cognitive functions such as language (aphasia), motor skills (apraxia), and perception (agnosia);
   impaired activities of daily living and altered patterns of behavior;
   family history of similar disorders, particularly if confirmed neuropathologically; and
   laboratory results of;
      normal lumbar puncture as evaluated by standard techniques.
      normal pattern or non-specific changes in EEG, such as increased slow-wave activity, and
      evidence of cerebral atrophy on CT with progression documented by serial observation.

III. Other clinical features consistent with the diagnosis of PROBABLE Alzheimer disease, after exclusion of causes of dementia other than Alzheimer disease, include:
   plateaus in the course of progression of the illness;
   associated symptoms of depression, insomnia, incontinence, delusions, illusions, hallucinations, catastrophie verbal, emotional, or physical outbursts, sexual disorders and weight loss; other neurologic abnormalities in some patients, especially with more advanced disease and including motor signs such as increased muscle tone, myoclonus, or gait disorder;
   seizures in advanced disease; and
   CT normal for age.

IV. Features that make the diagnosis of PROBABLE Alzheimer disease uncertain or unlikely include:
   sudden, apoplectic onset;
   focal neurologic findings such as hemiparesis, sensory loss, visual field deficits, and incoordination early in the course of the illness; and
   seizures or gait disturbances at the onset or very early in the course of the illness.

V. Clinical diagnosis of POSSIBLE Alzheimer disease:
   may be made on the basis of the dementia syndrome, in the absence of other neurologic, psychiatric, or systemic disorders sufficient to cause dementia, and in the presence of variations in the onset, in the presentation, or in the clinical course;
   may be made in the presence of a second systemic or brain disorder sufficient to produce dementia, which is not considered to be the cause of the dementia; and
   should be used in research studies when a single, gradually progressive severe cognitive deficit is identified in the absence of other identifiable cause.

VI. Criteria for diagnosis of DEFINITE Alzheimer disease are:
   the clinical criteria for probable Alzheimer disease and
   histopathologic evidence obtained from a biopsy or autopsy.

VII. Classification of Alzheimer disease for research purposes should specify features that may differentiate subtypes of the disorder, such as:
   familial occurrence;
   onset before age of 65;
   presence of trisomy-21; and
   coexistence of other relevant conditions such as Parkinson's disease.

(Reprinted from McKhann et al., 1984)

families have a point mutation in the amyloid precursor protein (APP) (Goate et al., 1991), while others are linked to chromosome 21, but not with the APP gene (St. George-Hyslop et al., 1987; Goate et al., 1989). In addition, there are families with no apparent linkages to any of these three chromosomes. No careful comparison of the different forms of familial Alzheimer disease, as defined by chromosome locus, has been performed to date.

Recently, an association between both late onset familial and late onset sporadic forms of Alzheimer Disease have been genetically associated with a specific allele of apoprotein E (the type 4 allele designated as APOE-ϵ4). With an increasing dosage of APOE-ϵ4 not only does the risk of Alzheimer increase, the age of onset decreases. It is suggested the APOE-ϵ4 protein may play a role in the binding of B-amyloid (Saunders et al., 1993; Corder et al., 1993).

Other subtypes include patients with prominent extra-pyramidal components (Hansen et al., 1990), myoclonus (Hansen et al, 1990), and a group which presents with primary language dysfunction as the initial symptom (Mesulem, 1982). Careful clinical pathologic correlation with serial functional imaging studies of these possible clinical subtypes remains to be done.

Why the interest in amyloid?

Research relative to Alzheimer disease goes through fads. Ten years ago the emphasis was on possible cholinergic mechanisms (Fibiger, 1992). Now the role of the protein, amyloid, is the 'hot' topic (Selkoe, 1991).

Interest in amyloid originally centered around the demonstration of β-amyloid in the core of plaques. The form of amyloid in plaques, and in blood vessels (referred to as βA4 protein or β-amyloid protein) is a 42-amino acid segment of a precursor protein (referred to as amyloid precursor protein (APP)).

The accumulation of plaques containing β-amyloid is seen in three conditions, Alzheimer disease, Down's syndrome, and to a lesser extent, normally aging brain. Determination of the amino acid sequence of β-amyloid and APP led to cloning of the gene for the precursor protein (APP) and its localization on chromosome 21. Interest was further heightened when the pathology of older persons with Down's Syndrome, who lose cognitive skills with age, was found to closely resemble that of Alzheimer disease—suggesting that Alzheimer might be a chromosome 21 disease. As mentioned above, this is true—but only for small number of families, and of these, even fewer have a mutation involving the amyloid gene (Kosik, 1992).

There are three possible mechanisms by which amyloid might accumulate in Alzheimer brain:

(1) A defect in the processing, or degradation of the precursor protein, APP, a mechanism similar to the dysfunction of lysosomal enzymes in ganglioside storage diseases. Such a defect in an APP proteinase has yet to be established. However, a recent report suggests a different pattern of amyloid fragments in blood of Alzheimer patients as compared with age-matched controls, suggesting such a defect in amyloid protein processing (Bush et al., 1992).

(2) Over-production of the precursor protein APP. Such might be the case in Down's syndrome with trisomy 21 and a subsequent gene dosage effect.

(3) Genetically determined defects in the APP structure, as has been found, as mentioned above, in a few families.

Those who espouse the amyloid theory of Alzheimer disease, suggest that accumulation of the β-amyloid, regardless of the cause, is toxic to the nervous system (Selkoe, 1991). This hypothesis could be tested in transgenic animals containing either the mutant amyloid gene, or, alternatively, animals designed to overexpress amyloid. Hence the great interest in developing such animals.

Therapy of Alzheimer disease

Table III outlines some of the approaches which have

Table III

Approaches to therapy of Alzheimer disease

Modifications of neurotransmitters
  Choline hypothesis
    Precursor
    Alteration of acetylcholinesterase
    Alteration of acetylcholine release
  Other Neurotransmitters
    Norepinephrine
    Serotonin
    Dopamine
    Glutamate
    γ-Aminobutyric acid
    Somatostatin
    Substance P
  Trophic hypothesis
    NGF
    Other growth factors
  Excitatory amino acids
  Amyloid
  Neuroprotective agents
    Gangliosides

---

either been tried or are contemplated for the therapy of Alzheimer disease. Time and space do not allow a full discussion of the rationale of these various approaches.

The cholinergic hypothesis (reviewed in Fibiger, 1992) is based on the finding of lowered levels of acetylcholine, and its synthetic enzyme, choline acetyltransferase (CAT) in the brain of Alzheimer patients both at autopsy and on biopsy (DeKosky et al., 1992), the demonstration of drop-out of cholinergic neurones in the nucleus basilis, and the cognitive effects of antimuscarinic drugs. This hypothesis has resulted in a variety of therapeutic attempts to increase acetylcholine in brain. Based on the precursor strategy, successfully employed in the use of L-dopa in Parkinson's disease, cholinergic precursors, such as lecithin, have been tried, but without significant success.

In September 1993, the Food and Drug Administration (FDA) of the United States approved Tacrine (tetrahydroaminoacridine) THA [Cognex®, Parker-Davis, Warner-Lambert Company, Morris Plains, NJ]. This approval is based primarily on a prospective double-blind, placebo-controlled study on 480 patients (Farlow et al., 1992). The results suggest that Tacrine reverses signs of dementia in some patients with Alzheimer Disease. Tacrine is a centrally active anticholinesterase inhibitor, but it may have other effects as well. These results are encouraging over a short term. (The patients were followed for only 12 weeks). Whether Tacrine, or similar choline-increasing agents will be of value over the long term remains to be proven. Conceivably, an identifiable subgroup of patients will be found to be responsive (Small, 1992).

Other transmitters

Further neurochemical studies have indicated that Alzheimer disease is not a deficiency of a single transmitter. Table III lists other neurotransmitters which are decreased, in at least some patients. It has been suggested that a 'cocktail' of precursors of neurotransmitters be tried. These studies have not been performed.

Trophic hypothesis

Based on the observation that NGF is involved in maintenance of cholinergic neurons, it has been suggested that NGF be tried therapeutically. NGF is now available for studies using pharmacologic dosages thanks to molecular biological production techniques. However, it is anticipated that NGF would penetrate into the brain poorly, thus necessitating either direct intraventricular injection or use of an NGF analogue with greater CNS permeability. Such studies are in progress or planning phases (Phelps et al., 1989a,b).

Excitatory amino acids

Excitatory amino acids, particularly glutamate, are invoked as toxic compounds in a variety of CNS insults such as ischemia, anoxia, prolonged epileptic seizures and trauma. Whether such compounds play a role in neurodegenerative diseases has been proposed for some time. The ability to block the actions of these compounds has been discussed by Costa (1993, this volume).

Amyloid

As discussed previously, the possible cellular toxicity of β-amyloid has been proposed. Thus, possible therapeutic agents which either ameliorate the accumulation of amyloid or result in its removal are being considered (Selkoe, 1991).

Neuroprotective agents

As discussed by Svennerholm (1993, this volume), gangliosides have potential as neuroprotective, or neurorehabilitative, agents. Some of these functions have been discussed by other speakers.

Therapeutic guidelines

A review of previous attempts at therapy suggest several principals of therapy.

(1) At this stage of knowledge, therapeutic trials with smaller numbers of well studied patients are preferable to studies of large numbers of poorly characterized patients. Such studies should include not only clinical evaluation, serial tests of cognitive function, but imaging studies of cortical function such as SPECT.

(2) Studies should include patients early in the course of the disease process. Attempts to evaluate therapeutic benefit in end-stage Alzheimer disease are similar to attempting evaluation of a therapeutic agent in end-stage motor neuron disease or cerebellar degeneration. Once neurones are lost, there is no longer a substrate for a therapeutic agent.

(3) Studies should be of longer duration, at least 6 months of therapeutic trial. Alzheimer disease may progress not only slowly but also in an intermittent fashion. An adequate period of trial is necessary.

(4) Treatment failures should be carefully catego-

rized. The syndromes of 'Parkinsonism Plus', such as progressive supranuclear palsy, striato-nigral degeneration, and Shy-Drager syndrome became more clearly defined when they were recognized as treatment failures among the larger pool of L-DOPA responsive patients with classical Parkinson's disease.

(5) At the present time there is no animal model of Alzheimer disease. Thus, therapeutic strategies must be evaluated in humans. As in the cancer field, or in other neurodegenerative diseases of the nervous system, desperate clinical situations justify bold approaches. For example, bone marrow transplantation, possible gene therapy, or the use of novel neurotrophic or neuroprotective factors.

In summary, our basic understanding of the pathophysiology of Alzheimer disease has made significant progress. However, we have yet to translate that knowledge to the practical problems of early diagnosis, objective measures of disease progression, and most importantly, therapy.

Acknowledgment

Previous speakers have acknowledged Professor Svennerholm's major contributions in the chemistry, biology, and physiology of gangliosides. I would like to add that Professor Svennerholm never forgot his roots in clinical medicine, and constantly bridged the gap between basic chemistry and biochemistry and clinical application. In addition, I must acknowledge the presence of Dr. Elizabeth Svennerholm, who spent her career in the area of geriatric medicine, and could just as appropriately have given this lecture as I.

References

Alzheimer, A. (1907) Über eine eigenartige Erkrankung der Hirnride. *Allg. Z. Psychiatr.,* 64: 146–148.
Breitner, J.C.S. and Folstein, M.F. (1984) Familial Alzheimer dementia: a prevalent disorder with specific clinical features. *Psychol. Med.,* 14: 63–80.

Bush, A.I., Whyte, S., Thomas, L.D., Williamson, T.G., Vantiggelen, C.J., Currie, J., Small, D.H., Moir, R.D., Li, Q.X., Rumble, B., Monning, U., Beyreuther, K. and Masters, C.L. (1992) An abnormality of plasma amyloid protein-precursor in Alzheimer disease. *Ann. Neurol.*, 32: 57–65.

DeKosky, S.T., Harbaugh R.E., Schmitt, F.A., Bakay, R.A.E., Chui, H.C., Knopman, D.S., Reeder, T.M., Shetter, A.G., Senter, H.J., Markesbery, W.R. and Intraventricular Bethanecol Study Group (1992) Cortical biopsy in Alzheimer disease: diagnostic accuracy and neurochemical, neuropathological, and cognitive correlations. *Ann. Neurol.*, 32: 625–632.

Corder, E.H., Saunders, A.M., Strittmatter, W.J., Schmechel, D.E., Gaskell, P.C., Small, G.W., Roses, A.D., Haines, J.L. and Pericak-Vance, M.A. (1993) Gene dose of apolipoprotein E Type 4 Allele and the risk of Alzheimer Disease in late onset families. *Science*, 261: 921–923.

Evans, D.A., Funkenstein, H.H., Albert, M.S., Scherr, P.A., Cook, N.R., Chown, M.J., Hebert, L.E., Hennekens, C.H. and Taylor, J.O. (1989) Prevalence of Alzheimer disease in a community population of older persons. *J. Ann. Med. Assoc.*, 262: 2551–2556.

Farlow, M., Gracon, S.I., Hershey, L.A., Lewis, M.S., Sadowsky, C.H. and Dolan-Ureno, J. (1992) A controlled trial of Tacrine in Alzheimer Disease. *JAMA*, 268(18): 2523–2529.

Fibiger, H.C. (1992) Cholinergic mechanisms in learning, memory, and dementia: a review of recent evidence. *Trends Neurosci.*, 14: 220–223.

Goate, A.M., Haynes, A.R., Owen, M.J., Farrall, M., James, L.A., Lai, L.Y.C., Mullan, M.J., Roques, P., Rossor, M.N., Williamson, R. and Hardy, J.A. (1989) Predisposing locus for Alzheimer disease on chromosome 21. *Lancet*, 1: 352–355.

Goate, A.M., Chartier-Harlin, M.C., Mullan, M., Brown, J., Crawford, F., Fidani, L., Giuffra, L., Haynes, A., Irving, N., James, L., Mant, R., Newton, P., Rooke, K., Roques, P., Talbot, C., Pericakvance, M., Roses, A., Williamson, R., Rossor, M., Owen, M. and Hardy, J. (1991) Segregation of a missense mutation in the amyloid precursor protein gene with familial Alzheimer disease. *Nature*, 349: 704–706.

Hansen, L., Salmon, D., Galasko, D., Masliah E., Katzman, R., DeTeresa, R., Thal, L., Pay, M.M., Hofstetter, R., Klauber, M., Rice, V., Butters, N. and Alford, M. (1990) The Lewy body variant of Alzheimer disease: a clinical and pathological entity. *Neurology*, 40: 1–8.

Kosik, K.S. (1992) Alzheimer plaques and tangles: advances on both fronts. *Trends Neurosci.*, 14: 218-219.

Markesbery, W.R. (1992) Alzheimer disease. In: A.K. Asbury, G.M. McKhann and I.M. McDonald (Eds), *Diseases of the Nervous System: Clinical Neurobiology*, Saunders, Philadelphia, pp. 795–801.

Masliah, E., Terry, R.D., DeTeresa, R.M. and Hansen, L.A. (1989) Immunohistochemical quantification of the synapse-related protein synaptophysin in Alzheimer disease. *Neurosci. Lett.*, 103: 234–239.

Masliah, E., Terry, R.D., Alford, M., DeTeresa, R. and Hansen, L.A. (1991) Cortical and subcortical patterns of synaptophysin-like activity in Alzheimer disease. *Am. J. Pathol.*, 138: 235–246.

McKhann, G.M., Drachman, D.A., Folstein, M.F., Katzman, R., Price, D. and Stadlan, E. (1984) Clinical diagnosis of Alzheimer disease: report of the NINCDS-ADRDA work group under the auspices of the Department of Health and Human Services Task Force on Alzheimer disease. *Neurology*, 34: 939–944.

Mesulem, M.M. (1982) Slowly progressive aphasia without generalized dementia. *Ann. Neurol.*, 11: 592–598.

Morris, J.C., McKeel, D.W., Fulling, K., Torack, R.M. and Berg, L. (1988) Validation of clinical diagnostic criteria for Alzheimer disease. *Ann. Neurol.*, 24: 17–22.

Phelps, C.H., Gage, F.H., Growdon, J.H., Hefti, F., Harbaugh, R., Johnston, M.V., Khachaturian, Z., Mobley, W., Price, D., Raskind, M., Simpkins, J., Thal, L. and Woodcock, J. (1989a) Potential use of nerve growth factor to treat Alzheimer disease (letter). *Science*, 243: 11.

Phelps, C.H., Gage, F.H., Growdon, J.H., Hefti, F., Harbaugh, R., Johnston, M.V., Khachaturian, Z.S., Mobley, W.C., Price, D.L., Raskind, M., Simpkins, J., Thal, L.J. and Woodcock, J. (1989b) Potential use of nerve growth factor to treat Alzheimer disease. *Neurobiol. Aging*, 10: 205–207.

Roses, A.D., Bebout, J., Yamaoka, P.C., Gaskell, P.C., Hung, W.-Y., Alberta, M.J., Clark, C., Welch, K., Earl, N., Heyman, A. and Pericak-Vance, M.A. (1990) Linkage studies in familial Alzheimer disease (FAD): application of the affected pedigree member (APM) method. *Neurology*, 40 (Suppl. 1): 275.

Saunders, A.M., Strittmatter, W.J., Schmechel, D.E., St. George-Hyslop, P.H., Pericak-Vance, M.A., Joo, S.H., Rosi, B.L., Gusella, J.F., Crapper-MacLachlan, D.R., Alberts, M.J., Hulette, C., Crain, B., Goldgaber, D. and Roses, A.D. (1993) Association of apolipoprotein E Allele ε4 with late-onset familial and sporadic Alzheimer's Disease. *Neurology*, 43: 1467–1472.

Schellenberg, G.D., Boehnke, M., Wijsman, E.M., Moore, D.K., Martin, G.M. and Bird, T.D. (1992a) Genetic association and linkage analysis of the apolipoprotein CII locus and familial Alzheimer disease. *Ann. Neurol.*, 31: 223–227.

Schellenberg, G.D., Bird, T.D., Wijsman, E.M., Orr, H.T., Anderson, L., Nemens, E., White, J.A., Bonnycastle, L., Weber, J.L., Alonso, M.E., Potter, H., Heston, L.L. and Martin, G.M. (1992b) Genetic linkage evidence for a familial Alzheimer disease locus on chromosome 14. *Science*, 258: 668–671.

Selkoe, D.J. (1991) The molecular pathology of Alzheimer disease. *Neuron*, 6: 487–498.

Small, G.W. (1992) Tacrine for treating Alzheimer's Disease. *JAMA*, 268(18): 2564–2565.

St. George-Hyslop, P.H., Tarzi, R.E., Polinsky, R.J., Haines, J.L., Nee, L., Watkins, P.C., Myers, R.H., Feldman, R.G., Pollen, D., Drachman, D., Growdon, J., Bruni, A., Foncin, J.-F., Salmon, D., Frommelt, P., Amaducci, L., Sorbi, S., Placentini, S., Stewart, G.D., Hobbs, W.J., Conneally, P.M. and Gusella, J.F. (1987) The genetic defect causing familial Alzheimer disease maps on chromosome 21. *Science*, 235: 885–890.

St. George-Hyslop, P.H., Haines, J.L., Farrer, L.A., Polinsky, R.,

382

Vanbroeckhoven, C., Goate, A., Mclachlan, D.R.C., Orr, H., Bruni, A.C., Sorbi, S., Rainero, I., Foncin, J.F., Pollen, D., Cantu, J.M., Tupler, R., Voskresenskaya, N., Mayeux, R., Growdon, J., Fried, V.A., Myers, R.H., Nee, L., Backhovens, H., Karlinsky, H., Rich, S., Heston, L., Montesi, M., Mortilla, M., Nacmias, N., Gusella, J.F. and Hardy, J.A. (1990) Genetic linkage studies suggest that Alzheimer disease is not a single homogeneous disorder. *Nature*, 347: 194–197.

Swearer, G.M., O'Donnell, B.F., Drachman, D.A. and Woodward, B.M. (1992) Neuropsychological features of familial Alzheimer disease. *Ann. Neurol.*, 32: 687–694.

L. Svennerholm, A.K. Asbury, R.A. Reisfeld, K. Sandhoff, K. Suzuki, G. Tettamanti and G. Toffano (Eds.)
Progress in Brain Research, Vol. 101

CHAPTER 29

Neuropathological changes in Alzheimer disease

Robert D. Terry

Department of Neurosciences, University of California, San Diego, 9500 Gilman Drive, La Jolla, California, 92093-0624, U.S.A.

Introduction

Information about many aspects of the pathology of Alzheimer disease has become surprisingly widespread in both lay and scientific communities. Interest on the part of the public has become remarkably intense, and partially as a result of this fact the number of autopsies on demented patients has increased sharply. On the other hand, the diagnostic skills of the clinicians having also improved, there is rarely a need for brain biopsy. Although this has a limiting effect on some kinds of investigations, many advances are being made.

Gross Changes

Cerebral atrophy is usually noticeable in most areas with only relative sparing of the occipital pole and of the primary motor and sensory gyri on either side of the central sulcus. No part of the cerebral cortex can be regarded as normal control tissue. While the atrophy can be extreme, bringing brain weight to below 900 g, it is most often between 1000 and 1100 g. Occasional specimens weigh as much as 1400 g, and do not show gross atrophy. The hippocampal, parahippocampal and amygdaloid regions are usually particularly shrunken. The white matter is lessened, and the ventricles are mildly to moderately dilated as a result. The cortical ribbon is often thinned by 10–20%. The basal ganglia and thalamus are intact,

as is the substantia nigra. The locus ceruleus, on the other hand, is usually very pale especially in its rostral half. Atrophy of the cerebellar vermis is sometimes noted, probably as an effect of malnutrition. Focal and or lobar atrophy is not unknown in Alzheimer disease, and sometimes one sees a case closely resembling the circumscribed fronto-temporal atrophy characteristic of Pick's disease, but with florid Alzheimer lesions visible only to the microscopist.

Neurofibrillary Tangles

Tangles (NFT) seem to appear earliest in layer two of the entorhinal cortex and, somewhat later, are found in the hippocampal pyramidal cells of CA1 (Braak and Braak, 1991). Later in the course of the disease when symptoms are readily apparent, these lesions are usually widespread in the larger neurons of the cerebral neocortex and are also present in the basal nucleus of Meynert (bnM), dorsal raphe nucleus and locus ceruleus. In the entorhinal cortex and in the hippocampal pyramidal layer, tangles are often found without the other neuronal components, these having apparently died leaving the tangles behind. Such extraneuronal tangles are very rare in the neocortex. Some specimens seem to be free or almost entirely free of neocortical tangles in the presence of entorhinal lesions, cortical plaques and a full-blown clinical picture (Terry et al., 1987b). While tangles are often difficult to detect in the usual hematoxylin and eosin or cresyl violet preparations, they are very apparent with several silver impregnations and with amyloid stains such as thioflavine and Congo red. At the ultrastructural level NFT are made up of clusters of paired helical filaments (PHF) with a half period of about 80 nm (Terry, 1963; Kidd, 1963), each filament of the pair being about 10 nm wide. Some particularly high resolution electron micrographs reveal 8 protofilaments in the form of a twisted or constricted tubule (Miyakawa, 1990).

The principal component of the NFT is hyperphosphorylated tau protein (Brion et al., 1985). At least two kinases are known to phosphorylate tau and can generate PHF type protein from tau. One is tubulin-associated kinase (Ishiguro et al., 1988). Biernat et al. (1992) reported that a serine-proline directed kinase 'turns normal tau into a PHF-like state'. Casein kinase-II binds to PHF (Iimoto et al., 1990) but does not generate PHF protein from the tau (Baum et al., 1992). The β-2 form of protein kinase C is decreased in the Alzheimer cortex (Cole et al., 1988).

Senile plaques

There are several forms of plaques, and the nomenclature is not standardized at this time. Growing opinion would have it that there are three major types. The first and most frequent is the *diffuse* plaque characterized by extracellular unformed but immunoreactive β/A-4 protein with only a few submicroscopic wisps of filamentous extracellular amyloid (Yamaguchi et al., 1991). The neuropil within the region of the diffuse plaque is free of alteration or displays only minimal change (Yamaguchi et al., 1988). Diffuse plaques are to be found in the disease and sometimes in normal aging throughout the neocortex, the basal ganglia, especially the putamen, and even in the cerebellum.

The *neuritic* plaque is made up of bundles and masses of filamentous amyloid surrounded by dystrophic neurites (Terry et al., 1964). In normal aging these abnormal neurites contain numerous neurofilaments, degenerating mitochondria and lysosomes, but do not have PHF. In Alzheimer disease the dystrophic neurites often contain clusters of PHF, as well as neurofilaments, mitochondria and lysosomes. In either situation the amyloid may or may not be massed into a dense central core. Those plaques without a core are called primitive, while those with the dense core are named mature or compact. Microglia and fibrous astrocytes are also to be found associated with the plaque.

The final type is a *burned-out* plaque, and is made up of a dense mass of β/A-4 fibrous protein surrounded by reactive astrocytic processes containing glycogen (Terry and Wisniewski, 1970). Neurites are absent, and are presumed to have died off. These lesions are not common in the human.

Diffuse and neuritic plaques are present usually in small numbers in normal aging and in significantly greater numbers in the disease. They are to be found in cortex, hippocampus, deeper layers of the entorhinal cortex and in the basal nucleus. Smaller numbers lie in the pontine and mesencephalic tegmentum. There is a common assumption that diffuse plaques progress to the mature form, but this is assuredly not always so.

Amyloid

Amyloid and its precursor protein, called APP, are the targets of most bench research in the field of Alzheimer disease. β/A-4 amyloid is the type common to the Alzheimer brain. It was first isolated and sequenced by Glenner and Wong (1984), and it was not very long after their report that other investigators found and sequenced the APP (Kang et al., 1987) and located the encoding gene on the long arm of chromosome 21 (Goldgaber et al., 1987). The latter finding was facilitated by the long known fact that patients with Down's syndrome (trisomy 21) invariably, by the time they are 35–40 years old, develop significant deposits of amyloid in plaques identical to those of Alzheimer disease. One familial Alzheimer disease (FAD) gene is on the same chromosome, but is located more proximally on the long arm and well separated from the APP gene (St. George-Hyslop et al., 1987). The amyloid itself is a 40–43 amino acid peptide, not entirely specific to Alzheimer disease, but very different from amyloids deposited systemically in relation to other disorders. In AD the amyloid comprises the core of the plaque and is frequently found in the media and adventia of cortical and meningeal vessels. It is also identifiable in extraneuronal tangles.

APP exists in three major forms, the most common of which has 695 amino acids, and is a glycosylated protein with a single hydrophobic transmembrane region (Kang et al., 1987). It is commonly diagrammed as traversing the plasma membrane with a relatively short intracytoplasmic carboxy terminus and a long extracellular amino terminus. The amyloid component begins in the membrane itself and extends for about half its length outside the transmembrane region (Kang et al., 1987). In fact, however, immunocytochemical procedures reveal that most APP is entirely intracellular rather than crossing the plasma membrane (Kawai et al., 1992). The second most prominent form of APP has 751 amino acids and includes within the 'extracellular' region a Kunitz serine protease inhibitor epitope accounting for the additional length (Kitaguchi et al., 1988). Normally the APP is cleaved outside the membrane such that the amyloid component is broken and cannot form the β-amyloid or be deposited in fibrillar form (Esch et al., 1990). Intensive search is underway to find a protease which would cleave this extracellular portion outside the amyloid peptide. Other investigations involve an alternative pathway through lysosomal processing (Cole et al., 1989). APP is synthesized primarily in the neuronal soma and is transported by the mechanisms of rapid axoplasmic transport toward the axonal terminals (Koo et al., 1990). Several mutations have been found in codon 717 of APP and are associated with the dominant early onset form of the disease (Chatier-Harlin et al., 1991), but these mutations account only for a very small minority of familial cases. Despite the efforts of many laboratories the connection between amyloid deposition and dementia is not at all clear. A few other proteins are associated with the amyloid including especially α1-antichymotrypsin (Abraham et al., 1988).

Three rodent transfection studies with the APP gene purport in two cases to demonstrate overexpressed amyloid (Quon et al, 1991; Wirak, 1991). The third claimed to produce plaques, tangles and neuron loss (Kawabata, 1991). None of the three models has been confirmed. Neither in vitro nor in vivo amyloid toxicity has been definitively demonstrated.

Neuropil threads

Neuropil threads are argentophilic fibers in the neuropil apart from the plaques (Tomlinson et al, 1970). They have also been called curly fibers and, in fact, are dystrophic neurites, usually dendritic (Kowall and Kosik, 1987). Small numbers may be present in the normal elderly, but there are large numbers in the diseased cortex. In that situation they contain PHF, neurofilaments and ubiquitin as well as a newly described protein which is called neuropil thread protein (NTP) and is closely related to pancreatic exocrine thread protein (de la Monte, 1990). McKee et al. (1991) report that the quantity of neuropil threads correlates closely with the severity of dementia.

Neuronal Numbers

Just as total brain weight declines in normal aging, there is also a decrease in the number of pyramidal cells in hippocampus (Ball, 1977) and in neocortex. While some investigators have reported a loss of total cortical neurons in the normal aged cortex (Brody, 1955), our own studies show that the pyramidal cells shrink into smaller size classes so that there is not a total loss of neurons but rather a shrinkage of the larger ones (Terry et al., 1987a). This implies a loss of axonal caliber, and this would require extensive myelin remodeling as well as a diminished number of synapses (vide infra). It must be kept in mind that even in normal aging where total neuron density is unchanged, the whole cerebrum shrinks so that there must be an overall neuronal loss. In Alzheimer disease there is a much greater loss of the large neurons and a total loss of neurons as well (Terry et al.,

1981). Formation of neurofibrillary tangles in the neocortex can not account for neuronal loss in AD, but these lesions are more clearly responsible for cell death in the entorhinal cortex.

In that Alzheimer disease is principally a disorder of the elderly, its pathology is additive to that of normal aging. Therefore, neuron loss is less in the old Alzheimer patients than it is in the younger ones (Hansen, et al., 1988). Similarly, the concentration of plaques and tangles in the older AD patients is less than in the younger ones. One might infer from these findings that there is a threshold of brain damage which must be reached in order that the patient display significant cognitive loss. Older patients need less additional pathology to reach this threshold than do younger individuals. A greater reserve of neurons and synapses is probably protective (Katzman et al., 1988), and is apparently augmented by education (Zhang et al., 1990)

Neurotransmitters

Many transmitters have been found to be reduced by biochemical assay. Immunocytochemical studies reveal that neurites of several transmitter classes are present in the plaques (Armstrong et al., 1989). The losses seem to be related to neuronal degeneration and can be correlated to some extent with neuron counts in these areas. For example, the loss of neurons and the presence of tangles in the basal nucleus are associated with the decreased choline acyteltransferase (Whitehouse et al., 1981). The locus ceruleus loses large numbers of cells, especially its forebrain-projecting rostral portion and this corresponds to the loss of noradrenergic activity (Bondareff, 1981). Serotonin decrements go along with degeneration of cells in the dorsal raphe (Yamamoro and Hirano, 1985). Somatostatin (Davies et al., 1980) and substance P (Crystal and Davies, 1982) are decreased in the cortex. Glutamate is deficient at least in the entorhinal cortex where there is a massive loss of neurons, especially in the glomerular clusters of the second layer (Hyman et al., 1984). This effectively isolates the hippocampus from much of its input.

Synapses

Synaptic measurements in the neocortex have been made by electron microscopic assay of brain biopsies. DeKosky and Scheff reported their ultrastructural study as demonstrating a major loss of synapses in prefrontal cortex (1990). They also showed that the zone of contact between pre- and post synaptic terminals was elongated as the number of synapses declined. More recently techniques have been developed which permit immunocytochemical display of synapses and subsequent quantification by microdensitometry (Masliah et al., 1990) and by confocal microscopy (Masliah et al., 1993). Immunochemical techniques have been used involving the same antibody, which is usually a monoclonal antibody against synaptophysin (Jahn et al., 1985). This is a 38 kDa protein integral to the membrane of synaptic vesicles. The protein is more concentrated in the small clear synaptic vesicles but is also present in the dense core vesicles. Therefore the antibody reacts with all presynaptic terminals which contain vesicles and is displayed as a granule representing a cluster in the presynaptic terminal. In normal aging there is an apparent steady decline in synaptic population density in the dorsal prefrontal region between ages 16 and 98 in the 25 cases studied by Masliah et al. (Masliah et al., 1993). The correlation coefficient of normal age versus synaptic density was 0.7, significant at $P < 0.01$. Using either optical density on paraffin sections of AD brain or confocal microscopy on vibratome sections we have found a major decline in synapses in three association areas — midfrontal (MF), inferior parietal (IP) and superior temporal (ST). There is little or no overlap between the synapse concentration in the normals and those found in the Alzheimer cases. As discussed below there is a very strong statistical correlation (Terry et al., 1991) between the concentration of synapses and the severity of the dementia as measured by three common clinical tests of global cognition — the Blessed IMC (Blessed et al., 1968) as modified by Fuld (1978) for American use, the Mini-Mental State Examination (MMSE) of Folstein (1975) and the Dementia Rating Scale (DRS) of Mattis (1976).

Membranes

In addition to the ganglioside abnormalities discussed elsewhere, there have been demonstrated several abnormalities of membrane phospolipids in Alzheimer disease. Pettegrew et al. have reported an increase in phosphomonoesters (PME) in AD as compared with both disease and normal controls (Pettegrew et al., 1988). They suggested that this might be related to a metabolic block in phospholipid synthesis at the level of cytidine triphosphate: phosphocholine cytidyltransferase, or that it might be due to decreased breakdown of PME by phospholipase D. More recently, Nitsch et al. have reported that levels of phosphatidylcholine and phosphatidylethanolamine are both significantly decreased in the Alzheimer cortex (Nitsch et al., 1992). Stokes and Hawthorne (1987) previously showed decreases in phosphatidyl inositol levels.

Clinico-pathologic correlations

In 1968, Blessed, Thomlinson and Roth published correlations between plaque concentrations and the Blessed Information-Memory-Concentration Test and with their test of activities of daily living (Blessed et al., 1968). They presented a correlation coefficient of 0.7, but their data included many cases without Alzheimer disease. If only their AD cases are used, the correlation with plaques falls to 0.4 which explains less than 20% of the variance. Our own data concerning more than 70 Alzheimer patients, judged on the basis of their clinical status and their histologic changes, provide an even less significant correlation between plaque counts and any of the three global tests. Other reports would have it that if one counts only neuritic plaques, then the correlation with dementia scores becomes stronger (Delaere et al., 1989), but our own series of 16 specimens did not confirm this.

When neurofibrillary tangles are present in the neocortex the patient is almost invariably demented, while in many normal elderly one can find quite numerous plaques in that location even in patients who were fully tested psychometrically shortly before death and found to be cognitively normal (Dickson et al., 1991). Therefore, although less specific to Alzheimer disease than are plaques, tangles are the more sensitive indicator. A problem arises, however, in the fact that 20–30% of the cases of AD greater than age 70 have very rare or no tangles in the neocortex (Terry et al., 1987). Many of these are also free of cortical Lewy bodies. There is some debate as to whether these cases should be called Alzheimer Disease, but we find that these specimens are identical to cases of AD with cortical tangles in respect to brain weight, plaque number, choline acetyltransferase concentration, neuron number and clinical course. These specimens do have tangles in the entorhinal cortex and in the hippocampal pyramidal layer. When these 'tangle-rare' cases are included with typical cases in a correlation of tangle number with the psychometrics, the correlation is barely significant. Even excluding the tangle-rare-cases, the correlation is not much better. It is interesting that neuritic plaques can be very frequent in the neocortex even in the absence or extreme rarity of neocortical tangles. Despite all these caveats, there can be no question but that there is a highly significant difference between Alzheimer specimens and normals as to the number of plaques or tangles. Such a difference exists despite considerable overlap as to plaque concentrations in some elderly cases. Therefore, the amyloid burden presented by plaques may be a contributory factor to the dementia, but is not the major one. In our series of 76–79 cases where plaques in midfrontal, inferior parietal and superior temporal areas were correlated with the IMC score, there was no significance. As to tangles, the same series showed a 0.01 significance in midfrontal, and 0.02 in parietal and temporal areas, with coefficients of 0.30, 0.28 and 0.28 respectively, accounting for less than 10% of the variance.

Neuron concentrations also correlate insignificantly with the severity of the dementia, but again there is a highly significant difference in their concentration between normals and AD cases.

In regard to structural changes in the neocortex, by far the strongest correlation lies between the severity of dementia on the one hand, and the concentration of synapses on the other. In the basal nucleus the strongest correlation is with tangles (Samuel et al., 1992). Midfrontal synapse concentration in 15 cases correlated with the total Dementia Rating Scale at $r = 0.68$ and $P < 0.007$. In the same series the correlation of these frontal synapses with the IMC was very similar with $r = 0.73$ and $P < 0.002$. The correlation with the MMSE yielded a coefficient of 0.71 (Terry et al., 1991).

The IMC test deals very largely with memory and therefore the basal forebrain is an important factor. Of bnM features, tangles provided the strongest correlate with the IMC with a coefficient of 0.73. The MMSE deals with the broader range of cognitive functions, and that test correlates best with frontal synapses and secondarily with bnM tangles. Stepwise regressions concerning data from the frontal cortex and from the basal nucleus reveal, in regard to the MMSE, a final coefficient of 0.99 after five-steps with frontal cortex synapses accounting for 50% of the variance and bnM tangles an additional 24%. The same assay in reference to the Blessed IMC went through three-steps and displayed a final r^2 of 86% with bnM tangles providing 53% and midfrontal synapses 26%. Therefore, the syndrome of dementia as tested globally in Alzheimer disease is related to both the cortex, especially frontal, and the basal nucleus. The memory components seem to be mostly related to basal nucleus, while other functions are primarily cortical. Synaptic loss in the cortical area is of particular importance and interest. The tangle counts in the bnM probably reflect the degree of magnocellular cholinergic cell loss.

Conclusions

The classical, diagnostic lesions of AD — the plaque and tangle — do not account satisfactorily for the clinical symptoms of global dementia. Other recently found factors are clearly of great significance. The neuronal and synaptic loss may well be related to abnormalities of the plasma membrane, and thus gangliosides and phospholipids are strongly implicated.

Acknowledgement

This work was supported by National Institutes of Health Grants (USA) AG08201, AG08205, AG10689-01.

References

Abraham, C.R., Selkoe, D.J. and Potter, H. (1988) Immunochemical identification of the serine protease inhibitor alpha antichrymotrypsin in the brain amyloid deposits of Alzheimer's disease. *Cell*, 52: 487–501.

Armstrong, D.M., Benzing, W.C., Evans, J. et al. (1989) Substance P and somatostatin coexist within neuritic plaques: implications for the pathogenesis of Alzheimer's disease. *Neuroscience*, 31: 663–671.

Ball, M.J. (1977) Neuronal loss, neurofibrillary tangles and granulovacuolar degeneration in the hippocampus with aging and dementia. A quantitative study. *Acta Neuropathol.*, 37: 111–118.

Baum, L., Masliah, E., Iimoto, D.S. et al. (1992) Casein kinase II is associated with neurofibrillary tangles but is not an intrinsic component of paired helical filaments. *Brain Res.*, 573: 126–132.

Biernat, J., Mandelkow, E.-M., Schröter, C., et al. (1992) The switch of tau protein to an Alzheimer-like state includes the phosphorylation of two serine-protine motifs upstream of the microtubule binding region. *E.M.B.O.J.*, 11: 1593–1597.

Blessed, G., Tomlinson, B.E. and Roth, M. (1968) The association between quantitative measures of dementia and of senile change in the cerebral gray matter of elderly subjects. *Br. J. Psychiatry*, 114: 797-811.

Bondareff, W, Mountjoy, C.Q. and Roth, M. (1981) Selective loss of neurons origin of adrenergic projection to cerebral cortex (nucleus locus ceruleus) in senile dementia. *Lancet*, I: 783–784.

Braak, H. and Braak, E. (1991) Neuropathological staging of Alzheimer-related changes. *Acta Neuropathol.*, 82: 239–259.

Brion, J.P., Passareiro, H., Nunez, J. et al. (1985) Immunological detection of tau protein in neurofibrillary tangles of Alzheimer's disease. *Arch. Biol.*, 96: 229–235.

Brody, H. (1955) Organization of the cerebral cortex. III. A study of aging in the human cerebral cortex. *J. Comp. Neurol.*, 102: 511–556.

Chartier-Harline, M.C. Crawford, F., Houlden, H. et al. (1991) Early onset Alzheimer's disease caused by mutations at codon 717 of the beta amyloid precursor gene. *Nature*, 353(6347): 844–846.

Cole, G., Dobkins, K.R., Hansen, L.A. et al. (1988) Decreased levels of protein kinase C in Alzheimer brain. *Brain Res.*, 452: 165–174.

Cole, G.M., Huynh, T.V. and Saitoh, T. (1989) Evidence for lysosomal processing of amyloid beta protein precursor in cultured cells. *Neurochem Res.*, 14: 933–939.

Crystal, H.A. and Davies, P. (1982) Cortical substance P-like immunoreactivity in cases of Alzheimer's disease and senile dementia of the Alzheimer type. *J. Neurochem.*, 38: 1781.

Davies, P., Katzman, R. and Terry, R.D. (1980) Reduced somato-statin-like immunoreactivity in cerebral cortex from cases of Alzheimer disease and Alzheimer senile dementia. *Nature*, 288: 279–280.

DeKosky, S.T. and Scheff, W.S. (1990) Synapse loss in frontal cortex biopsies in Alzheimer's disease: correlation with cognitive severity. *Ann. Neurol.*, 27: 457–464.

Delaere, P., Duysckaerts, C., Brion, J.P. et al. (1989) Tau, paired helical filaments and amyloid in the neocortex: morphometric study of 15 cases with graded intellectual status in aging and senile dementia of Alzheimer type. *Acta Neuropath.*, 77: 645–653.

de la Monte, S.M., Ozturk, M. Wands, J.R. (1990) Enhanced expression of an exocrine pancreatic protein in Alzheimer's disease and the developing human brain. *J. Clin. Invest.*, 86(3): 1004–1013.

Dickson, D.W., Crystal, H.A., Mattiace, L.A. et al. (1991) Identification of normal and pathological aging in prospectively studied nondemented elderly humans. *Neurobiol. Aging,* 13: 179–189.

Esch, F.S., Keim, P.S., Beattie, E.C., et al. (1990) Cleavage of amyloid β protein during constitutive processing of its precursor. *Science*, 248: 1122–1124.

Folstein, M.F., Folstein, S.W. and McHugh, P.R. (1975) "Minimental state": a practical method for grading the cognitive state of patients for the clinician. *J. Psychiatr. Res.*, 12: 189–198.

Fuld, P.A. (1978) Psychological testing in the differential diagnosis of dementias. In: R Katzman, R.D. Terry and K.L. Bick, (Eds), *Alzheimer's Disease: Senile Dementia and Related Disorders* Raven Press, Newark, pp. 185–193.

Glenner, G.G. and Wong, C.W. (1984) Alzheimer's disease and Down's syndrome: Sharing of a unique cerebrovascular amyloid fibril protein. *Biochem. Biophys. Res. Commun.*, 122: 1131–1135.

Goldgaber, D., Lerman, M.I., McBride, O.W. et al. (1987) Characterization and chromosomal location of a cDNA encoding brain amyloid of Alzheimer's disease. *Science*, 235: 877–880.

Hansen, L.A., De Teresa, R., Davies, P. et al. (1988) Neocortical morphometry, lesion counts, and choline acetyltransferase levels in the age spectrum of Alzheimer's disease. *Neurology* 38: 48–54.

Hyman, B.T., Van, H.G.W., Damasio, A.R., et al. (1984) Alzheimer's disease: cell-specific pathology isolates the hippocampal formation *Science*, 225: 1168–1170.

Iimoto, D.S., Masliah, E., De Teresa, R, et al. (1990) Aberrant casein kinase II in Alzheimer's disease. *Brain Res.,* 507: 273–280.

Ishiguro, K., Ihara, Y., Uchida, T., et al. (1988) A novel tubulin-dependent protein kinase forming a paired helical filament epitope on tau. *J. Biochem.*, 104: 319–321.

Jahn, R., Schiebler, W., Quimet, C., et al. (1985) A 38000 dalton membrane protein (p^{38}) present in synaptic vesicles. *Proc. Natl. Acad. Sci. U.S,A.*, 82: 4137–4141.

Kang, J., Lemaire, H.-G., Unterbeck, A., et al. (1987) The precursor of Alzheimer's disease amyloid A4 protein resembles cell-surface receptor. *Nature*, 325: 733–736.

Katzman, R., Terry, R., De Teresa, R. et al. (1988) Clinical, pathological, and neurochemical changes in dementia; a subgroup with preserved mental status and numerous neocortical plaques. *Ann. Neurol.,* 23: 138–144.

Kawabata, S., Higgins, G.A. and Gordon, J.S. (1991) Amyloid plaques, neurofibrillary tangles and neuronal loss in brains of transgenic mice overexpressing a C-terminal fragment of human amyloid precursor protein. *Nature*, V354: 476–478.

Kawai, M., Cras, P., Richey, P., Tabaton, M. et al. (1992) Subcellular localization of amyloid precursor protein in senile plaques of Alzheimer's disease. *Am. J. Pathol.,* 140: 947–958.

Kidd, M. (1963) Paired helical filaments in electron microscopy in Alzheimer's disease, *Nature,* 197: 192–193.

Kitaguchi, N., Takahashi, Y., Tokushima, Y., et al. (1988) Novel precursor of Alzheimer's disease amyloid protein shows protease inhibitory activity. *Nature*, 331: 530–532.

Koo, E.H., Sisodia, S.S., Archer, D.R., et al. (1990) Precursor of amyloid protein in Alzheimer disease undergoes fast anterograde axonal transport. *Proc. Natl. Acad. Sci. U.S.A.*, 87: 1561–1565.

Kowall, N.W. and Kosik, K.S. (1987) Axonal disruption and aberrant localization of tau protein characterize the neuropil pathology of Alzheimer's disease *Ann. Neurol.*, 22: 639–643.

Masliah, E., Terry, R.D., Alford, M. et al. (1990) Quantitative immunohistochemistry of synaptophysin in human neocortex: An alternative method to estimate density of presynaptic terminals in paraffin sections. *J. Histochem. Cytochem.* 38: 837–844.

Masliah, E., Mallory, M., Hansen, L. et al. (1993) Quantitative synaptic alterations in the human neocortex during normal aging. *Neurology*, 73: 192–197.

Mattis, S. (1976) Mental status examination for organic mental syndrome in the elderly patients. In: L. Bellack, T.B. Karasu, (Eds.), *Geriatric Psychiatry,* Grune and Stratton, New York, pp. 77–121.

McKee, A.C., Kosik, K.S. and Kowall, N.W. (1991) Neuritic pathology and dementia in Alzheimer's disease. *Ann. Neurol.,* 30: 156–165.

Miyakawa, T. (1990) Ultrastructure of neurofibrillary tangles in Alzheimer's disease. *Trans. XIth Internatl. Con. Neuropathol.,* p. 13.

Nitsch, R.M., Blusztrajin, J.K., Pittas, A.G. et al. (1992) Evidence for a membrane defect in Alzheimer disease brain. *Proc. Natl. Acad. Sci. USA*, V89 N5: 1671–1675.

Pettegrew, J.W., Mossy, J., Withers, G. et al. (1988) Nuclear magnetic resonance study of the brain in Alzheimer's disease. *J. Neuropathol. Exp. Neurol.,* 47: 235–248.

Quon, D., Wang, Y., Catalano, R., et al. (1991) Formation of beta-amyloid protein deposits in brains of transgenic mice. *Nature*,

352(6332): 239–241.

Samuel, W., Terry, R.D., Masliah, E. et al. (1993) Frontal cortex and nucleus basalis pathology in Alzheimer dementia. *Arch. Neurol.,* in press.

Stokes, C.E. and Hawthorne, J.N. (1987) Reduced phosphoinsitide concentrations in anterior temporal cortex of Alzheimer-diseased brains. *N. Neurochem.,* 48: 1018–1021.

St. George-Hyslop, P.H., Tanzi, R.E., Polinsky, R.J. et al. (1987) The genetic defect causing familial Alzheimer's disease maps on chromosome 21. *Science,* 235: 885–890.

Terry, R.D. (1963) The fine structure of neurofibrillary tangles in Alzheimer's disease. *J Neuropathol. Exp. Neurol.,* 22: 629–642.

.Terry, R.D. and Wisniewski, H. (1970) The ultrastructure of the neurofibrillary tangle and the senile plaques. In: G.E.W. Wolstenholme and O'Connor, M. (Eds.), *Ciba Foundation Symposium on Alzheimer's Disease and Related Conditions.* J.A. Churchill, London, pp. 145–168.

Terry, R.D., Gonatas, N.K. and Weiss, M. (1964) Ultrastructural studies in Alzheimer's presenile dementia. *Am. J. Pathol.,* 44:269–297.

Terry, R.D., De Teresa, R. and Hansen, L.A. (1987a) Neocortical cell counts in normal human adult aging. *Ann. Neurol.,* 21: 530–539.

Terry, R.D., Hansen, L.A., De Teresa, R. et al. (1987b) Senile dementia of the Alzheimer type without neocortical neurofibrillary tangles *J. Neuropathol. Exp. Neurol.,* 146: 262–268.

Terry, R.D., Peck, A., DeTeresa, R. et al. (1981) Some morphometric aspects of the brain in senile dementia of the Alzheimer type. *Ann. Neurol.,* 10: 184–192.

Terry, R.D., Masliah, E., Salmon, D. et al. (1991) Physical basis of cognitive alterations in Alzheimer's disease: synapse loss is the major correlate of cognitive impairment. *Ann Neurol.,* 30: 572–580.

Tomlinson, B.E., Blessed, G. and Roth, M. (1970) Observations of the brains of demented old people. *J. Neurol. Sci.,* 11: 205–242.

Whitehouse, P.J., Price, D.L., Clark, A.W. et al. (1981) Alzheimer's disease: evidence for selective loss of cholinergic neurons in the nucleus basalis. *Ann. Neurol.,* 10: 122–126.

Wirak, D.O., Bayney, R., Ramabhadran, T.V. et al. (1991) Deposits of amyloid β protein in the central nervous system of transgenic mice. *Science,* 253: 323–325.

Yamaguchi, H., Hirai, S., Morimatsu, M. et al. (1988) A variety of cerebral amyloid deposits in the brains of the Alzheimer type dementia demonstrated by β protein immunostaining. *Acta Neuropathol.,* 76: 541–549.

Yamaguchi, H., Nakazato, Y., Shoji, M., et al. Ultrastructure of diffuse plaques in senile dementia of the Alzheimer type: comparison with primitive plaques. *Neurol. Aging,* 12: 295–312.

Yamamoro, T. and Hirano, A. (1985) Nucleus raphe dorsalis in Alzheimer's disease: neurofibrillary tangles and loss of large neurons. *Ann. Neurol.,* 17: 573–577.

Zhang, M., Katzman, R., Jin, H. et al. (1990) The prevalence of dementia and Alzheimer disease (AD) in Shanghai, China; Impact of age, gender and education. *Ann. Neurol.,* 27: 428–437.

L. Svennerholm, A.K. Asbury, R.A. Reisfeld, K. Sandhoff, K. Suzuki, G. Tettamanti and G. Toffano (Eds.)
Progress in Brain Research, Vol. 101

CHAPTER 30

Ganglioside loss is a primary event in Alzheimer disease Type I

Lars Svennerholm

Department of Clinical Neuroscience, Section of Psychiatry and Neurochemistry, University of Göteborg, Göteborg, Sweden

Introduction

In an extensive study of the biochemical changes in brain in dementia disorders of Alzheimer type (Gottfries et al., 1983) we found gangliosides to be significantly reduced in caudate nucleus. The brain gangliosides had been suggested by us to be an optimal marker for neuronal plasma membranes as early as 1957 (Svennerholm, 1957) and we therefore interpreted the findings to indicate a significant loss of nerve endings in Alzheimer brains. In our continued examinations we found that the diminution of the ganglioside concentration was insignificant in most brain disorders of older subjects, and was only pronounced in the brains from Alzheimer cases of early onset.The marked diminution of gangliosides in early onset Alzheimer disease (AD) indicated an extensive loss of neuronal processes and we were interested in seeing whether the ganglioside loss was accompanied by demyelination and therefore determined the lipid composition of the cerebral white matter. The most specific myelin lipid (Svennerholm, 1956) cerebroside was not reduced in the brains from the early onset cases with the marked ganglioside loss, but rather in the brains from the cases with a late onset AD (Gottfries et al., 1985). In order to prove that the diminution of myelin lipids, cerebroside, cholesterol and phospholipids was caused by demyelination, myelin was isolated from brains of late onset cases and age-matched controls (Svennerholm et al., 1988). The study showed that in the Alzheimer brains the

yield of myelin was only 50% that of the controls, but the composition of the myelin from control and Alzheimer brains was the same. These results corroborated our view that there was pronounced demyelination in late onset Alzheimer disease (AD) and that the assay of myelin lipids, particularly cerebroside, is a good quantitative marker for myelin in AD.

The simultaneous assay of gangliosides in various brain areas of grey matter and myelin lipids in adjacent white matter indicated to us that AD could not be a single disease entity when the losses of such fundamental structures as nerve endings and myelin differed significantly between early and late onset cases. The aim of our future work was to elaborate clinical and biochemical methods which could be applied to patients with dementia disorders and would allow us to differentiate the Alzheimer patients into two or more subgroups. When the patients then died with a definite clinical diagnosis, we would perform determinations of the major membrane lipids in a large number of well defined areas of the brain. The same determination would also be performed on a carefully selected control material, which would serve as a library for the physiological changes of the human brain lipids during development and ageing. There was already a number of comprehensive studies of human brain lipids during development and ageing (Brante, 1949; Bürger, 1957; Rouser and Yamamoto, 1969; Svennerholm et al., 1978). The general lipid technology, however, has been improved. Particularly the extraction of strongly acidic lipids like gangliosides and inositol phosphoglycerides has been quantitative (Svennerholm and Fredman, 1980). Furthermore, the other key compound, cerebroside was determined in previous studies by the assay for hexose, a procedure which could include other neutral glycosphingolipids lead to falsely high values. There were thus several methodological reasons for making a reexamination of the brain membrane lipids during development and ageing. Another important factor was that the selection of relevant brain specimens had not always been well-controled in previous studies.

Membrane lipids in the aging human brain

One of the main difficulties associated with the study of the biochemistry of the normal human brain is obtaining specimens from individuals who have not suffered from any neurological or psychiatric disorder. It is equally important that the subjects have not suffered from a long agonal stage. Bowen et al., (1977) have paid particular attention to this problem. They did not find any effects of a long agonal stage on the ganglioside level, but in our own studies the concentration of myelin lipids seemed to diminish during a long agone (own unpublished results). We therefore decided to collect all our brain specimens from the Department of Forensic Medicine from individuals who had suffered an acute death of arteriosclerotic heart disease or accidents not affecting the brain. The brain specimens were dissected by the pathologist, who also divided the specimen for microscopic and biochemical examinations. All microscopic examinations have been performed by the same pathologist, Dr. Kerstin Boström, and when there have been problems with regard to the evaluation of the neurofibrollary tangles and of senile plaques in the specimens from the oldest cases, Dr. Patrick Sourander has been consulted.

Macroscopic examination of the brain allowed slight or moderate arteriosclerotic changes of the basal arteries without any obliteration of the lumen. No pathological change would be observed on macroscopic inspection of the brain or after sectioning. Microscopic examination was performed on the following areas: frontal cortex and white matter (Brodmann area 9), temporal cortex and white matter (Brodmann area 21 and 22), parietal cortex and white matter (Brodmann area 7), caudate nucleus, putamen, medial thalamus, hippocampus, amygdala nucleus, brainstem and cerebellar folie and white matter. The sections were fixed in phosphate buffered formalin, embedded in paraffin, stained with hematoxylin, Palmgren's silver staining and Luxol Fast Blue and incubated with antiserum against fibrillary acidic protein. The microscopic examination, not revealing any sign of organic brain disorder, was a criterion for the brain tissue to be accepted for this study.

After careful homogenization, the dissected material for biochemical examination was assayed with the methods used in our laboratory for more than 20 years—the only modification being the extraction procedure for the lipids (Svennerholm and Fredman, 1980). In the first report the lipid composition of brain from 21 men and 18 women, 60–97 years old, was communicated (Svennerholm et al., 1991). The percentage of solids diminished during the whole period—in the cortex the diminution was particularly marked after 90 years of age. Most of the cases with the highest ages were females which might explain why the correlation with age was found to be higher in females ($r = -0.61$ and $r = -0.58$) in frontal and temporal cortices, respectively than in males ($r = -0.24$ and $r = -0.29$). There was marked individual differences in the percentages of dry weight. They parallelled the phospholipid concentration, which indicates that the hydration of the brain varies markedly under physiological conditions. Variation in the hydration of brain leads to a simultaneous variation in brain volume and may be taken into account when, for example, atrophy is measured from X-ray examinations.

Phospholipids and cholesterol diminished slightly less than dry weight with ageing in frontal and temporal cortices, while the concentration of gangliosides diminished slightly more than the dry weight in the male cases. The diminution of dry weight and the concentration of major lipids were more marked in the white matter and equally pronounced in males and females—the correlation with ageing varied between $r = -0.37$ and $r = -0.78$ and was on the average $r = -0.6$. Since the number of brains assayed were relatively small and the age of the females 6 years above that of the males, the conclusions have to be preliminary. When the graphs of lipid concentration versus ageing are examined it seems that the major membrane lipids diminish linearly up to 90 years of age, when a marked diminution occurs, particularly pronounced for the gangliosides in the cortices and for cerebroside in the white matter. These results indicate a rapid loss of plasma neuronal membranes (nerve endings) and myelin in frontal and temporal lobes of human forebrain after 90 years of age.

Clinical differences

The syndrome described by Alzheimer in 1907 (Alzheimer, 1907) has until recently been regarded as an exclusively presenile disorder. The disease was given Alzheimer's name by Kraepelin in 1910. At that time AD designated presenile dementia (Sjögren et al., 1952; Katzman, 1986) and was considered to be a relatively rare condition (Rotschild and Kasanin, 1936). At autopsy, brains from Alzheimer cases presented gross atrophy and the microscopical examination showed the type of changes now known as neurofibrillary tangles and senile plaques. Patients with onset of dementia after the age of 65 and without symptoms of cerebrovascular disorder were considered to suffer from senile dementia. When histopathological examinations were continuously performed in cases with senile dementia, a major group was differentiated which contained significantly more of neurofibrillary tangles and senile plaques than age-matched controls. This group of senile dementia was then designated senile dementia of Alzheimer type (SDAT). In the USA, the presenile AD and SDAT were considered to represent a single homogeneous entity. From a clinical point of view, AD stands for progressive dementia in which no obvious cause (e.g. inflammation, brain tumor, multiple infarcts) of the symptoms is found on clinical examination, irrespective of age, symptomatogy, or family history of the disease (McKhann et al., 1984; McKhann, this volume).

The decision to consider the two forms of dementia, AD and SDAT, as a homogeneous entity was surprising. In Sweden, Sjögren et al. (1952), Sourander and Sjögren (1972), and Gustafson and Nilsson (1982) had given tentative criteria for the clinical differentiation of the two dementia forms. In AD the onset is often before 65 years of age, but later onset can occur. Early spatial disorientation or visuospatial dysfunction is present. During the second stage—the aphasic—apractic—agnostic stage (Sjögren 1950)—the parietotemporal focal deficits are marked. In SDAT the onset is generally after 65 years of age, the parietal symptoms are less marked, but hallucinations and delusions are common. In the second phase the patients often fail at every kind of cognitive task.

Histopathological and neurochemical examinations have also revealed major differences between Alzheimer cases with early and late onset (Rossor et al., 1984; Bowen and Davison, 1986; Roth, 1986). It is particularly relevant that the neurotransmitter deficits which have been the subject of hundreds of studies are more severe and widespread in younger cases (Rossor and Iversen, 1986). Previous studies thus indicated a heterogeneity of AD and although the designation of the two forms of dementia as AD have been met with great acclaim by the general public and particularly the relatives of demented patients, it is important that the heterogeneity is carefully studied. The heterogeneity might be looked upon as by Sir Martin Roth (1986). 'There is the same basic pathogenetic defect in the early and late onset forms, leading to the formation of plaques and tangles, but the intensity of the histopathological changes are greater in the early onset form'. Another explanation to the differences between early and late onset forms is the one suggested by me: that the early and late onset forms of AD have different primary pathogenesis and the neurofibrillary tangles and plaques are only secondary phenomena.

In order to create optimal conditions for investigations of the pathogenesis it is important to elaborate quantitative clinical parameters and biological markers which are able to differentiate the two forms. For this reason, Dr. Kaj Blennow (1990) in our department was given the research assignment of examining 75 Alzheimer patients, 31 with onset before the age of 65 and 44 with onset after this age, with our present clinical parameters and neurochemical tests. The controls consisted of 50 age-matched individuals without symptoms or signs of psychiatric or neurological disease. The clinical examination showed that a differentiation could not be made between the patients in the early demented stage and of those severely demented. There were then 19 patients left with early onset and 22 with late onset. The subgroup of patients with memory disturbances and predominant parietal symptoms was referred to as AD Type I, and patients with memory disturbances and general cognitive symptoms, but with no or only mild parietal symptoms were referred to as AD Type II.

AD Type I largely corresponds to early onset form; only 1 of the 19 early onset AD cases was classified as AD Type II, and 5 of 22 late onset form were referred to as AD Type I. White matter lesions analysed using computerized tomography were more frequent in AD Type II (91%) than in AD Type I (33%). The cerebrospinal fluid serum albumin ratio was significantly higher ($P < 0.0001$) in AD Type II than in controls and AD Type I. Cerebrospinal fluid ganglioside GM1 was significantly higher ($P < 0.001$) in AD Type I than in AD Type II (Blennow et al., 1991a). Cerebrospinal fluid 5-hydroxiindolacetic acid was significantly lower ($P < 0.0001$) in AD Type I than in controls (Blennow et al., 1991b). The level differed from that in AD Type II not only in being lower but also in showing the lowest value for stage 1. Sir Martin Roth (1986) has already used the designations Type I and Type II of AD, but he has used Type I for the late onset form, which we think is very unfortunate. The name Alzheimer disease was given to the early onset form, and so it is natural for this form to be called Type I. If future studies support our view that there is different pathogenesis for AD Type I and AD Type II, it will be awkward to use the name of Alzheimer disease for the late onset form. Our designation of Type I for the early onset form is also in accordance with the designation of diabetes Type I and II, where Type I stands for the early onset form.

Membrane lipids in Alzheimer disease Types I and II

The AD patients who had participated in our longitudinal study and who have died have been subjected to autopsy. Neuropathological and biochemical examinations have been performed on a large number of brain regions. The clinical diagnosis and the results of the neuropathological examination have been unknown for the technicians who performed the biochemical analyses.

Brain specimens were available from 12 cases with AD Type I and 21 cases with AD Type II. The clinical diagnosis was made after the principles elaborated by Blennow (1990). In the autopsy room, 20 cases were chosen to constitute an age-matched control material. Their clinical records did not show any signs or symptoms of neurologic or psychiatric disease. They had suffered from various somatic illnesses, and had had similarly long agonal stage as the AD cases. In the beginning of the study, white matter was not analysed, since we had already examined this tissue in previous investigations (Gottfries et al., 1985; Svennerholm et al., 1988). Therefore, the number of specimens of white matter is less. On a few occasions, lipid extracts were lost by accident and no further tissue specimen was available. The number of specimens from the five areas has thus varied — the exact number is shown in Figs 1–6. If the loss of specimen from the Alzheimer cases has changed the mean age of the cohort, the number of controls has also been changed so that age matching has been maintained. The procedure for lipid extraction and the assay methods are the same as used for the study of membrane lipids in the ageing human brain (Svennerholm et al., 1991). Protein-bound sialic acid was determined on the tissue residue after lipid extraction (Svennerholm, 1958).

Statistical analysis with regard to group comparisons was performed using the Wilcoxon 2-sample test. The study was approved by the Ethics

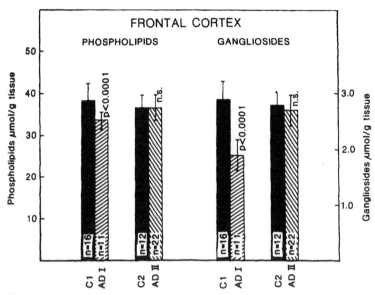

Fig. 1. Phospholipid and ganglioside concentrations in frontal cortex of AD cases and age-matched controls. n = number of cases examined, AD I = AD Type I, C1 = age-matched controls to AD Type I cases, AD II = AD type II. C2 = age-matched controls to AD Type II cases.

Committee, University of Göteborg. Next of kin gave their consent for the autopsy examination. The lipid composition has been determined in five regions, the frontal and temporal cortices, caudate nucleus, hippocampus, and frontal white matter. Significant diminutions of the membrane lipids in the four grey matter areas were found almost exclusively on AD Type I. The gangliosides were significantly reduced in all the four areas ($P < 0.0001$) but also the phospholipids ($P < 0.01$ to $P < 0.0001$) (Figs 1–4), while

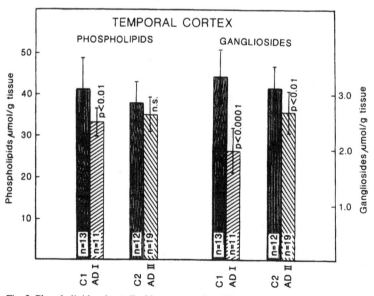

Fig. 2. Phospholipid and ganglioside concentrations in temporal cortex of AD cases and age-matched controls. Abbreviations as in Fig.1.

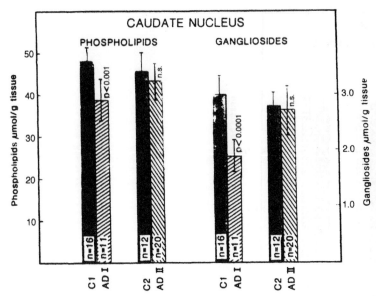

Fig. 3. Phospholipid and ganglioside concentrations in caudate nucleus of AD cases and age-matched controls. Abbreviations as in Fig. 1.

cholesterol was only significantly reduced in caudate nucleus ($P < 0.05$) and hippocampus ($P < 0.001$). Protein-bound sialic acid was only found to be significantly reduced in caudate nucleus ($P < 0.01$). In AD Type II there was no significant reduction of phospholipids in any of the four grey matter areas. Cholesterol was significantly reduced in temporal cortex ($P < 0.05$) and gangliosides in temporal cortex ($P < 0.01$) and hippocampus ($P < 0.05$) (Figs 1–4).

A completely different pattern was found in the

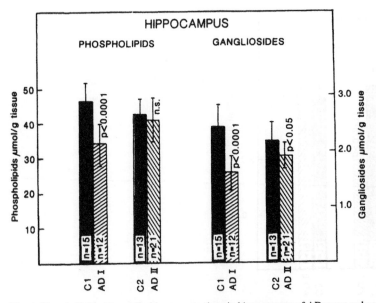

Fig. 4. Phospholipid and ganglioside concentrations in hippocampus of AD cases and age-matched controls. Abbreviations as in Fig. 1.

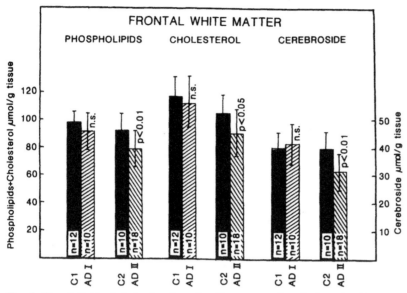

Fig. 5. Phospholipid, cholesterol and cerebroside concentrations in frontal white matter of AD cases and age-matched controls. Abbreviations as in Fig. 1.

white matter (Figs 5 and 6). A large reduction of myelin lipids was only found in AD Type II. Phospholipids ($P < 0.01$), cholesterol ($P < 0.001$) and particularly cerebroside ($P < 0.001$) and sulfatide ($P < 0.01$) were all significantly reduced. The gan-

glioside concentration was also significantly reduced in AD Type II ($P < 0.05$). In AD Type I frontal white matter only gangliosides were reduced ($P < 0.01$).

We also did a significant analysis of the differences in white and grey matter lipids between AD

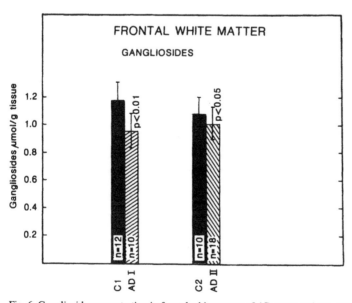

Fig. 6. Ganglioside concentration in frontal white matter of AD cases and age-matched controls. Abbreviations as in Fig. 1.

398

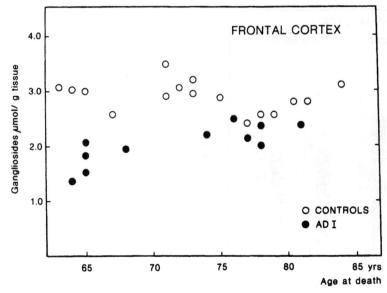

Fig. 7. Ganglioside concentration vs. age in frontal cortex of AD Type I cases and age-matched controls.

Type I and Type II. In frontal white matter phospholipids ($P < 0.05$), cholesterol ($P < 0.001$) and cerebroside ($P < 0.001$) were significantly lower in AD Type II than in AD Type I. To a certain extent the differences found depends on the increase in age of the Type II cases, since the lipids in white matter declines with ageing (Gottfries et al., 1985; Svennerholm et al., 1991). The significance analysis in the grey matter areas showed some remarkable results. Phospholipids were significantly lower in AD

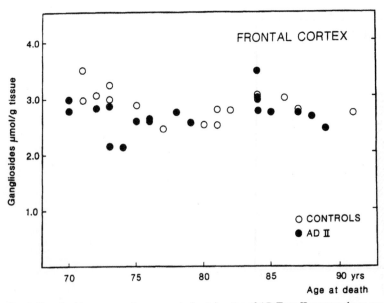

Fig. 8. Ganglioside concentration vs. age in frontal cortex of AD Type II cases and age-matched controls.

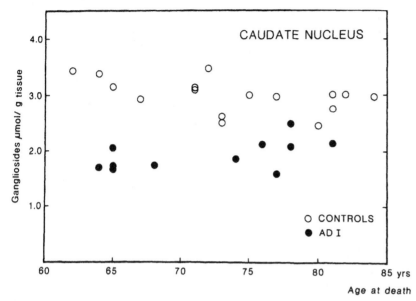

Fig. 9. Ganglioside concentration vs. age in caudate nucleus of AD Type I cases and age-matched controls.

Type I than in AD Type II in frontal cortex ($P <$ 0.01), caudate nucleus ($P < 0.05$) and hippocampus ($P < 0.05$). The differences for the gangliosides were still more pronounced. Gangliosides were significantly lower in Type I than in Type II in frontal cortex ($P < 0.0001$), temporal cortex ($P < 0.001$), candate nucleus ($P < 0.0001$) and hippocampus ($P < 0.01$). The difference in loss of gangliosides between AD Type I and AD Type II was particularly pronounced in frontal cortex and caudate nucleus. In AD Type I

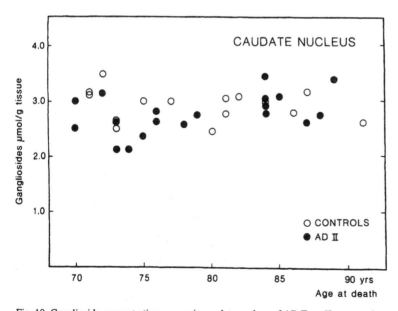

Fig. 10. Ganglioside concentration vs. age in caudate nucleus of AD Type II cases and age-matched controls.

the loss was 30% , in the younger cases 50%, while there was no loss in AD Type II (Figs 7–10). These differences are still more pronounced if one takes into account that the concentration of all the lipids diminish with age (Svennerholm et al., 1991).

Ganglioside pattern

The ganglioside patterns were also determined in all five areas. Significantly large variations from controls were only seen in AD Type I brains. The major differences in comparison with the controls was an increase in the proportions of ganglioside GD3, GM2 and GM1, and a decrease in the proportions of GD1a and GT1b, the diminution of GD1a being largest in hippocampus ($P < 0.01$) and temporal cortex ($P < 0.05$) and of GT1b in frontal cortex ($P < 0.001$), temporal cortex ($P < 0.001$) and caudate nucleus ($P < 0.01$) (Svennerholm and Gottfries, 1993). The proportion of GM2 was increased in frontal cortex ($P < 0.05$), caudate nucleus ($P < 0.05$) and hippocampus ($P < 0.05$). The only significant change of the ganglioside pattern in AD Type II grey matter was a diminution of the proportion of GM1 ($P < 0.05$) and GD1a ($P < 0.01$) in temporal cortex. In white matter only AD Type I cases showed differences in the ganglioside pattern from the controls. The proportion of GD3 was significantly increased ($P < 0.001$), and the proportions of GD1a ($P < 0.05$) and GT1b ($P < 0.05$) were diminished.

Fatty acid pattern of ethanolamine phosphoglyceride

Ethanolamine phosphoglyceride is the major phosphoglyceride of the human brain, with a large proportion of polyunsaturated fatty acids (Svennerholm, 1968). It was therefore of interest to study whether there was any significant loss of phosphoglycerides with a large proportion of polyunsaturated fatty acids. The proportions of 22:6 (n-3) and 20:4 (n-6) were slightly higher in the Alzheimer brains than in age-matched control brains, but the differences were not significant. The proportions of 18:0 were significantly lower in frontal cortex ($P < 0.01$) temporal cortex ($P < 0.001$) and caudate nucleus ($P < 0.05$) and of 18:1 higher in frontal cortex ($P < 0.01$), temporal cortex ($P < 0.01$) and hippocampus ($P < 0.05$) in AD Type I. No fatty acid change was found in AD Type II (Svennerholm and Gottfries, in manuscript).

Discussion of the results for brain membrane lipids in AD Type I and Type II

The axonal and dendritic processes of the neuron constitute at least 50% of the grey matter, and their volume of a large neuron is from 100 to over 1000 times that of the cell body. The neuronal cell body and its processes are surrounded by the plasma membrane which contains three major lipid classes; phospholipids, cholesterol and gangliosides — the neural processes contribute to more than 99% of the neuronal plasma membrane lipids, and the cell body is of no importance from a quantative point of view. Except for neuronal processes the grey matter contains glial cells both astrocytes and oligodendrocytes, blood vessels and other connective tissue and myelin sheath, surrounding the axons. Astrocytes and connective tissue is relatively lipid poor particularly in gangliosides and the cholesterol/phospholipid ratio is approximately 0.5:1. The myelin and the oligodendrocyte membranes are also ganglioside poor as compared with the neuronal processes, but they have a cholesterol/phospholipid ratio of 1.2:1 (Svennerholm et al., 1978). In the present study the cholesterol/phospholipid ratio in frontal and temporal cortices and caudate nucleus was 0.6:1.0 (the ratio was higher in hippocampus, because it is difficult also to remove myelin visible with the naked eye). Crude synaptosomal fraction isolated from the three areas mentioned above had a cholesterol/phospholipid ratio of 0.5:1.0. In the AD Type I brains the reduction of phospholipids was twice as large as that of cholesterol (Svennerholm and Gottfries, 1993), which suggests that the diminution of phospholipids in AD Type I was dependent on a selective loss of neural processes that were partly replaced by astrocytes, which are rich in GD3 (Gottfries et al., 1991). In AD Type II the concentration of cholesterol was equally much reduced or slightly more reduced than phospholipids expressed in μmol/g tissue in grey matter. This finding suggests that predominantly myelin was lost in grey matter. All the major myelin lipids — cerebroside, cholesterol and phospholipids — showed the same proportional diminution in frontal white matter of AD Type II cases which further support our view that the reduction of cholesterol and phospholipids in grey matter areas of this type depends on a myelin loss.

We have previously suggested that senile dementia of Alzheimer type might be a white matter disease (Gottfries et al., 1985), but the present study has shown the neuronal processes (the axons) to be rather well preserved and it is better to designate AD Type II as a demyelination disorder. The lipid changes found in our AD Type II brains — the mean ages of the cases were 80 years — are also similar to those found in the frontal white matter of 10–15 year older individuals without any sign of psychiatric and neurological disease except for cognitive disturbances (Svennerholm et al., 1991). From a lipid biochemical point of view the changes found in AD Type II are best described as early onset physiological brain ageing.

Suzuki et al. (1965) were the first to demonstrate a significant decrease of gangliosides in AD. The cerebral tissue was obtained by surgical biopsy of the three patients aged 52, 62 and 63 years, which suggests that they all belonged to AD Type I. Suzuki et al. (1965) also found that the concentration of cerebroside and phospholipids was diminished in the white matter and that the ganglioside pattern was normal, and they therefore drew the conclusion that the decrease in total gangliosides was secondary to a neuronal loss. Bowen and Davison (1986) have done extensive biochemical studies of Alzheimer brains. They have, however, missed the significant differences in the ganglioside concentration between AD Type I and AD Type II presumably because of their

early finding of no ganglioside loss in frontal lobe of old AD cases (Bowen and Davison, 1973). In later studies all ganglioside work was concentrated to temporal lobe in which ganglioside was reduced in both AD Type I and AD Type II as shown by us and a high precision of the ganglioside method has been required for detection of the difference between the two types.

There have also been two recent articles in which the ganglioside pattern has been examined in AD brains. Crino et al. (1989) reported a significantly lower concentration of gangliosides in 5 AD brains preferentially affecting gangliosides of the b-series. Their findings with regard to ganglioside concentration of individual gangliosides in control brains were, however, completely different from previously published values, which demonstrates that the assay of perbenzoylated gangliosides did give erroneous results in their hands. Kracun et al. (1992) found a decrease of all major gangliosides in temporal and frontal cortex and nucleus basalis of Meynert in 5 AD cases when the values were expressed per mg DNA. Only frontal cortex and white matter showed a decrease of ganglioside concentration when the values were expressed in μmol ganglioside per g fresh weight. The brains were said to have been collected from the early onset form and it is therefore strange that they did not find any ganglioside decrease in temporal cortex and hippocampus when the values were expressed in fresh weight. The variations in the ganglioside patterns in the different areas also differed from the similarity in ganglioside pattern changes found in our study. Our study is thus the first one in which significant biochemical differences between two major forms of AD have been demonstrated. In AD Type II the predominant feature is a demyelination with a ganglioside loss only in temporal cortex and hippocampus — areas known to have high contents of neurofibrillary tangles and neuritic plaques. In AD Type I there is an extensive reduction of gangliosides as a sign of a profound loss of nerve endings also in grey matter areas (caudate nucleus) known to have small to moderate numbers of plaques and tangles.

Terry et al. (1991; this volume) recently demonstrated similar findings as ours in AD. Their statistical data showed only weak correlations between psychometric indices and plaques and tangles, but the density of neocortical synapses revealed very powerful correlations with three psychological assays. The study included 15 patients who ranged in age from 59 to 91 years, and averaged 79 years of age at death, why it seems reasonable that the majority of the cases belonged to AD Type II. The morphometric determinations with counting of synapses were performed on specimens taken from midfrontal, superior temporal and inferior parietal areas of neocortex — the first two areas close to those examined in our biochemical study. In light of our finding that there was no diminution of gangliosides in the frontal area in AD Type II, we had expected the majority of the older cases not to have shown any loss of nerve endings, but similar histological pictures as the controls. Further work will show whether the immunochemical method used by Terry et al. (1991) will show negative staining of the nerve endings before they disappear or whether the nerve endings in AD Type II contain increased concentrations of gangliosides.

GM1 treatment in Alzheimer disease

Our study of membrane lipids in AD has shown a pronounced loss of the lipids which form the neuronal cellular processes, particularly the synaptosomes, while the neuronal cell bodies and the axons seemed to be rather well preserved. It would, then, be possible to stimulate the outgrowth of the nerve endings, by using nerve growth factor (NGF) as suggested by Marx (1990) and Svennerholm (1990). The proposed therapy was supported by the finding of Hefti and Mash (1989) of decreased NGF receptors in Alzheimer brains, but a recent study has shown no loss of NGF receptors in Alzheimers brains (Mufson and Kordower, 1992). The gangliosides have also been shown to possess neuritogenic and neuronotrophic properties (Ledeen, 1984) and of the gangliosides, GM1 has shown the best amelioratory effect in the treatment of neuronal damage in the CNS (Toffano et al., 1983; Karpiak et al., 1986). In

1988 we therefore began a study in which we administered 100 mg GM1 ganglioside subcutaneously or intramuscularly to 16 patients for 3 months (Svennerholm et al., 1990). The relatives of the patients saw a positive therapeutic effect on 8 of the 16 patients, but there was no objective sign of any improvement. Brain biopsy studies of patients to whom tritium labelled GM1 ganglioside was administered for 1 week before biopsy revealed that less than 0.05% of the given dose was recovered as GM1 ganglioside in brain. This amount of GM1 is less than 20% of the GM1 released from CNS to the cerebrospinal fluid (Davidsson et al., 1991) if the daily production of spinal fluid is 700 ml.

For this reason we decided to administer the gangliosides directly into the lateral ventricles, whose fluid is in direct contact with the brain intercellular fluid. Shunt catheters were implanted by stereotactic surgery in the frontal part of the two lateral ventricles and connected to a Rickham reservoir which allows sampling of cerebrospinal fluid during the treatment. A fine teflon catheter from the reservoir was tunnelated under the skin to a subcutaneous pocket under left arcus for the micropump (SynchroMed™ Infusion System, Medtronic Inc., Minneapolis, MN, U.S.A.). The pump has a reservoir of 17 ml which will be filled every 30 days with 17 ml of GM1 solution (60 mg/ml) (Fidia, Abano Terme). The patient will receive, by continuous infusion, 0.5 ml of GM1 (30 mg)/day. This means that 15 ml of the ganglioside solution is infused in the patient over a 30-day period. The solution remaining in the pump (approx. 2 ml) has been used for endotoxin examinations and bacteria cultivation on blood plates under aerobic and anaerobic conditions for 14 days. Five patients have been given GM1 ganglioside for more than 6 months. There has been no adverse effect in any patient, and it is remarkable how few negative effects the patients have from the inserted catheters.

Serum and CSF examinations are performed on the seventh day before, on the day of surgery, and 7, 30, 90, 180 and 360 days after initiation of ganglioside infusion. The serum examinations include all routine clinical chemical tests, thyroid function tests, immunoglobulins and antiganglioside antibodies. No patient has had any increased anti-GM1 antibody titer during the whole treatment period. Determinations on CSF from the ventricles and lumbar puncture include cytological examinations, CSF/serum albumin ratio, immunoglobulins, antiganglioside antibodies, gangliosides and several transmitter substances and their metabolites. None of the five patients treated for more than 6 months has shown any pathological biochemical finding in CSF and serum. The ganglioside concentration of lumbar CSF has varied between 32 and 45 µmol/l compared to the preoperative values of 16–39 nmol/l. The concentration of the major metabolites of dopamine and serotonin has increased during the treatment.

The clinical status of the patients will not be discussed until the patients have been followed for 1 year. Suffice it to say here that the disease process has stopped in all the five patients, and basal functions as appetite, sleep and sexual activity increased after 3 months. The patients have shown a significant increase in their power of initiative, and their emotional lives have improved.

Acknowledgements

The studies referred to in the present review are collaborative studies carried out at the Department of Clinical Neuroscience. I wish to express my warmest thanks to the following researchers who have participated in these studies; Lars-Erik Augustinsson, M.D., Ph.D., Kaj Blennow, M.D., Ph.D., Christian Blomstrand, M.D., Ph.D., Kerstin Boström, M.D., Ph.D., Görel Bråne, Ph.D., Pam Fredman, Ph.D., Carl-Gerhard Gottfries, M.D., Ph.D., Ingvar Karlsson, M.D., Ph.D., Annika Lekman, Ph.D., Jan-Eric Månsson, Ph.D. and Anders Wallin, M.D., Ph.D., Carsten Wikkelsö, M.D., Ph.D. Chief technician for the studies has been Birgitta Jungbjer. I wish also to acknowledge grants from the Swedish Medical Research Council (003X-627), The Bank of Sweden, Tercentenary Foundation, (86/326:1), the Greta and Johan Kock's Foundation, the Eivind and Elsa K:son Sylvan's Foundation and Fidia Research Laboratories.

References

Alzheimer, A. (1907) Über eine eigenartige Erkrankung der Hirnrinde. *Allg. Z. Psychiatr. Psych.-Gerichtl. Med.*, 64: 146–148.

Blennow, K. (1990) *Heterogeneity of Alzheimer's Disease*, Thesis, Medical Faculty, University of Göteborg, 199pp.

Blennow, K., Davidsson, P., Wallin, A., Fredman, P., Gottfries, C.-G., Karlsson, I., Månsson, J.-E. and Svennerholm, L. (1991a) Gangliosides in cerebrospinal fluid in probable Alzheimer's disease. *Arch. Neurol.*, 48: 1032–1035.

Blennow, K., Wallin, A., Gottfries, C.-G., Lekman, A., Karlsson, I., Skoog, I. and Svennerholm, L. (1991b) Significance of decreased lumbar CSF levels of HVA and 5-HIAA in Alzheimer's disease. *Neurobiol. Aging*, 13: 107–113.

Bowen, D.M. and Davison, A.N. (1986) Biochemical studies of nerve cells and energy metabolism in Alzheimer's disease. *Br. Med. Bull.*, 42: 75–80.

Bowen, D.M., Smith, C.B. and Davison, A.N. (1973) Molecular changes in senile dementia. *Brain*, 96: 849–856.

Bowen, D.M., Spillane, J.A., Curzon, G., Meier-Ruge, W., White, P., Goodhardt, M.J., Iwangoff, P. and Davison, A.N. (1979) Accelerated ageing or selective neuronal loss as an important cause of dementia? *The Lancet*, 1: 11–14.

Brante, G. (1949) Studies on lipids in the nervous system. *Acta Physiol. Scand. Suppl. 63*, Uppsala. 218 pp.

Büger, M. (1957) Die chemische Biomorphose des menschlichen Gehirns. *Abhandlungen der sächsischen Akademie der Wissenschaften zu Leipzig*, Akademie Verlag, Berlin, Band 46, Heft 6, pp.1–62.

Cherayil, G.D. and Cyrus Jr., A.E. (1966) The quantitative estimation of glycolipids in Alzheimer's disease. *J. Neurochem.*, 13: 579–590.

Crino, P.B., Ullman, M.D., Vogt, B.A., Bird, E.D. and Volicer, L. (1989) Brain gangliosides in dementia of the Alzheimer type. *Arch. Neurol.*, 46: 398–401.

Davidsson, P., Fredman, P., Månsson, J.-E. and Svennerholm, L. (1991) Determination of gangliosides and sulfatide in human cerebrospinal fluid with a microimmunoaffinity technique. *Clin. Chim. Acta*, 197: 105–106.

DeKosky, S.T. and Bass, N.H. (1982) Aging, senile dementia, and the intralaminar microchemistry of cerebral cortex. *Neurology*, 32: 1227–1233.

Gottfries, C.-G., Adolfsson, R., Aquilonius, S.-M., Carlsson, A., Eckernäs, S.-A., Nordberg, A., Oreland, L., Svennerholm, L., Wiberg, Å. and Winblad, B. (1983) Biochemical changes in dementia disorders of Alzheimer type (AD/SDAT). *Neurobiol. Aging*, 4: 261–271.

Gottfries, C.-G., Karlsson, I. and Svennerholm, L. (1985) Senile dementia — a white matter disease? In C.-G. Gottfries (Ed.), *Normal Aging, Alzheimer's Disease and Senile Dementia*, Editions de l'Université de Bruxelles, Brussels, pp. 111–118.

Gottfries, J., Fredman, P., Månsson, J.-E., Hansson, E. and

Svennerholm, L. (1991) Ganglioside characterization of rat astroglial cells in primary culture: detection of the ganglio and lacto series gangliosides. *Neurochem. Int.*, 19: 227–233.

Gustafson, L. and Nilsson, L. (1982) Differential diagnosis of presenile dementia on clinical grounds. *Acta Psychiatr. Scand.*, 65: 194–209.

Hefti, F. and Mash, D.C. (1989) Localization of nerve growth factor receptors in the normal human brain and Alzheimer's disease. *Neurobiol. Aging*, 10: 75–87.

Karpiak, S.E., Li, Y.S., Aceto, P. and Mahadik, S.P. (1986) Acute effects of gangliosides on CNS injury. In: G. Tettamanti, R.W. Ledeen, K. Sandhoff, Y. Nagai and G. Toffano (Eds.) *Ganglioside and Neuronal Plasticity*, Vol. 14, Fidia Research Series, Liviana Press, Padua, pp. 567–577.

Katzman, R. (1986) Alzheimer's disease. *N. Engl. J. Med.*, 314: 964–973.

Kracun, I., Kalanj, S., Talan-Hranilovic, J. and Cosovic, C. (1992) Cortical distribution of gangliosides in Alzheimer's disease. *Neurochem. Int.*, 20: 433–438.

Ledeen, R.W. (1984) Biology of gangliosides: neuritogenic and neuronotrophic properties. *J. Neurosci. Res.*, 12: 147–159.

Marx, J. (1990) NGF and Alzheimer's: Hopes and fears. *Science*, 247: 408–410.

McKhann, G., Drachman, D., Folstein, M., Katzman, R., Price, D. and Stadlan, E.M. (1984) Clinical diagnosis of Alzheimer disease: report of the NINCDS-ADRDA Work Group under the auspices of department of health and human services task force on Alzheimer's disease. *Neurology*, 34: 939–944.

Mufson, E.J. and Kordower, J.H. (1992) Cortical neurons express nerve growth factor receptors in advanced age and Alzheimer's disease. *Proc. Natl. Acad. Sci. USA*, 89: 569–573.

Rossor, M. and Iversen, L.L. (1986) Non-cholinergic neurotransmitter abnormalities in Alzheimer's disease. *Br. Med. Bull.*, 42: 70–74.

Rossor, M.N., Iversen, L.L., Reynolds, G.P., Mountjoy, C.Q. and Roth, M. (1984) Neurochemical characteristics of early and late onset types of Alzheimer's disease. *Br. Med. J.*, 288: 361–364.

Roth, M. (1986) The association of clinical and neurological findings and its bearing on the classification and etiology of Alzheimer's disease. *Br. Med. Bull.*, 42: 42–50.

Rotschild, D. and Kasanin, I. (1936) Clinicopathological study of Alzheimer's disease: relation to senile conditions. *Arch. Neurol. Psychiat.*, 36: 293–321.

Rouser, G. and Yamamoto, A. (1968) Curvi linear regression course of human brain lipid composition changes with age. *Lipids*, 3: 284–287.

Sjögren, H. (1950) Twenty four cases of Alzheimer's disease. *Acta Med. Scand. Suppl.*, 245: 225–233.

Sjögren, T., Sjögren, H. and Lindgren, A. (1952) Morbus Alzheimer and morbus Pick: A genetic, clinical and pathoanatomical study. *Acta Psych. Neurol. Scand. Suppl.*, 82: 66–115.

Sourander, P. and Sjögren, H. (1972) The concept of Alzheimer's disease and its clinical implications. In: G.E.W. Wolstenholme and M. O'Connors (Eds), *Alzheimer's Disease*, Ciba Foundation Symposium, Churchill, London, pp. 11–36.

Suzuki, K., Katzman, R. and Korey, S.R. (1965) Chemical studies on Alzheimer's disease. *J. Neuropathol. Exp. Neurol.*, 24: 211–224.

Svennerholm, L. (1956) The quantitative estimation of cerebrosides in nervous tissue. *J. Neurochem.*, 1: 42–52.

Svennerholm, L. (1957) Quantitative estimation of gangliosides in senile human brain. *Acta Soc. Med. Upsaliensis*, 62: 1–16.

Svennerholm, L. (1958) Quantitative estimation of sialic acids. III An anion exchange resin method. *Acta Chem. Scand.*, 12: 547–554.

Svennerholm, L. (1968) Distribution and fatty acid composition of phosphoglycerides in normal human brain. *J. Lipid Res.*, 9: 570–579.

Svennerholm, L. (1990) Gangliosides and nerve growth factors in Alzheimer's disease. In: A. Nordberg and B. Winblad (Eds), *Novel Therapeutic Strategies for Dementia Diseases. Acta Neurol. Scand. Suppl.*, 129: 21–22.

Svennerholm, L. and Fredman, P. (1980) A procedure for the quantitative isolation of brain gangliosides. *Biochim. Biophys. Acta*, 617: 97–109.

Svennerholm, L and Gottfries, C.-G. (1993) Membrane lipids selectively diminished in Alzheimer brains suggest synapse loss as primary event in early onset form (type I) and demyelination in late onset form (type II) *J. Neurochem.*, in press.

Svennerholm, L., Vanier, M.-T. and Jungbjer, B. (1978) Changes in fatty acid composition of human brain myelin lipids during maturation. *J. Neurochem.*, 30: 1383–1390.

Svennerholm, L., Gottfries, C.-G. and Karlsson, I. (1988) Neurochemical changes in white matter of patients with Alzheimer's disease. In: G. Serlupi Crescenzi (Ed.), *A Multidisciplinary Approach to Myelin Disease*, Plenum Publishing Corporation, New York, 1988, pp. 319–328.

Svennerholm, L.,Gottfries, C.-G., Blennow, K., Fredman, P., Karlsson, I., Månsson, J.-E., Toffano, G. and Wallin, A. (1990) Parental administration of GM1 ganglioside to presenile Alzheimer patients. *Acta Neurol. Scand.*, 81: 48–53.

Svennerholm, L., Boström, K., Helander, C.-G. and Jungbjer, B. (1991) Membrane lipids in the aging human brain. *J. Neurochem.*, 56: 2051–2059.

Terry, R.D., Masliah, E., Salmon, D.P., Butters, N., De Teresa, R., Hill, R., Hansen, L.A. and Katzman, R. (1991) Physical basis of cognitive alterations in Alzheimer's disease: synapse loss is the major correlate of cognitive impairment. *Ann. Neurol.*, 30: 572–580.

Toffano, G., Savoini, G., Moron, F., Lombardi, M.G., Calza, L. and Agnati, L.F. (1983) GM1 ganglioside stimulates the regeneration of dopaminergic neurons in the central nervous system. *Brain. Res.*, 261: 163–166.

Subject Index

Printed and bound by CPI Group (UK) Ltd, Croydon, CR0 4YY

03/10/2024

01040329-0015